Fores

Techniques in Ecology and Conservation Series

Series Editor: William J. Sutherland

Bird Ecology and Conservation: A Handbook of Techniques
William J. Sutherland, Ian Newton, and Rhys E. Green

Conservation Education and Outreach Techniques
Susan K. Jacobson, Mallory D. McDuff, and Martha C. Monroe

Forest Ecology and Conservation
Adrian C. Newton

Forest Ecology and Conservation

A Handbook of Techniques

Adrian C. Newton

This book has been printed digitally and produced in a standard specification in order to ensure its continuing availability

OXFORD
UNIVERSITY PRESS

Great Clarendon Street, Oxford OX2 6DP
United Kingdom

Oxford University Press is a department of the University of Oxford.
It furthers the University's objective of excellence in research, scholarship,
and education by publishing worldwide. Oxford is a registered trade mark of
Oxford University Press in the UK and in certain other countries

© Oxford University Press 2007

The moral rights of the authors have been asserted
Database right Oxford University Press (maker)

Reprinted 2013

All rights reserved. No part of this publication may be reproduced,
stored in a retrieval system, or transmitted, in any form or by any means,
without the prior permission in writing of Oxford University Press,
or as expressly permitted by law, or under terms agreed with the appropriate
reprographics rights organization. Enquiries concerning reproduction
outside the scope of the above should be sent to the Rights Department,
Oxford University Press, at the address above

You must not circulate this book in any other form
and you must impose this same condition on any acquirer

British Library Cataloguing in Publication Data
Data available

Library of Congress Cataloging in Publication Data
Data available

ISBN 978-0-19-856745-5

To my father, Alan Newton, for all his support and encouragement over the years.

Preface

For the past 25 years, forests have been the focus of international conservation concern. High rates of deforestation and forest degradation are common in many parts of the world, but it was the rapid loss of tropical rain forests that particularly captured the attention of the world's media from the early 1980s onwards. More recently, it has been increasingly recognized that many other ecologically important forest types, such as temperate rain forests and tropical dry forests, are also being lost at an alarming rate. In response, particularly following the United Nations Conference on Environment and Development in 1992, major international efforts have been devoted to forest protection and sustainable management. There have been some notable successes during this time, yet still the widespread loss of forests continues.

Despite the growth in the number of forest conservation and development projects, as well as in the scientific discipline of forest ecology, practitioners are often unsure how best to tackle the problems that they face. A lack of access to information about appropriate techniques is hindering both the development of the science and its application to forest conservation. This book was written in response to this need, and is part of a series providing information on methods in ecology and conservation, focusing on different species and habitats. The target audience is ecologists involved in forest research or conservation projects, including both established professionals and those just starting out on their careers. It is hoped that the book will also be of value to practising foresters. Although foresters have traditionally been trained primarily in management of forests for timber, the profession has undergone something of a revolution in recent years. The individual forest manager is now often expected to be familiar with social, economic, and ecological aspects of forests, as well as timber production. Hopefully this book will be of value to those practitioners aiming at the elusive goal of truly sustainable forest management.

Forest ecology and conservation is an enormous subject. I have therefore had to be highly selective in selecting material for this book. Inevitably, this choice has been influenced by my own interests and experience, and for this bias, I apologize. Although it is recognized that different forest types differ substantially in their ecology and composition, the book is designed to be relevant to all kinds of forests. This is undoubtedly an ambitious goal, but I am comforted by the fact that in my own experience I have been more struck by the similarities between different forests than by their differences, particularly regarding the conservation problems that they face. Many of the techniques described here have been applied to forests growing in very different parts of the world, although perhaps with some adaptation. These methods should therefore be applied flexibly, not as rigid protocols. There is no substitute for common sense!

It is important to remember that techniques are not fossils. This is a living discipline, in every sense. This means that there is scope to improve on all of the methods described here. Refining a method, or developing a new approach, is a worthwhile focus of research in its own right. Users of this book are therefore encouraged not to consider the techniques presented as a finished article, but rather as a starting point for further experimentation and innovation. I have deliberately provided extensive references, to provide examples of these methods being used in practice, and to encourage readers to investigate their chosen techniques in greater depth. Citing these examples illustrates the fact that different workers use techniques in different ways, and in many cases the best way of doing something is an issue still open to both critical appraisal and debate.

In preparing this book, I particularly thank the many wonderful postgraduate students and research assistants with whom I have had the privilege of working, and who have grappled with many of the methods described: Theo Allnutt, Claudia Alvarez Aquino, Siddhartha Bajracharya, Sarah Bekessy, Niels Brouwers, Philip Bubb, Elena Cantarello, Cristian Echeverría, Duncan Golicher, Jamie Gordon, Carrie Hauxwell, Gus Hellier, Valerie Kapos, Tracey Konstant, Fabiola López Barrera, Rizana Mahroof, Elaine Marshall, John Mayhew, Francisco Mesén, Lera Miles, Khaled Misbuhazaman, Gill Myers, Simoneta Negrete, Theresa Nketiah, Daniel Ofori, Tanya Ogilvy, Ashley Robertson, Patrick Shiembo, Tonny Soehartono, Kerrie Wilson.

While writing the text I became increasingly aware of how much I owe the people that taught me as a student. It was surprising to discover how many of the techniques described here were introduced to me when studying at Cambridge more than 20 years ago. It is easy to take a good education for granted. I was very fortunate to be taught by some eminent plant ecologists, and I here record a debt of gratitude to all of those who so generously shared their knowledge and expertise, particularly David Briggs, David Coombe, Peter Grubb, Bill Hadfield, Donald Pigott, Oliver Rackham, Edmund Tanner, Max Walters, and Ian Woodward.

Many thanks also to everyone who responded positively to a request for photographs, and to my wife Lynn for checking the text.

Forests are magnificent places. I deeply respect those individuals who dedicate their lives to forest conservation, and I very much hope that this book will be of some value in supporting their efforts. Please let me know if the book proves to be of use, and more importantly, how it could be improved.

Adrian C. Newton
School of Conservation Sciences
Bournemouth University
anewton@bournemouth.ac.uk
May 2006

Contents

Abbreviations	xiv

1. Introduction — 1

1.1 Defining objectives	2
1.2 Adopting an investigative framework	5
1.3 Experimental design	8
1.4 Achieving scientific value	9
1.5 Achieving conservation relevance	12
1.6 Achieving policy relevance	16
1.7 Defining terminology	21
1.8 Achieving precision and accuracy	26
1.9 Linking forests with people	27

2. Forest extent and condition — 32

2.1 Introduction	32
2.2 Aerial photography	33
2.2.1 *Image acquisition*	34
2.2.2 *Image processing*	36
2.2.3 *Image interpretation*	38
2.3 Satellite remote sensing	39
2.3.1 *Image acquisition*	42
2.3.2 *Image processing*	47
2.3.3 *Image classification*	49
2.4 Other sensors	54
2.5 Applying remote sensing to forest ecology and conservation	55
2.5.1 *Analysing changes in forest cover*	55
2.5.2 *Mapping different forest types*	60
2.5.3 *Mapping forest structure*	62
2.5.4 *Mapping height, biomass, volume, and growth*	63
2.5.5 *Mapping threats to forests*	66
2.5.6 *Biodiversity and habitat mapping*	66
2.6 Geographical information systems (GIS)	68
2.6.1 *Selecting GIS software*	71
2.6.2 *Selecting data types*	73
2.6.3 *Selecting a map projection*	74
2.6.4 *Analytical methods in GIS*	75

	2.7 Describing landscape pattern	76
	2.7.1 *Choosing appropriate metrics*	78
	2.7.2 *Estimating landscape metrics*	82

3. Forest structure and composition   85

3.1 Introduction	85
3.2 Types of forest inventory	85
3.3 Choosing a sampling design	87
3.3.1 *Simple random sampling*	88
3.3.2 *Stratified random sampling*	89
3.3.3 *Systematic sampling*	90
3.3.4 *Cluster sampling*	90
3.3.5 *Choosing sampling intensity*	91
3.4 Locating sampling units	91
3.4.1 *Using a compass and measuring distance*	91
3.4.2 *Using a GPS device*	92
3.5 Sampling approaches	93
3.5.1 *Fixed-area methods*	94
3.5.2 *Line intercept method*	95
3.5.3 *Distance-based sampling*	95
3.5.4 *Selecting an appropriate sampling unit*	98
3.5.5 *Sampling material for taxonomic determination*	102
3.6 Measuring individual trees	104
3.6.1 *Age*	104
3.6.2 *Stem diameter*	107
3.6.3 *Height*	109
3.6.4 *Canopy cover*	111
3.7 Characterizing stand structure	113
3.7.1 *Age and size structure*	113
3.7.2 *Height and vertical structure*	115
3.7.3 *Leaf area*	116
3.7.4 *Stand volume*	118
3.7.5 *Stand density*	120
3.8 Spatial structure of tree populations	121
3.9 Species richness and diversity	125
3.9.1 *Species richness*	125
3.9.2 *Species diversity*	131
3.9.3 *Beta diversity and similarity*	133
3.10 Analysis of floristic composition	135
3.10.1 *Cluster analysis*	136
3.10.2 *TWINSPAN*	138
3.10.3 *Ordination*	139
3.10.4 *Importance values*	142
3.11 Assessing the presence of threatened or endangered species	142
3.12 Vegetation classification	144

4. Understanding forest dynamics — **147**

  4.1 Introduction — 147
  4.2 Characterizing forest disturbance regimes — 148
      4.2.1 *Wind* — 149
      4.2.2 *Fire* — 151
      4.2.3 *Herbivory* — 153
      4.2.4 *Harvesting* — 159
  4.3 Analysis of forest disturbance history — 161
  4.4 Characterizing forest gaps — 164
  4.5 Measuring light environments — 167
      4.5.1 *Light sensors* — 167
      4.5.2 *Hemispherical photography* — 170
      4.5.3 *Light-sensitive paper* — 174
      4.5.4 *Measuring canopy closure* — 174
  4.6 Measuring other aspects of microclimate — 178
  4.7 Assessing the dynamics of tree populations — 181
      4.7.1 *Permanent sample plots* — 181
      4.7.2 *Assessing natural regeneration* — 182
      4.7.3 *Measuring height and stem diameter growth* — 184
      4.7.4 *Measuring survival and mortality* — 185
      4.7.5 *Plant growth analysis* — 189
      4.7.6 *Factors influencing tree growth and survival* — 191
  4.8 Seed bank studies — 195
  4.9 Defining functional groups of species — 198

5. Modelling forest dynamics — **203**

  5.1 Introduction — 203
  5.2 Modelling population dynamics — 204
      5.2.1 *The equation of population flux* — 204
      5.2.2 *Life tables* — 205
      5.2.3 *Transition matrix models* — 205
  5.3 Population viability analysis — 213
  5.4 Growth and yield models — 220
  5.5 Ecological models — 221
      5.5.1 *Gap models* — 222
      5.5.2 *Transition models* — 226
      5.5.3 *Other modelling approaches* — 228
      5.5.4 *Using models in practice* — 230

6. Reproductive ecology and genetic variation — **235**

  6.1 Introduction — 235
  6.2 Pollination ecology — 235
      6.2.1 *Tagging or marking flowers* — 236
      6.2.2 *Pollen viability* — 236

	6.2.3 *Pollen dispersal*	237
	6.2.4 *Mating system*	239
	6.2.5 *Hand pollination*	243
	6.2.6 *Pollinator foraging behaviour and visitation rates*	244
6.3	Flowering and fruiting phenology	245
6.4	Seed ecology	250
	6.4.1 *Seed production*	250
	6.4.2 *Seed dispersal and predation*	252
6.5	Assessment of genetic variation	262
	6.5.1 *Molecular markers*	262
	6.5.2 *Quantitative variation*	279

7. Forest as habitat — 285

7.1	Introduction	285
7.2	Coarse woody debris	285
	7.2.1 *Assessing the volume of a single log or snag*	286
	7.2.2 *Survey methods for forest stands*	287
	7.2.3 *Assessing decay class and wood density*	294
	7.2.4 *Estimating decay rate*	296
7.3	Vertical stand structure	297
7.4	Forest fragmentation	300
7.5	Edge characteristics and effects	302
7.6	Habitat trees	307
7.7	Understorey vegetation	312
7.8	Habitat models	316
	7.8.1 *Conceptual models based on expert opinion*	317
	7.8.2 *Geographic envelopes and spaces*	320
	7.8.3 *Climatic envelopes*	321
	7.8.4 *Multivariate association methods*	321
	7.8.5 *Regression analysis*	322
	7.8.6 *Tree-based methods*	323
	7.8.7 *Machine learning methods*	323
	7.8.8 *Choosing and using a modelling method*	324
7.9	Assessing forest biodiversity	326

8. Towards effective forest conservation — 332

8.1	Introduction	332
8.2	Approaches to forest conservation	333
	8.2.1 *Protected areas*	334
	8.2.2 *Sustainable forest management*	338
	8.2.3 *Sustainable use of tree species*	344
	8.2.4 *Forest restoration*	347
8.3	Adaptive management	354
8.4	Assessing threats and vulnerability	357

8.5	Monitoring	363
8.6	Indicators	367
	8.6.1 *Indicator frameworks*	368
	8.6.2 *Selection and implementation of indicators*	369
8.7	Scenarios	374
8.8	Evidence-based conservation	377
8.9	Postscript: making a difference	377

References	379
Index of Authors and Names	431
Subject Index	437

Abbreviations

AAC	allowable annual cut
ACE	abundance-based coverage estimator
AFLP	amplified fragment length polymorphisms
ANOVA	analysis of variance
ATBI	all taxa biodiversity inventory
ATFS	American Tree Farm System
AVHRR	advanced very high resolution radiometer
C&I	criteria and indicators
CBD	Convention on Biological Diversity
CCA	canonical correspondence analysis
CI	cover index
CIFOR	Centre for International Forestry Research
CSA	Canadian Standards Association
CWD	coarse woody debris
dbh	diameter at breast height
DCA or DECORANA	detrended correspondence analysis
DEI	depth of edge influence
DEMs	digital elevation models
DIFN	diffuse non-interceptance
DN	digital number
DPSIR	drivers, pressure, state, impact, and response
DSS	decision support system
ENFA	ecological niche-factor analysis
ESUs	evolutionarily significant units
FAO	Food and Agriculture Organization of the United Nations
FCR	fluorochromatic reaction
FCS	favourable conservation status
FHD	foliage height diversity
FLDM	forest landscape dynamics model
FLEG	forest law enforcement and governance
FLR	forest landscape restoration
FMU	forest management unit
FPA	formalin/propionic acid/alcohol
FRIS	Forest Restoration Information Service
FSC	Forest Stewardship Council
GAM	generalized additive model
GCP	ground control point
GFRA	Global Forest Resources Assessment

GIS	geographical information system
GLCF	Global Land Cover Facility
GLM	generalized linear model
GPS	global positioning system
GRMU	gene resource management unit
HBLC	height to base of live crown
HCVF	high conservation value forest
HPS	horizontal point sampling
HSI	habitat suitability index
IALE	International Association for Landscape Ecology
ICE	incidence-based coverage estimator
IFF	International Forum on Forests
IPF	Intergovernmental Panel on Forests
ISI	self-incompatibility index
ITTO	International Tropical Timber Organization
IUCN	World Conservation Union
IUFRO	International Union of Forest Research Organizations
kNN	k nearest neighbour
LAI	leaf area index
LAR	leaf area ratio
LMR	leaf mass ratio
MBR	Maya Biosphere Reserve
MU	management unit
MWP	modified-Whittaker plot
NDVI	normalized difference vegetative index
NFI	national forest inventory
NGOs	non-governmental environmental organizations
NTFP	non-timber forest product
OECD	Organisation for Economic Co-operation and Development
OTU	operational taxonomic unit
PAR	photosynthetically active radiation
PCA	principal components analysis
PCO	principal coordinates analysis
PEFC	Programme for the Endorsement of Forest Certification
PIT	passive integrated transponder
PPFD	photosynthetic photon flux density
PRA	participatory rural appraisal
PRC	population recruitment curve
PSP	permanent sample plot
PSR	pressure–state–response
PVA	population viability analysis

QTL	quantitative trait loci
RAP	rapid assessment programme
RAPD	random amplified polymorphic DNA
RAPPAM	rapid assessment and prioritization of protected areas management
RFLP	restriction fragment length polymorphism
RGR	relative growth rate
RGRH	relative growth rate of height
ROC	receiver–operator characteristic
RPVA	relative population viability assessment
RRA	rapid rural appraisal
RTU	recognizable taxonomic unit
SFI	sustainable forestry initiative
SFM	sustainable forest management
SI	suitability index
SLA	specific leaf area
SLA	sustainable livelihoods approach
SSR	microsatellite
UNCED	United Nations Conference on Environment and Development
UNEP	United Nations Environment Programme
UNFCCC	United Nations Framework Convention on Climate Change
UNFF	United Nations Forum on Forests
UTM	Universal Transverse Mercator (projection)
WCPA	World Commission on Protected Areas
WDPA	World Database of Protected Areas
WSSD	World Summit on Sustainable Development
WWF	World Wide Fund for Nature

1
Introduction

This book describes techniques that may be used in ecological research, survey or monitoring work in support of forest conservation and management. Yet conservation is much more than simply a research endeavour. Rather, conservation management depends on understanding the interplay between social, economic, and political issues relating to a particular forest, and on appreciating the values held by different people with an interest in it. In practice, the scientific understanding of the ecology of a forest often plays a relatively minor role in determining how it is conserved or managed. In many cases, management decisions are based on political or economic expediency rather than the results of the latest ecological research. Yet even though ecological understanding alone never conserved a forest, such an understanding can play a crucial role in ensuring that conservation management is effective. The aim of this introductory chapter is to help achieve this objective, by placing the application of ecological techniques in a broader context.

The science of forest ecology has progressed enormously over the past two decades, assisted by rapid technological developments in areas such as remote sensing, GIS, and molecular ecology. Such techniques have transformed our understanding of forest distribution and ecological condition, and have afforded profound insights into how forests respond to environmental change at a variety of scales. Yet our understanding of forest ecology has its roots far deeper, having grown out of more than two centuries of forestry practice. While some of the methods described here are still evolving rapidly, others have proved themselves over many years of practical application. Ecological researchers often have much to learn from forestry professionals with respect to methods of forest mensuration and inventory, and any technique that has stood the test of time is worthy of consideration. New technology is no guarantee of improved measurements.

Deciding which technique is appropriate for use in a particular situation depends critically on the objectives of the research or survey work to be undertaken. Defining these objectives clearly at the outset is of paramount importance to any research or conservation programme. The objectives of a research ecologist may differ substantially, however, from those of a conservation practitioner or a forest manager. Many forest conservation projects are designed to help implement some policy objective, whether this be a policy statement developed by the organization responsible for developing the project, or some national or international policy goal. Even in the case of relatively 'pure' ecological research, funding organizations are increasingly inclined to direct their support towards research that is policy-relevant. This chapter includes a brief overview of recent developments in international forest policy, as this provides a basis for so much of the current research focus on forests, together with definitions of some of the key concepts involved. A summary is also provided of recent initiatives aiming to provide conservation assessments of forests; these provide a valuable basis for much current and future research.

Other issues that should be considered in the early stages of planning research or survey work include the choice of an appropriate investigative framework and experimental design, how to ensure high scientific value and rigour, and how to place the work in an appropriate socio-economic context. This chapter provides some guidance on these issues, together with some reflections on how ecological research can be linked effectively with the practice of forest conservation and management. A significant divide currently exists between conservation research and practice, and this is an issue of widespread concern. How this divide can be bridged is considered in greater depth in the final chapter of this book.

1.1 Defining objectives

Investigators embarking on their first piece of research or survey work often make a major mistake: they fail to adequately define what it is that they are hoping to achieve. The need to set precise objectives may seem self-evident, or even trivial, but it is not. A failure to define aims with sufficient precision inevitably leads to poorly focused research, the lack of a clear result, and potentially a great deal of wasted effort. As noted by Underwood (1997), if there are no clear goals, there will be no useful results. Before investing time and resources in collecting data of any kind, and before choosing appropriate methods for data collection, it is important to ensure that the reasons for collecting the information are as clearly defined as possible.

Anyone who has attempted to teach ecological research to undergraduate students will be well aware of what can go wrong. Students new to research are often overly ambitious in their aims, giving little chance of generating a clear answer. An important early lesson is that it is not difficult to collect large amounts of data, but that this is no guarantee of a successful outcome. Some students seem to have a compulsion to measure as much as possible, then struggle to extract a clear message from the clouds of numbers that have been generated, a process that can be deeply disheartening. This problem can be pre-empted by paying greater attention to developing clear, precise objectives at the outset.

Such problems are not unique to novices. Even experienced researchers frequently make mistakes. It is not unusual for major research programmes, costing vast amounts of public money, to provide few genuine insights at the end of the day. Often, when a research programme is complete, some key piece of information will prove to be lacking. Hindsight truly is a wonderful thing. But without the benefit of hindsight, or experience, how can appropriate objectives be identified to minimize the risk of failure?

Choosing an appropriate question to ask can be a daunting process. The range of possible objectives, even for a relatively simple forest system, is potentially infinite. An important first step is to define the kind of study that is being attempted. It is useful to differentiate between ecological research, survey, and monitoring (sometimes the word surveillance is also used for the latter):

- *Research* is generally undertaken to answer a specific question, or to test a hypothesis.
- A *survey* is typically a descriptive piece of work, which might be more open-ended in nature than a research project, and might not have such a clear outcome.
- *Monitoring* is a form of survey that is designed to be repeated over time, enabling trends in some variable of interest to be determined.

Many of the techniques described in this book are equally relevant to each of these different approaches. However, the nature of the study will have implications for how the methods are implemented, and above all for the design of the data collection process.

The international scientific community tends to place greater emphasis on research rather than survey and monitoring work, and this is reflected in the content of scientific journals. According to Peters (1991), because of its lack of relationship to relevant theory, survey work does not qualify as science, but might be better referred to as natural history. Yet the importance of natural history should not be underestimated. Much of our current ecological theory was developed on the basis of painstaking field observations made by generations of naturalists. Furthermore, survey and monitoring methods are of fundamental importance to the practice of conservation, providing information of value to priority setting and management. There is great merit in simply observing how species behave in their natural habitats, and such observations can contribute directly to defining appropriate management interventions (Marren 2002). It is striking how little is known about even our most important forest-dwelling 'flagship' species. As an example, it is salutary to note that we do not know precisely how many individuals remain of any of the great ape species, nor what their precise habitat requirements are (Caldecott and Miles 2005).

There are situations where some form of survey will be preferred to a research programme. In forest areas for which no prior information is available, a descriptive survey is the logical first step, perhaps with the simple aim of describing forest composition and structure. A survey might be undertaken to assess the conservation status or condition of a particular forest or associated species, or to determine the occurrence of some potential threat. Many conservation organizations are currently investing heavily in survey work of this nature, with the aim of identifying priorities for conservation. An initial survey can provide a basis for developing more tightly defined questions relating to ecological processes or functions, which could be addressed by subsequent research. Yet even in the case of a preliminary, descriptive survey, clear objectives should be defined at the outset.

A brief checklist is provided here to help guide the definition of objectives for a research, survey or monitoring programme:

- *Is it original?* Has the information already been collected by somebody else? This can be most readily determined by conducting a review of relevant literature, for example by using an appropriate search engine (such as ⟨*www.google.com*⟩) or citation database (such as the ISI Web of Science, ⟨*www.thomsonisi.com*⟩). However, most information relating to forests has never been formally published, but resides in internal reports, data archives, newsletters and other so-called 'grey' literature. Accessing such information can present an enormous challenge. There may be no substitute for personally contacting relevant institutions and individuals to ascertain what work has been carried out previously, and to find out what happened to the results. Although tracking down such information can take a great deal of time and effort, the rewards may be significant. There is a fine tradition of meticulous survey work among many forestry institutions, which can still be a source of valuable information.
- *Is it tractable?* In other words, is it possible to deliver an answer to the question set, given available time and resources? If not, then the objectives need to be more tightly focused, for example by limiting the spatial or temporal scope of the project more narrowly. It is important to remember that some ecological questions are impossible to answer.
- *Is it interesting?* Interest can be increased by choosing an issue that is topical or novel. For example, has there been recent media interest in the chosen subject? Might the results of the research generate media interest? Many new researchers are unaware of the extent to which different scientific themes go in and out of fashion, yet an awareness of current trends can be of great importance in successfully publishing results or securing funding for further research.
- *Can the objectives be phrased as a question?* Presenting the objectives in this way can be a great help in focusing the design of the research, and in obtaining a clear answer from the results. It can be helpful to define a set of sub-questions under a general aim, to help break the problem down into more manageable, clearly defined units.
- *Is it of practical value?* Although this criterion may not be of paramount importance to a 'pure' researcher, much ecological information is collected with a specific end use in mind. To ensure that appropriate data are collected in a suitable form, the objectives should be developed in consultation with the intended users of the information, such as conservation practitioners or forest managers.

Time spent refining objectives is never wasted. Remember that not everything that can be measured, should be (Krebs 1999). In practice, this means considering alternatives, attempting to rephrase and refine the wording, always with the goal of increasing precision (see Box 1.1 for an example). Consult textbooks (for example Begon *et al.* 1996) or monographs (for example Hubbell 2001) to identify theories worth testing. Seek advice from your peers, colleagues, and supervisors before embarking on the project. Observe and analyse how the objectives are described in published scientific papers. Avoid questions such as 'why' and 'how come', and focus instead on developing questions that begin with 'how much', 'how many', 'when' and 'where' (Peters 1991). Critically consider the possible answers to the objectives that you have set.

> **Box 1.1** Defining research objectives
>
> Research should be both tractable and interesting. In order to ensure that research is manageable, objectives should be tightly focused, for example by limiting their temporal and spatial scope. In the example below, this has been achieved by explicitly stating the area of forest to be considered, avoiding broad statements about forests in general that would be impossible to evaluate in a field survey. To ensure that the research is interesting, it should be topical, something that can be ascertained by reference to the international media, as well as to recent issues of scientific journals. In this example, whereas measuring forest biomass was an active area of research during the boom in systems ecology in the 1960s, today estimation of carbon sequestration is arguably a much more topical issue—even if the basic techniques have not changed.
>
> | | Not interesting | Interesting |
> | --- | --- | --- |
> | Not tractable | Do forests have high biomass? | Do forests sequester a lot of carbon? |
> | Tractable | What is the above-ground biomass of this 0.01 ha forest plot? | How much carbon does this 0.01 ha of forest sequester in a year? |

Many researchers set great store by the need to state hypotheses clearly at the outset. Referees of manuscripts submitted to international scientific journals often expect to see the objectives of a piece of research stated in this form. Yet not everyone agrees with this approach. The role of hypothesis testing continues to be the subject of intense philosophical debate regarding how science should be done. This is a debate in which anyone embarking on a research project can usefully engage, perhaps involving some lively discussion with colleagues. This book cannot pretend to be a philosophical treatise, but researchers should be aware that opinions vary regarding how science should be carried out. It is worth noting, however, that statistical tests are explicitly designed to test hypotheses, and if there is an intention to employ such tests in the analysis of the results, then the hypotheses to be tested in this way should be made explicit at the outset.

1.2 Adopting an investigative framework

Regardless of what the precise objectives actually are, any piece of research or survey work should be carefully planned and implemented according to an appropriate investigative framework. Adopting a clear logical procedure is important for

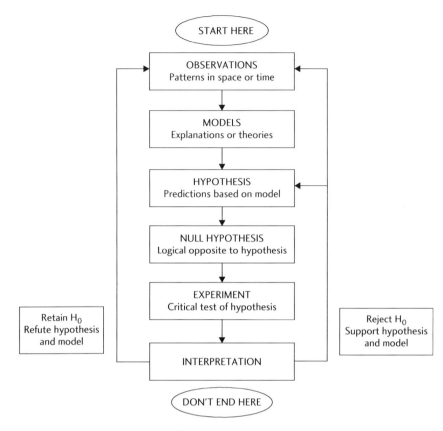

Fig. 1.1 Generalized scheme of logical components of a research programme. H_0 represents the null hypothesis. (From Underwood (1997). Experiments in Ecology: their logical design and interpretation using analysis of variance. Cambridge University Press.)

communicating the results to others, and to ensure that the information collected achieves the objectives set.

A framework is presented here based on that described by Underwood (1997) (Figure 1.1). Versions of this procedure are widely practised in ecological investigations. The framework comprises a series of logical steps, beginning with observations that are typically made in the field. Such observations can vary in spatial or temporal scale, and might be purely casual observations made during a visit to a particular forest, or the results of systematic survey work undertaken over a prolonged period. Usually some feature or pattern of potential interest will be detected, which might be worthy of further study. For example, it might be noticed that a particular species of tree appears only to occur in certain areas, perhaps along river banks or at particular altitudes.

The next step is to attempt to explain the phenomenon observed. Forest stands might be dominated by large, old trees, with little evidence of recent recruitment.

Why might this be? There might be several alternative explanations to an observation such as this: for example, failure of seed production or dispersal, destruction of juvenile trees as a result of fire or the activities of herbivores, or the lack of appropriate environmental conditions for seedling establishment. Typically there will be many possible explanations for the observations made, and research will be required to differentiate between them.

When undertaking any form of ecological investigation, it is helpful to differentiate between pattern and process. *Patterns* or *phenomena* are those things that we observe. Many ecological patterns are subtle and are difficult to detect, perhaps because they occur at a temporal or spatial scale that is difficult for us to perceive. Others might be more obvious but more difficult to explain. Such patterns are caused by ecological processes. Alex Watt, one of the founding fathers of forest ecology, was the first to explicitly separate pattern from process in considering the dynamics of vegetation in relation to its structure. His work on regeneration cycles in beech woodland in southern England laid the foundations of our current understanding of gap dynamics in forests, which has become such a powerful research paradigm in forest ecology. His classic paper (Watt 1947) is still worth consulting today, despite the fact that it is based purely on observation. A focus on ecological processes is a central feature of much modern ecological research.

Having detected and described an ecological pattern, how can we identify the many different *processes* that might have been responsible for its formation? How can we determine which of the many potential explanations for the pattern is correct? The formation of different logical hypotheses can help differentiate among explanations (Underwood 1997). A *hypothesis* can be defined as a prediction based on some explanation of the observations made. Once a hypothesis has been formulated, it can potentially be tested (Figure 1.1). This process of predicting an outcome by deducing what is logically consistent with a hypothesis, followed by its testing against observations made, is known as the *hypothetico-deductive* scientific method.

Often, objectives of research are expressed as a null hypothesis, which is the opposite of a hypothesis. This reflects the fact that it is easier to disprove something than to prove it (Underwood 1997). For example, it might be hypothesized that the abundance of an *Acacia* species is low in a particular area of savannah because it is preferentially browsed by giraffes. This would be difficult to prove, because there could always be some situation—another savannah, perhaps—that would provide an exception. So as an alternative, this could be expressed as a null hypothesis: the incidence of giraffe browsing has *no* effect on the abundance of the *Acacia* species. The null hypothesis could then be tested (or falsified), for example by experimentally altering the incidence of giraffe browsing and observing its effects on *Acacia* abundance.

An experiment is the only way of adequately testing a hypothesis. If the outcome of an experiment is to reject the null hypothesis, then the explanation or theory that it was designed to test is supported. If the experiment fails to falsify the null hypothesis, then the hypothesis is shown to be wrong, as its predictions were not correct. What happens next? If the explanation was supported, then it could be tested again—through an additional experiment—to see whether it applies to other situations, and if so, could then be considered as a general theory. If the hypothesis was not correct, then it needs to be revised in the light of the experimental results. There may be a need to collect additional observations. In either case, the process is a cyclic one (see Figure 1.1), and as Underwood (1997) points out, research can therefore be seen as a never-ending process—which might be comforting in terms of ensuring long-term job security!

Are there alternatives to this investigative framework? There is no doubt that it has its flaws, and is not supported by all ecological researchers. As an illustration, it can sometimes be difficult to determine whether or not the hypothesis should be retained or rejected, as a result of type I or type II errors (Underwood 1990).

An alternative way of approaching research is offered by the use of Bayesian methods. Bayesian inference involves the representation of beliefs or information in the form of probabilities. The knowledge or beliefs available before research is undertaken are represented as a likelihood distribution, known as the 'prior'. This can then be revised in the light of new information generated by research, through a process of statistical inference using Bayes' theorem. The revised probability distribution is known as the 'posterior' (Dennis 1996). The use of Bayesian methods in ecology was greatly stimulated by a series of papers published in the journal *Ecological Applications* in 1996. As noted by Dennis (1996), the application of Bayesian approaches in ecology is controversial, as it implies abandoning the scientific method based on testing hypotheses and the investigative framework described above. Protagonists of the Bayesian approach suggest that it makes better use of available data, allows stronger conclusions to be drawn from uncertain data, and is more relevant to environmental decision-making (Ellison 1996).

The application of Bayesian methods to conservation management is examined by Wade (2000), who highlights the value of presenting information in a form that decision-makers can readily understand, in a way that incorporates uncertainty directly into the analysis. Ghazoul and McAllister (2003) reached similar conclusions when considering the application of Bayesian methods to forest research. Whether or not Bayesian methods of analysis are adopted, it is helpful to make underlying models, paradigms, world views, and beliefs explicit, so that it is possible to determine their influences on what and how measurements were taken (Underwood 1997).

1.3 Experimental design

Any ecological technique must be applied according to an appropriate experimental or sampling design if the information generated is to be useful. A comprehensive treatment of the principles of experimental design is beyond the scope of this book. There are now a number of texts that provide a valuable introduction to the principles of designing surveys and experiments. I particularly recommend those by Ford (2000), Krebs (1999), Peters (1991), Southwood and Henderson (2000), and Underwood (1997). Dytham (2003) provides a highly practical guide to the principles of sampling and statistical analysis.

The design of the research or survey will depend on the objectives set, and the characteristics of the forest to be studied. Some general principles to remember are listed below:

- *Randomize.* Randomization is essential in order to avoid sample bias, and to ensure that samples are representative. Most statistical tests assume that samples are independent and free from bias, and this can most readily be achieved by sampling randomly. Stratified random sampling approaches are commonly adopted in forest ecology, where random samples are taken within a forest area divided into relatively homogeneous subareas classified on the basis of some environmental variable or forest composition. The number of samples taken should be proportional to the size of each subarea (Southwood and Henderson 2000). Remember that random samples should be truly random: use a random number generator provided by a pocket calculator or appropriate computer software (many statistical or spreadsheet packages have this feature).
- *Replicate.* Replication is essential in order to determine patterns of variation, so that the results obtained can be attributed to the experimental treatments or factors of interest. Replicate samples should be taken within each area of study, and should be genuinely independent, to avoid the risk of pseudoreplication (Hurlbert 1984).
- *Use appropriate controls.* Many experimental investigations fail because of inadequate selection of controls, which are characterized by the absence of the experimental factor or treatment of interest. By providing a basis for comparison, controls are of fundamental importance to effective experimentation. Many surveys and monitoring approaches fail to collect baseline information before the experimental treatments (or management interventions) are applied, and fail to include untreated controls. Only by including such controls can any effects detected be attributed to directly to the treatment of interest.
- *Perform a power analysis before starting the main investigation.* Before investing time and effort in intensive data collection, it is recommended that a pilot study be undertaken to test the methods and protocols. The data obtained during such a pilot study can then be used to perform a power analysis, which will enable estimates to be made of the number of samples required to detect effects of a given magnitude. Statistical power is influenced by the sample size, the variability of the population being sampled, and the magnitude of the effect of the experimental treatment or factor. Methods of power analysis are presented by Krebs (1999) (see also Chapter 8); software programs that perfrom such analyses are reviewed by Thomas and Krebs (1997).

1.4 Achieving scientific value

Once the objectives of a research or survey project have been defined, an investigative framework adopted, and an appropriate design put in place, then success in terms of delivering some valuable results might seem assured. However, failure to deliver results of genuine scientific value is more common than many researchers might care to admit. Most researchers have at least one file drawer full of results that have failed to see the light of day in terms of a scientific publication. Entire doctoral theses, representing years of honest endeavour, have been consigned to this form of oblivion. How can this failure be avoided, and results of real scientific value be achieved?

This raises the question of what constitutes scientific value. Although scientific performance is often assessed in terms of the number and quality of publications produced, this may be a poor measure of its real value. Most scientific papers are cited rarely, some not at all. Few have real impact in terms of influencing forest conservation policy or practice. But the problems may go deeper than how the results are disseminated: to the nature of ecological science itself.

Peters (1991) provides a comprehensive critique of the science of ecology and how it is currently practised. In many ways, this is an extraordinary book, and it is recommended reading for anyone interested in engaging in any form of ecological research. He concludes that:

> The weakness of the central constructs of contemporary ecology results because ecology compounds its single failings. Operational impossibilities spawn tautological discussions that replace predictive theories with historical explanations, testable hypotheses with the infinite research of mechanistic analysis, and clear goals for prediction with vague models of reality. The resulting mélange obscures appropriate research and attainable goals with sloppy, ineffective activity. As a result, the central constructs in ecology yield predictions with difficulty and these are often so qualitative, imprecise and specific that they are of little interest and less utility (Peters 1991).

Peters' book does not make comfortable reading for most ecologists. There is hardly a single area of ecological method that escapes some degree of censure. Needless to say, the conclusions are controversial, but at the very least the points raised are worthy of serious consideration and debate. But what are the practical implications of this critique? Some of these are summarized briefly below, but the reader is encouraged to consult the original text for a comprehensive consideration of these and many other issues.

- *Pluralism.* A single approach or technique will not be successful in all systems; rather, different approaches may be needed for different systems and different questions. This applies as much to the methods used in building theory as to those used to test it. Multiple working hypotheses should be encouraged.
- *Practicality.* Many ecological theories, it is argued, have little relevance to the real world. They are often based on abstract mathematical representations of phenomena rather than on empirical measurements. Researchers should focus on making practical observations, and on addressing little questions that can be answered rather than big questions that cannot. Theory should always be relevant, and inspired by pressing problems about nature rather than the search for scholasticism. Seek the simplest way of making testable predictions.
- *Sound variables.* Much ecological research suffers from use of concepts that are difficult or even impossible to measure. Even concepts as widely used as 'niche', 'ecosystem', and 'habitat' have been defined variously by different authors and are difficult to operationalize in practice. Variables should be simple, measurable,

and operationally defined, such as diversity, nitrogen concentration, biomass, population density, etc. Poorly defined variables should be avoided.
- *Empiricism.* Theories must be supported by data, or directly based on data. Patterns in these data should be identifiable by using simple statistical manipulations, such as regression. Focus on prediction rather than explanation. All predictive models are probabilistic; uncertainty should always be estimated or represented in such models.

Above all, Peters (1991) emphasizes the importance of testing predictions made on the basis of relevant theory, as an essential ingredient of high-quality ecological science. Investigators embarking on a new programme of research should therefore seek to identify relevant theory at the outset, and on the basis of such theory, develop hypotheses incorporating specific predictions that can readily be tested. Peters (1991) also notes that ecological research often encourages the construction of irrelevant theories and the collection of irrelevant data by proposing and testing underlying mechanisms that might contribute to an observed pattern, although they are neither necessary nor sufficient for that pattern. Theories that are scientifically relevant are those that can help resolve questions posed by the scientific community. Scientifically relevant data are those that can test the predictions from such theories (Peters 1991).

How can relevant theory be identified? Ecological science is not short of theoretical ideas, and a search of relevant textbooks or journal papers will soon unearth a variety of candidates. Selection of an appropriate theory will depend upon the characteristics of the problem that has been identified. Many problems relating to forest dynamics, for example, can be addressed by using theories relating to successional processes. In the conservation biology literature, island biogeography theory and metapopulation theory never seem to be very far away. New ideas are always worth seeking out and critically examining.

Problems relating to forest ecology have fortunately attracted the interest of some outstanding researchers, who have produced some highly original and stimulating works. To cite just two examples: the recent book by Stephen Hubbell, although it has implications far beyond the boundaries of forest ecology, was inspired by detailed analysis of community composition in tropical forests. His synthesis of island biogeography theory and theories of relative abundance is, in my view, the most important contribution to ecological science in the past three decades, and an outstanding intellectual achievement (Hubbell 2001). Read it and see whether you agree with me. Whether right or wrong, there are enough ideas in this book to keep generations of postgraduate students busy for decades. With respect to the forests of northern Europe, Vera (2000) has produced a remarkable book that seeks to overturn traditional views of forest succession, by critically examining almost a century's worth of accumulated empirical evidence. This book provides a salutary reminder that no theory in ecology is so well established that it could not be overturned by some appropriate research. As an aside, Vera has managed what many researchers aspire to but few achieve: a radical reappraisal of conservation practice, among practitioners themselves.

1.5 Achieving conservation relevance

Although scientific journals are bursting with research results that often appear to have important implications for forest conservation, most research seems to have little impact on how conservation is actually practised. This can be a source of great frustration to those who have worked so hard to obtain the results concerned. What is going wrong? Is it because practitioners do not read scientific papers? Or has the wrong research been done? Or does the problem go deeper—do practitioners fully understand the implications of research? Do researchers fully appreciate the problems faced by practitioners?

This divide between conservation research and practice is currently the focus of much concern, particularly among researchers. At the same time, policy-makers are increasingly requiring that management action be based on an 'evidence-based' approach—in other words, on the best scientific information available. How can this be achieved? The techniques that can be used to strengthen the links between conservation research and practice are described in Chapter 8. Here, I focus on more general issues that should be considered when planning any research or survey programme.

If research is to be relevant to conservation practice, then there is no substitute for effective communication between researchers and practitioners. A researcher may have a theory of how forests respond to environmental change, and be keen to test it. The practitioner's priorities may be very different: perhaps there is uncertainty about the potential impacts of a proposed management intervention, or how a forest is being affected by some newly emerging threat. Only a process of dialogue between both parties will ensure that the practitioner's needs are properly met by the proposed research. This dialogue can be great fun and enormously educational for both parties, if based on mutual respect.

Many conservation organizations appear to spend most of their time organizing workshops. Forest managers often feel as if they are spending more time attending meetings, discussing how things ought to be done, rather than getting out and doing it. There is now great emphasis on engaging in a process of 'stakeholder consultation' before implementing any conservation action. This reflects the growing recognition that for conservation to be effective, those people that have an interest or stake in the outcome need to be involved in the decision-making process from the outset. Ecological research or survey work often seems to play little role in this consultation process, and the idea of attending interminable meetings can be very off-putting to many researchers (as well as some practitioners). Discussions around a table can seem very remote from real-life conservation action. Yet, engagement in this process is often essential if research is to play its proper role in influencing conservation outcomes.

If you are a researcher who wants to make a difference, it is worth learning about how conservation decisions are made. Often the technical issues surrounding how a particular forest should be managed are relatively easy to solve. Yet the translation of research results into conservation practice can be a long and arduous process. It should be remembered that conservation is a highly political endeavour. It can be viewed as a struggle for competing values. Sometimes it even erupts into conflict.

Achieving conservation relevance | 13

One of the most striking trends in conservation over the past three decades has been the growth in size and influence of non-governmental environmental organizations (often referred to as NGOs). Collectively, these organizations have played a hugely significant role in placing forest conservation on the international agenda. Each NGO has its own particular objectives and mode of operation. Some, such as Greenpeace, focus exclusively on campaigning and direct action. Others, such as the World Wide Fund for Nature (WWF), actively develop and implement conservation projects on the ground. Some environmental NGOs are now large, influential organizations that work in partnership with government agencies and (sometimes controversially) the private sector in their conservation projects. Some NGOs, notably Conservation International and WWF, have invested heavily in developing their own research and assessment programmes, to provide a scientific basis to their campaigning and priority setting (Figure 1.2).

Fig. 1.2 The ultimate conservation priority? The island of New Caledonia is widely recognized as being of outstanding conservation importance, particularly because of the high diversity and endemicity of its flora. The island is home to many evolutionary primitive species as a result of being a remnant of Gondwanaland that has been isolated for a very long time. It is both a 'priority ecoregion' and a 'biodiversity hotspot', as defined by WWF and Conservation International respectively. The site pictured is the Chute de la Madelaine on the Plaine des Lacs, an exceptionally important site for endemic and extremely rare tree species, with six different conifer genera occurring within an area of about 10 ha. Two conifer species, *Dacrydium guillauminii* and *Retrophyllum minor*, are known only from this area. Although protected, the site is suffering from increasing visitor pressure. (Photo by Adrian Newton.)

14 | Introduction

Table 1.1 Some of the major assessments and campaigns relating to forests currently being undertaken by leading non-governmental conservation organizations.

Organization	Assessment or campaign	Comments	URL
Conservation International (CI)	Biodiversity hotspots	Hotspots are areas characterized by high endemism and high rates of habitat loss. To qualify as a hotspot, a region must contain at least 1500 species of vascular plant as endemics, and have lost at least 70% of its original habitat. A global assessment of biodiversity hotspots is available (Mittermeier *et al.* 2004). CI are also active in conserving high-biodiversity wilderness areas, many of which are forested	*www.conservation.org/xp/CIWEB/home* *www.biodiversityhotspots.org/xp/Hotspots*
Fauna and Flora International/ UNEP World Conservation Monitoring Centre	Global Tree Campaign	A collaborative programme involving campaigning and conservation action focusing on threatened tree species in different parts of the world	*www.globaltrees.org/*
Greenpeace	Ancient Forests	A conservation campaign focusing on 'the world's remaining forests which have been shaped largely by natural	*www.greenpeace.org/international/campaigns/forests*

Achieving conservation relevance | 15

		events and which are little impacted by human activities' (Greenpeace International 2002)	
IUCN	Red List	The most comprehensive assessments of the world's threatened species, undertaken and regularly updated by IUCN through its Species Survival Commission (SSC)	/www.redlist.org/
World Resources Institute / Global Forest Watch	Frontier forests	An assessment of the world's 'remaining large intact natural forest ecosystems' (Bryant et al. 1997), equivalent to the Ancient Forests featuring in the Greenpeace campaign	http://forests.wri.org/ www.globalforestwatch.org/ englishindex.htm
WWF	Ecoregions	An ecoregion is a large area of land or water that contains a geographically distinct assemblage of natural communities. WWF has identified 825 terrestrial ecoregions worldwide, and is targeting some 200 of them for conservation action (Olson and Dinerstein 1998, Olson et al. 2000, 2001). Thorough assessments are being undertaken of the biodiversity of these areas	www.worldwildlife.org/

NGOs can be powerful allies in disseminating research results and putting them into practice. Many have close links with the media and produce their own publicity material or technical publications. An awareness of their current priorities and activities is therefore useful. Conservation campaigns can also provide a context or justification for research. However, researchers should never forget their own capacity to set the agenda. Novel information about some conservation issue or problem might well be seized upon and become a focus of campaigning and eventual conservation action. The biodiversity hotspot concept developed by Norman Myers (Myers 1988, 1990, 2003), for example, has become the central focus of campaigning and action by Conservation International (Myers *et al.* 2000).

In recent years, some international NGOs have devoted substantial resources to undertaking biodiversity assessments, with a view to defining conservation priorities at the global scale. A summary of some of the key initiatives is presented in Table 1.1, together with some of the main campaigns relating to forests implemented by international NGOs. The ecoregion assessments produced by WWF provide a particularly informative account of the ecological characteristics of different forest areas, and the conservation issues affecting them, providing a very valuable source of reference (Burgess *et al.* 2005, Dinerstein *et al.* 1995, Ricketts *et al.* 1999, Wikramanayake *et al.* 2002). A valuable overview of current assessments and the conservation approaches of leading NGOs is provided by Redford *et al.* (2003). Although such assessments can provide useful context for research, they are not beyond criticism. In an important paper, Mace *et al.* (2000) point out the high degree of duplication between the conservation assessments that have recently been completed, and highlight the need for greater collaboration between conservation organizations. Whitten *et al.* (2001) go further and question the value of the conservation assessments that are currently being undertaken, as well as the relevance of scientific research to conservation practice.

Researchers should therefore be aware of the social, political, and institutional environment within which conservation action takes place. Recognize that all organizations have their own agenda and approach and, even if they appear to be working to a common goal of conservation, may have very different priorities or means of achieving it. It is important to keep abreast of developments and emerging issues, as illustrated by NGO campaigns. Consult websites, publications, and other media; attend conferences and meetings focusing on conservation practice as well as research. Remember that your research results may be of interest to a large community of conservation activists as well as practitioners, but be prepared to promote your findings if you believe them to be important. Conservation is not just a struggle about values, but a battle for ideas. And for resources.

1.6 Achieving policy relevance

Forest policy is something of a mystery to many researchers. They may dimly be aware of its presence, yet consider it of little relevance to their work. Cynics perceive international policy development fora as endless talking shops, which achieve little in terms of practical conservation action. The process by which policies are developed at national and subnational scales can similarly appear opaque and somehow divorced from the situation in the field. Yet, in reality, policy decisions and agreements made at a high level underpin many research and management actions, and have a major bearing on the availability of research funding to address specific problems. Does policy matter? Even if a research project is not designed to be policy-relevant, its results may be used in that way. It therefore pays to be aware of what is happening in the policy arena.

This is not the place for a comprehensive account of forest policy. The issue is covered in detail by other texts such as Mayers and Bass (2004) and Sample and Cheng (2004). Rather, the aim here is to encourage researchers to be aware of the policy context in which they are working. Some guidance is given regarding how to keep abreast of policy developments, and how to make the link between policy and research.

Some key recent developments in international forest policy are listed in Table 1.2. It is important to remember that policy-makers are just as subject to fashion as are many other elements of our increasingly globalized society. New issues can suddenly emerge and within a relatively short time come to dominate debate. Interest then gradually subsides as some other issue comes to the fore. Keeping abreast of policy developments presents a significant challenge to the average forest practitioner or researcher; after all, attending the international circuit of policy meetings is a full-time job for professionals dedicated to the (rather thankless) task. Fortunately, access to information has improved enormously with the development of the Internet, and most international forest policy processes now provide ready access to the many documents that they generate via their websites. Some relevant URLs are provided in Table 1.2. The Forest Policy Experts (POLEX) electronic list server ⟨www.cifor.cgiar.org/docs/_ref/polex/index.htm⟩ is a particularly useful way of keeping up to date on developments in forest policy. Another effective way to stay in touch is to monitor the websites of the leading environmental NGOs active in forest conservation. Many of these engage closely in policy processes and have teams of staff dedicated to the task, who report regularly via their own organization's websites.

Since the United Nations Conference on Environment and Development (UNCED) in 1992, the issue of sustainable forest management has been at the centre of the international policy debate relating to forests, and underpins many national policy initiatives. Much of this discussion has focused on how sustainable forest management can be defined and assessed. Considerable effort has been devoted to the development of criteria and indicators (C&I) that might assist in this process. These developments have primarily occurred under the auspices of international 'C&I processes' (see Table 1.2). It was recognized early on that forests in different parts of the world have very different characteristics, and therefore different sets of C&I would need to be developed. Although the idea of harmonizing or standardizing between these different indicator sets has been discussed from

Table 1.2 A summary of key international policy processes relating to forest conservation.

The Convention on Biological Diversity (CBD), agreed at the 1992 Earth Summit in Rio, is the main international convention focusing on biodiversity conservation and sustainable use. The CBD has developed a thematic programme specifically focusing on forest biodiversity (⟨*www.biodiv.org/programmes/areas/forest/*⟩) with an associated programme of work, which details what Parties to the Convention should actually be doing in this area.

The 'Forest Principles' and Chapter 11 of Agenda 21 are a set of non-legally binding principles relating to the conservation and sustainable development of forests that were agreed at the 1992 Earth Summit (UNCED) (⟨*www.un.org/esa/sustdev/documents/agenda21/*⟩).

The United Nations Forum on Forests (UNFF) was established in 2000 to 'promote the management, conservation and sustainable development of all types of forests and to strengthen long-term political commitment to this end' (⟨*www.un.org/esa/forests/*⟩). The UNFF provides an important forum for international dialogue about forests. It is the successor to two prior initiatives, the Intergovernmental Panel on Forests (IPF) and the International Forum on Forests (IFF), which together recommended more than 270 proposals for action to be adopted by the international community, specifically relating to implementation of the Forest Principles and Chapter 11 of Agenda 21. Implementation of these proposals is currently being assessed by the UNFF.

International C&I processes include ITTO, the Pan-European (or 'Helsinki') Process, the Montreal Process, and the Tarapoto, Lepaterique, Near East, Dry Zone Asia, and Dry Zone Africa processes, which have each generated sets of C&I (Castañeda 2001). Currently, around 150 countries are participating in these processes.

---

time to time, this is something that is no longer actively being pursued. The C&I processes like to keep their independence.

Many of the international C&I processes have developed indicators that are appropriate for use at the national scale, rather than at the local scale, and are used for the development and updating of national and international policy instruments (Castañeda 2001). However, these processes are increasingly driving the collection of information about forests at local scales, which can then be aggregated for reporting at higher scales. The Global Forest Resources Assessment (GFRA), coordinated by the FAO, has now structured the reports that it solicits from individual countries around C&I. These C&I therefore provide an important context for much of the data collection relating to forests.

At the scale of forest management units, the development of indicators for sustainable forest management has primarily been driven by the growth of interest in forest certification. Forest certification is essentially a tool for promoting responsible forestry practices, and involves certification of forest management operations by an independent third party against a set of standards. Typically, forest products (generally timber but also non-timber forest products) from certified forests are labelled so that consumers can identify them as having been derived from well-managed sources. There are now many different organizations certifying forests against a variety of different standards. Examples include the Sustainable Forestry Initiative (SFI) Program and the Forest Stewardship Council (FSC). At least at a general level, the standards developed by certification bodies can be viewed as supporting sustainable forest management, although not all certifying organizations use this precise terminology.

More recently, other key policy developments have come to dominate international discussion. Increased concern about widespread illegal logging has led to the development of regional Forest Law Enforcement and Governance (FLEG) processes, as well as action by the G8 group of countries and the World Summit on Sustainable Development (WSSD) ⟨*www.illegal-logging.info*⟩. Forests are also of concern to the United Nations Framework Convention on Climate Change (UNFCCC), particularly as the Kyoto Protocol potentially provides a mechanism for financing forest establishment and conservation. Under the Protocol, industrialized countries that lack options for expanding forests may partly compensate for their greenhouse-gas emissions by paying for the establishment and maintenance of forests in other countries ⟨*http://unfccc.int/*⟩. Forests do not feature so prominently in other recent international policy initiatives, such as the United Nations Millennium Development Goals and the WSSD held in 2002 ⟨*www.un.org/esa/sustdev/*⟩. However, the 2010 biodiversity target adopted by the Convention on Biological Diversity (CBD) and endorsed by the WSSD has become a central policy objective in conservation, aiming to achieve a 'significant reduction of the current rate of biodiversity loss at the global, regional and national level'. This potentially offers a significant opportunity to further conservation efforts worldwide, and its implementation is already taxing the scientific community (Balmford *et al.* 2005).

How can research be linked with policy? Many funding organizations now require that research be 'policy-relevant'. What does this actually mean? Simply put, research should strive to assist the process of policy implementation, without necessarily being policy-prescriptive. Put another way, policy-makers often do not like being told what to do, but recognize that research can play a role in helping to achieve policy goals. Researchers should not forget, however, that they have the capacity to significantly influence, or even lead, the policy agenda. Issues such as climate change, invasive species, and deforestation have all become the focus of international attention from policy-makers, partly as a result of the research that has been carried out on their actual or potential impacts.

How can research help implement policy? A key area is in helping to operationalize policy concepts. If many concepts in ecological science are difficult to define and measure precisely, as pointed out by Peters (1991), then the problem with forest policy is even more acute. Policy-makers seem to delight in coining terms whose meaning is difficult to pin down. Sustainable forest management is a case in point: what does this mean, exactly? Biodiversity is another term that means different things to different people. Of course, this obfuscation is partly deliberate: the use of vague terminology is designed to help provide politicians with room for manoeuvre, as they rarely enjoy being held to account. The use of poorly defined concepts is rightly an anathema to many ecological researchers, and this perhaps helps explain the antipathy that many researchers feel towards the area of policy.

It is practitioners who are typically at the sharp end of having to implement forest policy, and who often struggle with translating policy goals into practice. Forest managers are often assailed by poorly defined terminology: conceivably they might be asked to achieve sustainable forest management by using an ecosystem approach, by implementing multi-purpose forestry while adopting the precautionary principle, while not forgetting to consult stakeholders throughout the process. This kind of jargon is enough to task even the most hardened forestry professional. It is hardly surprising if these lofty policy goals sometimes fail to affect forest management on the ground.

Researchers can assist in the operationalization of policy concepts, by interpreting policy terms in the form of environmental variables that can be accurately and precisely measured. Researchers can also help determine whether policy goals are being achieved. There is a real concern that despite all of the policy interest in sustainable forest management, little is actually changing on the ground. Whether or not policy implementation is being successful is a worthy area of research itself, yet this is something that has been neglected by researchers. Available information suggests that the effects of certification and application of C&I have been limited to date (Rametsteiner and Simula 2003); many organizations that have certified forests are often those that were managing forests responsibly in any case (Leslie 2004). Why has application of C&I not been more successful in producing changes in forest management? Perhaps it requires the research community to engage more closely with the process, to help inform policy-makers and practitioners how best to define, measure and achieve progress towards policy goals; this is an important role that is often overlooked. Many of the indicators that have been proposed to date are difficult to implement in practice; often they are stated in vague or imprecise terms (Stork *et al.* 1997).

Forest ecologists who really want to make a difference to conservation may seek to see their results reflected in policy. How can this be achieved? Some suggestions:

- By engaging in a dialogue with policy makers, and by disseminating research results through policy fora such as the CBD. There are often mechanisms for researchers to present their results in this way, for example through preparation of an information note for delegates to the Convention's meetings.
- By presenting their results in a form that can be readily assimilated by policy-makers, for example by publishing a policy brief.
- By collaborating with NGOs who are continually campaigning for policy change.

- By publicizing their results in popular media, an approach of proven effectiveness in bringing issues to the attention of politicians, and an approach continually being adopted by campaigning NGOs, and even UN agencies.
- By publishing their results in scientific journals with a high impact factor, which can be remarkably successful in attracting media attention and increasing awareness among politicians.

1.7 Defining terminology

As mentioned above, one of the principal challenges to interpreting policy concepts relates to defining the terms used. Many terms used in forest conservation are open to a variety of different interpretations, and this can present a significant obstacle to clear communication. It is therefore important to define terms precisely at the outset of any investigation. In addition, make sure that your collaborators and partners share the same definition and understanding of the terms involved that you do.

The problems that can arise are usefully illustrated by reference to that most fundamental of definitions: what is a forest? Simply put, a forest is a type of vegetation dominated by trees. This might seem self-evident, and the problem of defining a forest might seem trivial, but the issue has been the subject of serious debate at the international level. Problems arise because of variation in the use of different terms, such as forest, woodland, savannah, and parkland, in different areas and among different communities of people. In some areas, the word 'forest' has specific legal connotations, defining rights of access and use. An example is provided by the royal hunting forests of northern Europe, which were traditionally used by the monarchy for exploiting populations of large vertebrates, and which often include extensive areas with low tree cover. Another aspect of the problem centres around how many trees are required within a given area in order for the vegetation to qualify as forest rather than some other vegetation type, such as savannah.

This issue has direct implications for conservation, most notably in the case of estimating deforestation rates and assessing the conservation status of particular forest types. If we cannot agree on what a forest is, how can we meaningfully analyse how much forest is being lost? This point was illustrated by Matthews (2001), who pointed out that the definition of forest adopted by the FAO has changed over time, and this has had a major impact on estimation of deforestation rates. For example, in the 1990 GFRA produced by the FAO, developed country forests were defined as land with tree crowns covering more than 20% of land area. In the FRA 2000, the definition was standardized to 10% for all countries. Matthews (2001) notes that a threshold of 10% is low enough to include land that might otherwise be described as tundra, wooded grassland, savannah, or scrubland rather than forest. In Australia, this change in definition led to an increase in estimated forest area from 40 million ha in the 1990 assessment to 158 million ha in the FRA 2000—enough to significantly influence estimates of change in global forest area (Matthews 2001). Another vexed question is whether plantations should be considered together with natural forests in estimations of global forest cover. Most forest ecologists recognize that plantation forests have characteristics substantially different from those of natural forests, and that expansion in area of the former does not adequately offset losses of the latter. This did not stop FAO considering both together when developing estimates of change in global forest cover (Matthews 2001). The lesson is: state your definitions clearly. They may be challenged, and you may be required to justify them.

Forest ecologists, managers and conservationists seem to delight in inventing concepts that are difficult to define exactly, or to apply in practice. For example, forests characterized by a relatively low level of human influence have been variously described as pristine, old-growth, primary, antique, climax, and ancient. Such terms should not be used uncritically, but should be exposed to rigorous scrutiny and defined precisely before being invoked. It is worth continually asking the question: can this variable be measured? And if so, how?

An example is provided by the concept of naturalness. This is considered to be very important by many conservationists, and reflects belief in a rather intangible property of 'natural' or 'wild' forests that plantations patently do not possess. But the issue is not clear-cut. Forests that were originally planted but have been left to regenerate naturally over a prolonged period of time can be very difficult to differentiate (in terms of structure and composition) from forests that have never been felled. Similarly, 'natural' forests that have been 'enriched' through localized planting of particular tree species can also retain most, if not all, of the characteristics of truly 'natural' forest. Naturalness therefore has a lot to do with the history of a particular forest, which can have a profound influence on its ecological characteristics. Unfortunately, information on the history of how a site has developed is often lacking, and consequently attempts are often made to infer the degree of naturalness from measurements of the current ecological characteristics of a forest, a process fraught with difficulty. A thoughtful consideration of how different types of naturalness might be defined is provided by Peterken (1996).

The problem of adopting poorly defined terminology can be further illustrated by the example of 'authenticity'. This is a term that has been used by the World Conservation Union (IUCN) and WWF as a way of describing the 'quality' of forest habitat. The term is defined as 'the extent to which an existing forest has a balanced ecology and a full range of species … a fully authentic forest is a forest in which all the expected ecosystem functions can continue to operate indefinitely' (IUCN/WWF International 1999). Measuring authenticity presents a considerable, perhaps insurmountable challenge. For example, what is meant by 'a balanced ecology' and 'a full range of species'? How might the indefinite operation of ecosystem functions be assessed? An inability to operationalize a concept such as this fatally undermines its value to ecological science and conservation management.

In response to this terminological confusion, the FAO has coordinated a process to harmonize forest-related definitions through an ongoing series of international meetings, for which proceedings are available (FAO 2005). These provide a useful reference point in terms of selecting definitions for use in any particular investigation (Table 1.3), and provide some valuable background with respect to the usage of different terms. However, even these definitions should not be applied uncritically. For example, many of them employ the word 'natural', which as noted above, has itself been subject to a variety of interpretations.

Some of the terms commonly used by forest conservation organizations are defined in Table 1.3. To these should be added the concept of *high conservation value forest* (HCVF), which was first defined by the FSC as part of their principles relating to forest certification, and is increasingly being used in conservation and natural resource planning and advocacy, most notably by WWF (Jennings *et al.* 2003). The concept focuses on the values that make a forest particularly important in conservation terms, rather than the definition of particular forest types (primary, old growth, for example) or methods of forest management. HCVF may therefore have widespread value as a tool for forest conservation planning and management, but this again depends on how the concept is operationalized. The values by which HCVF is defined include measures of biodiversity value (such as endemism, endangered species, refugia), the occurrence of 'naturally occurring species . . . in natural patterns of distribution and abundance' and presence of 'rare, threatened or endangered ecosystems', as well as provision of environmental services to people (such as watershed protection, erosion control, subsistence, health) (Jennings *et al.* 2003). Some practical guidance is now available for forest managers and conservation practitioners to support implementation of the concept (Jennings *et al.* 2003).

The aim of this section is to encourage increased precision in the use of terms relating to forest conservation, and not to provide a comprehensive survey of the terms in current use. However, there are four further concepts that merit further elaboration, given their widespread inclusion in policy documents and their current importance to those involved in practical forest conservation. These are the ecosystem approach, ecosystem management, the precautionary principle, and adaptive management. Definitions of these concepts are provided in Box 1.2.

Development of these concepts reflects growing recognition of the many ecological services provided by forests, and a shift away from managing forests purely for timber. To a degree, they could all be viewed as different perspectives on the same theme, but their overlapping and uncertain definitions can be the source of great confusion. IUCN *et al.* (2004) provide a comparison of some of these terms, noting some linkages between them: for example, the ecosystem approach advocates use of the precautionary principle. Perhaps the main difference between the ecosystem approach and sustainable forest management is that the former places greater emphasis on negotiation to solve problems, whereas the concept of

Table 1.3 Proposed definitions for some key concepts relating to forests (Carle and Holmgren 2003, FAO 2005). (Note that not all of the definitions listed below have been formally 'harmonized' or agreed by the FAO process).

Natural forest	Forest stands composed predominantly of native tree species established naturally. This can include assisted natural regeneration, excluding stands that are visibly offspring/descendants of planted trees.
Semi-natural forest	A managed natural forest which, over time, has taken on a number of natural characteristics (such as layered canopy, enriched species diversity, random spacing, etc.) or planted forests which acquire more natural characteristics over time.
Planted forest	Forest stand in which trees have predominantly been established by planting, deliberate seeding or coppicing, where the coppicing is of previously planted trees.
Primary forest	A forest that has never been logged and has developed following natural disturbances and under natural processes, regardless of its age.
Secondary forest	A forest that has been logged and has recovered naturally or artificially.
Old-growth forest	Stands distinguished by old trees and related structural attributes that may include tree size, accumulations of large dead woody material, number of canopy layers, species composition, and ecosystem function.
Forest management	The formal or informal process of planning and implementing practices aimed at fulfilling relevant environmental, economic, social, and/or cultural functions of the forest and meeting defined objectives.
Sustainable forest management	The stewardship and use of forests and forest lands in a way, and at a rate, that maintains their biodiversity, productivity, regeneration capacity, vitality, and their potential to fulfil, now and in the future, relevant ecological, economic, and social functions, at local, national, and global levels, and that does not cause damage to other ecosystems (definition from the Ministerial Conference on the Protection of Forests in Europe, now adopted by FAO).

sustainable forest management tends more towards the application of professional judgement (IUCN *et al.* 2004). Practitioners should certainly be aware of these concepts, and are encouraged to consider critically how they might be put into practice.

Box 1.2 The ecosystem approach, ecosystem management, the precautionary principle, and adaptive management.

- The *ecosystem approach* has been adopted by the CBD as a central strategy in the implementation its goals. It can be described as 'a strategy for the integrated management of land, water and living resources that promotes conservation and sustainable use in an equitable way'. It is described by the CBD (*www.biodiv.org*) as being 'based on the application of appropriate scientific methodologies focused on levels of biological organization which encompass the essential processes, functions and interactions among organisms and their environment. It recognizes that humans, with their cultural diversity, are an integral component of ecosystems'. With respect to forests, the CBD states that 'the ecosystem approach requires adaptive management to deal with the complex and dynamic nature of forest ecosystems and the absence of complete knowledge or understanding of their functioning... the conservation of their structure and functioning should be a priority target'.
- The goal of *ecosystem management* is the simultaneous use of biological resources and the maintenance of the integrity of the ecosystems that produce the resources. It can be considered as the basis of sustainable forest management and the ecosystem approach, and focuses on managing ecological units in an integrated and holistic way (IUCN *et al.* 2004). The term has been particularly used in the USA, where it is was adopted by the US Forest Service in the 1990s. A useful overview of the application of the concept to forests is provided by Johnson *et al.* (1999).
- The *precautionary principle*, or precautionary approach, is increasingly being employed in environmental policy and management. The principle has been interpreted differently by various workers, and a number of different definitions exist, leading to some confusion about what it actually means. Cooney (2004) provides a comprehensive account of the development of the principle, and how it has been applied to conservation, together with the different definitions that have been used. As a minimum, the precautionary principle requires that scientific certainty of environmental harm is not required as a prerequisite for taking action to avert it (Cooney 2004). When applied according to a relatively 'strong' definition, the principle may lead to prohibition of any activities that pose an environmental threat, and for this reason application of the principle is often controversial. Application of the concept to forest management and conservation is considered in depth by Newton and Oldfield (2005).
- *Adaptive management* approaches focus on acquiring knowledge from experience, monitoring and research, and integrating this information into more effective management practices (Lindenmayer and Franklin 2002). Further information and resources on adaptive management approaches are provided in section 8.3.

1.8 Achieving precision and accuracy

This book provides information about techniques that can be used to make measurements of ecological phenomena. If the measurements are to be useful, and an appropriate degree of scientific rigour is to be achieved, then they need to be as effective, precise, and accurate as possible. How can this be achieved? Some suggestions are given below (Ford 2000):

- *Choose an effective technique.* Make sure that the chosen technique actually measures the thing that you are interested in. Simply because an instrument is available that generates numbers does not mean that these numbers are relevant to the problem that you have identified.
- *Practice.* Making measurements is a skill that can be developed with practice, which can increase the degree of familiarity with the instruments or techniques being applied. Always perform pilot experiments or take preliminary measurements before applying a new technique.
- *Avoid bias.* Accuracy refers to how closely the measurements made represent reality. In order to be accurate, the methods used must be free from bias. This can be assessed by comparing the results obtained by different methods in different situations. To improve accuracy, use multiple techniques to measure the variable of interest. Ensure that instruments are properly calibrated both before and during use.
- *Repeat measurements.* Precision relates to the repeatability of values measured under the same conditions and using the same technique. This is influenced by sampling intensity and design, and by the performance of the instrument or method used. Repetition of measurements improves precision and enables the degree of precision to be estimated, which has an important bearing on the interpretation of results. Results should be presented to a level of precision that is consistent with the method used. Simply reporting a high degree of precision in the results does not ensure that the information is accurate, and could potentially be misleading.
- *Measure variability.* Many ecological phenomena are highly variable, and as a result, making accurate measurements is often difficult. Variation can be due to intrinsic properties of the ecological system being investigated, or can be introduced as a result of the technique employed. Measurements of photosynthesis, for example, might be highly variable over short timescales because of variation in sunlight, and the response time of the instrument being used. Adequately assessing variability is fundamental to ecological science. Observations can best be portrayed as frequency distributions of the variable being measured (Underwood 1997).

It is important to make sure that results obtained are reported with an appropriate degree of precision. It is incorrect to report results with more significant digits than were observed during measurement. For example, if a tree height was measured with a precision to the nearest metre, then the result should not be presented as 36.0 m, as this suggests that the measurement is more precise than it actually is. It is important to consider the number of significant digits to take and record during original measurements, and care should be taken in presenting results. If numbers are multiplied or divided, the number with the fewest significant figures limits the number of significant digits in the result. For example, if measurements 756.83 and 42.1 are multiplied, the product is 31862.543. However, only the first three figures in the product (3, 1, and 8) are significant, because the number 42.1 was presented to only three significant digits. The first three figures in the product are the only ones that are reliable (Husch *et al.* 2003). Similarly, in the addition of 253.026 + 1.4 = 254.426, the result has only four significant digits and should therefore be expressed to only one decimal place, i.e. 254.4.

Using greater precision than is needed is a waste of time, effort and money. Some suggestions (Husch *et al.* 2003):

- Do not try to make measurements to a greater precision (more significant digits) than can reliably be indicated by the measuring process or instrument. For example, it would not be appropriate to attempt to measure the height of a tree to within a few centimetres with an Abney level.
- The precision needed in original data may be influenced by how large a difference is important in comparing results. For example, if the objective is to compare different approaches to forest management in terms of how much biomass is produced, and estimates are to be compared to the nearest tenth of a cubic metre, there is no need to estimate biomass with any greater degree of precision than this.
- The degree of precision chosen is influenced by the variation in the population sampled and the size of the sample. If variation within the population is high, or if the sample size is not large, then high measurement precision is worth while.

1.9 Linking forests with people

One of the main developments in forestry practice over the last three decades has been its evolution from a practical discipline with a primary, or even exclusive, focus on management of forests for timber, to a more holistic approach recognizing that forests provide a wide range of environmental and social services and that provision of these should form an objective of management. The development of concepts such as forest ecosystem management and multi-purpose forestry are symptomatic of this process. The importance of forests to people has been increasingly recognized, as illustrated by the widespread implementation of forest management approaches explicitly aimed at or involving local communities, such as community forestry and social forestry. The importance of actively involving local communities and other stakeholders is consistently an element of approaches to sustainable forest management.

In a similar way, the practice of conservation has shifted from a primary focus on the conservation of individual species, to a broader approach in which the importance of meeting the needs of local people is explicitly recognized. Many conservation projects now integrate conservation actions with rural development approaches.

What has this meant for forest managers and conservation practitioners? Often, they are now expected to fulfil roles that they were not trained for. For example, many foresters who were trained primarily in practical silviculture and the principles of forest management now find themselves responsible for managing a wide range of habitats, and having to justify their management decisions to a wide range of individuals and organizations. Multi-purpose forests require multitasking managers. Similarly, many conservation professionals, who might have trained as biologists or ecologists, now find themselves spending more time dealing with people than with the habitats and species that they were trained to manage.

As a result, it is now rare for ecological techniques to be applied in a conservation context that does not involve some link with people. Concepts such as the ecosystem approach and sustainable forest management place great emphasis on the need to involve the public, consult stakeholders, develop partnerships with local communities, and understand the role of forests in supporting livelihoods. These are all laudable aims, but there is no doubt that this shift in focus represents an enormous challenge to those involved in practical forest management and conservation.

Ideally, the social components of projects should be carried out by specialist professional staff who have received appropriate training and possess the required skills. The number of people working in forest conservation with a background in social science or rural development training is, however, very small. Often managers with a technical training in forestry, environmental management, or some other biophysical subject are required to extend their activities to include social elements. However, it is also worth noting that many young people entering into a career in conservation today recognize the importance of social issues and are keen to develop skills in this area to complement their biophysical training. Perhaps the new generation of conservation managers will overcome the traditional barriers between subjects and employ whichever techniques and methods seem appropriate to the task in hand, whether they be social or biophysical in origin. In my experience, I have seen many students take on integrated projects of this nature and do them with alacrity and success, and in my mind, this form of integration is definitely to be encouraged. On the other hand, someone trained in ecological science may find their first meeting with a social scientist extremely challenging: the world views of these schools of thought can be profoundly different.

Techniques for social science are often a crucially important part of the toolkit of practitioners involved in forest conservation. A wide variety of methods are available, and a considerable body of experience has now been accumulated regarding their application in situations relevant to forest management and sustainable use. Widely used methods include *participatory rural appraisal* (PRA), *rapid rural appraisal* (RRA), and the *sustainable livelihoods approach* (SLA). Tools that are commonly used as part of these methods include:

- structured or semi-structured interviews with key informants, group interviews, workshops
- focus group discussions
- preference ranking and scoring
- mapping and modelling
- seasonal and historical diagramming, use of timelines
- direct observation, foot transects, familiarization, participation in activities
- biographies, local histories, case studies
- ranking and scoring.

Detailed description of social survey techniques is beyond the scope of this book. Some excellent guides are now available that provide an overview of the methods. Particularly recommended are those by Chambers (1992, 2002), Holland and Campbell (2005), McCracken *et al.* (1988), Pretty *et al.* (1995), and Theis and Grady (1991). An example of how these methods can be applied in practice, and integrated with biophysical research methods, is provided by the CEPFOR Project (Box 1.3).

One approach that has proved particularly valuable for understanding the complex issues surrounding the role of forests in rural development (see Box 1.3) is the sustainable rural livelihoods framework (Ashley and Carney 1999, Carney 2002). This approach views the livelihoods of people as depending on the availability of certain assets, namely:

- *Natural capital.* This includes the natural resource stocks (forest resources) from which products and services useful for livelihoods are derived.
- *Physical capital.* This comprises the basic infrastructure and producer goods needed to support livelihoods (shelter and buildings; tools and equipment used for farming or forest management; transportation, energy, and communications; etc.).

Box 1.3 Example of an integrated forest conservation and development research programme: the CEPFOR Project.

CEPFOR was an international collaborative research project that examined the commercialization of non-timber forest products (NTFPs) in Mexico and Bolivia (Marshall *et al.* 2006). NTFPs have recently been the focus of a great deal of interest among forest conservation and development organizations, as they seem to offer a 'win–win' solution by enabling local communities to benefit financially from the sale of forest products, thereby increasing the economic value of forests and acting as an incentive for their conservation. However, in practice attempts at commercializing NTFPs have often failed to deliver the expected benefits. The CEPFOR Project was designed to find out why.

Fig. 1.3 Interviewing mushroom collectors in Cuajimoloyas, Mexico. An example of one of the social survey techniques used in the CEPFOR project investigating the use of non-timber forest products by local communities. (Photo by Elaine Marshall.)

The commercial development of forest resources is a complex issue. Clearly, success has a lot to do with economics: there must be demand for the product and a market within easy reach. There are many social factors that can also influence success, such as the way a community is organized and how it collects, processes, and trades the product. Ecologically, the main issue is how the forest resource is managed and whether extraction of the product is sustainable. To address these different issues, CEPFOR employed a multidisciplinary team of specialists, with expertise in economics, social science, and forest ecology. Social scientists employed a variety of participatory methods, including structured and semi-structured interviews, group interviews, workshops, ranking and scoring, and development of narratives based on personal experience. Interviews were used to collect much of the economic information, which was analysed by using traditional econometric approaches.

One of the biggest challenges to an interdisciplinary project such as this is how to integrate the different types of data that are generated, including both quantitative and qualitative information. Unlike some purely social research projects, CEPFOR explicitly aimed to test a series of hypotheses identified at the outset of the project on the basis of a thorough literature review, and this provided a valuable focus for analysing the many different forms of data collected. In addition, CEPFOR found that the SLA provided a very valuable analytical framework, which successfully enabled research findings to be integrated and related directly to the livelihoods of poor people. Bayesian approaches were also used to integrate the research results, and to present them in the form of a decision-support tool, designed to support practical decision-making (Marshall *et al.* 2006).

- *Human capital.* This includes the skills, knowledge, ability to work, and health that people need to pursue different livelihood strategies and achieve their objectives.
- *Financial capital.* This includes the financial resources that people use to achieve their livelihood objectives, including savings in various forms, access to credit, earnings, remittances, and any debt burdens.
- *Social capital.* This refers to the social resources that people draw upon to help meet their livelihood objectives, including networks and connections between people, memberships; relationships of trust; and the rules, norms, and sanctions associated with different institutions.

The 'livelihoods framework' concept considers the impact of different environmental, socio-economic and political factors on the availability of these different assets that are required for living. A livelihood is considered sustainable when it can cope with and recover from stresses and shocks and maintain or enhance its assets into the future, while not undermining the natural resource base (Chambers and Conway 1992). The ways in which people combine and use their assets to achieve their goals are referred to as their livelihood strategies, which might include harvesting particular forest products (Box 1.3). The livelihoods framework therefore provides a useful way of considering how different environmental, socio-economic, and political factors might affect the livelihoods of people, and their livelihood strategies, by influencing the availability of different assets. Further details of this approach are available at ⟨*www.livelihoods.org/*⟩.

2
Forest extent and condition

2.1 Introduction

Tools for mapping the extent of forest cover are of central importance to forest conservation and management planning. This chapter describes the use of remote sensing technologies and geographical information systems (GIS), which together have revolutionized our ability to map patterns of forest distribution and estimate rates of forest loss and recovery. Increasingly, remote sensing and GIS methods are also being used to assess forest condition and the environmental services provided by forests. This can include assessment of forest composition, the degree of canopy cover, tree density, the pattern and intensity of natural and anthropogenic disturbance, and forest fragmentation, among many other variables. Many of these variables can be measured by using techniques other than remote sensing and GIS, as described in subsequent chapters.

The application of remote sensing and GIS technologies to forest assessment is the focus of ongoing research, and the techniques continue to develop rapidly. In order to keep abreast of developments, the reader is encouraged to monitor relevant scientific journals such as the *International Journal of Remote Sensing*, *Remote Sensing of the Environment*, and *Forest Ecology and Management*. Other textbooks that provide a useful introduction to remote sensing include Campbell (1996) and Lillesand and Kiefer (1994). Recent books that explicitly focus on applying remote sensing methods to forests include Corona *et al.* (2003), Franklin (2001, 2006), and Wulder and Franklin (2003). Horning (2004) has provided an exceptionally useful online resource that provides practical guidance to the use of remote sensing methods in biodiversity conservation, with a particular focus on the use of Landsat imagery ⟨http://cbc.rs-gis.amnh.org/guides/⟩. Cohen and Goward (2004) provide a more general review of the role of Landsat imagery in ecology.

Remote sensing data can be collected from a wide variety of ground-based, airborne, or satellite sensors, which vary markedly in their spectral characteristics, resolution, and scale. Rather than provide a comprehensive survey of these different methods, this chapter focuses on those techniques that are most likely to be of use to forest ecologists and conservationists. First, the use of aerial photography is described, then the most widely used forms of satellite remote sensing are considered. Most other ground-based and airborne sensors have a relatively limited, specialist use, and are not considered here in depth, although mention is made of lidar, which is increasingly proving of value to ecologists (Lefsky *et al.* 2002).

The application of these methods to forest ecology and conservation is described, with reference to practical examples. An introduction to GIS methods is then provided, followed by a description of the methods used to assess the spatial characteristics of forest landscapes.

2.2 Aerial photography

Aerial photography has been widely used for assessment of forests for more than 50 years; it therefore has the benefit of being a tried and tested technique (Lachowski et al. 2000). Black-and-white, colour, and infrared aerial photographs are routinely collected over many forest areas, and are used for forest mapping, assessment of forest condition, forest management planning, and conservation assessments (Figure 2.1).

Despite the development of satellite technologies, aerial photographs are still the most common form of remote sensing used to assess and map forests, primarily because they can provide high-resolution images at relatively low cost, and are relatively easy to use. They are also flexible: photographs are available at a range of scales, and can be produced by using a variety of different films, lenses, and cameras (Franklin 2001). The most significant advantage of aerial photographs

Fig. 2.1 Aerial photograph of part of the New Forest National Park in southern England. Such photographs are an extremely useful tool for assessing forest extent and distribution. On this image, the boundaries of forest fragments and even the location of individual trees can readily be determined. Comparison of such images taken at different times enables vegetation change to be assessed. (Courtesy of Getmapping plc, with permission.)

over satellite images is that they can be interpreted with little or no processing, which greatly increases their practical value and reduces their cost. With relatively little practice or training, most people are able to readily interpret many of the features illustrated in a typical aerial photograph. Useful introductions to the use of aerial photography in forest assessment are provided by Franklin (2001) and Hall (2003). Details of the methods used for estimating stand variables from aerial photography, including appropriate algorithms, are provided by Stellingwerf and Hussin (1997). Wolf and Dewitt (2000) provide a detailed account of the principles of photogrammetry (or the methods by which information can be derived from remote sensing imagery, including photographs) and its links with GIS.

2.2.1 Image acquisition

If financial resources are available, it may be possible to commission an organization or company to provide the photographs required. Specialist companies are now available in many areas that will acquire specific imagery on request. However, it is often the case that aerial photographs are already available for the area of interest. National forest services or conservation agencies may possess extensive archives of air photo imagery, although it is important to remember that interest in such photographs goes well beyond the forestry and conservation sectors. Land planning and rural development agencies, local or regional government administrations, hydrological surveys, and agricultural departments may all have commissioned aerial surveys at various times. Failing that, national military or defence organizations can usually be relied upon to possess comprehensive air photo coverage, which in some cases is made available to the public sector. Some companies, such as those involved in mining or the construction of oil, gas, or water pipelines, also invest in developing extensive archives of air photos. Increasingly, comprehensive coverage is offered by specialist companies who can provide specific images to order, from archives that they have already developed. The Internet resource Google Earth ⟨*www.earth.google.com*⟩ is a particularly useful source of such imagery.

Whatever the source of the imagery, a key consideration is scale, which determines the area on the ground that the photograph can detect. The *spatial resolution* or resolving power of a photograph describes the degree of detail on the ground that can be observed, and is influenced by the properties of the camera lens and film used in taking the photograph, as well as the characteristics of the object itself, such as its degree of contrast with its surroundings. The resolution of the image is often expressed in the form of line pairs per millimetre, values of which can be calculated based on the finest set of parallel lines that can be clearly resolved when the image is examined (Wolf and Dewitt 2000). These values can be converted into an effective spatial resolution at a given scale, giving an indication of the size of objects on the ground that can be differentiated. Resolution values that are obtained for photography systems used to assess forests generally fall within the range 0.25–0.8 m at a scale of 1 : 20 000 (Hall 2003), although higher-resolution imagery can sometimes be obtained.

The *spectral sensitivity* of an aerial photograph refers to the sensitivity of the film to different parts of the electromagnetic spectrum, which depends on the type of film used. The main types of film used are black-and-white, black-and-white infrared, colour, and colour infrared (Lillesand and Kiefer 1994). Radiometric resolution refers to the image contrast or density on the film, and is influenced by the dyes and metallic silver used in the manufacture of the film, and its degree of exposure. Films also differ in terms of their types of emulsion and how they are processed, which can affect the graininess of the image, and therefore its spatial resolution. The characteristics of films used in aerial photography can most readily be obtained by reference to the websites of the relevant manufacturers (Hall 2003). The quality of the image produced can also be influenced by the type of paper used in printing the image from the negative.

Often, aerial photographs are digitized by using a scanner, to create digital images that can be viewed on a computer and incorporated within a GIS. It is important that the scanner used has sufficient geometric and radiometric resolution, as well as high geometric accuracy, as otherwise the scanning process can introduce artefacts into the image. A number of scanners specially designed for photogrammetry are commercially available (Wolf and Dewitt 2000). Although relatively low-cost desktop scanners can also be used, they are generally less accurate and may distort the image. The minimum radiometric resolution of the scanner should be 8-bit (256 levels), although most modern scanners are able to capture images at 10-bit (1024 levels) or higher. Minimum pixel sizes should be of the order of 5–15 mm, and the positional accuracy should be around 2–3 mm (Wolf and Dewitt 2000). Scan resolution is often given in the form of dots per inch (dpi). The size of a single pixel on the ground can be calculated by expressing the scale as 1 cm = x m, then using the following simple formula (Hall 2003):

$$\text{pixel size (m)} = \frac{2.54 \times \text{scale (m)}}{\text{dpi}}$$

As a general rule, images should be scanned so that the pixel size is no larger than 20–25% of the size of the object to be resolved. For example, at a map scale of 1 : 20 000, 1 mm on the map is equivalent to a distance of 20 m on the ground. In order to be able to map this level of precision with a scanned image, the pixel size should be no larger than 4–5 m (Hall 2003).

It is important to check the date when the photograph was taken. The time of year influences the solar angle and therefore affects how objects within the photograph are illuminated by the sun, and the size of shadows that are cast. The phenology of the vegetation changes through the growing season, even in evergreen forests, and this influences the characteristics of the vegetation on the photograph. Details of the flightpath may also be useful as an aid to interpretation.

It is also important to note that aerial photographs vary in terms of the angle above the ground at which they are taken. Whereas vertical photographs are taken with the axis of the camera arranged vertically, oblique photographs will result if the camera axis is tilted. In forest ecological work, vertical air photographs are almost always preferred. If obliques are available, they may be of some value in terms of interpreting the characteristics of a forest area, but are likely to be of limited value with respect to development of forest maps unless they can be adequately converted through some form of digital processing (orthorectification).

2.2.2 Image processing

Much useful information can be gained by examining aerial photographs visually, for example by using a hand lens or binocular microscope. Stereoscopes can be used to view stereo pairs of photographs, enabling a three-dimensional image to be viewed. This is an important technique for photogrammetry, and is described in detail by Wolf and Dewitt (2000). However, at least for most ecological applications, photographs are generally viewed on a computer screen following digitization (it should be noted that instruments for on-screen stereoscopic examination of digitized images are now available; Hall 2003).

Once in the digital domain, the photograph will often need to be rectified if it is to be used as a basis for spatial analysis or mapping. *Image rectification* refers to the process of producing an image that is geometrically corrected, removing any distortions introduced during the photographic process. The process of correcting for distortion caused by variation in topography is called *orthorectification*, and photographs that have been processed in this way are often referred to as *orthophotos*. Orthorectification is almost always necessary, unless photographs are obtained that have already been processed in this way.

There are two main methods by which an aerial photograph can be orthorectified. First, *ground control points* (GCPs) are taken at selected locations within a landscape; these may be obtained from field surveys, perhaps by using a *global positioning system* (GPS), or directly from a published map. These points are then located on the image and their coordinates entered. At least 3–5 GCPs must be established in this way, but this is a minimum, and more accurate results are obtained if a larger number of GCPs is used. An alternative method is to use *digital elevation models* (DEMs), which may be derived from digital maps, some remote sensing data (such as lidar) or from stereoscopic models by photogrammetric methods. As with GCPs, a relationship is determined between the map coordinates in the real world and locations on the digitized aerial photograph, and the digital image is then resampled to create the rectified image. This resampling involves warping the image so that distance and area measurements made on the image are closely related to those in the real world. Orthorectification is generally done by using specialist software; some widely used software packages are listed in Table 2.1.

Whether or not a photograph is rectified, it will certainly need to be referenced if it is to be used as a basis for producing maps. The process of *georeferencing* (sometimes called ground registration) involves processing an image so that it is aligned according to a ground coordinate system (Wolf and Dewitt 2000). Although the process shares some similarities with orthorectification, it is important not to confuse the two processes: whereas orthorectification corrects for distortion in the image, georeferencing enables the image to be related to existing map coordinate

Table 2.1 Selected computer software packages appropriate for processing aerial photographs.

Product	Comments	URL
Aerial Image Corrector (AIC)	A relatively cheap product that enables aerial images to be rectified, referenced and mosaiced.	www.tatukgis.com
ArcView Image Analysis	An extension to the widely used ArcView GIS software that enables imagery to be manipulated and viewed. Can be used for data visualization, data extraction/creation, and analysis.	www.esri.com/software/arcview/extensions/imageanalysis/
ER Mapper	Very powerful, relatively expensive, high-specification software package, enabling a wide range of digital image analyses including orthorectification.	www.ermapper.com/
ERDAS IMAGINE	Another powerful, and relatively expensive, but widely used software package. The same company also markets simpler versions of IMAGINE software, and other sophisticated products such as the Leica Photogrammetry Suite.	http://gis.leica-geosystems.com/
Orthoengine	A flexible software package specifically designed for orthorectification; can also be used to produce digital elevation models (DEMs).	www.pcigeomatics.com/product_ind/orthoengine.html
SmartImage	An application that provides integrated spatial analysis and visualization, including rectification, aimed at MapInfo or ArcView users.	www.mappingandbeyond.com/
3D Mapper	A fully featured, desktop soft photogrammetry package solution that enables users to capture 3D vector data and orthophotos from (scanned) digital photography.	www.3dmapper.com/3dmapper.htm

systems. Again, GCPs are selected in the image for which coordinates on the ground are available. These are then entered, usually by clicking on the appropriate points on a computer screen using a mouse. Most GIS software packages have the capability to georeference images in this way, although it should be noted that their ability to warp images (as required during orthorectification) is often limited. As a result, a GIS package cannot be relied upon to provide all of the tools that may be required to process a photograph to the degree necessary for its use in mapping activities; additional image-processing software (Table 2.1) may also be needed.

Once an image has been georeferenced it can be combined with other spatial data in a GIS with the same coordinate system. Measurements of variables such as distance and area can now be made, enabling different forms of spatial analysis to be carried out. The accuracy of these measurements depends on the accuracy of the georeferencing process, as well as the characteristics (such as spatial resolution) of the image, so care should be taken throughout the processing procedure, and attention paid to the levels of accuracy achieved.

2.2.3 Image interpretation

Aerial photographs are usually interpreted visually, a skill that can be developed through training and practice. Guidance on the interpretation of air photographs is provided by Avery (1968, 1978) and Avery and Berlin (1992). Areas or objects can be differentiated by inspection of characteristics such as tone, texture, pattern, size, shadows, shapes or associations (Lillesand and Kiefer 1994). These may be defined as:

- *Image tone or colour*. Many objects have a characteristic colour or tone, depending on the specific signatures of electromagnetic radiation that are reflected or emitted. Different types of vegetation may therefore vary in how they appear on either black-and-white or colour images, according to their species composition, phenological state and canopy characteristics. Coniferous tree canopies, for example, often appear darker than those of broadleaved tree species. Usually, similar objects emit or reflect similar wavelengths of radiation. The types of camera and film used can also influence how objects appear on an image. For example, on colour infrared images forest canopies tend to appear pink or red rather than the usual tones of green.
- *Texture*. Vegetation canopies differ in their surface texture, or whether they appear rough or smooth. Texture can readily be used to differentiate between different types of land cover, such as forest and cropland; for example, agricultural crops often appear to have a smoother, more homogeneous texture than most natural forest canopies. Forest canopies with many strata or canopy gaps can appear rougher than even-aged stands. Texture, as with the size of the objects being observed, is related to the scale of the image.
- *Pattern*. How objects are arranged within an image can help aid their identification. Plantation forests or orchards, for example, tend to be characterized by a more regular pattern of tree distribution than natural forests.

- *Size.* It is important to note the scale of the image during visual inspection, as this has a major bearing on how objects are interpreted. Both the relative and absolute sizes of objects are important in their identification. The absolute size of an object is determined by reference to the scale of the image.
- *Shadow.* The presence of shadows can greatly complicate image interpretation, as shaded features generally appear to be dark and difficult to discern. On the other hand, shadows can be used to help interpret features: for example, the length of a shadow cast by an individual tree can give an indication of its height relative to other objects on the image. Shadows can also display the shape of an object on the ground.
- *Shape.* The shape of objects can be highly diagnostic. Roads, rivers, and urban development can readily be identified because of their characteristic shapes, but it is also possible to differentiate some individual tree species on the basis of crown shape, as well as different kinds of disturbance to which a forest has been subjected—a fire or logging coup may produce a canopy gap with a very different shape to one caused by a natural windthrow.

Visual interpretation of aerial photographs can be used to map a wide variety of forest features, including variation in species composition, degree of crown closure, height class and density, and pattern of disturbance. Typically, forests are mapped as stands, which may be defined as areas with relatively homogeneous characteristics. Determining the boundaries between forest stands can, however, be difficult in practice. As with all other aspects of visual interpretation, decisions made by the analyst are subjective and may therefore be subject to a degree of error that can be difficult to quantify. However, experienced practitioners are able to discern features with a very high degree of accuracy, which can potentially be validated by reference to field surveys, and for this reason aerial photography is still the main technique of choice for producing maps in support of forest management. For example, in Finland, data for forest management planning are generally gathered by field surveys of forest stands that are first delineated by reference to air photos (Pekkarinen and Tuominen 2003). Many studies have shown that digitized aerial photographs can do better than satellite remote sensing data for a variety of forest mapping applications (Franklin 2001, Hyyppä *et al.* 2000, Poso *et al.* 1999).

2.3 Satellite remote sensing

There is no doubt that satellite remote sensing techniques have revolutionized the mapping, assessment and monitoring of forests, enabling precise estimates of forest extent and condition to be made at a range of scales, from local to global. However, there are two main challenges to the effective use of these methods. First, the analysis of satellite imagery is a highly technical field, with its own specialist methods and associated literature. Use of such imagery at anything other than a superficial level has generally required access to high-specification computing facilities and specialist software, which typically requires a high level of technical expertise to be used effectively. Second, the development of satellite remote sensing methods has been driven largely by rapid technological advancement supported by a great deal of scientific research. It can be argued that these developments have been led by producers of the imagery and the research community, rather than by its potential users. In reality, only a tiny proportion of available imagery has been put to any practical use; much of it resides in very extensive data archives, some of which are only rarely accessed. These problems have sometimes been compounded by a lack of understanding among practitioners of the strengths and weaknesses of satellite imagery, leading to unrealistic expectations of what such imagery can deliver. There are many examples of forest projects that have made substantial investment in satellite imagery, only for the results to be disappointing and of little practical value (Franklin 2001).

The barriers to effective use of satellite imagery in forest assessment are increasingly being overcome. Software tools are now available that enable images to be processed and classified relatively easily, and combined with other spatial data to produce highly visual outputs that can easily be understood by forest managers. Dialogue between the developers and potential users of this technology has helped improve the practical application of remote sensing methods in ways that can genuinely support decision-making. Experience of applying different methods to particular problems has enabled ready identification of the situations where use of satellite imagery is most likely to be valuable, and those where results are less likely to be satisfactory. As a result, it is now feasible for someone with no prior experience of the methods (such as a postgraduate research student, perhaps) to be producing highly accurate analyses of satellite remote sensing data with just a few weeks of appropriate training.

On the other hand, even though satellite imagery and the requisite analytical software are increasingly becoming available at little or no cost, use of these methods still requires substantial investment of time and effort, as well as capital expenditure. Careful consideration should be given to whether satellite imagery offers the most cost-effective way of addressing the issue at hand, or whether some cheaper, alternative method might be available. A valuable critique of the use of remote sensing in forest planning is provided by Holmgren and Thuresson (1998), which highlights a series of limitations in the technique, such as the difficulty of differentiating more than a small number of different forest types, and the high levels of inaccuracy that are often associated with vegetation classifications based on remote sensing imagery. Satellite sensors are only able to detect the canopy from above, and cannot directly measure the age, structure, height, or volume of forest stands, particularly when crown cover is complete. These methods should not be viewed as a quick technical fix; they can require a great deal of technical expertise and resources to be used effectively. It is worth remembering that better and less expensive ways of obtaining the necessary information might often be available (Holmgren and Thuresson 1998). Just because aerial photographs are a relatively mature technology does not mean they are no longer of value: they may often be the most cost-effective method available. Yet there may be no realistic alternative to the use of satellite data, particularly where the forest areas to be assessed are large and inaccessible (Figure 2.2).

As noted by Franklin (2001), satellite remote sensing should not be viewed as a panacea, and cannot be expected to meet all needs relating to forest monitoring and assessment. However, it can be considered as one of the most important sources of information available to those involved in forest conservation and management, and has the potential to be of even greater value in the future. A key challenge is to integrate remote sensing information with field observations, and to identify the appropriate role for remote sensing methods. Perhaps the most useful approach is to consider how field observations, aerial photography and satellite data can be used in a complementary way (Franklin 2001).

Fig. 2.2 A map of forest cover produced using satellite remote sensing (raster) data. The image is the island of Borneo and was created using MODIS data at a spatial resolution of 500 m. The depth of shading on the image relates to the density of tree cover; areas in white are largely absent of trees. These MODIS data have been used in a variety of different forest conservation assessments (e.g. Miles et al. 2006, DeFries et al. 2005). (Data from Hansen et al. 2003.)

How can the risk of failure be minimized? The following sections are designed to help identify some of the pitfalls and limitations in use of satellite imagery. A clear definition of the problem or objective in question is very important. Consider the level of accuracy and precision that is actually required in the analysis; it may be that relatively low-resolution data will be adequate, which could save much time and resources. Above all, discuss your intended approach with experienced practitioners if you possibly can, or better still, collaborate with someone who has experience of using these methods.

In which situations might satellite remote sensing be preferable? Common situations where this method is used include (Franklin 2001):

- mapping spatial distribution of forest cover and different types of forest
- assessing forest condition and forest health
- monitoring changes in forest extent and structure, for example in response to some management intervention or another form of environmental change; some believe that the greatest strength of remote sensing techniques lies in their use for environmental monitoring
- assessing forests at multiple scales and resolutions, enabling plot-based data to be scaled up to landscapes and regions
- assessing landscape pattern and structure by using quantitative methods
- assessing forest growth and biomass, which can be of value for estimating the environmental services provided by forests, such as carbon sequestration
- monitoring threats to forests (such as fire, flooding, or deforestation).

Remote sensing data derived from satellite-borne sensors first need to be acquired, then subjected to various forms of image processing. Typically, the image is then classified by using additional data, such as information derived from a field survey or forest inventory. Each of these steps in the use of satellite imagery is considered in separately in the following sections. The limitations and common problems associated with using satellite imagery for forest assessment are then profiled. Examples of applying satellite remote sensing to practical problems relating to forest ecology and conservation are considered in a subsequent section (2.5).

2.3.1 Image acquisition

The provision of satellite remote sensing data is highly dynamic, with new data sources continually becoming available as new satellites are placed in orbit. Traditionally, the main source of such data has been the national and international space agencies involved in developing and launching earth observation satellites, such as NASA, the European Space Agency, and the space agencies of countries such as Russia and India. Data can be purchased directly from these sources, but increasing amounts of data are now being made available for free download over the Internet. Researchers in academic institutions are particularly well placed to take advantage of educational discounts and academic agreements when accessing data, and much information is now available to researchers free of charge. Increasingly, however, remote sensing data are being provided by private corporations. For example, the very high-resolution data provided by Ikonos and Quickbird are available from Space Imaging and Digital Globe, respectively, typically at relatively high cost. Such high-resolution imagery is now so detailed that it can compete effectively with aerial photography, and can be used to map individual trees (see, for example, Koukal and Schneider 2003). The cost of obtaining remote sensing imagery is typically a major factor influencing purchasing decisions. Although an increasing number of archives are offering free access to satellite imagery, much of it is at low (less than 250 m) resolution and therefore of limited value for application at the scale of forest stands or landscapes. Before purchasing imagery, it is worth checking whether anyone else (perhaps a researcher in an institute or university, the forest service or other government agency) has already analysed imagery for the area in which you are interested—they may be prepared to share their data.

Some widely used sources of satellite imagery are listed in Table 2.2. The choice of which satellite data are most appropriate for a particular task depends on the nature of the problem being addressed, and the characteristics of available data. Different sources of satellite data vary in their resolution, which refers to the ability of the sensor to acquire data with specific characteristics and comprises the following components (Franklin 2001):

Table 2.2 Selected sources of satellite imagery.

Product	Comments	URL
EOMOnline	Sells a variety of imagery, such as Landsat 5, IRS Imagery and Landsat 7	www.eomonline.com/
Terraserver		www.terraserver.com/
ResMap	Provides free online access to satellite data, including over 10 terabytes of Landsat imagery	www.resmap.com
Ikonos		www.spaceimaging.com/
SPOT		www.spotimage.fr/
Landsat 7	Available from USGS	http://landsat.usgs.gov/
Orbimage		www.orbimage.com/
AVHRR	Available from USGS	http://edc.usgs.gov/products/satellite/avhrr.html
Quickbird		www.digitalglobe.com/
Global Land Cover Facility (GLCF)	Provides access to Landsat TM images for much of the world for download	http://landcover.org/

- *Spectral resolution.* The number and dimension of specific wavelength intervals in the electromagnetic spectrum to which a sensor is sensitive. Particular wavelength intervals are selected for specific applications; for example, the red region of the spectrum can be related to the chlorophyll content of leaves. Hyperspectral sensors detect very many narrow (2–4 nm) intervals (Figure 2.3).
- *Spatial resolution.* A measure of the smallest separation between objects that can be distinguished by the sensor. A system with higher spatial resolution can detect smaller objects. Spatial resolution can best be understood as the size of a pixel in terms of dimensions on the ground, and is usually presented as a single value representing the length of one side of a square (for example, a spatial resolution of 30 m indicates that one pixel represents an area of 30×30 m on the ground).
- *Temporal resolution.* The image frequency of a particular area recorded by the sensor. Satellite sensors vary in the time interval between visits to a particular area.
- *Radiometric resolution.* The sensitivity of the detector to differences in the energy received at different wavelengths. Greater resolution enables smaller differences in radiation signals to be discriminated.

Some of the most important sources of satellite remote sensing data are listed in Table 2.3, together with an indication of their spatial resolution. There are often trade-offs between different forms of resolution; for example, an increase in the number of bands that the sensor is able to detect increases spectral resolution, but often decreases spatial resolution. There may also be trade-offs in the design of sensors between spectral and spatial resolution (Franklin 2001). Imagery should be

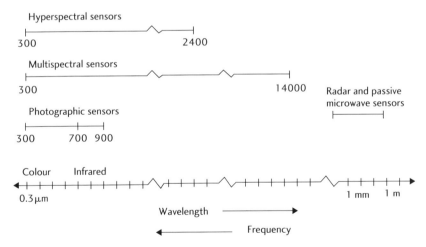

Fig. 2.3 Electromagnetic spectrum with regions of interest for remote sensing of forests. The main regions of interest are the optical/infrared and microwave portions of the spectrum. (Copyright 2001, From *Remote sensing for sustainable forest management*, by S. E. Franklin. Reproduced by permission of Taylor and Francis, a division of Informa plc.)

selected with a spatial resolution (pixel size) that is appropriate to the scale of the pattern being assessed. Collection of an unnecessary degree of detail can hinder interpretation of the imagery and slow the analytical process. Imagery should also be selected with a spectral resolution that is appropriate to measurement of the phenomenon of interest.

Variation between sources of remote sensing imagery in their resolution influences the scale to which they can be usefully applied. *Scale* refers to the area over which a pattern or process is mapped or assessed. According to cartographic convention, 'small-scale' refers to a large area of coverage, whereas 'large-scale' refers to a small area of coverage and consequently greater detail. The scale on a map or image refers to the difference in size between features represented in the image and

Table 2.3 Characteristics of selected sources of satellite remote sensing data that are of value for forest assessment and monitoring (adapted from Franklin 2001).

Satellite name	Sensor	Number of bands	Spatial resolution (m)
Landsat-5	TM	7	30–120
	MSS	4	82
Landsat-7	ETM+	7	15–30
SPOT-2	HRV	4	10–20
SPOT-4	HRV	5	10–20
	VI	4	1150
SPOT-5	HRG	4	2.5
	HRS	4	10
	VEGETATION 2	4	1150–1700
IRS-1B	LISS	4	36–72
IRS-1C, -1D	LISS	4	23–70
	PAN	1	5.8
ERS-1, -2	AMI (SAR)	1	26
	ATSR	4	1000
Space Imaging	Ikonos-2	5	1–4
NOAA-15	AVHRR	5	1100
NOAA-14	AVHRR	5	1100
NOAA-L	AVHRR	5	1100
Terra (EOS AM-1)	ASTER	14	15, 30, 90
	MODIS	36	250, 500, 1000
	MISR	4	275
DigitalGlobe	Quickbird2	5	0.61–2.44
Orbview-3	Orbview	4	1–4
IRS P6	LISS 3	4	23
	AWiFS	3	56

on the ground, and is usually expressed as a ratio of image distance over ground distance. For example, a scale of 1 : 100 000 indicates that a distance of 1 cm on the map represents a distance of 100 000 cm (1 km) on the ground. On an image of this scale, an area of 5 × 5 cm would represent an area of 2500 ha on the ground.

The relation between the resolution of imagery and the scale of phenomena that can be detected can be summarized as follows (following Franklin 2001):

- *Low spatial resolution* (for example NOAA, AVHRR, MODIS, SPOT VEGETATION, Landsat data). Usual applications are the study of patterns varying over hundreds or thousands of metres and for mapping at the small scale, for example the distribution of forest types within a region.
- *Medium spatial resolution* (for example Landsat, SPOT, IRS). Most relevant for assessing patterns that vary over tens of metres and for mapping at the medium scale, such as tree density, size of forest stands, and the characteristics of forest patches.
- *High spatial resolution* (for example Ikonos, Quickbird). Applicable to the assessment of phenomena that vary over distances of centimetres to metres and for large-scale mapping, such as mapping individual trees, detailed assessment of forest structure, or mapping the distribution of individual species within forest stands.

Other issues that might be considered when selecting imagery include (Franklin 2001, Horning 2004):

- *Spectral bands (channels)*, characterized by the bandwidth (the range of wavelengths detected), the placement of the bands (the portion of the electromagnetic spectrum that is used), and the number of bands: sensors with more than one band are *multispectral* and images with very many bands (more than 100) are called *hyperspectral*.
- *The history of the satellite sensor*. Analysis of deforestation, for example, requires a time series of images for comparison. Landsat imagery is available from the early 1970s, but most other imagery is only available from much more recent dates, particularly in the case of very high-resolution imagery.
- *The surface area covered by an image*, which varies between different satellite sensors, and influences the number of scenes that you will need to obtain. If possible, it is preferable to use just a single image to cover the entire study area, as tiling together different images to cover a particular area can pose problems (for example, if the images differ in quality).

How can the most appropriate source of imagery best be selected? The most effective way of making an informed decision is to discuss the alternatives with experienced practitioners, who can help identify successful approaches and potential pitfalls (Horning 2004). There are many relevant discussion fora accessible via the Internet, and specialists within university departments, research institutions, and government agencies might also be able to offer advice. It is also worth consulting research publications providing examples of the kind of investigation that you want to undertake. For example, Corona *et al.* (2003) provide a large number of recent examples involving application of different remote sensing data to forest management and biodiversity assessment. Most satellite data can be inspected before purchase, enabling appropriate images to be selected. A common problem in selecting imagery for forest assessment is that clouds may obscure the areas of interest; being able to browse the imagery before purchase enables cloud-free images to be selected. Sensors such as Terra MODIS that revisit areas relatively frequently (every 1–2 days) increase the chances of locating imagery that is cloud-free.

How can an appropriate spatial resolution be selected? There may be no substitute for trial and error. As a rule of thumb, try to select imagery that has a resolution a factor of 10 higher than the area of the features you are identifying (Horning 2004). For example, in order to differentiate forest patches with a minimum size of 1 ha (100 × 100 m), imagery with a spatial resolution of 30 m or finer would be preferred. In order to map tree crowns of around 3 m diameter, imagery with a resolution of at least 1 m should be used.

2.3.2 Image processing

Satellite remote sensing data need to be subjected to at least two forms of image processing in order to be of use: *radiometric processing*, which seeks to ensure that the measured data accurately represent the spectral properties of the target feature, and *geometric processing*, which aims to relate the locations of pixels to coordinates on the Earth's surface according to a chosen map projection (Franklin 2001). Radiometric processing removes the effects of sensor errors and/or environmental factors such as atmospheric interference, often using additional data collected when the image was obtained. Geometric processing corrects any errors caused by the curvature of the Earth and movement of the satellite, and involves registering an image to points on a map that has already been rectified.

After radiometric and geometric processing, the image may be further processed in a variety of ways. Unprocessed satellite imagery is generally stored in a grey-scale format, in which the numerical values of each pixel (known as the digital number, DN) are stored as numbers representing shades of grey from black to white, which represent the variation in energy detected. Image-processing techniques can be used to increase the degree of contrast between shades of grey, enabling different objects to be differentiated more readily. Usually, the image is converted from grey-scale to colour, which is achieved by assigning specific DN values to different colours. Colours chosen may be 'true colour', representing the spectral bands that were detected, or 'false colour', where colours are chosen that are different from the spectral bands detected (Horning 2004).

Many satellites produce multispectral data, or data collected from a number of different spectral bands. Use of multiple spectral bands enables land-cover types to be differentiated more easily, and therefore images from different sensors are often combined into what is known as a composite image. Different spectral bands can be mapped with different colours to help visualize the data. For example, a composite image can be produced from Landsat ETM+ data by selecting three spectral bands and assigning one to each of the three primary colours used in an RGB display—red, green, and blue (Horning 2004). The level of each colour in each pixel represents the measured value of each appropriate spectral band. The final composite image could potentially include thousands of different colours produced by different combinations of red, green, and blue, illustrating variation in the different spectral bands detected. The choice of colours used in producing an image can have a major influence on which patterns and objects are visually detectable.

Satellite remote sensing data are often used to map forest cover by using the *normalized difference vegetative index* (NDVI), which is calculated from the ratio between measured reflectivity in the red and near infrared parts of the electromagnetic spectrum (Franklin 2001). These bands are particularly sensitive to the density of green vegetation and the absorption of chlorophyll, and provide a high degree of contrast between vegetation cover and the soil surface. The NDVI is produced by using mathematical algorithms to transform raw data produced by the satellite sensor. The transformed data are used to produce a new image using the calculated colour values of each pixel. NDVI is calculated according to the following formula, where R = red and NIR = near infrared:

$$\mathrm{NDVI} = \frac{(\mathrm{NIR}-\mathrm{R})}{(\mathrm{NIR}+\mathrm{R})}$$

The *advanced very high resolution radiometer* (AVHRR) sensor, mounted on NOAA meteorological satellites, is a widely used source of data for calculation of NDVI and has been widely used in forest mapping over large areas. Landsat TM and ETM data can also be used to produce images based on calculation of NDVI, at substantially higher resolution than images derived from AVHRR data (i.e. 30 m rather than 1 km).

Some form of computer software is needed to process satellite remote sensing imagery. Although images can be manipulated to a basic degree by using standard graphics software, such as Adobe Photoshop (Horning 2004), typically specialist software is required that includes the algorithms needed for radiometric correction and geometric processing and for producing transformed or composite images. Some of the software packages listed in Table 2.1 can be used for this purpose; for example, ERDAS IMAGINE (Leica Geosystems) is popular software for this task. Increasingly, however, GIS software packages are offering at least some of the functionality that is required to process remote sensing imagery (see section 2.6). Some freely available software that can be used to process remote sensing imagery is listed in Table 2.4.

It is important to note that image processing is demanding in terms of computing resources; as the data files tend to be very large, high-speed processors, large

Table 2.4 Selected freely available software suitable for image processing of satellite remote sensing data (adapted from Horning 2004).

Product	Comments	URL
OpenEV	An Open Source project available for a range of platforms to display and analyse vector and raster geospatial data.	*http://openev.sourceforge.net/*
MultiSpec	Developed at Purdue University, MultiSpec was originally designed as a teaching tool but is now widely used by remote sensing practitioners, and has some sophisticated features.	*www.ece.purdue.edu/~biehl/MultiSpec/*
OSSIM	An integrated, Open Source tool for image processing and GIS. A graphical user interface is available, called ImageLinker.	*www.ossim.org/*
IDV	An Open Source Java-based program for visualizing and analysing geoscience data, developed by Unidata.	*http://my.unidata.ucar.edu/content/software/metapps/index.html*
SPRING	Developed by Brazil's National Institute for Space Research (INPE), combining GIS and image processing capabilities. Documentation available in Spanish and Portuguese.	*www.dpi.inpe.br/spring/*

amounts of RAM, and substantial hard disk space are required. Furthermore, specialist software is often technically demanding, and it can take a long time to learn how to use it effectively. The choice of an appropriate software package depends primarily on the objectives of the investigation and the financial and computing resources available, although the technical skills available among project staff is also an important determinant.

2.3.3 Image classification

Forest managers and ecologists generally want to know about the distribution of forest within a particular area. In order to obtain this information from satellite remote sensing data, some form of image classification is needed. How this classification is done depends largely on the types of land-cover class that need to be mapped and on the particular definition of forest adopted (see Chapter 1), as well as on how different forest types are defined. Ideally, the classes used should be defined by their characteristics visible during field survey, for the images and maps to be of practical value (Franklin 2001).

Images derived from satellite data can be classified visually, in much the same way as an aerial photograph, using characteristics of the image such as tone, colour, contrast, and shape of different objects. For this kind of classification, a true-colour image would generally be used. More usually, images are classified by using appropriate computer software that classifies each pixel on the basis of its spectral characteristics. As with image processing, the process of image classification may require specialist software, although increasingly many GIS software packages provide tools for this purpose. The key objective of image classification is to relate the spectral characteristics of individual pixels to the classes of interest, such as different forest stand types or communities. In general, multispectral images are used for this purpose.

Classifications are of two principal types: *supervised* and *unsupervised*. A supervised classification is one where some knowledge about the distribution of the classes within an image is used to assist the classification process. Locations for which some information is available, perhaps obtained from field survey or forest inventory, are used as 'training sites' to help identify the spectral characteristics of each class of interest, a task carried out by appropriate software. Once these spectral 'signatures' have been identified for each class, the software can be used to produce a map of different forest classes by relating the information available for each pixel to the signatures that have been determined. Each pixel is assigned to the forest class that it resembles most closely. An unsupervised classification does not make use of any external information, but involves classification of the pixels purely on their spectral characteristics by using some form of statistical procedure. Once groups of pixels have been identified by this approach, they can be related to classes of interest.

The choice of whether to use a supervised or unsupervised approach to image classification depends on whether or not appropriate information is available with which to train the data. Either approach can have advantages in different situations. A supervised approach might be preferred if there are classes that need to be mapped (for example, some forest type of particular conservation importance), regardless of how well the data actually represent them. The accuracy of the map will depend on how closely the remote sensing data are related to the classes of interest (Franklin 2001). In contrast, an unsupervised approach might be preferred if few field survey data are available, or if there is a need to identify variation in forest characteristics within an area where there are few preconceptions about the kinds of forest that might be present. The main risk of this approach is that the classes identified as a result of the classification process can sometimes be difficult to relate to the forest characteristics observable in a field survey. It should be remembered that remotely sensed imagery detects variation in forests from above the canopy, whereas forest classifications used in ecology or management are instead developed from below the canopy. For this reason, they sometimes do not coincide.

If a supervised approach is employed, a decision needs to be made regarding which types of forest are to be mapped. Many different ecological classifications of different forest types are available, some applicable only at local or national scales, others designed for application at regional or global scales (see section 3.12). Decisions also need to be made regarding whether various non-forest land-cover types should be included, such as agricultural land, urban development, or wetland areas. Consideration should therefore be given to how the maps produced will be applied in practice, while choosing an appropriate classification. The definitions used in the classification should be clearly documented.

It is also important to note the difference between the terms 'land cover' and 'land use'. Although they are sometimes used interchangeably, they refer to different things: whereas land cover is a description of what covers the Earth's surface, land use refers to how this land surface is being used by people. This is a particular issue in relation to forests, where different forest types (such as coniferous or broadleaved forest) might be differentiated as land-cover types. Each of these forest types might be used in different ways, such as a protected area, a recreational area, or a forest concession from which timber is being extracted. Whereas different ecological communities of forest trees can be considered as different land-cover types, forestry can be considered as a form of land use, and this has led to some confusion in the classifications adopted in forest mapping. Land use or land-cover types can usefully be differentiated by using different colours, patterns, or symbols. Most importantly, each class included on any map should be clearly defined and the distinctions clearly drawn between classes.

Many different methods can be used to classify remote sensing imagery, and new methods are continually being developed. Some of the principal methods that have been used in support of forest mapping and assessment are listed in Table 2.5. The choice of which classification method to use will largely be determined by the objectives of the study, the available software, and the skills and experience of the person undertaking the task. Typically, classification is undertaken by using some form of automated method, in which an algorithm is used to assign individual pixels or groups of pixels to an appropriate category. The algorithm can be applied to the entire image by using data from multiple satellite bands, in some cases together with other spatial data such as elevation, slope, and aspect. A variety of different algorithms are available, particularly for supervised classifications, which differ in how a particular pixel is assigned to a land-cover category.

Supervised image classifications may result in a pixel being unequivocally assigned to a particular land-cover class. Such 'hard' classifications are suitable when the signature data are gathered from training sites that are homogeneous, and where pixels with mixed land-cover classes are rare. However, many forests are spatially complex, characterized by the occurrence of mixed stands and high spatial heterogeneity. In this situation, an area represented by an individual pixel can contain more than one land-cover class, and a 'hard' classification method may be inappropriate. As an alternative, some form of 'soft classification' approach might be adopted, such as Bayesian or fuzzy methods (Chirici *et al.* 2003, Foody 1996).

Table 2.5 Classification methods used to classify satellite remote sensing data for use in forest mapping and assessment (adapted from Horning 2004, Franklin 2001, and Franklin et al. 2003).

Method	Description
Unsupervised classification	One of the most commonly used unsupervised classification algorithms is ISODATA, which is used to identify clusters of pixels or classes that are then labelled to produce a forest cover map. Before ISODATA is run, several variables need to be defined, such as the number of clusters that will be produced, the number of iterations, and how the clusters will be combined and divided.
Supervised statistical classification	A number of statistical algorithms are available, including linear discriminant analysis, minimum distance, and maximum likelihood. These algorithms differ in how the similarity between pixels is determined. The maximum likelihood method has been particularly widely used in forest classification, in which classes are defined on the basis of their mean and covariance matrix. However, if the data are not normally distributed, then non-parametric methods should be used instead, such as a parallelepiped classifier, neural networks, or decision trees.
Bayesian and fuzzy approaches	Fuzzy classification methods assign pixels to multiple classes, but based on different degrees of membership, expressed as a likelihood. Fuzzy approaches are becoming increasingly popular as many ecological classifications of forests do not have sharp boundaries between classes, and because many forest stands are heterogeneous in composition. Bayesian methods, and modifications such as Dempster–Shafer theory, similarly establish probabilistic relationships between pixels and the classes used.
Artificial neural net classification	Neural networks have developed from the field of artificial intelligence, and can be used to classify images through a process of machine learning. This method is more technically challenging than statistical approaches, and can require a degree of experience in order to be performed effectively. Extensive training data are required, but the results can be very accurate.
Decision tree classification	Another machine learning tool based on a set of rules governing how land cover classes should be assigned to pixels. This method involves defining nodes in a decision tree by using a set of rules, generally implemented by means of specialist software. Useful for combining remote sensing and other variables.
Image segmentation	A method that enables contiguous pixels to be grouped into relatively homogeneous areas (segments). This can be done before classifying an image. The method employs algorithms that analyse the value of a pixel in relation to the values of neighbouring pixels, to determine whether or not they should be grouped together. After being 'segmented' in this way, an image can be classified at the level of segments rather than pixels, reducing the amount of time required. This method is increasingly being implemented in remote sensing applications relating to forests, particularly with high-resolution imagery.

Such methods provide an estimate of the degree to which individual pixels belong to different land-cover classes, which may be illustrated on separate maps. This method of mapping each class as a separate layer indicating the percentage coverage of that particular cover type in each pixel is referred to as *continuous fields mapping*.

However a map is produced, it will contain errors. There are two main types of error in forest maps derived from satellite imagery: *position error* and *thematic error*. Position error refers to inaccurate placement of a feature or object on a map, even if its size and shape are correct. Thematic error refers to a situation when an object or feature on a map is identified incorrectly. For example, if an area labelled as mangrove forest on a particular map is in fact shrubland or some other vegetation type, then this would be classified as a thematic error. If it were correctly classified as shrubland, but was located incorrectly, then this would be a position error. These different types of error are often difficult to separate in practice. However, estimation of these errors is an essential part of quality control and should form part of any mapping project using such data.

Accuracy can be influenced by the characteristics of the area being mapped, on the quality of the data used to assist in the classification, and the classification method used. Accuracy estimates of 80% or less are common. The accuracy of a land-cover map can be assessed by selecting sites of different cover type on the image, then doing a field survey to determine whether or not the forest has been classified correctly. The accuracy of the classification can be assessed with respect to each land-cover class. The data can be compared statistically, typically by using contingency tables, sometimes known as a classification confusion matrix (Lillesand and Kiefer 1994). Care must be taken to ensure that a representative sample of locations is selected for comparison. A comprehensive description of the analysis of accuracy, including the use of contingency tables, is provided by Czaplewski (2003). Such analyses can be used to estimate the errors of *omission* (the probability of excluding a pixel that should have been included in a particular class) and *commission* (including a pixel in a class when it should have been excluded) (Franklin 2001, Horning 2004). Accuracy estimates may be based either on the pixels used for training during the (supervised) classification process, or on a set of pixels that were not used during this training. Usually, the latter approach is adopted, as the former tends to give an artificially high estimate of accuracy. Another widely used statistic is kappa (κ), which enables two or more contingency tables to be statistically compared (Horning 2004). The accuracy of a classification can often be significantly increased by inclusion of ancillary information, such as a DEM.

One of the problems inherent in all maps is that boundaries between areas appear to be discrete and well defined, whereas in reality different types of land cover may vary continuously and merge into each other. As a result, the boundaries on maps produced from remote sensing data may be something of an abstraction, and be difficult to relate to patterns observable on the ground. Although images produced from satellite data can look impressive, they may give a very misleading impression of how different forest types are actually distributed within an area, and therefore they should always be interpreted with caution. Interpreting an image visually, rather than making some form of quantitative classification, can save a great deal of time and resources, and may even produce a more useful result at the end of the day (Horning 2004).

2.4 Other sensors

Although a variety of other sensors are currently being used for assessing forests, principally from airborne platforms, most of these have narrow, specialist uses and have not been widely applied to questions relating to ecology and conservation. However, it is likely that the importance of such techniques to conservation applications will increase in future. For example, thermal imagery can be used in reconstruction of surface temperature patterns that can be related to the incidence of water stress in forests, and has been used to provide an indicator of biodiversity (Bass *et al.* 1998).

Ground-based and airborne radar sensors have been applied to forest assessment, primarily for estimates of biomass or timber volume extraction (Franklin 2001). Radar has particular value in areas of persistent cloud cover, where its ability to produce imagery despite the presence of clouds may offer an advantage over other remote sensing techniques. Airborne digital multispectral scanners, frame cameras, and videography are other technologies at a relatively early stage of development that are increasingly being used to assess forests (Franklin 2001, Wulder and Franklin 2003).

Lidar (light detection and ranging) devices use lasers to scan terrain to produce high-resolution topographic data, which are being increasingly used for the development of DEMs. Lidar can be used to measure the three-dimensional structure of vegetation, unlike conventional optical sensors. Application of this technique to forest assessment has been relatively limited to date, but recent technological developments are leading to a growth in interest (Dubayah and Drake 2000). Lidar has now been used to estimate forest stand characteristics such as tree height, timber volume, tree density, and canopy structure (Lefsky *et al.* 2002, Naesset 1997, Naesset and Bjerknes 2001). For example, Hirata *et al.* (2003) describe the use of a helicopter-borne lidar instrument for high-resolution measurement of canopy structure in a mixed deciduous forest, including assessment of the vertical structure of the different canopy layers. Similarly, Lefsky *et al.* (1999a) describe use of lidar to assess the canopy structure of conifer forests, and Hinsley *et al.* (2002) used the same method to characterize woodland structure and habitat quality for birds. Lidar systems applied to forest assessment vary in the size of their 'footprint' or spatial resolution, a large-footprint system with a diameter of 10–25 m covering a group of standing trees, and a small-footprint system with a diameter of less than a metre enabling the canopy surface of an individual tree to be surveyed in detail (Hirata *et al.* 2003).

Lidar has a substantial advantage over other remote sensing techniques in enabling three-dimensional images to be produced. It is possible to envisage a number of potential applications to forest ecology and conservation; for example, the recent growth in interest in the ecology of forest canopies might benefit from high-resolution mapping of canopy surfaces. Lidar data generally require less processing than satellite data, although filtering may be required to determine the structures of interest (Hirata *et al.* 2003). The main challenge to the use of lidar is the large amount of data involved, requiring a lot of computer memory capacity and processing power.

2.5 Applying remote sensing to forest ecology and conservation

As described in the previous sections, a wide variety of methods are available for analysing remote sensing data. Choosing an appropriate method for a particular task can be a daunting exercise. The best way of making an appropriate choice is to learn from the experience of others, particularly in a rapidly developing field such as this. This section describes the approaches that have been adopted in a range of published studies, to illustrate how remote sensing methods have been applied to address specific questions relating to forest ecology and conservation. Problems that may be encountered when using particular techniques and general limitations of these methods are also highlighted.

The selection of an appropriate method is governed by the overall objectives of the investigation, and how the data are to be used. Once a map has been produced from remote sensing data, it can be used as simply as a visual aid for forest planning purposes or to inform a stakeholder consultation process. Alternatively, a range of different quantitative analyses can be carried out, including species habitat modelling, deforestation modelling, landscape pattern and forest fragmentation analysis, and conservation priority-setting. Some of these methods are considered in subsequent chapters of this book. In the current chapter, methods are presented that are useful for assessing changes in forest area and in forest condition.

2.5.1 Analysing changes in forest cover

The use of remote sensing imagery to assess changes in forest cover is of great interest to forest conservationists. Such analyses can be used to estimate deforestation rates and patterns, which can be of value in identifying conservation priorities and potential sites for forest restoration. In addition, analysis of changes in forest cover can be used to infer changes over time in the availability of habitat for forest-dwelling species.

Aerial photography can be used for very detailed assessment of rates and patterns of change in forest area (Price 1986); as photographs are often obtained relatively easily, this is frequently the method of choice. Aerial photographs have the added advantage of being potentially available over longer time periods than other types of imagery, enabling forest changes over longer timescales to be evaluated. The main problem with this method is the difficulty in mapping the boundaries of vegetation types, which do not always coincide with those observed during field surveys (Franklin 2001). Despite this problem, aerial photography continues to be very widely used for assessments of forest change (see, for example, Lowell *et al.* 1996).

Typically, satellite imagery is used to evaluate changes in forest cover by producing classified maps that illustrate the distribution of different change classes, such as forest to non-forest, forest unchanged, non-forest to forest, and non-forest unchanged (Horning 2004). The commonest way of producing a map of changes in forest cover is to compare two classified images produced for different dates. If the amount of change is large then this approach can be highly effective (Franklin 2001). A variety of different algorithms can be used to calculate the difference between two images, ranging from simple subtraction to more complex statistical manipulations, which are available in image-processing software. Details of these algorithms are provided by Gong and Xu (2003).

The main problem with this post-classification approach is the fact that errors associated with each of the individual land cover maps are accumulated into the final map illustrating forest cover change, which therefore tends to be less accurate than either of the individual land-cover maps (Horning 2004). As an alternative, the images from the two different dates can be combined into a single image, which can then be classified by using the approaches described earlier. This method can potentially enable classes of land-cover change to be mapped directly, producing lower errors than the post-classification method. The main problem with this approach is the potential difficulty of identifying changes in forest cover if there is variation within the images that is not directly related to changes in forest cover (Horning 2004).

Other methods that can be used include (Horning 2004):

- *Image difference or ratio.* This involves the analysis of individual bands or single-band image products, such as vegetation indices (such as NDVI). Images from different dates can be compared by subtracting them, and the values analysed to determine changes in forest cover. Although rapid, this method does enable changes between specific land-cover types (for example, conversion of forest cover to different non-forest land-cover types) to be determined.
- *Spectral change vector analysis.* This involves assessing changes in spectral composition and intensity of pixels between different dates. For example, conversion of forest cover to bare ground is likely to result in an increase in the brightness of an image as well as a change in colour. This method is typically employed with multispectral imagery.
- *Manual methods.* Visual interpretation, supported by on-screen digitizing, can be used to manually produce maps of changes in forest cover. For example, polygons can be drawn on screen by using suitable GIS software representing different classes of land-cover change. Although relatively simple, the work can be arduous, and is subject to biases introduced by the analyst.

- *Hybrid approaches.* These incorporate elements of both the automated and manual methods described above. For example, an image might first be produced by using an automated method, then edited visually to produce the final map.

One of the biggest challenges to the analysis of changes in forest cover is the fact that the characteristics of the imagery often differ over time. Satellite remote sensing data are only available from the early 1970s onwards, and therefore detection of change that has occurred over longer timespans than the past 30 years requires other imagery, such as aerial photographs. Even in this case, earlier photographs are likely to be black-and-white rather than colour images, which can complicate interpretation and comparative analysis. Comparison between aerial photographs and satellite data can be achieved by using visual methods; this is most readily achieved by digitizing (scanning) and georeferencing the photograph and displaying it together with the satellite image in a GIS.

In the case of satellite imagery, a common problem is encountered when comparing Landsat images from different dates. Earlier images obtained from the Landsat MSS sensor are at a coarser resolution than more recent Landsat TM imagery, and the number of spectral bands is fewer. To overcome this problem, the lower-resolution data set can be resampled to match the resolution of the other data set so that the pixel sizes for the two data sets are equal, making automated analysis possible (Horning 2004). Some software packages now enable imagery with different resolutions to be combined, avoiding the need for resampling.

Another key problem relates to validating the results obtained. Whereas current maps of forest cover can be validated relatively easily by comparing the results with recent field surveys or forest inventories, validating images from the past can present significant difficulties. Determining the pattern and distribution of different forest types even just a few decades previously can be highly problematic, particularly in areas undergoing rapid deforestation. A range of different sources of information might be used to help validate the maps produced, such as aerial photographs, interviews with local people familiar with the area, evidence from cut stumps or logging records, and information from field plots or forest inventories that originate from the time in question. Sometimes it is possible to infer the kind of forest a particular site is likely to have supported given the current environmental characteristics of the site, such as soil type, altitude, aspect, and drainage. None of these methods is without problems, and inferring historical patterns of forest cover is always likely to be subject to a high degree of error. This uncertainty should obviously be taken into account when applying the results of the analyses.

Any assessment of forest cover change must carefully consider the type of change classes that are to be mapped. For example, what are the possible land-cover types that might have replaced forest? Is it possible that some land-cover types have reverted to forest through a process of succession, and if so, how might successional vegetation types be classified? Have natural forests been replaced by plantation forests, and if so, are their spectral characteristics likely to differ? It is important to select images that adequately cover the period of interest, and that provide information at an appropriate spatial and spectral resolution to allow the detection of significant changes in forest cover. Often, investigations are significantly limited by the available imagery, and the quality of the analyses depends primarily on making the best of what can be obtained. Some of the issues that should be considered when selecting imagery for analysing changes in forest cover are listed in Table 2.6.

It should be noted that remote sensing imagery can be used to assess forest recovery as well as loss. Regeneration surveys based on field survey, supported by aerial photography, are a routine approach in many forest areas, for example to determine the density of young trees and the success of planting initiatives. Ground-based and airborne sensors have been used for direct estimation of cover, seedling, and stem counts. However, satellite remote sensing of forest regeneration assessment is much less common. One approach is to consider the changes in reflectance characteristics as the forest stand develops over time. For example, in Tanzania, Prins and Kikula (1996) reported that detection of coppicing from roots and stumps in miombo woodland (*Brachystegia* and *Julbernardia*) was possible by using Landsat MSS data acquired during the dry season.

Here are some examples of the use of satellite remote sensing methods to detect changes in forest area:

- Alves *et al.* (1999) used Landsat imagery to assess tropical deforestation by comparing separately classified images from 1977, 1985, and 1999.
- Cushman and Wallin (2000) used Landsat imagery to assess landscape change in the central Sikhote-alin mountains of the Russian Far East. Maximum likelihood classification of the satellite imagery identified four broad cover types (hardwood, conifer, mixed, and non-forest); multitemporal principal components analysis was used to describe the magnitude and direction of landscape change in six watersheds.
- Parmenter *et al.* (2003) analysed land-cover change in the Greater Yellowstone Ecosystem in the USA. Classification tree regression analysis was used to define land use and land-cover classes in the landscape, and to produce maps from Landsat TM scenes.
- Hayes and Sader (2001) used three dates of Landsat TM imagery to assess land-cover change in Guatemala's Maya Biosphere Reserve (MBR). Three change-detection methods were evaluated: NDVI image differencing, principal components analysis, and RGB–NDVI change detection. A technique to generate reference points by visual interpretation of colour composite Landsat images, for kappa-optimizing thresholding and accuracy assessment, was employed. The highest overall accuracy was achieved with the RGB–NDVI method (85%).
- Zhang *et al.* (2005) used Landsat TM and MSS to assess deforestation in central Africa.
- Mayaux *et al.* (2005) provide an overview of the results of recent research that has employed satellite remote sensing data to assess tropical deforestation.

Table 2.6 Some of the variables that should be considered when selecting imagery for assessing changes in forest cover (adapted from Horning 2004 and Franklin 2001).

Sensor characteristics (spatial and spectral resolution)	Ideally, images from different dates that are selected for comparison should be obtained from the same sensor, so that sensor characteristics are consistent between the images. However, even if imagery from the same sensor is used, this is no guarantee that sensor characteristics will be directly comparable, as sensors change over time. Such changes may need to be corrected by applying published radiometric correction factors or by procuring radiometrically corrected imagery.
Solar illumination	Images should be selected that are similar in terms of solar illumination angles, to ensure that areas under shadow are similar in all images that are to be compared. This can be achieved by selecting images acquired at the same season and time of day. It is also possible to use a DEM to correct for the influence of different illumination angles.
Atmospheric conditions	Images to be compared should ideally have been acquired under similar atmospheric conditions, although this is often hard to achieve. Selecting images acquired at the same season and time of day can help, but again there may be a need to use some form of correction algorithm.
Soil moisture	Variation in soil moisture can greatly complicate comparison of different images, particularly when image bands sensitive to water (such as Landsat TM band 5) are used in the analysis. Variation in soil moisture availability can also influence the spectral characteristics of vegetation.
Acquisition date and frequency	Select imagery acquired at a time of year when the features of greatest interest can be accurately differentiated from other features. For example, if there is a need to map areas of deciduous forest, it might be preferable to use images obtained when the forests are leafless, enabling them to be differentiated from evergreen vegetation types. However, images acquired during seasons when refoliation or leaf senescence occurs can be difficult to compare over time, because the forest type of interest is changing rapidly. Usually, images obtained at the same time of year are used as the basis of comparison. Lambin (1999) emphasizes the need for long-term data sets for monitoring forest degradation in tropical regions.

2.5.2 Mapping different forest types

A forest ecologist is typically interested in mapping the distribution of different forest communities or ecosystems, and perhaps different types of forest stand. Assessment of the distribution or status of particular forest types often forms an important part of any forest conservation project. Some forest types, such as tropical montane cloud forests or tropical dry forests, are considered as globally threatened and a high priority for conservation. Determination of where such forests occur within a particular area may therefore be a high priority. Alternatively, particular species of conservation concern may be associated with particular types of forest, and estimation of the extent and distribution of potential habitat for such species may therefore form an important part of conservation planning. Can analysis of remote sensing data enable different types of forest community to be resolved?

Remote sensing methods can be used to map different forest types according to a range of different classification methods, based on consideration of different attributes. However, some forest attributes of particular interest to forest ecologists and conservationists are poorly differentiated by remote sensing methods, and therefore the potential use of these methods should be critically considered before implementation. It should also be noted that the classifications typically used in assessments of land cover may have limited value in terms of illustrating the distribution of ecological communities. A land-cover type is not necessarily equivalent to an ecological community. Many existing forest maps were not developed with ecological objectives in mind, and may therefore be of limited use for applications relating to forest ecology or conservation.

Mapping the distribution of different ecological communities, defined in terms of species composition, is generally done by field survey, which can often usefully be supported by interpretation of aerial photographs (Avery 1968). Forest stands, or areas with relatively homogeneous species composition, can often be differentiated on photographs through differences in colour and texture. In forests with a relatively high diversity of tree species, this method is less reliable. Although estimates of accuracy are rarely provided, the use of aerial photographs to assist in the process of identifying tree species and mapping forest communities is well accepted. Experienced human practitioners can be very effective at interpreting photographic images, in a way that is difficult to duplicate with automated procedures (Franklin 2001). In some areas, keys have been developed to assist in the identification of tree species from air photos. For example, in the Dominican Republic, Hudson (1991) described how tree species can be identified from such photographs by using criteria such as crown shape, crown margin (smooth or serrate), tone (light or dark grey), and texture (rough or smooth).

Identification of individual tree species can also be achieved by the use of other remote sensing methods, such as field spectroradiometric techniques and airborne digital imagery, by examining illuminated tree crowns at different times to detect phenological differences between species. Satellite imagery has proved to be less useful for mapping the distribution of individual tree species; in low- to medium-resolution imagery, such as Landsat, the pixel size is too large to differentiate the characteristics of individual trees. However, the increasing availability of high-resolution satellite data, such as Ikonos and Quickbird, should facilitate mapping individual tree species and forest communities in future. Analytical methods for analysing such imagery are still at a relatively early stage of development, but it can be amenable to the visual interpretation methods used with aerial photographs.

Potentially, if appropriate field data are available, a satellite image could be classified according to the communities of tree species present in the canopy. A key point made by Horning (2004) is that the accuracy of any classification tends to decline as the number of classes increases. In other words, the higher the precision for the class definitions, the lower the accuracy for the individual classes. Horning (2004) provides a useful general guideline of how classification accuracy varies with the number of classes used, based on experience with Landsat imagery:

- With a simple classification, such as forest/non-forest, water/no water, soil/vegetated, accuracies of over 90% can be achieved.
- If conifer and hardwood forest types are differentiated, accuracies decline to 80–90%.
- Classifications based on presence of different tree genera give accuracies of 60–70%.
- If individual species are included in the classification, accuracies are likely to fall in the range 40–60%. Landsat imagery is limited to detecting the dominant tree species in forest canopies.
- Classification accuracies can be improved if a DEM is used in the classification (Franklin 2001, Horning 2004). This can help define ecological communities associated with different topographical characteristics, such as slope or aspect.

One method that appears to offer particular promise for mapping forest composition by using satellite imagery is *spectral mixture analysis*. Many forest ecosystems show small-scale heterogeneity, which results in many individual pixels representing forest areas that are mixed in terms of their species composition. Mixture modelling approaches assume that the reflectance of a pixel is a combination of the spectral reflectance of different cover-type objects (or endmembers). The resulting spectra are thus a composite of the endmembers of pure spectra of objects in a pixel, weighted by their area proportion (Corona *et al.* 2003). Methods of spectral mixture analysis are described in detail by Asner *et al.* (2003). As an example, Köhl and Lautner (2001) found that for a test site in the Ore mountains, Germany, spectral mixture analysis provided accurate results for the assessment of mixture proportions of deciduous and coniferous trees, enabling classification of stand types and differentiation of tree species groups by using a maximum likelihood algorithm.

Examples of studies that have used satellite remote sensing data to map different forest types include:

- Ramirez-Garcia *et al.* (1998) mapped 10 land-cover classes in Mexico, including two mangrove communities, with over 90% accuracy by using a Landsat

TM image, a supervised maximum likelihood classifier, and approximately 80 field plots.
- Lobo (1997) accurately differentiated four forest types (non-flooded alluvial plains forest, lowland seasonally flooded forest, palm forest, and swamp forest) in the Bolivian Amazon (Chimanes) from Landsat TM imagery, by using a segmentation algorithm that works in a similar way to the visual interpretation of aerial photographs, by identifying areas with similar tone and texture.
- Adams *et al.* (1995) used spectral mixture analysis to classify Landsat TM imagery into broad categories of land cover in Amazonian Brazil, including primary forest and different communities of regrowth vegetation. The classification considered the amount of shade cast by different vegetation types to improve accuracy, which was over 90%.
- Foody *et al.* (1996) analysed tropical forest using Landsat TM imagery. They were able to separate six types of forest according to age class and species composition with an accuracy above 80%. The method used was a segmented image classification approach based on spectral and textural similarities. A number of other examples are available of studies that have successfully differentiated tropical forest types, including Hill (1999), Hill and Foody (1994), Tuomisto *et al.* (1994, 1995).
- Cayuela *et al.* (2006) classified the spatially heterogeneous montane forests of Mexico by using the Dempster–Shafer method of 'soft' classification applied to Landsat data. The use of such 'soft' approaches is likely to increase in future, now that analytical tools are more readily available (in software packages such as IDRISI, for example).

The classification of forest stands according to their successional stage can also be of great ecological and conservation value. For example, mapping the location of old-growth forest stands is often a conservation priority. In practice, old-growth forest can be difficult to differentiate with satellite imagery, particularly from other successional stages with large trees and high basal areas. Fiorella and Ripple (1993) found that old-growth stands appeared slightly darker in some Landsat TM bands, probably as a result of the higher number of large canopy gaps in old-growth forests. Cohen *et al.* (1995) were able to relate variation in satellite spectral response to broad age classes (<80, $80-200$, and >200 years) of coniferous forests in Oregon, again as a result of differences in illumination, absorption and shadows between different age classes.

2.5.3 Mapping forest structure

Variables of potential interest relating to forest structure include the degree of forest crown closure, diameter at breast height (dbh) or basal area of forest stands, timber volume, tree height, stem density, stand age, and stage of development. These structural characteristics have a major influence on the value of a forest stand in terms of provision of habitat for different species. Again, such measurements are generally obtained through field survey, but aerial photography has been widely applied to assessing variables such as tree height, canopy cover, and crown density (Löfstrand *et al.* 2003).

Satellite imagery has also been used to estimate structural variables, but in a different way from image classification. Generally, some form of empirical model estimation has been used, which follows this general procedure (Franklin 2001):

- Collect information on structural variables through a field survey at a number of sites.
- Acquire and process imagery for the area, and locate the sites where the field surveys were undertaken on the image.
- Extract the remote sensing data for these sites from the image.
- Develop a model relating the field and spectral data (for example, by using regression techniques).
- Use the model to predict the structural variables for all of the pixels classified as forest within the area of interest, based on the spectral data.

Generally, statistical regression procedures are used in this form of analysis, in which the remote sensing data are the dependent variables and the structural measures (such as stand volume or density) are the independent variable. The remote sensing data are then used as predictors of the structural variables so that they can be mapped across the landscape (Franklin 2001). Attempts have also been to estimate forest structure from remote sensing imagery by using structural indices that integrate a number of variables (Cohen and Spies 1992, Danson and Curran 1993). Details of these methods are given by Asner *et al.* (2003). Ingram *et al.* (2005) describe the use of NDVI derived from Landsat ETM+ for assessing the structure of littoral forests in south-eastern Madagascar. Strong relationships were identified among individual bands and field-based measurements of basal area, but not stem density measurements. An artificial neural network was used to predict basal area from radiance values in four bands and to produce a predictive map of basal area for the entire forest landscape.

2.5.4 Mapping height, biomass, volume, and growth

It is possible to estimate tree height from aerial photographs, for example by using parallax or shadows; details of these methods are given by Stellingwerf and Hussin (1997). With respect to other sensors, lidar measurements have greatest potential for measurement of tree height. In general, satellite remote sensing has not been successfully used for this purpose, with a few exceptions (see, for example, Shettigara and Sumerling 1998).

Estimation of forest biomass (or total organic matter content) is currently of great interest with respect to measurement of the carbon sequestration potential of forests. Traditionally, estimates of stand biomass have been derived from stem volume estimates obtained from forest inventories or other forms of field survey (Aldred and Alemdag 1988). Relatively few attempts have been made to estimate biomass from satellite remote sensing data, reflecting the difficulty of relating biomass estimates to spectral response. Here are some examples:

- Brown *et al.* (1999) mapped biomass for forests in the eastern USA by combining estimates of biomass derived from an inventory with AVHRR satellite data based on 4 ×4 km grid cells.
- Kesteven *et al.* (2003) produced biomass estimates by integrating remote sensing data with process-based models of forest development.
- Sader *et al.* (1989) estimated biomass accumulation in regenerating tropical areas by using Landsat data.
- Foody *et al.* (2003) showed that the use of vegetation indices, multiple regression and feedforward neural networks with Landsat TM data can be used for biomass estimation in tropical forests, although the relations defined had limited predictive power.

Ground-based and airborne sensors, such as radar and lidar, have also successfully been used to estimate forest biomass (Lefsky *et al.* 1999b, Patenaude *et al.* 2005).

Estimation of timber volume can be achieved by using the same method as for estimating biomass. Volume and dbh can be estimated from aerial photography, usually by using regression equations relating these variables to tree height and crown diameters, which can be measured from the photograph directly (Hall *et al.* 1989b, Stellingwerf and Hussin 1997). Use of satellite data for this purpose requires a relation to be established between the spectral characteristics of the imagery and variables such as crown size and closure, which may be related to volume or dbh estimates. In general, such relations are relatively weak, although some attempts have been made at using remote sensing for this purpose (see, for example, Trotter *et al.* 1997 in New Zealand).

Although forest managers are very interested in mapping and measuring forest growth, this is not something that is readily obtained directly from remote sensing imagery. However, it may be inferred from other structural variables that can be measured. Most commonly *leaf area index* (LAI) is used for this purpose; this is the leaf area per unit ground area, usually defined in units of $m^2 m^{-2}$. LAI can be measured in the field by using a range of different methods (see section 3.7.3) and can be used as input to process-based models that estimate growth and productivity across the landscape (see Chapter 5). LAI can be estimated from remote sensing imagery by using an empirical modelling approach, involving correlation of spectral indices such as NDVI with field estimates (Curran *et al.* 1992). Estimates can be derived for areas for which no field data are available, by means of regression approaches. Guidance on how remote sensing data can be used to estimate LAI is provided by Asner *et al.* (2003), Fassnacht *et al.* (1997), Fournier *et al.* (2003), and Franklin (2001).

There is increasing research interest in the integration of remote sensing data with forest inventories to provide estimates of forest cover, volume, biomass, and condition over large areas, which cannot readily be met by remote sensing data alone

(Corona *et al.* 2003). An example of this approach is provided by experience in Sweden, involving integration of Landsat TM and AVHRR data with information from the Swedish National Forest Inventory (NFI) (Fazakas and Nilsson 1996). Landsat TM imagery and NFI field plots were analysed by using regression methods, enabling timber volume at the stand level to be predicted. AVHRR data (NDVI values) were then used to produce additional regression models, enabling maps to be generated for all of southern Sweden and associated estimates of timber volumes and biomass to be estimated at the regional scale (Fazakas *et al.* 1999). This provides a useful illustration of how relatively coarse satellite imagery, such as AVHRR, can be used to scale up from analyses carried out at a higher spatial resolution, for example by using Landsat TM data. A similar approach was used by DeFries *et al.* (2005) in their analysis of rates of deforestation in the vicinity of protected areas in tropical forests.

Nilsson *et al.* (2003) provide a detailed consideration of how remote sensing and forest inventory data can usefully be combined, and highlight a number of issues that should be considered, including the following:

- The presence of geometric errors in an image, together with errors in the positioning of field plots, affects how accurately field data and satellite images can be matched, which in turn affects the estimation or classification accuracy. Issues relating to the accuracy of field plots are considered by Curran and Williamson (1985).
- Attempts to relate relatively small field plots to data from medium-resolution sensors such as Landsat TM are subject to a high degree of error, particularly if local variation in the forest landscape is high.
- Accuracy is also influenced by the fact that the correlation of satellite spectral data with variables such as stem volume or basal area is stronger in young stands than in old ones (Horler and Ahern 1986).
- Care must be taken to ensure that the data from different inventories that are combined in this way are comparable, for example in terms of the definitions of variables measured.
- Some forest types are spectrally very similar to other land-cover classes, making them difficult to differentiate and map; problems related to mixed pixels also negatively affect the classification accuracy.

One method that has attracted particular attention in this context is the *k nearest neighbour* (kNN) method, which is a technique designed to help relate spectral information to data collected in the field (Kilkki and Päivinen 1987, Tomppo 1993). Field survey data are related to corresponding pixel values provided by the imagery. For those pixels for which field data are lacking, the *k* closest field sample plots are selected (in terms of spectral distance rather than physical distance). The unknown variables are then estimated by calculating a weighted average of the values of the *k* sample plots (a decaying exponential weight is usually used) (Holmgren *et al.* 1999). This enables maps to be produced by interpolating values for which no field information is available.

2.5.5 Mapping threats to forests

Assessing the factors responsible for changes in forest extent and condition is of fundamental importance to conservation planning, as well as to forest management. Forests are subject to a wide variety of different causes of environmental change, of both natural and anthropogenic origin, which may threaten the viability of forest ecosystems and the species that reside within them. The methods by which such threatening processes can be analysed are considered in more detail in section 8.4. Here it is sufficient to note that remote sensing imagery can be an important source of information about such threats and their impacts on forests. Examples of relevant studies include:

- floods (Michener and Houhoulis 1997)
- winds (Mukai and Hasegawa 2000, Ramsay *et al.* 1998)
- wildfires (Eva and Lambin 2002, Koutsias and Karteris 2000, Salvador *et al.* 2000)
- insect attack (Franklin and Raske 1994, Leckie *et al.* 1992)
- defoliation, for example that caused by aerial pollution (Brockhaus *et al.* 1992, Olthof and King 2000).

Detection of fires continues to be one of the most important applications of remote sensing to forests. Global observations of fire occurrence are available on a daily basis from AVHRR, SPOT VEGETATION, or EOS MODIS sensors, and remote sensing is also being used to provide rapid assessments of fire outbreaks at a regional level, such as in Amazonian Brazil. Remote sensing imagery can also be used to map fire history in forest areas, and to ascertain its impact on spatial variation in forest composition and structure (see, for example, Kushla and Ripple 1998).

A number of studies have used remote sensing methods to assess the impact of forest harvesting on forests, particularly in relation to mapping clearcuts. For example, Cohen *et al.* (1998) mapped cutovers between 1972 and 1993 in 1.2 million ha of forest in central Oregon by using Landsat images, with an accuracy of over 90%, illustrating the value of this method for surveying very extensive forest areas. Other examples are provided by Fransson *et al.* (1999) and Hall *et al.* (1989a, 1991). Few studies have assessed silvicultural interventions other than clearcuts. As the degree of canopy disturbance declines, the potential for using satellite imagery for its detection is reduced. However, the advent of high-resolution imagery may increase the potential role of satellite imagery in assessing harvesting impacts. For example, Furusawa *et al.* (2004) used Ikonos data to successfully assess impacts of selective logging in the Solomon Islands.

The use of remote sensing imagery to assess the degree of forest fragmentation, which can itself be considered a threatening process, is considered in section 2.7.

2.5.6 Biodiversity and habitat mapping

There is currently great interest among conservationists and ecologists in the use of remote sensing methods to detect and map biodiversity. However, progress has been relatively limited to date (Foody 2003, Kerr and Ostrovsky 2003, Nagendra 2001, Turner et al. 2003). Although remote sensing imagery can be used to map the composition of tree species within a forest, as noted above, mapping the distribution of other forest-dwelling species is more problematic. The integration of field survey data and remote sensing imagery to assess the status distribution of species is at a relatively early stage. Nagendra (2001) outlines three main approaches that have been used to date:

- direct mapping of individual plants or associations of single species in relatively large, spatially contiguous units
- habitat mapping by means of remotely sensed data, and predictions of species distribution based on habitat requirements
- establishment of direct relations between spectral radiance values recorded from remote sensors and species distribution patterns.

To these may be added the correlation of spectral imagery with maps of species richness (see, for example, Jorgensen and Nohr 1996). A number of studies have successfully related field-based estimates of species richness to NDVI. For example, Fairbanks and McGwire (2004) identified relations between NDVI and species richness in vegetation communities of California, and Gould (2000) described similar results in the Canadian Arctic.

Methods of habitat mapping are considered in section 7.8. In this approach, environmental data are incorporated in a GIS to produce ecological land classifications that can be used for mapping and modelling the distribution of habitats and the potential distribution of associated species. Models may include variables in addition to forest cover, including topography, hydrology, climate, and field data on the presence of target organisms (Dettmers and Bart 1999). Issues relating to the use of remote sensing data for habitat mapping include (Franklin 2001) the following:

- Spectrally distinct land-cover classes may differ from interpreted habitat classes.
- Ancillary data such as DEMs and biophysical land classifications can be of great value in producing habitat maps from remote sensing imagery.
- Field verification of the maps produced is often very difficult, because of the need for intensive surveys of the species of interest.
- There is a need to ensure consistency in mapping approaches adopted across management units.

Here are some examples of other recent approaches using satellite imagery to map biodiversity:

- Amarnath *et al.* (2003) used remote sensing data (IRS LISS III) together with field survey data to do a GIS-based analysis of the evergreen forests of the Western Ghats, India. Spatial analysis was used to delineate homogeneous large patches of evergreen forest, and to identify the relationship between species richness and forest fragmentation, to assist in the prioritization of sites

for restoration and conservation. Similar work is described for elsewhere in India by Roy *et al.* (2005) and Roy and Tomar (2000).
- Behera *et al.* (2005) used satellite image interpretation with field survey and spatial analysis in the eastern Himalayas to assess patterns of species endemism and disturbance to natural forests.
- Foody and Cutler (2003) analysed data from field plots and Landsat TM imagery by using feedforward neural networks, to derive predictions of biodiversity indices from the imagery, when comparing logged and unlogged forest in Borneo.

2.6 Geographical information systems (GIS)

During the past two decades, GIS has grown from a specialist technique with only a small number of practitioners to become one of the most important tools in environmental science and management. Use of GIS in forest ecology and conservation is now very widespread. Forest managers use GIS to manage and display inventory data, and as a basis for management planning and monitoring. Ecological researchers use GIS to visualize and analyse spatial patterns in ecological communities, and increasingly to investigate the spatial processes responsible for generating such patterns (Figure 2.4). The increased ability to collate and process spatial data afforded by GIS has led to the development of landscape ecology as a subdiscipline in its own right, with its own specialist journals (*Landscape Ecology*), organizations (the International Association for Landscape Ecology (IALE), ⟨www.landscape-ecology.org/⟩ and textbooks (Gutzwiller 2002, Turner *et al.* 2001). GIS is also widely used to support conservation planning and priority setting, and for mapping biodiversity.

GIS can be defined most simply as a tool for the collection, integration, processing, and analysis of spatial data (DeMers 2005). A common feature of GIS is the ability to present data in different layers, which can be overlaid on top of each other (I once heard GIS described, by no less than a professor of geography, as 'glorified tracing paper'). Most importantly, GIS can be used to produce maps, which differentiates it from most computer-aided drafting (CAD) systems that otherwise share some features with GIS. The technologies associated with GIS, both hardware and software, continue to develop rapidly. For a thorough introduction to GIS, specialist textbooks should be consulted. Many such books are now available (Borrough and McDonnell 1998, DeMers 2005, Longley *et al.* 2005). Johnston (1998) provides a useful introduction to the use of GIS in ecology. Informative resources are also available via the Internet ⟨*http://gislounge.com*⟩.

Although GIS software is becoming more accessible and user-friendly, it is also becoming more powerful, which can increase the length of the learning curve. As noted by Johnston (1998), ecologists should not assume that mastering GIS software will be as rapid as learning word processing or spreadsheet software. Problems that arise can be difficult to solve without access to an experienced practitioner. Such people can be difficult to find, even in academic environments, because they

Geographical information systems (GIS) | 69

Fig. 2.4 Example of application of GIS to support a forest conservation project in the Isle of Wight, southern England. In this example, a map of woodland cover has been produced by creating a separate polygon (vector data) for each individual woodland (coloured black on the figure) in the landscape. This was achieved by digitizing woodland boundaries identified on an aerial photograph. To achieve this, the photograph was georeferenced and included as a separate data layer in the GIS database. The white points were created as an additional data layer, and represent the locations of a threatened insect species determined in a field survey using a GPS device. (Courtesy of Niels Brouwers.)

can usually earn a lot more in the private sector (Johnston 1998). Extensive resources are available via the Internet that can help solve specific problems, but there is no substitute to discussing a technical issue with a colleague, so before embarking on a GIS project it is worth spending time searching for people who might be able to help. Johnston (1998) provides some further advice for those about to start using GIS (Table 2.7).

Implementation of GIS requires appropriate software and computer hardware on which to run it. As spatial data sets can be very large and require a lot of processing power, the computer should be equipped with large amounts of RAM and hard disk memory, as well as a reasonably fast processor. Most commercially

Table 2.7 Advice that should be heeded before embarking on a GIS project (adapted from Johnston 1998).

Keep it simple	Begin with relatively simple data and software. Although high-specification GIS software may appear attractive, a simpler package may be adequate for the task at hand, and is likely to be easier to learn.
Read the documentation	Manuals provided with the software should be carefully consulted. Extensive online help is now available, for example on the websites of the companies producing the software, via user groups for particular software packages, and from online teaching resources developed by educational establishments.
Use existing data	Much spatial data can now be accessed online, and downloaded free of charge in many cases. Develop collaborative partnerships with others investigating the same study area so that data can be shared, duplication of effort avoided, and considerable time saved.
Plan ahead	GIS analysis usually requires a series of steps to be undertaken, which should be planned beforehand
Keep good records	This is a crucial and often neglected aspect of GIS work. Each data set used should be carefully described and documented, and the source noted. (Such information about data is often referred to as metadata.) Keep detailed notes of each step undertaken in the compilation and analysis of the data.
Check results	It is important to apply quality control throughout the process; inspect the data and compare different data sets against each other to check on their accuracy. Check that simple operations, such as distance and area operations, are giving correct results.
Consult with experts	Before starting any GIS project, consult experienced practitioners for advice on database management and GIS procedures. If no experts are readily available, form a user group with interested colleagues to share tips and techniques.

available GIS software will perform well on most modern desktop personal computers. Facilities for data storage and backup are an important part of any GIS computer system; writable CD and DVD drives and additional hard disks are all potential options that should be explored. Once appropriate hardware is available, the main choices that need to be made concern selection of appropriate software, types of data, map projection, and the analytical procedures to be used. These are each considered in the following sections.

2.6.1 Selecting GIS software

Several different GIS software programs are available (Table 2.8), which vary in their specifications and ease of use. How to choose which is most appropriate? This will depend on the precise objectives of the task set, and also the previous experience of the user. The best approach is to try using demonstration copies of different software packages, which are often available as a free download via the Internet. Information about the products provided by the manufacturers should also be consulted. Most GIS software programs provide the same basic facilities, for example the ability to enter geographic objects such as lines, polygons, and points, together with their attributes, and to overlay different data layers. Where software programs differ most markedly is in their ability to process and analyse spatial data, rather than basic mapping. Also, some are much easier to use than others: pay careful attention to the documentation, tutorials and help facilities provided.

Some points to consider when selecting GIS software include the following:

- Will the software run on the hardware that is available? Is it compatible with the computer's operating system? Does the hardware have sufficient capacity (in terms of RAM, hard disk memory, and processor speed) to run the software adequately?
- Which file types are supported? Digital data are available in a very wide variety of formats, which are changing all the time. If there are specific data that you wish to use, check which format they are available in, and ensure that you choose software that can import data in this format. Although some software products provide facilities to convert from one file type to another, these do not always work entirely satisfactorily.
- Which type of software is being used by others with whom you wish to collaborate or share data? Given that tools for converting between file types are widely available, it may not necessarily present a problem if you wish to share data with others using a different type of software. However, a lot of potential problems can be avoided if a common type of software is used.
- Do you wish to carry out image-processing operations, such as orthorectification of aerial photographs, or classification of satellite remote sensing imagery? Although some GIS packages provide these facilities, many do not. Although most provide tools for georeferencing images, the ability to warp images is often very restricted.

Table 2.8 Selected GIS software programs that are widely used for mapping and analysis of forests. Further GIS resources are available at ⟨http://gislounge.com⟩

Product	Comments	URL
AGISMap	A simple, easy-to-use GIS package that is distributed as shareware via the internet ESRI	www.agismap.com/
ArcGIS, ArcView, ArcInfo	ESRI (Environmental Systems Research Incorporated) is the market leader in GIS software, and is widely used to support practical forest management and planning. The main products provided by ESRI are ArcView and ArcInfo, which are now combined in a single software package called ArcGIS. Additional functionality is provided by a suite of extensions and additional tools; see website for further details. A free version of the software (ArcExplorer) is also available, which allows basic mapping and spatial querying.	www.esri.com/
GRASS	A well-known and widely used program that can be downloaded free of charge. Can be used for analysis and presentation of both vector and raster data, and for image processing. Not easy to use at first.	www.geog.uni-hannover.de/grass/index.php
Idrisi32	Another popular GIS package, with a large community of users interested in forests. Although it can be used to analyse both vector and raster imagery, it is particularly designed for the latter, and is widely used by the remote sensing and modelling community for this reason. Possesses many powerful analytical features. For example, Idrisi Kilimanjaro includes tools for spatial modelling of forest cover change, such as GEOMOD.	www.clarklabs.org/
MapInfo	A leading GIS software package, with a wide range of basic functions. Although marketed towards the business sector, and lacking some of the more sophisticated analytical functions of IDRISI or ArcGIS, this package is widely used by land managers and planners.	www.mapinfo.com/
Map Maker	A low-cost and easy to use programme, yet powerful enough to produce sophisticated maps. An excellent option for those new to GIS, wanting to learn the basic techniques quickly. A basic version is available for free download, and a more powerful version (Map Maker Pro) is available free of charge to non-profit organizations, educational establishments, and African students.	www.mapmaker.com

- How do you wish to analyse your data? Does the GIS software offer appropriate tools for the analysis method that you have in mind? Most GIS programs allow different data layers to be integrated and simple analytical procedures to be performed, such as buffering around objects, calculation of areas and distances, or querying of the data. However, some offer much more sophisticated analytical tools, including tools for geostatistics, spatial analysis, and modelling. In some cases, these tools are available as additional software modules or extensions that can be purchased separately. An alternative approach is to export the data from the GIS software to a statistical program (such as SPSS, Statistica, SAS, or S Plus), which can be used to further analyse and process the data.

2.6.2 Selecting data types

A key principle relating to the use of GIS is that digital data are generally available in one of two forms: raster and vector.

- *Raster data* represent an area as a grid of (usually square) cells. Various properties or attributes may be assigned to these cells to produce maps. These cells are equivalent to the pixels that form a digital image, such as those generated from satellite remote sensing data or digitized air photos. It is important to note that when features are represented in this way, information about variation within the pixel is lost. Selection of a particular grid or pixel size therefore has an important bearing on how information on some object of interest is represented within the GIS.
- *Vector data* focuses on mapping objects as points (or vertices), lines, or polygons, which are areas bounded by points that are joined by lines. The term *polyline*, which is widely used, refers to a curved line represented by a series of straight lines joined together.

With respect to mapping forests, both types of data have their uses. Forest stands can usefully be mapped as polygons, perhaps derived from an aerial photograph, enabling precise estimates of distances and area to be obtained. Such measurements are less reliable if derived from raster data, because of reduced spatial accuracy. A vector-based map of forest stands might also look more presentable, or more accurate, than a map of forest stands based on raster data. Yet in practice the boundaries of forest stands are often not sharp, and may in fact be more accurately represented as raster data (for example, if there is a gradient in species composition of a forest across an area). The main advantage of raster data is that many analytical functions, including overlay operations, spatial analysis, and modelling are far more easily done with raster than with vector data. The fact that satellite remote sensing data are provided in raster format is another main reason why raster data are widely used in forest mapping. In general, file sizes tend to be significantly larger for raster data than for vector data, and this has implications for data storage and the time required for analytical procedures.

Many GIS software packages now provide tools for the display and analysis of both types of data, and even enable data to be converted from one type to another.

However, most software programs are biased towards one or other data type, in terms of the functionality provided. Consideration should therefore be given to whether data are to be represented as raster or as vector data layers within the GIS, and the ability of the software to process the selected data type should be checked. If satellite remote sensing data are to be used as a basis for forest mapping, for example, it would make sense to choose GIS software that is explicitly designed to analyse and present raster data.

2.6.3 Selecting a map projection

Given that the Earth's surface is curved, any representation of features on the Earth's surface as a two-dimensional map inevitably results in some form of distortion. A wide variety of different map projections are available that differ in how the Earth's surface is represented on a flat surface. These projections distort features in different ways, and therefore the choice of map projection has a major influence on the results of any calculations or analysis performed on the mapped data. A fundamental issue is that areas and angles cannot be preserved at the same time: if a projection is selected that accurately represents angles, then measurements of area will be inaccurate. Conversely, if a projection is used that preserves area, then measurements of angles will be inaccurate.

Many GIS software packages enable data to be mapped according to a range of different projections. Choice of an appropriate map projection therefore depends on the type of calculation that is to be made, and can be summarized as follows (DeMers 2005):

- If the objective is to make measurements of area, for example when analysing changes in forest area over time, then some form of *equal area projection* should be selected. Examples include Alber's equal area and Lambert's equal area projections. The size of the area to be mapped will influence how much angular distortion occurs; small areas display less angular distortion than large ones when equal area projections are used. This is important if both the area and shape of forest areas are of interest. The Universal Transverse Mercator (UTM) projection is widely used in the production of large-scale maps; it preserves the shape of mapped features (Johnston 1998) and provides accurate estimation of distances.
- If the objective is to analyse motion or the changing direction of objects, for example when the movements of animals are detected by using radio telemetry, then a *conformal projection* is most appropriate. This type of projection is preferred whenever angular information is important, such as in navigational maps and with topographic data. Examples of this type of projection include the Mercator, Lambert's conformal conic, and conformal stereographic projections.
- *Azimuthal projections* are used are used when the determination of shortest routes is required, particularly over long distances. Examples include Lambert's equal area, azimuthal equidistant, and gnomonic projections.

2.6.4 Analytical methods in GIS

Typically, spatial data incorporated in a GIS will be georeferenced, enabling maps to be produced according to an appropriate coordinate system. GIS software packages usually provide georeferencing tools that enable coordinates to be entered for an image or any data that have been collected, as described in section 2.2.2. However, an appropriate coordinate system must be selected. Cartesian coordinate systems are widely used, which represent coordinates as pairs of x, y values that indicate the position on a grid. For example, the UTM projection uses a Cartesian coordinate system, in which easting (x) and northing (y) distances are measured in metres relative to the origin of the coordinate system, which lies at the intersection of the equator and the central meridian of each 6° zone of the Earth's surface (Johnston 1998). Although Cartesian coordinates are appropriate for relatively small areas of the Earth's surface, they cannot be used for the entire planet, because of its (approximately) spherical shape. The latitude–longitude system provides Earth coordinates according to a spherical system, with lines of longitude defined as circles that pass through both poles, and lines of latitude defined as concentric circles around the poles (DeMers 2005). Longitude and latitude coordinates according to this system are presented as degrees from the prime meridian and equator, respectively.

Generally, a coordinate system will be selected that is typically used on published maps available for a particular area. Individual countries use different coordinate systems based on different datums, which are ways of describing the shape of the Earth. Awareness of different datums and their associated coordinate systems is particularly important with respect to the use of GPS (see section 3.4.2), which typically provide options for location data according to a large number of different datums. It is therefore essential to know which datum is used for georeferencing a map. The same datum and associated coordinate system must be used for each map included in a GIS if they are all to be located at precisely the same places on the Earth's surface (DeMers 2005). A widely used datum is the world geodetic system, WGS84.

Once data have been incorporated within a GIS, a wide range of analytical procedures can be carried out. The methods available depend on the software being used, but a number of basic procedures that are common to most GIS packages are listed below. Many of these operations can be carried out on either raster or vector data.

- *Querying.* The data incorporated in a GIS are generally organized in a database, which permits a variety of operations to be performed on the data, including filtering and sorting. Querying refers to the process of selecting specific database entries according to their characteristics, and this may be done by using a variety of approaches: interactively, by using a look-up table, by specifying numerical thresholds, or through Boolean logic (Johnston 1998). For example, different forest areas could be classified according to the tree species present by using these different approaches. An interactive query might select those areas

where a particular species tree is present; a look-up table might relate the presence of particular species to ecological community types; numerical thresholds could be used to classify forest areas on the basis of relative abundance of different species; rules based on Boolean logic could be used to classify forests on the basis of specific combinations of the species present.
- *Overlaying.* One of the characteristic features of GIS is the ability to superimpose different layers of spatial data over one another, to visualize how different features coincide. This is a very useful feature for producing customized maps incorporating specific features of interest. For example, a map of a forest area could be produced that illustrated the distribution of features such as protected areas, logging concessions, roads, and urban development, each feature having been derived from an individual data layer.
- *Combining data layers.* In addition to the simple graphical overlay of data layers, it is also possible to combine them in different ways to generate new data layers. This can be achieved by a variety of processes. *Clipping* involves cutting one data layer according to the features on another layer; *intersecting* creates a new feature from the area where features from different layers overlap; a *union* creates a new layer combining the features of other layers.
- *Buffering.* Most GIS software enables buffer zones to be generated around objects such as points, lines, or polygons. The radius of the buffer can be specified. This could be used, for example, to define the potential area of environmental impact associated with construction of a feature such as a new road, or to define areas of potential tree colonization around forest patches.

Although most GIS software can do these basic tasks, the ability to do so with different data types varies between software programs, and this is something to be considered when selecting software.

Many GIS software programs also offer a variety of other, more sophisticated, analytical tools, including various spatial analysis and modelling methods. The ability to carry out such advanced tasks is one of the main distinguishing features separating commercially available GIS software programs. Some of these methods are described in subsequent chapters of this book. The remainder of this chapter focuses on one specific issue, the description and analysis of the spatial characteristics of forest areas.

2.7 Describing landscape pattern

Forest fragmentation refers to the division of large, continuous expanses of forest into smaller discrete patches, which are separated by some other type of land cover (such as agricultural land) commonly referred to as the *landscape matrix* (Forman and Godron 1986). Fragmentation can be caused by natural forms of disturbance, and some forests are naturally fragmented because they are associated with particular edaphic conditions (such as soil type or climate) that are patchily distributed. However, it is the widespread forest fragmentation caused by human activities that is currently of such concern to conservationists, given that many species appear to be negatively affected by fragmentation of their habitat. Forest fragmentation is widely considered to be one of the major causes of biodiversity loss, and consequently the development of methods to estimate forest fragmentation has been the focus of much research interest. The issue is considered further in section 7.4.

Descriptions of the pattern of forest landscapes can be produced from maps of land cover derived from field survey, aerial photography, or satellite remote sensing imagery. Few attempts have been made to describe landscape pattern by using field-based approaches. Kleinn and Traub (2003) indicate that plot-based designs used in ecological surveys and forest inventories can be used to produce unbiased estimates of some spatial attributes, such as total forest area or total perimeter length. However, other attributes of interest, such as mean patch area or mean patch perimeter, cannot be directly derived. Typically, analyses are done in GIS, although other software tools such as statistical analysis packages can also be used for estimating some spatial patterns. A number of specialized software tools have been developed specifically for assessment of landscape pattern, and this has greatly encouraged use of these methods among the research community.

Turner *et al.* (2001) provide a valuable overview of how to describe landscape pattern, and highlight the following important caveats that should be borne in mind before using these methods:

- *Objectives.* A clear statement of objectives of the analysis, or an explicit hypothesis, can avoid misleading or confusing results. As many different metrics are available, it is important to select those most appropriate for addressing the objectives set.
- *Classification scheme.* The choice of the land-cover categories, or classes, used in the pattern analysis has a major influence on the results obtained. The classes selected, for example during the classification of remote sensing imagery, should be appropriate to the objectives set and must be rigorously applied across all the landscapes being compared.
- *Scale.* The spatial resolution and extent of the data will also have a major influence on the results obtained. Comparison of analyses done at different scales may be invalid because of scale-related artefacts. As the size of the study area declines, the risk of the map boundary truncating patches increases, producing additional artefacts. On the basis of an analysis of Landsat imagery, O'Neill *et al.* (1996) presented the following rule of thumb: the spatial resolution of the map should be 2–5 times smaller than the size of the features being analysed, and map extent should be 2–5 times larger than the size of the largest patches.
- *Patch identification.* Many of the available metrics are based on the concept of a forest patch. It is important to consider how this concept relates to reality, and how a patch should be defined. Methods used for analysis of raster imagery, for example, usually define a patch as a contiguous group of cells of the same mapped land cover category. But what does contiguous mean in this

context? Should only the four nearest cells be considered, as is often used in practice (the 'four-neighbour rule'), or should diagonal neighbours be included also? Also, different species have different perceptions of what constitutes a habitat patch, which may differ markedly from those perceived by people. Careful consideration should therefore be given to how patches are mapped and defined, as this will have a major bearing on the results obtained.

2.7.1 Choosing appropriate metrics

Many metrics have been developed that can be used to describe the spatial pattern and characteristics of forest landscapes (Franklin 2001, Jorge and Garcia 1997, McGarigal 2002, Trani and Giles 1999) (Table 2.9). These may be divided into two general categories: those that assess composition, or the variety and abundance of different types of land-cover patches within a landscape, and those that quantify their spatial configuration (McGarigal 2002, Wolter and White 2002).

Importantly, composition metrics are only applicable at the landscape level, as they are integrated over all patch types within a landscape (McGarigal 2002). Measures of landscape composition include (Turner *et al.* 2001):

- the proportion of the landscape that is occupied by a given land-cover type
- the number of land-cover types present, as a percentage of the total possible number of cover types
- diversity, or relative evenness, which refers to how evenly the proportions of cover types are distributed
- dominance, which is the deviation from maximum possible diversity.

Note that either dominance or diversity could be reported, but not both, as they are correlated. However, as similar values can be reported for these metrics for landscapes that are very different in character, the usefulness of these measures is limited.

Spatial configuration refers to the spatial characteristics and arrangement or orientation of land-cover patches within a landscape (McGarigal 2002). Configuration metrics are spatially explicit at the scale of individual patches, rather than the landscape, although measures of the spatial relationships between patches and patch types can also be derived.

The metrics that have been developed can be divided into the following principal groups (Echeverría 2005, Franklin 2001, McGarigal 2002):

- *Area metrics*. Metrics describing the area of patches, such as mean patch size, can be summarized for different patch types and for entire landscapes.
- *Edge metrics*. These measures of patch geometry represent the length of edge between land-cover types, and are useful for assessing the extent of edge habitats.
- *Shape metrics*. These refer to the shape of land-cover patches and are most commonly represented as the relative amount of patch perimeter per unit

area, or as a fractal dimension. The edge-to-area ratio indicates whether patches are compact and simple or elongated and complex in shape.
- *Core metrics.* These refer to the interior of patches, after a user-specified degree of edge buffer has been subtracted. This represents the part of the patch that is unaffected by edge effects. Core area metrics effectively integrate patch size, shape and distance from the edge into a single measure, which has been used to assess the extent of large forest patches in a landscape (Wolter and White 2002).
- *Isolation/dispersion/proximity metrics.* These describe whether patches are regularly distributed or are clumped, and also describe how isolated patches are from each other. Calculation of these metrics is based on nearest-neighbour distance, which is defined as the distance from a patch to a neighbouring patch.
- *Contagion and interspersion.* Contagion refers to the tendency of patches to be spatially aggregated, and measures the extent to which cells (in a raster grid) of a similar type are aggregated. Interspersion refers to the extent to which different types of patch are spatially intermixed, and is calculated on patch adjacencies.
- *Connectivity.* This generally refers to the connections between patches, which can be based on strict adjacency (i.e. patches that are touching), a threshold distance, a decreasing function of distance that reflects the probability of connection at a given distance, or a resistance-weighted distance function.

As no single metric captures all aspects of fragmentation, a suite of selected metrics tends to be used to characterize landscapes (Baskent and Jordan 1995, Hansen *et al.* 2001, Staus *et al.* 2002, Wolter and White 2002). The metrics should be selected carefully, attempting to avoid redundancy (Armenteras *et al.* 2003) and considering the characteristics of both the patches and the matrix (Lambin *et al.* 2001). Area metrics, particularly mean patch size, patch size distribution, and number of patches, have been used most widely in forest fragmentation studies. For example, the size of largest patch was used to analyse forest fragmentation in relation to representativeness of protected areas in Colombian montane forest (Armenteras *et al.* 2003). Although this metric is simple, it is restricted to assessing forest fragmentation in areas where large patches occur. In landscapes dominated by patches of a wide range of sizes, metrics such as mean patch size or number of patches may be more suitable (Echeverría 2005). Largest patch size, edge density, and mean core area have been shown to be useful descriptors in the context of forest interior-dependent species (Riitters *et al.* 1995). In contrast, some studies have found patch density, number of patches, and perimeter-to-ratio metrics to behave erratically over time (Hargis *et al.* 1998, Millington *et al.* 2003, Trani and Giles 1999).

Edge density has been used to analyse patch edges in several landscapes (Hansen *et al.* 2001, Staus *et al.* 2002), and fractal dimension and perimeter-to-area ratio metrics to examine patch shape have been widely used (Imbernon and

Table 2.9 Selected metrics used for assessing the spatial characteristics of forest landscapes (adapted from Baskent 1999, Echeverría 2005, Franklin 2001, McGarigal 2002).

Category of metric	Metric	Description
Area	Total landscape area	
	Largest patch	Percentage of area accounted for by the largest patch.
	Number of patches	Number of patches per unit area.
	Patch density	
	Number of patch types	
	Mean patch size	
Edge	Total edge	Total length of all patch edges.
	Edge density	Length of patch edge per unit area.
	Total edge contrast index	The degree of contrast between a patch and its immediate neighbourhood.
	Mean edge contrast index	The average contrast for patches of a particular type.
Shape	Mean shape	Mean patch perimeter/area ratio for a patch type.
	Fractal dimension	The complexity of patch shape in a landscape.
	Mean patch fractal dimension	Length of diagonal of smallest enclosing box divided by the mean width.
	Elongation	
	Landscape shape	Measures of landscape compared to a standard.
Core	Core area	Area of interior habitat of patches defined by specified edge buffer width.
	Number of core areas	Number of core areas per unit area.
	Core area density	
	Mean core per patch	

Isolation/dispersion/proximity	Similarity	The size and proximity distance of all patches, regardless of type, whose edges are within a specified search radius of the focal patch.
	Proximity	The size and proximity distance over all patches of the corresponding patch type, whose edges are within a specified radius of the focal patch.
	Mean proximity	For a class or for the landscape as a whole.
	Mean nearest-neighbour distance	For a class or for the landscape as a whole.
	Spatial autocorrelation	Patch type spatial correlation; patch type distribution.
	Dispersion	Degree of fragmentation/complexity of patch boundaries.
Contagion and interspersion	Clumpiness	The frequency with which different pairs of patch types (including pairs of the same patch type) appear side-by-side on the map.
	Aggregation	The number of like adjacencies in a landscape, in which each class is weighted by its proportional area in the landscape.
	Splitting	The number of patches obtained by dividing the total landscape into patches of equal size in such a way that this new configuration leads to the same degree of landscape division as obtained for the observed cumulative area distribution.
	Contagion	The tendency of land cover types to clump within a landscape.
	Interspersion	The number of pixels in a square that are of a land cover type different from that of the central pixel.
Connectivity	Patch cohesion	The physical connectedness of the patch type.
	Connectance	The number of functional joinings between patches of the same type, where each pair of patches is either connected or not, based on a user-specified distance criterion.
	Traversability	Degree of resistance to movement of organisms.

Branthomme 2001, Jorge and Garcia 1997), as have isolation metrics (Cumming and Vernier 2002, Ranta *et al.* 1998). In contrast, metrics of connectivity are relatively uncommon in forest fragmentation analyses despite the emphasis given to habitat connectivity in the conservation literature. Connectivity is difficult to quantify (McGarigal 2002), and some confusion exists regarding the most appropriate way to measure it (Tischendorf and Fahrig 2000b). Different conclusions about the connectivity of forest patches may be obtained when using different metrics (Tischendorf and Fahrig 2000a).

2.7.2 Estimating landscape metrics

Ideally, the choice of metrics should reflect some hypothesis about the observed landscape pattern and what processes might be responsible for generating it. Also, metric selection should include consideration of the initial pattern of a landscape. If a landscape contains forest patches, or if the potential for patch formation exists, then the use of patch-level metrics is warranted (McGarigal 2002). Conversely, if the landscape comprises extensive areas of contiguous forest cover, then other metrics (forest interior, percentage forest cover, or contiguity, for example) may be preferred (Trani and Giles 1999). It is also important that the choice of metrics should be informed by an understanding of the ecological processes relevant to the investigation. The size and degree of isolation of forest patches, for example, might most usefully be analysed in relation to the specific habitat needs of selected species of conservation concern. Many landscapes have been described by using the metrics available with little reference to the needs of individual species, and this has limited the value of such studies.

Although some analyses can be done using appropriate GIS software, several software tools have been developed by researchers explicitly for generating landscape spatial metrics (Table 2.10). The most widely used of these is FRAGSTATS (McGarigal and Marks 1995). Note that some of these tools require specific GIS software in order to run, and some are limited to analysis of specific types of data.

There are two important issues regarding the interpretation of metrics. Firstly, as noted earlier, they are sensitive to scale. Consequently, landscape metrics calculated by using different types of satellite imagery might not be comparable because of differences in pixel size, which can influence the computation of individual metrics (McGarigal 2002). However, it has been shown that patch size, some edge metrics, and a patch diversity metric appear to be relatively insensitive to variation in spatial resolution between 30 and 1100 m (Millington *et al.* 2003). To minimize the risk of erroneous interpretation, it is recommended that imagery with the same scale and spatial resolution be used for analysis (Franklin 2001). This may be difficult to achieve when comparing images obtained from different dates. To address this problem, some researchers have converted the pixel size resolution in all images to a common standard (Imbernon and Branthomme 2001).

The second constraint is that many landscape metrics are highly correlated. This reflects the fact that only a few primary measurements can be made on land-cover patches; most metrics are then derived from these primary measures. Several

Table 2.10 Software tools for generating spatial metrics for forest landscapes (adapted from Echeverría 2003 and Turner et al. 2001).

Product	Comments	URL
Spatial Analyst	An ArcView extension that provides tools to create, query, analyse, and map cell-based raster data and to perform integrated vector–raster analysis.	www.esri.com/software/arcview/extensions/spatialanalyst/
Patch analyst v3.1	An ArcView extension software that facilitates the spatial analysis of landscape patches and modelling of attributes associated with patches. Patch Analyst (Grid) extends these capabilities to gridded data. Both require the ESRI Spatial Analyst extension to ArcView, and neither works with ArcGIS software. Available as a free download.	http://flash.lakehead.ca/~rrempel/patch/
FRAGSTATS	Has been widely used to assess landscape structure, offering a comprehensive choice of landscape metrics including area metrics, patch density, edge, core metrics, etc. (McGarigal and Marks 1995). Version 3 includes a graphical user interface and the addition of several new landscape metrics, and analysis capabilities (McGarigal et al. 2002). Works with raster data. Available as a free download.	www.umass.edu/landeco/research/fragstats/fragstats.html
LEAPII (Landscape ecological analysis v 2.0)	Designed to explore, monitor, and assess a landscape for its ecological status. Both raster and vector data can be imported. Freeware.	www.ai-geostats.org/software/Geostats_software/leap.htm
ATTILA v2.0	An ArcView extension designed to generate common landscape indicators. Does not calculate landscape metrics per se, but allows the user to combine classes and obtain class areas and neighbours relatively easily. Requires the Spatial Analyst extension to ArcView.	http://epamap1.epa.gov/emap/ca/pages/nca_at_frame.htm
APACK	A stand-alone analysis package for rapid calculation of landscape metrics on large-scale data sets, developed at the University of Wisconsin-Madison. Works with raster data. Freeware.	http://landscape.forest.wisc.edu/projects/apack
r.le	A suite of programs that interface with GRASS GIS software.	http://grass.itc.it/gdp/terrain/r_le_22.html

investigators have attempted to identify a parsimonious suite of independent metrics (Li and Reynolds 1993, Riitters *et al.* 1995, Trani and Giles 1999) by using correlation matrices along with factor or variance analysis, which should be considered for any forest area under investigation. Some studies have evaluated large numbers of metrics: 55 and 61 metrics were used to explain landscape pattern change by Riitters *et al.* (1995) and Imbernon and Branthomme (2001), respectively. By means of statistical analysis, it was demonstrated that only 5 and 6 metrics, respectively, were needed to capture most of the information. An alternative is to use metrics that display similar behaviour through time and across different sites (Millington *et al.* 2003). Some studies have selected those metrics most related to the ecological questions being addressed (Armenteras *et al.* 2003, Hansen *et al.* 2001, Wolter and White 2002), and this is surely the best approach (Tischendorf 2001).

3
Forest structure and composition

3.1 Introduction

Assessments of forest structure are of fundamental importance to forest management, providing information on the size distribution of trees on which harvesting plans can be developed. Measurements of forest stands also provide much information relevant to forest ecology and conservation, enabling the regeneration characteristics of different tree species to be identified, providing insights into the processes of forest dynamics (Chapter 4), and indicating the potential value of the stands as habitat for other organisms (Chapter 7). If there is one thing that foresters know about, it is how to measure trees. The sections presented here on forest mensuration techniques therefore draw heavily on the forestry literature.

Assessments of the species composition of forest stands are also of crucial importance to effective conservation planning, enabling different communities of tree species to be differentiated, sites with high species richness to be identified, and the presence of rare or threatened species to be determined. The classification of forest communities has a long history in ecological science, and a wide range of different classification systems are currently in use. Rather than attempting to provide a comprehensive overview of these systems, this chapter describes methods that can be used to characterize communities of tree species based on a field survey, with a focus on those methods that are most commonly used today. Information is also provided on techniques for estimating the species richness and diversity of forest communities, which has received increasing attention in the wake of the Convention on Biological Diversity.

3.2 Types of forest inventory

Forest inventory refers to the process of collecting information about the extent and condition of forest resources within a specified area. Traditionally, forest inventories were primarily carried out to determine the quantity of available timber, but increasingly the scope of such inventories has been expanded to include ecological variables such as measures of the quality of habitat provided for different species. A forest inventory may therefore be carried out to obtain a range of different information. For estimating potential timber supplies, the key variable to estimate is timber volume, which requires forest area to be measured together with a sample of some or all of the trees in the area. Other information relevant to the development of forest management plans includes descriptions of forest ownership, access and transport infrastructure, topography, hydrology, and soils.

Inventory methods used by forestry professionals have been refined over many years of use in a wide range of forest types. They are therefore 'tried and tested' methods. Recent books describing inventory techniques used by foresters include Avery and Burkhart (2002), Husch *et al.* (2003), Philip (1994), Reed and Mroz (1997), and West (2004). The design of the inventory or forest survey (sometimes referred to as a cruise or enumeration) will vary according to the specific information needs. For example, ecologists may be interested in variables such as the structure and composition of the stand, the extent and pattern of natural regeneration, and those variables that describe the quality of the forest as habitat for wildlife, such as the amount of deadwood (Section 7.2).

National forest inventories (NFIs) have been implemented in many countries, particularly those with extensive forest resources, to provide information on forests at the national scale in support of forest planning and policy development. NFIs are usually established according to a systematic grid or network of plots across the whole country, with field plots arranged in clusters to make the process of data collection more efficient. Examples are provided by Brändli *et al.* (1995) for Switzerland and Ranneby *et al.* (1987) for Sweden (Figure 3.1). Typically the field data are supported by remote sensing data (Chapter 2). The use of these and other new technologies in NFIs is described by Kleinn (2003).

Many developing countries still lack comprehensive NFIs. This is something currently being addressed by the FAO through provision of capacity-building and support. According to the methods developed by FAO, the country is first stratified into relatively homogeneous regions, then tracts are established within which field sample plots are located (FAO 2003). The number of tracts may vary between 50 and 500, depending on country size and homogeneity. Four field plots are

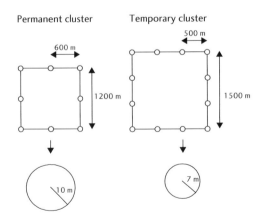

Fig. 3.1 Design of cluster plots as implemented in the NFI in Sweden. (From Ranneby *et al.* 1987.)

established per tract, which are marked permanently on the ground. For example, in Guatemala, the country was divided into three strata according to national ecological zones, namely lowland, southern coastal plain, and central high mountain areas. One hundred and eight tracts (1 × 1 km) were systematically distributed throughout the country, at higher density in the heterogeneous central stratum and at lower density in the relatively homogeneous northern and southern strata. In each tract four sample plots (25 × 20 m^2) were established, oriented south–north, west–east, north–south, and east–west. Some variables were measured in subplots located within these plots, to save effort (FAO 2003).

It is unlikely that any of the readers of this book will ever be in the position of needing to design an NFI—if so, useful guidance is provided by de Vries (1986) and Shiver and Borders (1996). However, NFIs can be of great value as a source of information for forest ecology and conservation. For example, Soehartono and Newton (2000, 2001) used the Indonesian NFI to estimate the population size of *Aquilaria* spp. (gaharu), threatened tree species of great commercial importance for which limited information on conservation status was previously available. Field visits were undertaken to check the accuracy of the NFI data used in these analyses, and a number of sources of error were discovered, including inaccurate descriptions of plot locations and taxonomic confusion caused by the use of local, rather than scientific, names of tree species. Despite such problems, this study demonstrated that NFIs can provide information of value to conservation that is difficult to obtain in any other way, particularly when the species concerned are widespread and occur in inaccessible areas. Another example is provided by ter Steege (1998), who used NFI data in Guyana for development of a national protected area strategy. However, the limitations of NFIs should also be noted: they often lack information of great interest to conservationists, such as production of non-timber forest products. Suggestions regarding the use of NFIs for generating biodiversity information are provided by Newton and Kapos (2002).

Most inventories are implemented at the scale at which forest management decisions are typically taken, the *forest management unit* (FMU) or stand level. Regardless of which scale the inventory is designed to address, and which variables are selected for measurement, an appropriate sampling design needs to be implemented, and a choice must be made regarding which type of sampling unit to use. These issues are addressed in the following sections. The methods described here are those used for field survey; typically these will be supported by use of aerial photographs or other remote sensing data (see Chapter 2).

3.3 Choosing a sampling design

It is rarely possible to measure all of the trees in a particular forest. Therefore, a *sample* must be taken from the complete *population* of all sample units. The most important basic principle is to ensure that the sample is representative. Otherwise, the information obtained will be biased in some way, and the inferences drawn from the data are likely to be invalid. The first step is to divide the forest area to be

88 | Forest structure and composition

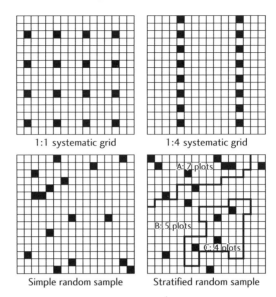

Fig. 3.2 Different sampling designs, based on four possible arrangements of 16 samples in a population composed of 256 square plots. (From Avery and Burkhart 2002. *Forest Measurements*, 5th ed. with permission of McGraw-Hill.)

surveyed into sampling units, for example by dividing it into grid squares or forest stands or patches. The next step is to decide which sample units to include in the survey, according to an appropriate design. Four basic sampling designs are used in forest surveys: *simple random, stratified random, systematic sampling* and *cluster sampling* (Figure 3.2). Each of these is considered below.

Whichever method is used, it is important that the survey is as accurate as possible. If a sample is taken from a population, the values of measurements obtained from the sample may differ from those of the population. This difference is referred to as the *sampling error*, and is expressed as the standard error of the mean. Sampling error can be reduced by increasing the sample size. Other (non-sampling) errors may arise as a result of mistakes or inaccuracies in data collection, bias in the estimates because of a lack of independence in the sample units, inaccurate production of maps and calculation of areas, and poor data processing or management (Husch *et al*. 2003). Attempts should be made to document and minimize such sources of error.

3.3.1 Simple random sampling

This method requires that there be an equal chance of selecting every possible combination of sampling units. It is important to note that this is not the same as each sampling unit having an equal chance of being selected (Avery and Burkhart 2002). The method involves ensuring that the selection of any unit is completely independent of selection of any other unit. This can be achieved by assigning every unit in the population a number, then selecting a sample of these according to randomly generated numbers. Alternatively, random numbers can be used to select intersection points on a sample grid. Random numbers can be obtained from statistical tables, or can be generated by some pocket calculators as well as some spreadsheets or statistical analysis software. It is important to use numbers that genuinely are randomly generated in this way, and not simply plucked from the air in some arbitrary fashion.

The main problem with this approach is that it may be difficult to accurately locate the selected sample points, and it may be difficult to define the most suitable route between two points, making the process of collecting information less efficient than some other methods (Reed and Mroz 1997). It is also important to note that randomly distributed locations will tend to be clustered. As a result, some parts of the surveyed area will not be included in the sample.

3.3.2 Stratified random sampling

In this approach, the forest to be surveyed is first divided into relatively homogeneous areas (or strata). Sample units are then randomly selected from each stratum (usually at least two from each), using the same approach as for simple random sampling. There are a number of key advantages to this approach. Forests are usually spatially heterogeneous, as a result of variation in environmental variables such as topography, soils, aspect, and altitude, as well as in patterns of natural disturbance and previous management. This variation can often be detected on aerial photographs or during preliminary field surveys. Strata can therefore be defined on the basis of such information, in a way that is relevant to the objectives of the survey. For example, it may be known that in different parts of the forest the stands are dominated by different tree species because of variation in soil conditions and drainage. In this case, if the objective is to assess stand structure in each of the different kinds of forest stand present, stratified approaches offer an advantage by enabling each type of forest stand to be adequately represented within the sample.

How should the number of sample units in each stratum be determined? There are two main options. Samples can be allocated among strata in proportion to their relative areas (*proportional allocation*); in other words, more samples are allocated to larger strata than smaller strata. Alternatively, sample units can be allocated to strata taking into account both the size and expected variance of the strata (*optimal* or *Neyman allocation*). In other words, more variable strata are sampled more intensively than less variable strata of the same size (Reed and Mroz 1997). The latter method results in a more precise estimate of the population mean, but requires prior estimates of the sample variance within the individual strata. However, optimal allocation is generally preferred if stratum areas and variances can be reliably determined (Avery and Burkhart 2002). Both of these methods provide separate estimates of mean values for each of the strata, which might be different forest types or administrative units.

3.3.3 Systematic sampling

In this method, the first sampling unit is selected randomly or arbitrarily located on the ground, and locations to be sampled are thereafter spaced at uniform intervals throughout the area to be surveyed. For example, if 10% of the area were to be sampled, every tenth sampling unit would be selected. Typically, sample units are established on a grid.

This method has been widely used in forest survey because the sampling units are easy to locate on the ground, and because the samples are distributed over the entire area, giving the impression that a representative sample has been obtained. The main problem is that it is less statistically powerful than random sampling methods. Specifically, it is not possible to obtain a genuinely valid estimate of the sample variance, because the sample units are not truly independent. This makes it difficult to estimate the precision of the measurements taken. This is particularly the case when there is a regular spatial pattern in the forest being surveyed, which may be caused by regular variation in soils, topography, or hydrology. The effects of topography can be minimized by referring to soil maps or by orientating grid lines to be surveyed up and down slope, rather than along the contours (Reed and Mroz 1997). However, in situations where estimations of precision are not required, systematic sampling may be preferred, primarily because it may be more efficient (in terms of information gained per unit effort expended) than random sampling approaches (Avery and Burkhart 2002).

3.3.4 Cluster sampling

A cluster is a group of smaller units (subplots) that taken together make up the sampling unit. Clusters can be arranged in many different ways, depending on the number of subplots included, the distance between them, and their spatial relations. As noted earlier, clusters are often used in NFIs, particularly in areas that are remote or difficult to access. This is because time and resources can be saved if information is collected from a number of locations within a particular area. However, cluster sampling is also sometimes used in regeneration surveys at a local scale.

Clusters are randomly selected from the population. However, they are not stratified, as in stratified random sampling. Cluster sampling may also be divided into a number of stages. For example, in two-stage clustering, clusters are randomly sampled and then, instead of each subplot within the cluster being surveyed, a subsample of these is randomly selected for survey (Reed and Mroz 1997). When implementing cluster sampling, the first step is to specify appropriate clusters. Ideally, the number of subplots within a cluster should be relatively small relative to population size, and the number of clusters should be relatively large. Cluster sampling will be more precise than simple random sampling if variation at the local scale is high relative to variation at the scale of the entire population (Avery and Burkhart 2002). It should be remembered, though, that the subplots within a cluster are not independent of each other, and therefore the independent sample unit is the cluster rather than the subplot.

3.3.5 Choosing sampling intensity

How many samples should be taken? Enough to obtain the level of precision required. It is therefore important to specify an acceptable level of precision before the survey is initiated. For example, in a survey designed to estimate the basal area of a forest stand, the forest manager might require an estimate within $\pm\ 5\ m^2\ ha^{-1}$ with a 95% confidence level. This corresponds to achieving estimated mean basal area that is within $5\ m^2\ ha^{-1}$ of the actual value 95% of the time (Reed and Mroz 1997).

There is a trade-off between precision and cost, because both increase as the number of sample locations increases. An index of efficiency can be calculated as the product of the squared standard error, which is a useful measure of precision, and the survey time (or expenditure) required (Avery and Burkhart 2002). This is based on the fact that to halve standard error four times as many sampling units are required. The required sampling intensity for a specified level of precision is given by the following formula:

$$n = (ts/E)^2$$

where n is the number, t is the t value (which can be obtained from statistical tables), s is standard deviation, and E is the desired half-width of the 95% confidence interval (Avery and Burkhart 2002). In order to apply this equation, an estimate is required of the expected variance that is likely to be achieved, as indicated by the confidence interval. This can best be obtained by carrying out a preliminary survey before the main investigation. Equations for calculating means, standard deviations, and confidence intervals for these different sampling approaches are given by Avery and Burkhart (2002), Cochran (1977), and Reed and Mroz (1997).

3.4 Locating sampling units

3.4.1 Using a compass and measuring distance

Sample location can be determined by using a hand compass to identify direction and by pacing to estimate distance. If using this method, it is helpful to calibrate the length of your stride over different terrain, by measuring the number of paces required to cover a set distance marked out with a measuring tape or marker poles. When performed by experienced practitioners, this method can be astonishingly accurate. I once accompanied a forest manager in Belize who managed to precisely relocate a sample plot, abandoned more than 50 years previously, after a couple of hours' walking through dense rainforest with only a compass and his carefully calibrated paces by which to navigate. Instruments such as pedometers are also available that can be used to measure distance. A sketch map, or better still an accurate map of forest stands, can be of great help in relocating sample locations in the future. Photographs can also help in this respect. Field notes on the location and orientation of the camera when the photograph is taken can assist interpretation of such photographs.

92 | Forest structure and composition

In ecological surveys, distances are often measured by using 30 m or 50 m tapes, which are usually made of steel, fibreglass cloth, or plastic. Other instruments that can be used to measure distances include *optical rangefinders*, which work on a similar principle to focusing a camera. A split image is created in the viewfinder by using mirrors or prisms, and a focusing knob is turned until the two images are coincident (Husch *et al.* 2003). Two types of optical rangefinder are available, fixed-based and fixed-angle, which differ with respect to whether the distance or the angle between the mirrors is altered. *Electronic rangefinders* are also available that measure the time taken for a laser pulse to travel to the target and return to the receiver. The range over which such rangefinders operate is influenced by the characteristics of the forest understorey and whether or not reflectors are used; effective range can vary from 20 to 100 m, with larger ranges obtained with the use of reflectors. Accuracy is also increased through the use of reflectors; accuracies of 0.4–0.5% can be achieved with handheld units (Husch *et al.* 2003). Alternatively, electronic rangefinders can be used that are based on the use of ultrasound, which have a maximum range of 20–30 m.

3.4.2 Using a GPS device

The development of satellite navigation systems has proved to be of enormous value to forest surveys (Kleinn 2003), enabling sampling locations to be located and relocated relatively easily. Most GPS units use the NAVSTAR-GPS system operated by the United States Ministry of Defense. The Russian government also operates a satellite navigation system called GLONASS. In Europe, a new programme called Galileo is currently being deployed and is expected to significantly improve coverage and precision when it becomes operational in 2008.

GPS receivers can now be obtained relatively cheaply, and are reasonably user-friendly. Although models made by different manufacturers vary in the details of their operation, typically they can be used to determine a location, to navigate from one location to another, and to store both individual locations and tracks. GPS is particularly useful where available maps are of poor quality, or where there are few features that can be identified from a map. As different models vary in their accuracy, the type of GPS unit used in collecting a set of information should be reported (Johnston 1998).

One of the main problems with use of GPS in forest survey is that it can sometimes be difficult to obtain an accurate location fix from underneath a forest canopy, which can prevent communication between the GPS unit and the satellites. To solve this problem, it may be necessary to locate the nearest canopy gap in order to obtain a measurement. The second main problem is the degree of accuracy obtained with a typical hand-held GPS unit, which is typically 5–10 m at best (Longley *et al.* 2005), especially under forest canopies (Figure 3.3). This may not be of sufficient accuracy to relocate a sampling unit. Sometimes a more accurate fix can be obtained if the GPS unit is left for several minutes, to increase the chances of detecting a relatively large number of satellites. Taking averages of repeated

Fig. 3.3 A differential GPS system. The antenna and GPS receiver are illustrated here. A reference station is also required (not illustrated), which consists of an additional GPS receiver and antenna. Such systems can provide location data with an accuracy of a few centimetres. However, accurate measurements may be difficult to obtain under a forest canopy. This instrument is manufactured by Leica Geosystems AG of Switzerland. (Photo by Harry Manley.)

readings can also increase accuracy (Johnston 1998). More expensive differential GPS systems are available that can provide a degree of accuracy of 1 m or better, but these still function less well under a forest canopy. Measurements of altitude made by GPS units are often prone to a high degree of error, and it may be preferable to use a conventional altimeter instead.

3.5 Sampling approaches

There are two main types of sampling unit that might be selected. The commonest approach is to use a *fixed area* method, which involves establishment of a field plot that may be square, rectangular, or circular in shape. Alternatively, a *distance-based* sampling approach might be adopted, where the area sampled varies but the number of individuals sampled is fixed. These two approaches are considered in the following sections, followed by an evaluation of their relative merits. In addition, line transects are briefly considered. Examples of the application of different sampling units to the assessment of tropical forest are provided by Dallmeier and Comiskey (1998a,b).

3.5.1 Fixed-area methods

- *Circular plots* are widely used because a single dimension (the radius) can be used to define the perimeter (Husch *et al.* 2003). They can be established relatively easily, and are readily marked through use of a single post, stake, or other form of marker (although they may be more difficult to relocate in future, for the same reason). Plots may range from 10 to 10 000 m^2 in area. In general, smaller circular plots are more efficient than larger ones (Husch *et al.* 2003). The main problem with this shape of plot is the accurate determination of the plot boundary. The best approach is to extend a tape measure from the centre of the circle, or use an electronic measuring device. Beers (1969) describes a method for establishing circular plots on a slope (which are elliptical in horizontal projection).
- *Square plots* offer an advantage over circular plots as the boundaries are straight lines, making it easier to determine whether or not an individual tree falls within the plot. To ensure that the corners of the plots are right angles, a compass or right-angle prism should be used. Plot limits can be established by measuring diagonals from the centre of the plot. As with circular plots, areas may vary from 10 to 10 000 m^2; for ecological surveys, plots of 1 ha in area or less are typically used.
- *Rectangular plots* are usually established by measuring distances from the central axis. Rectangular plots may be preferable in forests with difficult topography and large altitudinal variation (Husch *et al.* 2003). The word *strip* is sometimes used to refer to long rectangular plots. Although a strip may be divided into smaller subplots, the entire strip is equivalent to a single sampling unit, and for this reason separate smaller plots are generally preferred to strips as they provide greater statistical power for the same amount of effort.

Within fixed-area plots, different size classes of trees are sampled in proportion to their frequency. In natural forests, the number of smaller trees is often much larger than the number of large ones, and can take a great deal of time to measure. Generally some form of subplot is used to sample the smaller trees, which is typically nested within the larger plot. In circular plots, subplots can be established as concentric circles. For example, a plot design with concentric circles of 16, 64, 255, and 1018 m^2 can be used to measure seedlings, shrubs, small trees, and large trees, respectively (Husch *et al.* 2003). Similarly, in square plots, nested square subplots tend to be used. For example, within a 1 ha plot (100 $\times$ 100 m) all large trees might be measured, whereas young trees could be surveyed in a subplot 20 $\times$ 20 m, and seedlings within a subplot of 10 $\times$ 10 m or 5 $\times$ 5 m, typically located in the corner of the main plot. Subplots do not necessarily have to be the same shape as the larger plot, and can be arranged according to a variety of fixed designs (Husch *et al.* 2003).

Krebs (1999) points out that the ratio of the length of edge of a plot to the area enclosed inside it varies with shape, as follows:

$$\text{circle} < \text{square} < \text{rectangle}$$

The edge effect is important because it can lead to counting errors, which can arise because it is sometimes difficult to determine whether an individual tree is inside a plot or not. Such errors are fewer in circular plots. However, in some studies longer, thinner (rectangular) plots may be preferred because they include greater habitat heterogeneity.

3.5.2 Line intercept method

This method has been widely used by ecologists for measuring the cover of plants along line transects (Krebs 1999). A *line transect* can be established by pegging a measuring tape or cord across the area of interest. Estimates of cover are simply calculated as the fraction of the line length that is covered by the canopy of a particular species. Density or abundance estimates are obtained by measuring the longest perpendicular width w for each plant or sample unit intercepted. This width determines the probability that any individual plant will be bisected by the sampling line (Krebs 1999). Following Eberhardt (1978), population size can be estimated as:

$$N = \left[\frac{W}{n}\right] \sum_{i=1}^{k} \left(\frac{1}{w}\right)$$

where N is an estimate of population size, W is the width of the baseline from which transects begin, n is the number of transects sampled, w is the perpendicular width of plants intersected, and k is the total number of plants intercepted on all lines ($i = 1, 2, 3, \ldots, k$). To estimate the density of organisms for any shape of area, simply divide this estimate of numbers by the area studied (Krebs 1999). If a series of line intercepts are measured, each can be used to generate an estimate of population size, enabling confidence limits to be calculated, providing an estimate of variability.

3.5.3 Distance-based sampling

Distance sampling methods are used primarily for estimation of population densities and abundances (i.e. the number of individuals of a particular species occurring in an area), although these methods can also be used for estimating other variables such as tree heights, basal area, and canopy cover. The methods focus on sampling a certain number of individuals, rather than a fixed area or plot. A comprehensive description of different distance sampling methods is provided by Buckland *et al.* (2001) and Krebs (1999), although it should be noted that many of the methods described are more widely used for investigations of animal populations than for plants. A comparative analysis of the performance of various distance-based sampling approaches is provided by Engeman *et al.* (1994). Use of these methods in forestry is described by Payandeh and Ek (1986). In each of these methods, the sampling locations should be selected by using a random or stratified random design, although Hall (1991) used a systematic design, with sample points arranged on a grid.

Distance methods most commonly used by forest ecologists include:

- The *point-centred quarter method*, in which a sample is taken of four trees at each sample point by selecting the nearest tree within each of four 90° quadrants around the sample point (usually defined using compass bearings) (Morisita 1954). An example is provided by Haridasan and de Araújo (1988). Data for each sample point are pooled before analysis, by calculating the mean of the four distances from each sample point.
- The *nearest individual method*, where the nearest tree to the sample point is located and the distance between it and the sample point is measured. Density of trees can be calculated according to the equation

$$\text{density} = 1/(2\,D_2)^2$$

  where D_2 is the mean of the distances over all of the samples (Bullock 1996).
- The *multiple-nearest-tree technique*, which is characterized by sampling multiple nearest neighbours to each sample point, rather than just one or four (Williams *et al.* 1969). Application of this method is described by Hall (1991) in montane forest in Tanzania. In this case, a sample was taken of the nearest 20 trees $\geq$ 20 cm dbh occurring around each sample point, although results indicated that a sample of 15 trees would have provided sufficient precision. Sample points were located 200 m apart to avoid any chance of overlap between the samples. Distances from the sample points to the trees were estimated by extending 50 m measuring tapes from each point.
- The *T-square sampling method*, in which the distance is measured from the sample point to the nearest tree. A line is then drawn at right angles to the line from the sample point to the tree, and the nearest tree to the sample point positioned on the other side of this line is then measured (Figure 3.3) (Greenwood 1996).
- The *variable-area transect approach* involves extending a single rectangular plot until it includes the specified number of stems (Parker 1979). This method is considered by Engeman *et al.* (1994) to be the simplest and most practicable of the distance methods that they considered. Further refinements to the method are presented by Engeman and Sugihara (1998). Sheil *et al.* (2003) highlight some of the problems with this approach, such as the fact that transects may extend over large areas, complicating the analysis of relations between density measurements and site characteristics such as soil and topography. Sheil *et al.* (2003) present a refinement of the variable-area transect method developed for rapid assessment of diversity in tropical forests, in which the sample unit is a cluster of cells, each of which is a modified variable-area transect; a set of decisions is used to define the sampling effort on the transects.

The following formula is used to estimate population density from data collected by the point-quarter method (Krebs 1999):

$$N = \frac{4(4n-1)}{\pi \sum (r_{ij}^2)}$$

where N is the estimate of population density, n is the number of sample points from which observations are made, and r_{ij} is the distance from random point i to the nearest organism in quadrant j ($j = 1, 2, 3, 4; i = 1, \ldots, n$).

In methods where distances are measured between an organism and a fixed point, the following formula can be used to estimate density (Krebs 1999):

$$N = \frac{n}{\pi \sum (x_i^2)}$$

where N is the estimate of population density, n is sample size, and x_i is the distance from a random point i to the nearest organism.

For methods that calculate distances between the organism and nearest neighbours, the following formula can be used to measure density (Morisita 1957, Krebs 1999):

$$N = \frac{n}{\pi \sum (r_i^2)}$$

where r_i is the distance from an organism to its nearest neighbour.

For T-square sampling, the following formula can be used (Krebs 1999):

$$N = \frac{2n}{\pi \sum (z_i^2)}$$

where z_i is the T-square distance associated with a random point i.

Other equations for estimating densities for use with distance measures and corresponding estimates of variation are presented by Buckland *et al.* (2001), Husch *et al.* (2003) and Krebs (1999).

The use of distance (or plotless) methods to assess tree density is described by Bullock (1996), who recommends a minimum of at least 50 sample points for each estimate. He also points out one of the problems with distance measures: the sample may be biased, because more isolated trees are more likely to be sampled. The T-square method can overcome this problem, and may therefore be preferred (Greenwood 1996) (Figure 3.4). Bullock (1996) also notes that it takes longer to obtain samples by using the point-centred quarter method than for the nearest-individual method, but the latter gives a more variable estimate and therefore the sample size needs to be higher for the same degree of accuracy. The techniques also work less well when rare species, which occur at very low densities, are being surveyed. In such cases, it can take an enormous amount of effort to locate individuals and measure distances from sample points.

98 | Forest structure and composition

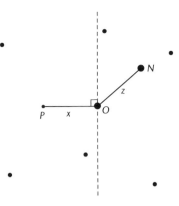

Fig. 3.4 The *T*-square method. The black circles represent individuals of the study species. *O* is the nearest individual to a random point *P*; *N* is *O*'s nearest neighbour on the opposite side of the dashed line (which is perpendicular to the line *OP*). (From Greenwood 1996.)

3.5.4 Selecting an appropriate sampling unit

Many factors influence the choice of sampling unit, including cost-effectiveness, required accuracy and precision, resource availability, and ease of data analysis and presentation (Sheil *et al.* 2003). Kenkel *et al.* (1989) provide an overview of such factors. Most importantly, the choice of sampling unit varies with the objectives of the investigation and the characteristics of the forest being surveyed. Establishment of field plots in forests that are dense, inaccessible, or located on steep topography can be physically very challenging; in such situations, either a small plot size or a distance-based method may often be preferred. Overall, distance-based methods tend to be faster and more efficient than establishment of fixed-area plots, but despite these advantages, the use of fixed-area plots is far commoner in forest ecology research.

Phillips *et al.* (2003) describe two standard plot-based methods that are widely used in tropical forest ecology, particularly for assessment of floristic diversity:

- *The 1 ha method.* This involves a one-time census of all stems ≥ 10 cm in diameter in a 1 ha plot, which is usually square in shape.
- *The 0.1 ha method.* This involves sampling all stems ≥ 2.5 cm diameter in 10 0.01 ha transects each of 2 × 50 m (as developed by Gentry 1982, 1988). This method has been applied mostly, but not exclusively, in the neotropics.

Phillips notes that more than 650 0.1 ha inventories have been established in tropical forests to date, which compares with more than 700 1 ha plots surveyed throughout the tropics.

Another standard plot design particularly used for assessing plant diversity and cover is the modified-Whittaker plot (MWP). This approach employs subplots with a variety of different uses nested within each other, and can provide accurate estimation of mean species cover and analysis of plant diversity patterns at multiple spatial scales (Stohlgren *et al.* 1995). Examples of this approach being used in forests are provided by Campbell *et al.* (2002) and Keeley and Fotheringham (2005). In the former study, which was done in lowland tropical forest, 0.1 ha MWPs were found to record species composition and abundance similar to that of 1 ha plots, and were equally effective at detecting rare species. In addition, MWPs were more effective at detecting changes in the mean number of species of trees ≥ 10 cm dbh and of herbaceous plants.

A key aspect that should be considered is the relative efficiency of different methods—in other words, the amount of information gained per unit effort expended. This issue has received surprisingly little attention from ecological researchers, although foresters have long been aware of its importance (Avery and Burkhart 2002). Phillips *et al.* (2003) repeatedly sampled forests in two regions of Amazonia by 1 ha and 0.1 ha plot-based methods, and compared their performance against the amount of effort required. Results indicated that the 0.1 ha method is more efficient for floristic assessment, but the authors also note that 1 ha plots still may be preferred in some situations, such as for ground-truthing remotely sensed measurements; they are also widely used in studies of forest dynamics (Condit 1998, Dallmeier and Comiskey 1998a, b). A key advantage of both of these methods is that their widespread use enables comparisons to be made with a range of other studies. On the other hand, the fact that they have been widely used by other ecologists does not in itself provide a strong justification for their use in a particular study. It may be that for the characteristics of a particular forest, or a specific set of objectives, some other approach might be more efficient.

This point is further illustrated by the work of Gordon (2005), who compared fixed area, distance-based, and *ad hoc* methods for the rapid inventory of tropical forest tree and shrub diversity in eight seasonally dry tropical forests sites in southern Mexico, with the aim of identifying priority sites for conservation. Results indicated that the 2×50 m protocol with 10 repetitions popularized by Gentry (1982) was relatively inefficient and lacking in statistical power. A 6×50 m protocol and fixed-count circular plots (equivalent to the variable area transect approach) were found to be more efficient, in terms of results obtained per unit effort. Preliminary surveys testing different methods are therefore to be recommended before committing substantial time and resources to a particular approach.

Hall *et al.* (1998) provide a detailed consideration of different sampling approaches for tropical forests. For determining stand density, many small plots are more efficient in terms of time required to establish and enumerate, and can provide similar precision to larger plots. There is a trade-off between total area to be surveyed and number and size of plots, which depends on the extent of variation between plots in comparison to the variation between sites that are to be compared. This trade-off point can best be estimated by doing a preliminary survey with different plot sizes (Hall *et al.* 1998). In Amazonian Peru, Stern (1998) compared fixed-area plots following the strip transect design of Gentry (1982) with variable-area transects; she used a total sample size of 50 stems for each of three size categories. Her conclusions were that the fixed-count plots were more flexible, particularly when different vegetation structures were encountered, but that strip transects had the advantage of being comparable to assessments from many other sites worldwide. Kint *et al.* (2004) also found that equal or higher sample sizes are needed for plot sampling than for distance sampling to obtain the same degree of accuracy, and that distance methods were generally more efficient. However, plot-based sampling was more efficient for estimating stand structure at low to medium accuracy. Kint *et al.* (2004) also demonstrated that, at least in the low-diversity forests in which they worked, minimum sample size is negatively correlated with tree density and is generally lower in large stands than in small ones.

Choice of plot size and shape can have a major influence on the results obtained from the survey (Laurance *et al.* 1998a). For example, Condit *et al.* (1996) found that in tropical forests 5–27% more species were found in rectangular plots than in square plots of the same area, with longer and narrower plots increasingly diverse. This result reflects the aggregated distributions of individual species, and indicates the importance of sampling the same number of stems if the objective is to make comparisons of diversity. In Mediterranean vegetation, however, Keeley and Fotheringham (2005) found no such difference between square versus rectangular plots. The size of the trees being measured can also have implications for plot design. For example, Gray (2003) found that in mature Douglas-fir forests, samples of at least 40% of a stand (4 subplots of radius 18 m) were required to reduce errors for estimated density of large trees (≥ 122 cm dbh) below 25% of true density at least 66% of the time. However, for trees < 75 cm dbh, the standard inventory sample of 0.07 ha with 4 subplots of radius 7.3 m met this degree of accuracy for estimates of density and mortality.

It is also worth noting that there is an alternative to the use of either fixed-area or distance methods: a systematic or *ad hoc* search of habitat. Searches are typically done by the surveyor walking around a site, looking for the target species or recording all of the species encountered. Usually, the search is timed so that the information gained can be corrected for the amount of survey effort expended. Timed searches can be further standardized by restricting the search to a specific area; an area might be subdivided into units and a sample of these selected randomly for surveying. This method is generally used to determine whether a particular species is present in a specific area (the absence of the species from the area, however, can be difficult to demonstrate conclusively) or to produce a checklist of the species present in an area. The approach is widely used in support of conservation planning, particularly where information is needed on the distribution of rare or threatened species, or where a rapid assessment of floristic composition is required (see, for example, Schulenberg *et al.* 1999).

Searches suffer from the problems of bias (Nelson *et al.* 1990) because survey effort will not be equally distributed over the forest area, regardless of how carefully the search is done. As a result, the data obtained in this way cannot be considered truly representative of the area that has been surveyed, and are not therefore amenable to statistical analyses based on the assumption of unbiased data. Comparisons between the information gathered from different areas are therefore difficult to make, and the method cannot readily be used to provide accurate estimates of abundance. However, searches require much less effort than the establishment of a field plot and can provide valuable information (Droege *et al.* 1998), particularly where information is needed on the presence of specific species, which can be difficult to gather by means of any other approach. Searches have been used to assess tree diversity by Gordon *et al.* (2004) and Hawthorne (1996), and checklisting was also recommended for this purpose by Gray (2003).

Given the range of different sampling units that are available, how can an appropriate method be selected? Husch *et al.* (2003) suggest that the following points should be considered:

- The overall aim should be to sample a large enough number of trees so that results of sufficient accuracy are obtained, but a small enough number so that the time required for measurement is not excessive.
- In general, the most efficient sampling unit is the one that samples proportionally to the variance of the variables being measured.
- Smaller sampling units are often more efficient than larger ones. In a relatively homogeneous forest, the precision obtained for a given sampling intensity tends to be higher for relatively small sampling units than for larger ones, because the number of independent sampling units is greater. However, in heterogeneous forests, high variation will be obtained with small sampling units, and therefore larger sampling units will be preferred.
- In general the cost of sampling, in terms of time and effort expended, is generally greater for a large number of small sampling units than for fewer plots of larger size.
- Each sampling unit should be large enough so that it adequately represents the composition and structure of the forest. As plot size is reduced, the probability that it is not representative of the forest increases.
- For dense stands of small trees, plots should be relatively small, but for widely spaced stands of large trees, plots should be relatively large.

Selection of an appropriate plot size therefore depends on the characteristics of the forest, and can usefully be determined by doing a preliminary survey testing a range of designs. Two methods are available for choosing the best plot size statistically (Krebs 1999). Wiegert (1962) suggested that the most appropriate plot size was one that minimizes the product of relative cost and relative variability. When comparing different plot sizes, relative cost for each plot size can be defined as:

$$\text{relative cost} = \frac{\text{time to measure a plot of a given size}}{\text{minimum time to measure a plot}}$$

Here, the minimum time refers to the least time taken to measure a plot, of all the plot sizes considered. Relative variance can be defined as:

$$\text{relative variance} = \frac{(\text{standard deviation})^2}{(\text{minimum standard deviation})^2}$$

In this case, the standard deviation of a variable measured (such as biomass) is calculated, and divided by the minimum standard deviation of the same variable for all of the plot sizes considered. The two values, relative cost and relative variance, are multiplied together to determine the plot size that produces the lowest value, which can be considered the most efficient.

Hendricks' method (Hendricks 1956) proposes that the optimal plot size is determined as follows:

$$\hat{A} = \left(\frac{a}{1-a}\right)\left(\frac{C_0}{C_x}\right)$$

where $\hat{A}$ is an estimate of optimal plot size, a is the absolute value of the slope of the regression of log variance on log (plot size), assumed to be between 0 and 1, C_0 is the cost of locating one additional plot, and C_x is the cost of measuring one unit area of sample. This method cannot be applied as generally as Wiegert's method, because it is based on a series of assumptions: that the absolute value of the slope of the regression between log of the variance and log of the plot size is between 0 and 1, and that the amount of time taken to survey a plot is directly proportional to plot size. It is important to note that a sampling strategy that is adequate but not optimal for all species or habitats may still save time and money (Krebs 1999).

3.5.5 Sampling material for taxonomic determination

One of the biggest challenges in working with trees is that the parts of the plant most useful for identification—the flowers, fruits, and leaves—are often out of reach. This can make taxonomic determination very difficult. It is good practice to retain voucher specimens of taxonomically critical species, or any specimen for which identification is uncertain, and deposit them as a reference collection in a suitable herbarium (Stern 1998) (Figure 3.5). Often trees may have to be climbed to collect samples for identification, although this should only be attempted by people who have received appropriate training and have access to adequate climbing equipment and a safety harness (Figure 3.6). Pruning poles provide a safer alternative, but their reach is limited to a few metres. Other methods that have been used to obtain samples for identification include catapults, shotguns, and even trained monkeys but, as noted by Richards (1996), these are not methods that have been used extensively. Vegetative characters, such as fallen leaves, bark, and wood characteristics, can often be used to identify species. It is important to avoid over-collecting. A simple rule of thumb is not to collect any specimen unless there

Fig. 3.5 When undertaking a field survey, it is often necessary to collect samples for taxonomic identification. Care should be taken to minimize the environmental impacts of such collecting, for example by collecting only small parts of a plant. Voucher specimens should be deposited in a herbarium for future reference. (Photo by Adrian Newton.)

are 20 individuals present, and one should not collect more than 1 out of 20 plants (Wagner 1991). For trees, it is usually sufficient just to collect samples of the foliage and reproductive structures, rather than the whole plant!

Research into the ecology of forest canopies is area of growing interest among ecologists, who have been very inventive in developing novel methods of accessing tree crowns, including the use of balloons, canopy towers, and cranes as well as relatively cheap and simple methods such as climbing ropes. These techniques are described by Houle *et al.* (2004), Lowman and Wittman (1996), Mitchell (1982, 1986), and Moffett (1993).

104 | Forest structure and composition

Fig. 3.6 Researchers are often tempted to climb trees to collect samples of leaves, flowers, or seeds for analysis, but tree climbing should only be attempted by people who have received appropriate training. Use of safety equipment, such as the harness and head protection pictured here, is essential. (Photo by Adrian Newton.)

3.6 Measuring individual trees

3.6.1 Age

For some conifers, age estimates can be obtained by counting the number of branch whorls, although this method is less accurate in older trees because it can be difficult to determine where branches have been abscised. In some broadleaved trees and tree ferns it is possible to estimate ages by counting the number of leaf scars on the terminal shoot, although again this method is only useful with younger trees and branches.

Measuring individual trees | 105

Fig. 3.7 Using an increment borer on a *Larix decidua* tree during a dendroecological study of mixed woodlands at the upper timberline of the central Italian Alps. (Photo by John Healey.)

Age estimates are generally obtained by counting the number of annual rings in a stem cross-section. Annual ring formation depends on the fact that wood formed earlier in the year tends to be more porous and lighter in colour than that formed later in the year. Total tree age is obtained by counting rings at ground level; if measurements are made higher up the stem, then the number of years that the tree takes to grow to this height should be added to the total (Avery and Burkhart 2003). Although this method works well in most temperate forests, it can also be used in some tropical forests, particularly where there is a pronounced dry season or where the trees are seasonally deciduous (Schweingruber 1988).

Ring counts are often taken on sawn sections of a tree trunk, which should be smoothed with a plane or knife and viewed with a hand lens or dissecting microscope in order for accurate counts to be obtained. If sawn sections are not available, an *increment borer* can be used (Figure 3.7), which consists of a hollow tube with a cutting bit that is screwed into the tree (Husch *et al.* 2003) and is obtainable from forestry suppliers. A reverse turn snaps the core of wood inside the tube, which is

106 | Forest structure and composition

Fig. 3.8 Extracting the stem-core from an increment borer for dendrochronological analysis of growth rings. Great care is needed when extracting and transporting the cores, as they can be very fragile. (Photo by John Healey.)

removed with an extractor (Figure 3.8). Accurate determination of tree age by this method requires coring through the centre or pith of the tree, which can be difficult to locate. This process can be assisted by employing two people, one of whom takes the core and the other of whom indicates the perpendicular axis of the tree from a distance of 2–3 m (Schweingruber 1988). Alternatively, a holding device can be used, which can be adjusted in the axial direction by a peg and screw.

The maximum length of an increment borer is around 50 cm, which determines the upper limit to the size of tree that can be aged using this method. Cores in standing trees are typically taken at breast height (1.3 m above ground level). The bore holes should be sealed with grafting wax (available from garden centres and forestry suppliers) to minimize the risk of introducing disease as a result of coring (Schweingruber 1988). Tree species with dense wood can be very difficult to core, and the wood cores or the borer itself can be difficult to extract intact. Extracted cores are fragile and should be stored in a plastic tube, drinking straw, or other appropriate container. They can be labelled with soft pencil when freshly collected. The cores can be glued into a groove in a block of wood with water-soluble glue, to assist preparation and inspection.

The visibility of tree rings in trunk disks or cored samples can be increased by cutting transverse radial strips with a sharp blade, such as a multiple-snap-off blade knife, or by polishing the sample with sandpaper of different grades. Samples displaying little contrast between tree ring boundaries can be mounted between two blocks and cut to a thickness of 0.25 mm, then examined in transmitted light under a microscope. Staining of tree rings with paper dye can also be used to improve visibility, but this is generally not successful (Schweingruber 1988). The surface of prepared samples can best be examined under a stereomicroscope with a spotlight, or a hand lens. A calibrated eye-piece is useful for making measurement of ring widths. Dedicated instruments are also available for detailed analysis of tree ring widths, linked to custom-designed computer software (such as WinDENDRO, ⟨*www.regeninstruments.com/*⟩).

Analysis of tree rings has been widely used in archaeology as a means of dating wood fragments, and has also been widely used to analyse past climate change. Details of the methods used in dendrochronology are described by Cook and Kairiukstis (1990) and Schweingruber (1988).

A number of problems may be encountered when obtaining ring counts (Husch *et al.* 2003), namely:

- In slow-growing trees, rings may be very close together and consequently difficult to count.
- In some species rings are indistinct, because there is little difference between wood formed in the spring and in the summer.
- Some species may form more than one ring in a growing season, for example during a period of dry weather or as a result of defoliation caused by insect attack. False rings often do not extend around the entire circumference of the tree, however.
- Tree ring counts are generally very difficult to obtain from tropical trees, except in areas with a pronounced annual dry season.

Where ring counts are not possible, radiocarbon dating can potentially be used, although this technique is limited to trees of great age (> 500 years old) (Martínez-Ramos and Alvarez-Buylla 1999). It is also possible to estimate tree age using models of growth increment, for those species where reliable long-term growth data are available (Chambers and Trumbore 1999).

3.6.2 Stem diameter

Measurements of stem diameter are widely used in both forest ecology and management to characterize the size distribution of forest stands and to estimate timber volumes. Diameter measurements are usually taken at a standard height, the *diameter at breast height* (dbh), which is defined as 1.3 m above ground level (or 4.5 ft in the USA). Measurements are complicated by the fact that tree stems are often not circular in cross-section, and may be leaning or surrounded by prominent buttresses, making them difficult to measure. The following standard procedure is recommended by Husch *et al.* (2003):

- When the tree is on a slope, measure dbh on the uphill side of the tree.
- When a tree is leaning, measure dbh on the high side of the tree, in a way that is perpendicular to the longitudinal axis of the stem.

- When the tree has a bulge, limb or some other abnormality at breast height, measure dbh above the abnormality; attempt to measure the dbh that the tree would have had if the abnormality were not present.
- When a tree is multistemmed at breast height, measure each stem separately; when a tree forks above breast height, measure it as a single stem. If the fork occurs at breast height, measure the dbh below the enlargement of the stem caused by the fork.
- When a tree has a buttress than extends higher than 1 m, measure the stem at a fixed distance (30 cm) above the top of the buttress.
- When the breast height point has been marked on the tree with paint, assume the point of measurement to be the top of the paint mark. Use of such paint marks can greatly improve accuracy when repeated measurements are made.
- If the tree stem is elliptical in cross-section, then measure the major and minor diameters separately, and produce an overall figure by calculating the mean of the two values.

The two most commonly used instruments for measuring tree stem diameters are calipers and diameter tapes. *Calipers* are usually used when the trees are less than 60 cm dbh; although larger calipers are available, they can be difficult or unwieldy to use in the field. Calipers are usually constructed out of metal, wood, or plastic, and enable the diameter to be read directly off a scale when the arms of the caliper are placed around the tree stem. The caliper arms should be pressed firmly against the tree stem with the main beam of the caliper placed perpendicular to the axis of the tree stem, and the arms parallel and perpendicular to the beam (Husch *et al.* 2003).

Standard *measurement tapes* can be used to measure stem diameter by placing the tape around the circumference of the tree at breast height (Figure 3.9). Measurements of circumference taken in this way can be converted to diameters (assuming a circular cross-section) by dividing the values by π. Diameter tapes can be obtained, however, that are graduated at intervals of π units (in cm or inches) enabling diameter to be measured directly. Care should be taken to ensure that the tape is positioned correctly: it should be in a plane perpendicular to the trunk of the tree, and pulled taut around the trunk so that accurate measurements are obtained (Husch *et al.* 2003). Although more accurate results can be obtained with calipers, measurements with tape tend to be more consistent if repeated measurements are made, because caliper measurements are more sensitive to the positioning of the instrument (Husch *et al.* 2003). A review of different methods for measuring tree diameters is provided by Clark *et al.* (2000).

Bark thickness can be determined by using a bark gauge, which consists of a steel shaft that is pushed through the bark. The thickness of the bark can be read directly off a scale with the instrument in place. A minimum of two readings should be taken (Avery and Burkhart 2003). When measurements of diameter are made, whether or not the bark was included in the measurement should be recorded.

Measuring individual trees | 109

Fig. 3.9 Measuring stand structure in the New Forest National Park, southern England. The student in the centre is measuring diameter at breast height of an oak tree (*Quercus robur*) using a diameter tape. The others are carrying a laser rangefinder (left) and a hypsometer (right). (Photo by Adrian Newton.)

A number of instruments are available for obtaining measurements of upper stem diameters, which can be useful for assessing the form of the tree or extent of stem taper. Options include optical forks, optical calipers, and fixed-based or fixed-angle rangefinders (Clark *et al.* 2000). Many of these instruments are expensive and are prone to inaccuracy (Avery and Burkhart 2003). Most commonly, a *relascope* is used, which is a form of optical rangefinder (Husch *et al.* 2003) (see also Section 7.2). However, calipers or diameter tapes provide more accurate measurements of upper stem diameters, if the upper parts of the tree can be accessed through use of climbing ropes or ladders.

3.6.3 Height
Total tree height can be defined as the distance along the axis of the tree stem from ground level to the top of the canopy. Other terms commonly used by foresters

Fig. 3.10 Measuring tree height using a hypsometer. The top of the tree is sighted through the instrument, and the distance from the tree is measured. Note that in deciduous forests, it is much easier to make measurements when the trees are leafless. (Photo by Adrian Newton.)

include *bole height*, which refers to the distance between ground level and the first crown-forming branch, and *crown length*, which is the distance on the axis of the tree stem between the first crown-forming branch and the top of the canopy.

The heights of relatively short trees can be readily measured by using a graduated pole. Height measurements of tall trees are generally made by means of *hypsometers*, which use trigonometric relations to estimate height (Figure 3.10). The user sights the top of the canopy of the tree being measured and takes a reading, then sights to the base of the tree and takes a second reading. Many hypsometers are scaled according to appropriate units, enabling the height to be calculated directly as the sum of the two readings. However, if measurements are made on a slope, and the observer's position lies below the base of the tree, tree height is derived by taking the difference between the two readings (Avery and Burkhart 2003) (Figure 3.11). Some hypsometers are graduated in degrees, and require the use of basic trigonometry for conversion to height measurements. In both cases, the distance between the point of measurement and the tree should be measured, typically with a measuring tape. Bole height and crown length can be measured, as well as total tree height, by measuring the heights of the appropriate locations on the tree stem. Commonly used types of hypsometer include the Abney level and the Suunto clinometer (Husch *et al.* 2003). More recently, electronic hypsometers have become commercially available, which use lasers to measure horizontal distances and calculate tree heights

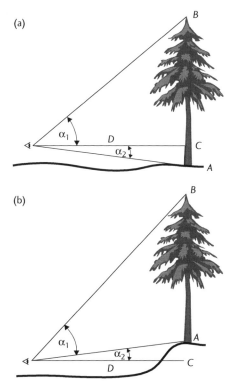

Fig. 3.11 Measuring tree height with a hypsometer, based on tangents of angles. (After Husch *et al.* 2003): (a) The total height of the tree may be determined as $AB = D(\tan\alpha_1 + \tan\alpha_2)$. (b) On steep ground, where the tree may be viewed from below, the height of the tree $BC - CA$ is given by $AB = D(\tan\alpha_1 - \tan\alpha_2)$.

from angular measurements. However, these are often bulkier than traditional hypsometers, and much more expensive.

The main challenge to obtaining accurate height measurements, regardless of the type of hypsometer used, is the difficulty of sighting the top of the tree canopy, particularly in closed forest stands. Large, flat-crowned trees are particularly difficult to measure, simply because it is not easy to see the crown apex. As a rule of thumb, tree heights should be measured at a distance approximately equivalent to the height of the tree (Husch *et al.* 2003). Leaning trees are also difficult to measure; in this case, height should be measured for the point on the ground that is vertically below the canopy apex. Accuracy of measurements can also vary between different users, so, ideally, repeated measurements should be made.

3.6.4 Canopy cover

Jennings *et al.* (1999) distinguish two basic types of measurement of forest canopies: *canopy cover*, which is the area of the ground covered by a vertical projection of the canopy, and *canopy closure* (or *canopy density*), which is the proportion of the sky hemisphere obscured by vegetation when viewed from a single point. These two terms are often confused in the literature. Canopy cover is an important variable for estimating stand variables from remote sensing data (see section 2.5), and in young forest stands may correlate closely with basal area; however, the relation between these variables is often less pronounced in more mature forest stands (Jennings *et al.* 1999). Canopy closure is likely to be more closely related to light regime and microclimate, as well as plant growth and survival, at the point of measurement (Jennings *et al.* 1999).

As noted in Chapter 2, crown diameters can be measured from high-resolution aerial photographs. Field measurements of tree crowns are complicated by their inaccessibility and irregularity. The commonest method to measure the size and shape of a tree crown is to project the perimeter of the crown vertically down, then measure it at ground level. Estimates of crown area are typically obtained by measuring the crown diameter at its widest point, then again at right angles to this measurement (Husch *et al.* 2003). Hand-held or pole-mounted mirrors, prisms, or pentaprisms may be used to achieve the vertical projection. Appropriate instruments and associated methods for this purpose have been described by Cailliez (1980) and Tallent-Halsell (1994). Crown area is estimated from these measurements by using the formula for calculating the area of a circle, from either the mean value of the two measurements made or the mean value of minimum and maximum crown diameters.

The method for estimating crown cover of a forest stand is described by Jennings *et al.* (1999). At each point of measurement, the observer looks vertically upwards and records whether or not the forest canopy obscures the sky. An estimate of forest canopy cover can be produced by calculating the proportion of points where the sky is obscured. Observations can be made without use of any instrumentation, although both accuracy and repeatability can be improved by doing so. Examples of instruments designed to ensure that sightings are truly vertical include the gimbal balance (Walter and Soos 1962) or the sighting tube, which often has an internal crosshair (Johansson 1985). Commercial versions of the latter incorporate bubble levels to ensure that the tube is positioned vertically, and 45° mirrors to ensure that the head posture is horizontal during use. Random or stratified random sampling approaches should be used when taking such measurements according to Jennings *et al.* (1999), who present formulae for calculating the confidence limits of such measurements by using a binomial distribution. Accurate estimates require large sample sizes; these authors suggest that at least 100 observations should be made in any forest area being surveyed.

Measures of canopy closure, rather than canopy cover, are generally to be preferred in ecological studies. This requires estimation of the light received at a particular point, including both direct solar radiation and the indirect radiation that arrives from all parts of the sky. The entire hemisphere surrounding the sample point should therefore be assessed, rather than just the sky immediately above the sample point (Jennings *et al.* 1999). Methods for assessing canopy closure are presented in section 4.5.4.

3.7 Characterizing stand structure

A forest stand may be defined as a group of trees that occupy a given area, which share some characteristics such as species composition, size, or age. *Stand structure* refers to the distribution of species or tree sizes within the stand. Canopy characteristics and leaf area can also be considered as structural variables.

3.7.1 Age and size structure

Stands are commonly differentiated into those that are *even-aged*, where all trees are of approximately the same age, and those that are *uneven-aged*, where trees display a variety of different ages. The age distribution of trees within a stand can provide some insights into the history of tree recruitment and patterns of previous disturbance. The age structure of a stand is generally characterized by sampling a subset of trees from the stand and determining their ages, as described in section 3.6.1. In even-aged stands, the age of the stand is usually determined as the mean age of those trees that dominate the canopy (i.e. the largest individuals), and sampling may be restricted to these. Age distributions are typically illustrated by plotting a frequency distribution or histogram of the number of trees within each age class.

The diameter (dbh) of a stand can be expressed a mean of the diameters of the trees present. Often, basal area is calculated to provide a measure of stand density and to provide a basis for calculating stand volume. Basal area can be calculated from measurements of stem diameter by using the following equation (assuming that trees are circular in cross-sectional area):

$$g = \frac{\pi d^2}{4}$$

where g is cross-sectional area of the tree and d is the diameter of the cross-section.

When measured using metric units, tree diameter is usually expressed in centimetres, and cross-sectional area in square metres. In this case, the equation is as follows (Husch *et al.* 2003):

$$g(m^2) = \frac{\pi d^2}{4(10\,000)} = 0.00007854\, d^2$$

Stand basal areas can be calculated by summing values obtained from dbh measurements of individual trees.

Stand structure is most commonly described by using measurements of tree diameters (section 3.6.2). These data may be illustrated as a stand table, which indicates the number of trees of each species per unit area belonging to each diameter class. Such data can also be illustrated as a frequency distribution or histogram of the number of trees within each size class. This method of illustrating stand structure is very widely used by forest ecologists, and can be used to infer the dynamics of the stands under investigation and their phase of development.

For example, dominance of a stand by many small-diameter trees would suggest that the stand has only recently been established, perhaps following some form of disturbance event. Alternatively, presence of some very large diameter trees with representation of trees in smaller diameter classes might be interpreted as mature or 'old-growth' forest, within which continuous recruitment is taking place (Spies 1997, Spies and Turner 1999). However, it should be noted that the size and age of trees are not necessarily closely related. Trees growing on adverse sites or subjected to a high intensity of browsing can grow very slowly, leading to much greater variation in age than in size. Stands that appear to be even-aged on the basis of their diameter distributions may in fact have been recruited over a prolonged period. For this reason, caution should always be exercised when interpreting diameter measurements in terms of ages; ideally both age and size should be measured.

Uneven-aged stands are typically characterized by the presence of a large number of trees in smaller-diameter size classes with decreasing frequency as the size class increased. This form of size–frequency relationship is often referred to as an 'inverse-J' shape (Figure 3.12) and is often an objective of approaches to sustainable forest management. If a particular tree species displays such a size distribution, then continuous recruitment can generally be inferred, suggesting that the population is viable as sufficient regeneration it taking place for the population to be maintained. Techniques for studying forest dynamics and the population viability of tree species are described in detail in Chapters 4 and 5.

Stand diameter distributions can be represented mathematically by *probability density functions*. A number of different functions have been used, including normal, exponential, binomial, Poisson, Pearl, Reed, Schiffel, and Fourier series (Husch *et al.* 2003). Details of the use of these functions are provided by Johnson

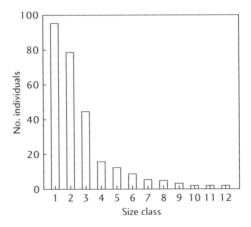

Fig. 3.12 Inverse-J structure of a forest stand, illustrated by plotting the number of individuals in each size class. The inverse-J structure is characterized by larger numbers of smaller individuals than of larger individuals. (After Peters 1994, http://www.panda.org.)

(2001) and Schreuder *et al.* (1993). The most widely used function in relation to analysis of forest stand data is the Weibull function (see, for example, Soehartono and Newton 2001), which can be expressed as follows (Husch *et al.* 2003):

$$f(D) = \frac{c}{b}\left(\frac{D-a}{b}\right)^{1/c} e^{-[(D-a)/b]^c}$$

where $f(D)$ is the probability density, a is a location parameter (theoretical minimum population value), b is a scale parameter, c is a shape parameter, and D is the diameter.

The Weibull function can exhibit a variety of different shapes depending on the value of c (Husch *et al.* 2003):

- $c < 1$, inverse J-shape
- $c = 1$, exponential decreasing
- $1 < c < 3.6$, positive asymmetry
- $c = 3.6$, symmetric
- $c > 3.6$, negative asymmetry.

The parameters of the Weibull distribution can be derived directly from diameter measurements. Parameter a is usually set to the smallest value of diameter observed. An algorithm for recovering the other two parameters of the function is provided by Burk and Newberry (1984).

3.7.2 Height and vertical structure

The vertical structure of a forest stand refers to its structural complexity, which is influenced by the presence of different plant life forms (such as vines and epiphytes), the arrangement of leaves on branches, and the amount and distribution of leaves, branches, and twigs, at different heights (Brokaw and Lent 1999). Vertical structure has a major influence on the provision of habitat for wildlife. Quantitative methods for assessing vertical structure in this context are described in Section 7.3.

Foresters traditionally classify the crown position of trees according to the following simple scheme (Oliver and Larson 1996):

- *dominant*, where tree crowns extend above the general canopy level and are not physically restricted from above
- *co-dominant*, where crowns form the general level of a forest canopy and are somewhat crowded by other adjacent trees
- *intermediate*, where trees are shorter, but their crowns extend into the general canopy that is primarily composed of the crowns of dominant and co-dominant trees
- *suppressed* (*overtopped*), where crowns lie entirely below the general level of the canopy and are physically restricted from above.

This classification is subjective, and can be difficult to apply in practice, but is widely used.

Ecologists have traditionally described the vertical structure of forest stands through *profile diagrams*. Some excellent examples are provided in Richards' classic work on tropical rain forests (Richards 1996). The method involves marking out a rectangular strip of forest, typically at least 60 m long with a width of 10 or 20 m. The positions of all trees above 5 m height are then mapped and their diameters recorded. Height and crown width are then recorded by using the methods described above. These measurements are used to produce a diagram of a vertical section through the forest. Richards (1996), who was one of the people responsible for developing the method, was well aware of its limitations: it is difficult to select a site that is truly representative of a particular forest, and it is of limited value as a source of quantitative information. However, profile diagrams do capture something of the vertical structure of forest stands, and are an effective way of illustrating its complexity.

Average canopy height can be obtained by calculating a mean value of all trees, or a sample of trees in the stand. Mean values can be weighted according to the position of trees in the canopy, for example whether they are classified as dominant or co-dominant, or according to their basal area (Husch *et al.* 2003).

3.7.3 Leaf area

Measurements of leaf area may be expressed as a quantity (m^2), or more typically, as *leaf area index* (LAI), which is the leaf area per unit ground area, usually defined in units of $m^2\,m^{-2}$. LAI is an important structural attribute of forest ecosystems because of its role in influencing exchanges of energy, gas, and water and physiological processes such as photosynthesis, transpiration, and evapotranspiration. It is therefore very widely used in ecophysiological investigations. As noted in Chapter 2, LAI can be estimated from remote sensing data, and is also an important component of many process-based models of forest dynamics (Chapter 4) (Running *et al.* 1989, Chen and Cihlar 1995).

In the field, LAI can be estimated by using a wide variety of different instruments and techniques. These have been reviewed by a number of authors (Chason *et al.* 1991, Larsen and Kershaw 1990), and most recently by Bréda (2003), who noted that LAI is difficult to quantify because of large spatial and temporal variability. Methods developed for estimating LAI may be grouped into direct and indirect methods.

Direct or semi-direct methods described by Bréda (2003) include:

- *Direct measurement of leaf area*, using either a commercially available leaf area meter or a defined relation between leaf area and some other measured variable. In the latter approach, leaf area is typically measured on a subsample of leaves and related to dry mass—via *specific leaf area* (SLA, $cm^2\,g^{-1}$), for example—then the total dry mass of leaves collected within a known ground-surface area is converted into LAI by multiplying it by the SLA.
- *Relation between foliage area and sapwood area*. This is based on the hypothesis that leaf area is in balance with the amount of conducting tissues, and

therefore allometric relations can be developed. Because of the difficulties of measuring conducting area, sapwood area is often replaced by more readily measured variables, such as dbh. Estimating allometric relations through destructive sampling is generally a reliable method of deriving LAI for a given experimental site, but different relations may need to be established for different years. Examples of this approach are provided by Medhurst and Beadle (2002) and Pereira *et al.* (1997).

- *Collection of leaf litter*. In deciduous stands, leaves can be collected in traps of known collecting area distributed below the canopy during leaf fall. Litter should be removed from the traps at least every second week to avoid losses and decomposition. Collected litter is dried (at 60–80 °C for 48 h) and weighed, and the dry mass of litter calculated as $g\,m^{-2}$. Leaf dry mass at each collection date is converted into leaf area by multiplying the collected biomass by the SLA. LAI is the leaf area accumulated over the period of leaf fall. As leaves can be sorted, litter collection enables the contribution of each species to total leaf area index to be assessed. The depth of fresh leaf litter can also be assessed by using a point quadrat method, by inserting a needle into the litter layer and counting the number of leaves that it touches.

Indirect methods infer LAI from measurements of the transmission of irradiance through the forest canopy, and are based on statistical descriptions of the arrangement of leaves. Two main approaches may be differentiated: *radiation measurement* methods, which assume that leaves are randomly distributed within the canopy, and '*gap fraction*'-*based* methods that are dependent on estimating leaf angle distributions (Bréda 2003). Radiation measurement methods require measurement of irradiance both incident on the canopy and below the canopy; LAI is calculated from an extinction coefficient which is influenced by the total leaf area present within the canopy, as well as by canopy architecture and stand structure. Further details of these methods, including the analytical equations used in LAI estimation, are presented by Bréda (2003). A number of commercial sensors are available that use gap fraction-based methods of estimating LAI (Table 3.1). Although it should be noted that these sensors were primarily developed for use with crop plants, they may also be adapted for use with forest canopies.

A further indirect method involves the use of hemispherical photographs, with supporting digital analysis. The use of hemispherical photographs to characterize forest light environments is considered further in section 4.5.2.

Comparisons between direct and indirect methods indicate a significant underestimation of LAI by the latter techniques in forest stands, mainly because of clumping of leaves and the contribution of stem and branches. Reports indicate that the degree of underestimation varies from 25% to 50% depending on the characteristics of the stands. The sampling strategy adopted also has a major influence on the accuracy of the results, as the spatial heterogeneity of forest canopies is often very large (Bréda 2003). However, direct methods are all very labour-intensive (Fassnacht *et al.* 1997) and require many replicates to produce a precise result; they are therefore costly in terms of time and money.

Table 3.1 Characteristics of four commercially available sensors that can be used for indirect estimation of LAI (adapted from Bréda 2003).

	Principle	Sensor	Company	URL
AccuPAR	Gap fraction or sunflecks	80 PAR sensors distributed along a 0.90 m rod	Decagon Devices, Pullman, USA	*www.decagon.com*
DEMON	Gap fraction zenith angles from the sun at different angles to the vertical	Detector sighted at the sun	CSIRO, Canberra, Australia	*www.cbr.clw.csiro.au/ pyelab/tour/demon.htm*
LAI-2000	Gap fraction for each zenith angle acquired simultaneously	Fish-eye sensors with five concentric rings of sensors	Li-Cor, Lincoln, Nebraska, USA	*www.licor.com*
SunScan	Gap fraction or sunflecks	64 PAR sensors distributed along a 1 m rod	Delta-T Devices Ltd., Cambridge, UK	*www.delta-t.co.uk*

3.7.4 Stand volume

Measurement of the volume of wood produced by a forest stand is of fundamental importance to forestry, and consequently foresters have developed a variety of methods for estimating it. Particular efforts have been directed towards developing functions for stem volume and taper that allow estimates to be made from relatively simple measurements. Details of these functions are provided by Avery and Burkhart (2003), Husch *et al.* (2003), and West (2004). Stem volume is less important to forest ecologists, but as it is used for estimation of stem biomass and carbon content, a brief overview of the principal methods used in volume estimation is presented here (see also section 7.2).

The main method used to measure tree stem volume is the *sectional method*, which involves measuring the stem in relatively short sections, determining the volume, and each then summing these values to produce an estimate of total volume (West 2004). The volume of a stem section is determined by measuring its length, and the stem diameter at the lower end of the section ('large end diameter'), the upper end ('small end diameter') and/or at the midpoint of the section. These measurements are used to estimated volume by using one of three formulae (West 2004):

- *Smalian's formula:*

$$V_S = \pi l (d_L^2 + d_U^2)/8$$

- *Huber's formula:*

$$V_S = \pi l d_M^2 / 4$$

- *Newton's formula:*

$$V_S = \pi l (d_L^2 + 4d_M^2 + d_U^2)/24$$

where V_S is the volume of a section of a stem, l is the length of the section, d_L is the stem diameter at the lower end of the section, d_U is the diameter at the upper end of the section, and d_M is the diameter midway along the section.

These formulae provide accurate estimates of stem volume so long as the stem is circular in section or the stem is shaped in the form of a *quadratic paraboloid*. A variety of stem taper functions have been developed to accurately describe stem shape, but for most tree species the quadratic paraboloid is a reasonable approximation to the actual shape of tree stems, and for this reason these three functions are still in widespread use (West 2004). Of the three, Newton's is generally the most accurate because it uses the most information in the calculation.

Most measurements of tree stem volume use section lengths of 0.5–1 m for large trees; shorter lengths are used for smaller trees. Stem diameters can be measured by the methods described in section 3.6.2.

'Importance sampling' or 'centroid sampling' methods offer an alternative to the sectional method. These methods require the stem dbh to be measured (D_0) together with total height (H). A further measurement of stem diameter is required high on the stem (D_1). The location of this measurement point can be selected by either importance sampling or centroid sampling approaches. These involve application of the following formulae.

First, the value K is determined as:

$$K = 2H(H_U - H_L) + H_L^2 - H_U^2$$

where H_L is the lower height and H_U is the upper height of the stem section for which the volume estimate is required. The height at which the required upper stem diameter is to be measured (H_S) is then determined as:

$$H_S = H - \sqrt{(H - H_L)^2 - NK}$$

If importance sampling is used, then N is a randomly selected value in the range 0–1. If the centroid method is used, then $N = 0.5$ (the centroid being the position along the stem section above which half of the section volume lies) (West 2004).

Once H_S has been determined, the diameter at that height must be measured (D_S), by using one of the methods in section 3.6.2 (such as an optical dendrometer). The stem volume (V_{LU}) between H_L and H_U can then be determined as:

$$V_{LU} = \pi K (D_S D_1 / D_O)^2 / [8(H - H_S)]$$

This method is relatively simple, as it requires few diameter measurements to be made. The entire stem can be treated as a single section, or it can be divided into a series of sections and estimates obtained for each. Care should be taken to measure upper diameters accurately, and measurements should be made above any buttresses or stem swelling present near the base of the tree.

3.7.5 Stand density

Stand density refers to the number of trees within a given area. This can be most simply obtained by counting the number of trees present within a stand and measuring its area, then dividing the former by the latter. However, basal area and the extent of crown cover can also be used as measures of stand density. Stand density is of great importance to forest management, primarily because of its importance in determining the volume of timber likely to be obtained from a particular stand. Consequently foresters have developed a range of metrics for describing it, including various measures of relative spacing, stand density indices, crown competition factors, and stocking diagrams. Details are provided by forest mensuration textbooks such as Avery and Burkhart (2003) and Husch *et al.* (2003).

As noted earlier (section 3.5), distance measures are often used to estimate stand density. For example, Patil *et al.* (1979) proposed a plant density estimator based on point-to-plant distances that produces consistent results. Barabesi and Fattorini (1995) showed for various spatial plant patterns that an improvement over simple random sampling can be achieved by estimating plant density by a ranked set sampling of point-to-plant distances.

Two main issues should be borne in mind when measuring plant density (Bullock 1996). First, it may be difficult to differentiate individuals of clonal plants; in such cases, the number of ramets (i.e. shoots or stems) tends to be counted, rather than the number of genets (i.e. distinct genotypes, which usually can only be differentiated if information from molecular markers is available; see section 6.5.1). Second, measures of density that fail to take into account differences in the size of individual plants may give ecologically misleading results. For example, if a herbaceous species were to be compared with a tree species, it might demonstrate a much higher density in terms of number of individuals, but be much less important in terms of its contribution to the structure of the forest stand. For this reason, combined measures of density and plant size are often preferred. The commonest of these is cover, which is a measure of the above-ground parts of the plant (such as the tree canopy) when seen from above (for methods of estimating canopy cover, see section 3.6.4; methods for assessing the cover of understorey vegetation are described in section 7.7).

The 3/2 law of self-thinning has attracted a great deal of interest from both foresters and ecologists, because it is one of those rare things in ecology—a straight line. Usually, the logarithm of mean tree volume or mass is plotted against the logarithm of the number of trees per unit area. For stands undergoing density-dependent mortality ('self-thinning'), the slope of the line is approximately $-3/2$ (Kershaw and Looney 1985).

3.8 Spatial structure of tree populations

In recent years, substantial progress has been made in developing methods for analysing the spatial pattern of plant populations. These techniques can be used to describe the spatial distribution of individuals within populations, and to develop testable hypotheses about the underlying processes responsible for generating these patterns (such as seed dispersal, competition, and herbivory) and their relationship to environmental heterogeneity. Dale (1999) and Fortin and Dale (2005) provide detailed accounts of methods of spatial pattern analysis used in plant ecology.

Often, spatial patterns are analysed by mapping the positions of individual plants within a study area or sample plot (see Figure 4.4). A number of methods are available that are explicitly used with the kind of data produced by this approach, which consist of points. Many of these methods focus on determining whether the individuals are randomly dispersed, clumped or overdispersed. *Random dispersion* refers to the situation where points occur independently of one another; when *clumped*, the presence of one point increases the probability of finding another nearby; when *overdispersed*, the presence of one point decreases the probability of finding another nearby (Dale 1999). Often, a key objective is to determine the scale of the spatial pattern—the size of any clumps present, and their spacing.

Many methods are based on analysis of the distance of each plant to its nearest neighbour. For example, the test described by Clark and Evans (1954) enables random, clumped, and overdispersed patterns to be differentiated by analysis of nearest-neighbour distances. However, such methods provide no information on the size or spacing of the clumps, and therefore have limited usefulness (Dale 1999). Alternatively, it is possible to analyse the distances between all possible pairs of plants, in what is termed *plant-to-all-plants distance analysis* (Galiano 1982). This method examines the frequency distribution of the distances between all pairs of plants in the area surveyed, but again the method does not always give a full picture of the characteristics of spatial pattern (Dale 1999).

More commonly, *second-order statistics* are used for the analysis of point patterns. These methods are based on the distribution of distances of pairs of points, and count the number of points located within a certain specified distance of each point. The methods are described in detail by Ripley (1981, 1988), who first developed them. Two commonly used second-order statistics are Ripley's K-function and the pair-correlation function g, which use the information on all interpoint distances within the data set being analysed (Ripley 1981, Diggle 1983). These statistics provide more information than those that use nearest-neighbour distances only, such as Diggle's nearest-neighbour functions G or F (Diggle 1983), which are also commonly used. These three statistics have different sensitivities to different types of spatial distributions: for example, K has a slightly lower power than F for aggregated patterns and a higher power than G for regular patterns

(Diggle 1979). K also presents the advantage of being density-independent, unlike the two other tests (Ripley 1981). In their analysis of the tropical palm tree *Borassus aethiopum*, Barot *et al.* (1999) found that use of F, G, and K together, as proposed by the developers of these methods (Ripley 1981, Diggle 1983), provided greater insight than did the use of individual methods in isolation. The functions for these statistics are presented by Barot *et al.* (1999).

During the past few years, methods based on Ripley's K-function have undergone a rapid development and are now widely used in plant ecology. This method is reviewed in detail by Wiegand and Moloney (2004), who cite a large number of studies that have used the technique. Recent examples of the use of Ripley's K function to assess the spatial structure of tree populations include Aldrich *et al.* (2003), Barot *et al.* (1999), Condit *et al.* (2000), He and Duncan (2000), and McDonald *et al.* (2003). Statistical significance is usually evaluated by comparing the observed data with Monte Carlo simulations of a null model, which is most commonly complete spatial randomness (Wiegand and Moloney 2004). The statistics can be calculated with appropriate statistical analysis software, such as the spatial statistics module of S-Plus software (produced by Insightful Corporation; ⟨www.insightful.com/⟩). Thorsten Wiegand (Department of Ecological Modelling, UFZ-Centre for Environmental Research, Leipzig, Germany) has developed a freely available software program (Programita) for analysis of point data, which can be used to calculate Ripley's L function (a transformation of Ripley's K) and the Wiegand–Moloney *O-ring statistic*, described by Wiegand and Moloney (2004). The *O*-ring statistic is a probability density function that is complementary to the K statistic and can detect aggregation or dispersion at a given distance. Wiegand and Moloney (2004) also provide a valuable step-by-step series of recommendations for the use of these methods.

Morisita's index (I_δ) is another measure of dispersion that has been widely used to examine the spatial pattern of trees. This may be calculated as follows (Dale 1999):

$$I_\delta = n\left\{1 + \frac{s^2 - \bar{x}}{\bar{x}^2}\right\}/(n-1)$$

where x_i is the number of individuals of a particular kind in the ith quadrat, n is the total number, and s^2 is the sample variance. Typically, values of the index are calculated from measures of density or presence–absence data collected from contiguous quadrats. Quadrats can be combined into squares of increasing size, and values of the index calculated for each. As the size of the area analysed increases, values of the index remain constant until the mean clump size is reached, and then it increases (Dale 1999). However, if there is more than one scale of pattern in the data, the method does not provide clear results (Dale 1999). On the other hand, the method does not require the positions of all trees to be mapped, a process that can be very labour intensive. Examples of use of this method to analyse the spatial pattern of trees are provided by Veblen (1979), Taylor and Halpern (1991) and Bunyavejchewin *et al.* (2003).

Increasing interest in the spatial ecology of plant populations has led to increasing awareness of *spatial autocorrelation* (Figure 3.13). This refers to the situation when the observed value of a variable at one locality is dependent on values of the variables at other localities (Johnston 1998). Importantly, spatial autocorrelation impairs the ability to perform standard statistical tests; for example, the assumption that different samples are independent may be invalid. Two measures are commonly used to assess spatial autocorrelation: Moran's I statistic and Geary's c statistic, both of which indicate the degree of spatial autocorrelation summarized for the entire data set (Johnston 1998). The formulae are given below.

- *Moran's I* is calculated for N observations on a variable x at locations i, j as:

$$I = (N/S_0)\sum_i\sum_j w_{ij}(x_i-\mu)(x_j-\mu)/\sum_i (x_i-\mu)^2$$

where μ is the mean of the x variable, w_{ij} are the elements of the spatial weights matrix, and S_0 is the sum of the elements of the weights matrix:

$$S_0 = \sum_{ij} w_{ij}$$

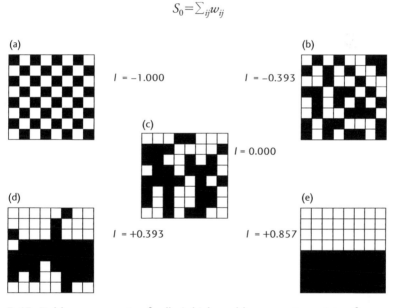

Fig. 3.13 Field arrangements of cells (which could represent a variety of measured variables, or the distribution of individuals) exhibiting (a) extreme negative spatial autocorrelation, (b) a dispersed arrangement, (c) spatial independence, (d) spatial clustering, and (e) extreme positive spatial autocorrelation. Values of Moran's I are given next to the figures. (From Longley et al. 2005, *Geographical Information Systems and Science*, 2nd ed., Copyright John Wiley and Sons Limited. Reproduced with permission.)

- *Geary's c statistic* is expressed in the same notation as:

$$c = (N-1)/2S_0 \left[\sum_i \sum_j w_{ij}(x_i - x_j)^2 / \sum_i (x_i - \mu)^2 \right]$$

Positive spatial autocorrelation is indicated by a value of Moran's *I* that is larger than its theoretical mean of $-1/(N-1)$, or a value of Geary's *c* less than its mean of 1 (Johnston 1998).

Another statistical approach that can be used for analysis of spatial data is *geostatistics*. This provides a set of techniques that can be used for the analysis and prediction of spatially distributed phenomena. The approach was originally developed for use in geology, but is increasingly being used by ecologists. Its use is likely to become more widespread among ecologists in future with the increasing availability and power of software tools. As noted in section 2.6, some GIS software programs are now able to do geostatistical analysis, either as part of the program itself (for example IDRISI) or as additional software modules or extensions (for example Geostatistical Analyst, which is an ArcGIS extension produced by ESRI; see section 2.5.1 for URLs). Some statistical software programs (such as the spatial statistics module of S PLUS, see above) can also be used to do geostatistics.

Geostatistical methods measure the similarity or dissimilarity between variables based on the spatial dependence of measurements taken at different locations, and then use this information for interpolation, extrapolation, or simulation. Geostatistics is based on the assumption that data values located closely together in space are likely to be more related than locations that are further apart (Johnston 1998). Spatial dependence decreases with increasing distances among sample points. Analyses are usually done by producing a semivariogram (or variogram for short), which illustrates the degree of dissimilarity between values at different intervals of distance and direction.

Once a variogram has been produced, interpolation methods can be used to estimate values at unsampled locations (Johnston 1998). The most commonly used geostatistical method of interpolation is *kriging*, which uses weighted average of sample measurements to estimate the value at non-sampled points. The weights are calculated from expected spatial dependence between estimated points and sampled points. In practice, kriging is done by selecting a mathematical function from a variety of alternatives, and fitting the function to the observed data points to obtain the best possible fit. The fitted variogram is then used to estimate values at locations of interest (Longley *et al.* 2005). A number of different options are available regarding the choice of mathematical function for the variogram; an appropriate option has to be determined by the analyst (Longley *et al.* 2005). Introductions to kriging are provided by DeMers (2005), Longley *et al.* (2005) and Johnston (1998).

Textbooks describing geostatistical techniques include those by Goovaerts (1997), Isaaks and Srivastava (1989), Wackernagel (2003), and Webster and Oliver (2000). Use of geostatistics in ecology is described by Johnston (1998),

Robertson (1987), and Wagner (2003). Examples of the use of geostatistics in forest ecology include:

- Köhl and Gertner (1997), who used these methods in forest damage surveys
- Bebber *et al.* (2002), who used geostatistics to determine the spatial relationships between canopy openness and seedling performance in secondary lowland forest in Borneo
- Hohn *et al.* (1993), who used three-dimensional kriging in space and time to predict defoliation caused by gypsy moth in Massachusetts
- Nanos *et al.* (2004), who used geostatistics to analyse the stand characteristics of a pine forest in Spain
- Schume *et al.* (2003), who carried out a spatiotemporal analysis of soil water content in a mixed Norway spruce–European beech stand in Austria.

3.9 Species richness and diversity

Species richness refers to the number of species in a particular area, whereas *species diversity* refers to a combination of richness and relative abundance. A comprehensive review of the methods used for estimating both of these variables is provided by Magurran (2004).

3.9.1 Species richness

The main problem with measuring species richness is that the result depends on the number of individuals recorded. Two main sampling methods are used in forest ecology. One approach is to sample individual trees selected at random within a field plot, and record sequentially the species identity of one tree after another. This is referred to as *individual-based sampling*. Alternatively, a series of subplots can be established in each plot, and the number and taxonomic identity of all of the trees within each subplot can be recorded, noting the increase in the number of species as additional subplots are surveyed. This is referred to as *sample-based assessment* (Gotelli and Colwell 2001). The key difference between these approaches is that the unit of replication is different between the two methods, being an individual tree in the former case and a sample of individuals in the latter. The choice of method has a major influence on the estimate obtained (Gotelli and Colwell 2001, Longino *et al.* 2002). Magurran (2004) mentions another method of assessing species richness: the use of incidence or occurrence data, representing the number of sampling units in which a species is present. Sample units that could potentially be used in this way include grid squares, quadrats, or point samples.

There are three main methods of estimating species richness from samples: extrapolation of species accumulation or species–area curves, use of the shape of the species abundance distribution, or use of a non-parametric estimator. Each of these is considered below. Further information on these methods is provided by Chazdon *et al.* (1998) and Colwell and Coddington (1994). Recent progress in interpolating and extracting species accumulation curves is described by Colwell *et al.* (2004) (see also Golicher *et al.* 2006).

Species accumulation curves plot the cumulative number of species recorded as a function of sampling effort, and illustrate the increase in the total number of species encountered during the process of data collection. Species–area curves can be considered as one form of species accumulation curve, in which species richness is related to an increase in the area sampled. Smooth curves can be produced if samples are added randomly and the process repeated a number of times (at least 100 is recommended). Species accumulation curves are often plotted on a linear scale on both axes, although Longino *et al.* (2002) suggest that the *x*-axis should be log-transformed to enable easier differentiation between asymptotic and logarithmic curves.

The total species richness within a particular area can be estimated by extrapolating from species accumulation curves. The best method of making this extrapolation has been the subject of some debate, as has the most appropriate way of obtaining samples (Magurran 2004). According to Colwell and Coddington (1994), random samples should be taken from areas of relatively homogeneous habitat. Rosenzweig (1995) states that a nested design should be used; in other words, subplots that are used to sample individuals for production of a species–area curve should be contiguous. However, in this case subplots are not statistically independent of one another; this could lead to the results being biased and statistical inferences being invalid (Crawley 1993).

Functions fitted to species accumulation curves may be either asymptotic or non-asymptotic. The equation most commonly used for fitting an asymptotic curve is the Michaelis–Menten equation (Magurran 2004):

$$S(n) = \frac{S_{max} n}{B + n}$$

where $S(n)$ is the number of species observed in n samples, S_{max} is the total number of species in the assemblage, and B is the sampling effort required to detect 50% of S_{max}.

Non-asymptotic curves that have been used include logarithmic transformations of the *x*-axis (a log–linear model) and of both axes (a log–log relation). However, Colwell and Coddington (1994) suggest that non-parametric methods (see below) are preferable to either of these.

If information on the *species abundance distribution* is available, namely the relation between the number of species and the number of individuals in those species, this can also be used to estimate species richness. A wide variety of different models have been used to characterize species abundance distributions, and these are considered in detail by Hubbell (2001) and Magurran (2004). Those with the greatest potential for estimating species richness are the log series and log normal distributions, which are evaluated by Colwell and Coddington (1994). Again, however, the use of non-parametric estimators is generally preferred to these methods (Magurran 2004).

A range of different *non-parametric estimators* are described by Chazdon *et al.* (1998) and Colwell and Coddington (1994). These include Chao 1, Chao 2, the *abundance-based coverage estimator* (ACE), the partner *incidence-based coverage estimator* (ICE), two methods based on the use of jackknife statistics (Jackknife 1 and 2), and a bootstrap estimator (see Box 3.1). These can all be calculated easily by using EstimateS software, which can be downloaded free of charge from ⟨*http://viceroy.eeb.uconn.edu/EstimateS*⟩ (Colwell 2004a, b). A detailed manual is provided with this software, which should be consulted carefully before use. Although this program is now widely used, it should be employed with caution because of potential errors (Golicher *et al.* 2006). Selection of which estimator is most appropriate depends on the characteristics of the forest being studied, including the sample size, the patchiness of the vegetation, and the total number of individuals in the sample (Magurran 2004). Relatively few comparative studies of

---

Box 3.1 Formulae for selected non-parametric estimators of species richness (after Magurran 2004)

- *Chao 1*, an abundance-based estimator of species richness

$$S_{Chao1} = S_{obs} + \frac{F_1^2}{2F_2}$$

- *Chao 2*, an incidence-based estimator of species richness

$$S_{Chao2} = S_{obs} + \frac{Q_1^2}{2Q_2}$$

- *Jackknife 1*; first-order jackknife estimator of species richness (incidence-based)

$$S_{Jack1} = S_{obs} + Q_1\left(\frac{m-1}{m}\right)$$

- *Jackknife 2*; second-order jackknife estimator of species richness (incidence-based)

$$S_{Jack2} = S_{obs} + \left[\frac{Q_1(2m-3)}{m} - \frac{Q_2(m-2)^2}{m(m-1)}\right]$$

- *Boostrap estimator* of species richness (incidence-based)

$$S_{boot} = S_{obs} + \sum_{k=1}^{S_{obs}} (1-p_k)^m$$

- *Abundance-based coverage estimator* (ACE) of species richness

$$S_{ace} = S_{comm} + \frac{S_{rare}}{C_{ace}} + \frac{F_1}{C_{ace}}\gamma_{ace}^2$$

- *Incidence-based coverage estimator* (ICE) of species richness

$$S_{ice} = S_{freq} + \frac{S_{ifreq}}{C_{ice}} + \frac{Q_1}{C_{ice}}\gamma^2_{ice}$$

where:

S_{est} = estimated species richness, where est is replaced in the formula by the name of the estimator

S_{obs} = total number of species observed in all quadrats pooled

S_{rare} = number of rare species (each with 10 or fewer individuals) when all quadrats are pooled

S_{comm} = number of common species (each with more than 10 individuals) when all quadrats are pooled

S_{ifreq} = the number of infrequent species (each found in 10 or fewer quadrats)

S_{freq} = number of frequent species (each found in more than 10 quadrats)

m = total number of quadrats

F_i = number of species that have exactly i individuals when all quadrats are pooled (F_1 is the frequency of singletons, F_2 is the frequency of doubletons)

Q_j = the number of species that occur in exactly j quadrats (Q_1 is the frequency of uniques, Q_2 is the frequency of duplicates)

P_k = the proportion of quadrats that contain species k

C_{ace} = sample abundance coverage estimator

C_{ice} = sample incidence coverage indicator

γ^2_{ace} = estimated coefficient of variation of the F_i for rare species

γ^2_{ice} = estimated coefficient of variation of the Q_j for infrequent species

these estimators have been made to date; examples include Chazdon *et al.* (1998), Colwell and Coddington (1994), Condit *et al.* (1996b), and Longino *et al.* (2002).

When estimating species richness, Gotelli and Colwell (2001) emphasize the importance of standardizing data sets that are to be compared to a common number of individuals. This can be achieved by using species accumulation curves and *rarefaction curves*, which are produced by repeatedly resampling the pool of individuals or samples at random and plotting the average number of species represented as the sample size increases (Figure 3.14) (Gotelli and Colwell 2001). Gotelli and Colwell (2001) make the following recommendations with respect to this approach:

- It is essential that a species accumulation curve or rarefaction curve is plotted when estimating species richness. Raw species richness counts can only be validly compared when such accumulation curves have reached a clear

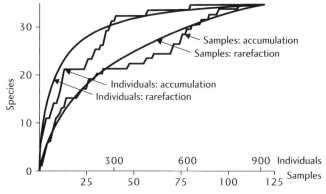

Fig. 3.14 Sample- and individual-based rarefaction and accumulation curves. Accumulation curves (jagged curves) represent a single ordering of individuals (solid-line, jagged curve) or samples (open-line, jagged curve), as they are successively pooled. Rarefaction curves (smooth curves) represent the means of repeated re-sampling of all pooled individuals (solid-line, smooth curve) or all pooled samples (open-line, smooth curve). The smoothed rarefaction curves therefore represent the statistical expectation for the corresponding accumulation curves. The sample-based curves lie below the individual-based curves because of the spatial aggregation of species. Curves were produced by using EstimateS (see text). (From Gotelli and Colwell 2001. Quantifying biodiversity: procedures and pitfalls in the measurement and comparison of species richness. *Ecology Letters*, Blackwell Publishing.)

asymptote. Estimates of species richness should be reported together with information about the sampling effort involved.
- When sample-based approaches are used, the number of species should be plotted as a function of the accumulated number of individuals, not the accumulated number of samples, because data sets may differ in the mean number of individuals per sample.
- Individual-based rarefaction analysis is based on the assumption that the spatial distribution of individuals in the environment is random, that sample sizes are sufficient, and that assemblages being compared have been sampled in the same way. If these assumptions are not met, misleading results may be obtained.
- It is invalid to simply divide the number of species encountered by the number of individuals included in the sample to correct for unequal numbers of individuals between samples. This is because such a correction assumes that richness increases linearly with abundance, which is rarely the case (Figure 3.15).

Rarefaction curves can be plotted by EstimateS software, as well as by other commercially available software packages such as *Species diversity and richness* (Pisces Software, ⟨www.pisces-conservation.com⟩). Kindt and Coe (2005) have

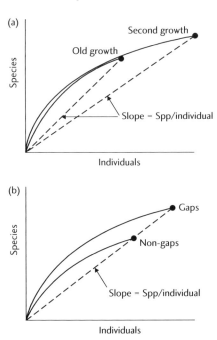

Fig. 3.15 Pitfalls of using species : individual ratios to compare data sets. In (a), an old-growth and a second-growth forest stand are compared. The two stands have identical individual-scaled rarefaction curves and thus do not differ in species richness. The second-growth curve extends farther simply because stem density is greater, so that more individuals have been examined for the same number of samples. However, when the species : individual ratio is computed for each, the ratio is much higher for the old-growth stand. In (b), species richness in treefall gap quadrats is compared with richness in non-gap (forest matrix) quadrats. In this case, species : individual ratios are identical, yet the true species richness is higher in gaps. (From Gotelli and Colwell 2001. Quantifying biodiversity: procedures and pitfalls in the measurement and comparison of species richness. *Ecology Letters*, Blackwell Publishing.)

produced a useful software program and accompanying manual that can also be used to calculations a number of measures relating to tree diversity, including rarefaction. These can be downloaded from ⟨*www.worldagroforestry.org/treesand-markets/tree_diversity_analysis.asp*⟩.

Many ecological studies assess the number of species present in a particular area, or the *species density*, which depends both on species richness and on the mean density of individuals. To compare species density estimates from different locations, the *x*-axis of individual-based rarefaction curves can be rescaled from the number of individuals to area, by using a measure of average density (Gotelli and Colwell 2001). However, when comparing species density data for two unequal areas it is invalid to divide the number of species by the area measured, because the number of species increases non-linearly with area (Gotelli and Colwell 2001). Again, species accumulation curves should be used to compare samples.

Which measure should be used, species richness or species density? For most conservation assessments, species density is more likely to be of value. Species richness may be preferred when testing ecological theories or models (Gotelli and Colwell 2001).

3.9.2 Species diversity

Species diversity is generally assessed by using some type of *diversity index*, which incorporates information on species richness and evenness. The term *evenness* refers to the variability in the relative abundance of species. A wide variety of different indices are available, which are comprehensively described by Magurran (2004). Measures of evenness (or heterogeneity) can be divided into two groups: those that are based on the parameter of a species abundance model, and non-parametric methods that make no assumptions about the underlying species abundance distribution. Examples of parametric measures of diversity include log series a, log normal l, and the Q statistic. Examples of non-parametric measures include the Shannon evenness measure, Heip's index of evenness, Simpson's index, Simpson's measure of evenness and the Berger–Parker index (Magurran 2004). Some of these are given in Box 3.2.

How should an appropriate diversity index be selected? The following advice is provided by Magurran (2004):

- Rather than simply calculating a range of diversity measures, it is preferable to define in advance which measure is most appropriate to the objective of the investigation; do not simply select the measure that provides the most appealing answer.
- Sample size must be adequate for the method selected.
- Replication is always recommended. Many small samples are generally preferable to a single large one. Replication permits statistical analyses, such as the calculation of confidence limits.
- Consider whether a diversity index is really necessary. A robust estimate of the number of species present, without any consideration of their relative abundance, may be sufficient to meet the objectives set.
- If some measure of evenness is required, consider using either Fisher's α or Simpson's index (Box 3.2). These measures are relatively well understood and are relatively easy to interpret. Fisher's α is relatively unaffected by sample size once the sample size is greater than 1000. Simpson's index provides an accurate estimate of diversity for relatively small sample sizes.
- The Shannon index is not recommended, despite its popularity, because of its sensitivity to sample size. Interpretation can also be difficult.
- The Berger–Parker index and the Simpson's evenness measure provide relatively simple and easily interpretable measures of dominance (i.e. weighted by the abundance of the commonest species).

Box 3.2 Selected measures of diversity (after Condit *et al.* 1998, Magurran 2004 and Southwood and Henderson 2000).

- *Shannon–Wiener function:*

$$H = -\sum_{i=1}^{S_{obs}} p_i \log_e p_i$$

where p_i is the proportion of individuals in the ith species.

- *Fisher's α:*

$$S = \alpha \ln\left(1 + \frac{N}{\alpha}\right)$$

where S is the number of species and N is the number of stems.

- *Simpson's index:*

$$D = \sum f_i^2$$

where f_i is the proportion of individuals in the ith species.

- *Simpson's measure of evenness:*

$$E_{1/D} = \frac{(1/D)}{S}$$

where D is Simpson's index, and S is the number of species. The measure ranges from 0 to 1 and is not sensitive to species richness.

- *Berger–Parker index:*

$$d = N_{max}/N$$

where N_{max} is the number of individuals in the most abundant species and N is the total number of individuals. This index has the great advantage of being very easy to calculate.

Based on their detailed assessment of diversity in highly diverse lowland tropical forests, Condit *et al.* (1998) made the following additional recommendations:

- Compare samples with approximately equal numbers of stems, but ignore area.
- Use the same sampling protocol across sites that are to be compared; use the same shape of field plot.
- Compare samples by using the same dbh limit.
- With samples of fewer than 3000 stems, do not use species richness as a diversity metric; instead use Fisher's α.
- Be aware of sampling error; confidence limits associated with diversity assessments in samples of fewer than 1000 stems extend about 30% above and below the estimate.

Condit *et al.* (1998) place particular emphasis on the importance of stem number in influencing diversity estimates in tropical forests, suggesting the following specific rules:

- Never use samples of fewer than 50 stems, and in very diverse forests use 100 stems or more. These are absolute minimum values; larger samples are preferable.
- In samples of more than 2000 stems, Fisher's α can be used to compare samples, even if they differ substantially in size.
- For samples between 50 and 20 000 stems, either subsample stems to provide a common number of stems for all sites to be compared, or apply a correction factor to Fisher's α to adjust for sample size (for details of this approach see Condit *et al.* 1998).

Magurran (2004) further emphasizes the importance of sampling approach for determining the outcome of diversity measures. Most importantly, the sample size must be adequate. However, this can be difficult to determine in practice. Assessment of the rate at which new samples are being encountered is one useful guide; the experience of knowledgeable field ecologists is another (Magurran 2004). In high-diversity sites, Sørensen *et al.* (2002) recommended the following rule of thumb: 30–50 individuals should be sampled per species. When using sample-based approaches, it is necessary to determine an appropriate number of replicates. A number of different studies have found 10 replicates to be adequate, but the optimum number varies with the scale of the sampling unit in relation to the size of the assemblage being surveyed (Magurran 2004). As additional samples are included in the analysis, their effects on the precision of diversity estimates can be measured. Unequal sample sizes should be avoided; rather, a consistent approach to sampling should be used across the entire investigation. It is also important to remember that samples should be independent; repeated samples of the same plot are not true replicates, and replicates should be located randomly and not grouped together (Crawley 1993).

3.9.3 Beta diversity and similarity

Beta diversity is a measure of the extent to which the diversity of two or more spatial units differs (Magurran 2004) and is generally used to characterize the degree of spatial heterogeneity in diversity at the landscape scale, or to measure the change in diversity along transects or environmental gradients. A variety of indices are available for describing beta diversity, most of which are based on the use of presence–absence data, although quantitative abundance data can also be used. One of the most widely used indices is *Whittaker's measure* (β_W), which is calculated as:

$$\beta_W = S/\bar{\alpha}$$

Box 3.3 Measures of similarity and complementarity (from Magurran 2004)

- *Jaccard similarity index* (C_J):

$$C_J = \frac{a}{a+b+c}$$

where a is the total number of species present in both quadrats or samples being compared, b is the number of species present only in quadrat 1, and c is the number of species present only in quadrat 2.

The statistic can be adapted to give a single measure of complementarity across a set of samples or along a transect:

$$C_T = \frac{\sum U_{jk}}{n}$$

where $U_{jk} = S_j + S_k - 2V_{jk}$ and is summed across all pairs of samples; V_{jk} is the number of species common to the two lists j and k (the same value as a in the formula above); S_j and S_k are the number of species in samples j and k, respectively; and n is the number of samples.

- *Sørenson's similarity index* (C_S):

$$C_S = \frac{2a}{2a+b+c}$$

(see above for definitions of a, b, and c).

- *Sørenson's quantitative index* (C_N):

This is a modified version of Sørenson's index that takes into account the relative abundance of species, introduced by Bray and Curtis in 1957:

$$C_N = \frac{2jN}{N_a + N_b}$$

where N_a is the total number of individuals in site a, N_b = the total number of individuals in site b, and $2jN$ is the sum of the lower of the two abundances for species found in both sites.

- *Morisita–Horn index:*

This is recommended because it is not strongly influenced by sample size and species richness (Henderson 2003):

$$C_{MH} = \frac{2\sum(a_i b_i)}{(d_a + d_b)(N_a N_b)}$$

where N_a is the total number of individuals in site a, N_b is the total number of individuals in site b, a_i is the number of individuals in the ith species in a, b_i = the number of individuals in the ith species in b, and d_a (and d_b) are calculated as follows:

$$d_a = \frac{\sum a_i^2}{N_a^2}$$

- *Percentage similarity:*

$$P = 100 - 0.5 \sum_{i=1}^{S} |P_{ai} - P_{bi}|$$

where P_{ai} and P_{bi} are the percentage abundances of species i in samples a and b respectively, and S = the total number of species.

where S is the total number of species recorded in the system, and α is the average sample species richness, where each sample is a standard size. This is one of the simplest and most effective measures of beta diversity (Magurran 2004). Where this measure is calculated between pairs of samples or between adjacent quadrats along a transect, values of the measure will range from 1 (complete similarity) to 2 (no overlap in species composition). Subtracting 1 from the answer enables results to be presented on a scale of 0–1. Often, beta diversity is assessed by using measures of similarity such as the Jaccard similarity index and Sørensen's measure (Box 3.3).

Clarke and Warwick (2001) note that similarity measures can be markedly affected by the abundance of the commonest species, and recommend that data be transformed before calculating the similarity measure. Data should either be square-root transformed or transformed to log $(x + 1)$. Vellend (2001) suggests that measures of beta diversity should be differentiated from measures of species turnover, which measure the extent of change in species composition along predefined gradients. According to Vellend (2001), Whittaker's measure of beta diversity should not be used to assess species turnover; rather, matrices of compositional similarity and physical or environmental distances among pairs of study plots should be used.

One of the main problems with using such similarity measures is that they are based on the assumption that sites being compared have been completely censused (Magurran 2004). Often, this is not the case. Recently developed methods focusing on the use of ACE are designed to address this, which estimate the number of unobserved shared species (Chao *et al.* 2000).

3.10 Analysis of floristic composition

Information on floristic composition collected from sample plots can be used in a variety of ways to define and analyse the distribution of ecological communities. If the objective is simply to assess the degree of similarity in floristic composition between samples collected from different locations, then a similarity measure such as those described in section 3.9 could be used. If the objective is to analyse the relationship between sites or samples in terms of the species present, or to relate species composition to environmental variables, then some form of multivariate analysis will need to be used. An extensive literature is available relating to multivariate techniques. Textbooks describing the methods in detail include those by Digby and Kempton (1987), Jongman *et al.* (1995), Kent and Coker (1992), and Legendre and Legendre (1998). Shaw (2003) provides a useful and accessible introductory account.

For multivariate analyses, specialist statistical or ecological analysis software generally needs to be used (Table 3.2). Typically, data sets describing species composition are arranged as a two-dimensional matrix with the different samples forming the columns and the species forming the rows. The cells of the matrix can represent the observed abundance of each species, an abundance score, or presence–absence information, represented by a 1 if the species was found in the sample and 0 if it was not (Henderson 2003). Such data can readily be entered by using a spreadsheet program such as Microsoft Excel, or directly into the software package that is to be used for analysis.

3.10.1 Cluster analysis

Cluster analysis can be used to identify groups of sites or samples that are similar in terms of their species composition. There are two main types: *hierarchical*, where samples are assigned to groups that are themselves arranged into groups, and *non-hierarchical*, where the samples are simply assigned to groups (Henderson 2003). Hierarchical methods are most commonly used. The results are usually illustrated by means of a *dendrogram* (Figure 3.16), in which sites or samples that are more similar in terms of species composition group more closely together. A wide variety of different algorithms can be used to produce dendrograms, which may produce different results (Shaw 2003). Cluster analysis can be done by many standard statistical analysis software packages, including SAS, SPSS, and GENSTAT, as well as specialist ecological software.

Two key decisions need to be made in cluster analysis: which measure of similarity (or distance) to employ, and which rules for cluster formation to adopt (Shaw 2003). The similarity measures described in section 3.9, such as the Jaccard and Sørensen indices, which are suitable for use with presence–absence data, can be used as a basis of cluster analysis. For continuous variables, Euclidean distance is typically used. With respect to the rules for cluster formation, the simplest method is to use single-link or nearest-neighbour clustering, in which clusters are joined based on the smallest distance between any pair of their component individuals (Shaw 2003). However, dendrograms produced by this method can sometimes be difficult to interpret. A number of alternative methods are described by Digby and Kempton (1987) and Shaw (2003). UPGMA is a widely used alternative method that allows clusters to be joined on the basis of the lowest value of the average distance between clusters.

Table 3.2 Software packages suitable for multivariate analysis of ecological data.

Product	Comments	URL
Canoco for Windows 4.5	A powerful, high-specification package able to perform a wide variety of multivariate analyses, but particularly suitable for CCA, as the software is designed by the developers of this method. Detailed supporting materials are available (ter Braak and Šmilauer 1998). Takes time to learn, but is the software of choice for many researchers.	*www.plant.dlo.nl/*
Community Analysis Package (CAP)	A reasonably powerful Package that is relatively easy to use. Developed and marketed by Pisces Software.	*www.pisces-conservation.com.*
DECORANA and TWINSPAN	Available as a free download from the Centre for Ecology and Hydrology in the UK, who originally developed them.	*http://science.ceh.ac.uk/ products_services/software/ corn.html*
PC-ORD	A set of Windows programs for multivariate analysis of ecological data, developed by Bruce McCune at Oregon State University. Widely used.	*http://oregonstate. edu/~mccuneb/ pcord.htm*
Primer 5	Developed by Plymouth Marine Laboratory (UK), capable of performing a wide range of multivariate analyses. Although developed by marine ecologists, the software is applicable to analysis of forest communities. Well supported.	*www.primer-e.com/*

The main problem with cluster analysis, as with other classification techniques, is that the results can easily be misinterpreted. Definition of groups does not necessarily reflect genuine ecological differences. Cluster analysis will always divide the data into groups; even randomly generated data can produce a dendrogram (Henderson 2003). Another problem is that different clustering techniques can provide very different answers, and as no statistical significance is obtainable for this form of analysis, it can be difficult to decide which output is 'correct'. Careful consideration of the results is therefore required, to examine whether they make

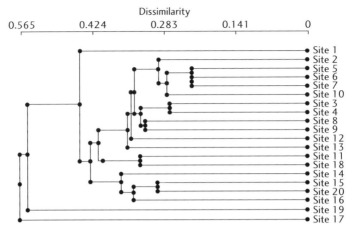

Fig. 3.16 A dendrogram showing the results of a cluster analysis. The dendrogram illustrates the degree of similarity between sites according to their species composition. The diagram was produced by using CAP (Table 3.2). (From Henderson 2003.)

sense and have some genuine ecological meaning. Cluster analysis should be viewed as a tool for data exploration and hypothesis generation, and not as an end in itself (Shaw 2003).

3.10.2 TWINSPAN

TWINSPAN (Two-Way INdicator SPecies ANalysis procedure) has been very widely used by plant ecologists since its development more than 30 years ago (Hill *et al.* 1975). This method can be used with presence–absence, percentage cover, or abundance data. TWINSPAN progressively divides a data set into groups based on all of the information content of the data, rather than merging individuals into groups as is done in cluster analysis (Shaw 2003). Results are presented in the form of a dendrogram, but the reciprocal averaging ordination method is used to order the samples or sites. As a result, the outputs of TWINSPAN analysis are less prone to misinterpretation than dendrograms produced by standard cluster analysis (Shaw 2003). A particularly valuable feature of TWINSPAN is that it allows indicator species that are characteristic of each group to be identified, and therefore the method is especially useful when there is a need to identify species that can be used to characterize particular communities (Henderson 2003). In addition, both the samples and species are ordered along a gradient in the TWINSPAN output tables, which can greatly help data interpretation (Shaw 2003). An unusual feature of the method is the use of artificial constructs called 'pseudospecies', which are used to convert continuous data (such as percentage cover of different species) into categorical variables. Full details of the method are provided by Hill (1979). Although TWINSPAN output can be rather complicated, the method has been very widely used and is recognized as a standard technique, with few criticisms levied against it. Its main limitation is that it only considers one axis of variation (Shaw 2003).

3.10.3 Ordination

A wide variety of different ordination methods are available that can be used to analyse the relationships between sites or samples in terms of their species composition. The best method is arguably the one that gives the clearest and most easily interpreted results, and therefore it may be worth trying a variety of different methods and comparing the results (Henderson 2003). Ordination methods that are commonly used in forest ecology are considered below. A word of caution is appropriate, regardless of which method is preferred. These techniques provide little useful information unless there is some genuine pattern to be detected and differences exist between individual samples of vegetation. A great deal of time can be wasted exploring multivariate analyses that produce little of value in the way of clear results. Ordination outputs can be dominated by outliers and swamped by noise. One solution to this problem is to use sample plots that are large enough to include sufficient stems in the sample, even in sites with low stem densities. Also, do not be too disheartened if no clear pattern emerges in the results: ecological data can just be like that.

Principal components analysis (PCA)

PCA is most appropriate for use with abundance data. The method enables the relationship between samples to be illustrated in a two- or three-dimensional space by producing a plot of the results (Figure 3.17). This can then be interpreted visually. The analysis is usually done on a correlation matrix that includes all of the correlation coefficients between the variables included in the analysis. If the species vary greatly in abundance between samples the data will probably need to be transformed before analysis, using either a square-root or logarithmic transformation. The latter procedure cannot handle zeros, and therefore 1 is often added to all of the observations before the transformation (Henderson 2003). However, as this can distort the output, it is generally preferable to use the square-root transformation.

Often it is necessary to normalize the data before analysis. This is because multivariate datasets often consist of very different variables: for example, the cover of different tree species might be measured on a scale of 0–100%, whereas soil potassium concentration might vary between 5 and 20 mg 100 g^{-1}. To analyse such data, they should be normalized to ensure that they are comparable. This is usually achieved by converting them to Z scores. Each observation (X_i) can be converted to a Z score (Z_i) as follows (Shaw 2003):

$$Z_i = \frac{X_i - \mu}{s}$$

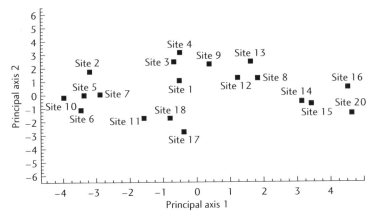

Fig. 3.17 The results of a principal components analysis to illustrate differences between sites according to their species composition. (Graphical output produced by using CAP; see Table 3.2). (From Henderson 2003.)

where μ is the mean of a sample and s is the standard deviation. The Z score has no units (it is the number of standard deviations each observation is from the mean of the sample), and therefore data that have been normalized in this way can be readily compared in the same analysis.

Outputs of PCA are usually visualized by plotting the sites or samples to be compared on the two or three main axes of variation detected by the analysis. Samples that are grouped more closely together are more similar in terms of their species composition. Attention should be paid to the proportion of the variability in the data set that is explained by these axes; the higher the proportion, the more robust the results. For data sets with more than 20 species, the first three main axes should account for more than 30% of the variance for the results to be of reasonable value (Henderson 2003).

PCA is very popular among plant ecologists because it is relatively easy to do. The outputs are relatively easy to interpret, but the patterns produced can sometimes be difficult to relate to environmental variables of interest. It is important to note that no test of significance is provided with the analysis: therefore it can only be used to explore data and generate hypotheses, rather than test them. However, the axis scores (which are generated during the analysis and describe the position of each sample along the principal axes) can be used as data themselves and subjected to further analysis, such as regression or analysis of variance (ANOVA) (Shaw 2003). For example, by using ANOVA it would be possible to test whether the axis scores associated with different forest communities were significantly different from one another (assuming that the scores are normally distributed). PCA cannot cope with missing values in the data and may be inappropriate where communities develop along environmental or temporal gradients (when DCA may be preferred; see below) (Shaw 2003).

Principal coordinates analysis (PCO) is sometimes confused with PCA, because of its similar abbreviation, but is in fact is a very different technique (sometimes referred to as multidimensional scaling). PCO uses a square matrix of distances between individuals and produces a map from the distances measured. This map can be used as a form of ordination to show the relative position of individuals sampled; usually Euclidean distance is used (Shaw 2003). The method is relatively little used by plant ecologists.

Detrended correspondence analysis (DCA or DECORANA)

DCA is an ordination technique designed specifically to assist with the exploration of ecological data (especially abundance data) and is very widely used by plant ecologists. It is particularly appropriate for use in situations where the sites that are sampled can be arranged along an environmental gradient, such as successional stages in vegetation. In such cases, use of PCA can lead to an artefact in the output (the so-called 'arch effect'), which DCA was designed to overcome. Ordinations of both species and sites can be plotted on the same figure, which enables the influence of different species on the ordination of the sites to be evaluated visually. The output of DCA is otherwise similar to that of PCA.

Canonical correspondence analysis (CCA)

This method was developed by ter Braak (1986), and is widely used by researchers interested in exploring the relations between community composition and environmental variables. CCA is based on a similar analytical technique to DCA and TWINSPAN, but differs in its inclusion of environmental data within the ordination itself so as to maximize their importance in the output. CCA therefore requires data on the environmental conditions at each site. These may be represented in the form of classificatory variables (such as 'high' or 'low' altitude) or continuous variables (such as temperature measured in degrees Celsius). The data matrix including the environmental variables measured must have the same number of rows (observations) as the species data, but need not have the same number of columns (variables) (Shaw 2003).

The outputs of the analysis enable the relation between environmental variables and the observed species communities to be evaluated. The method is sensitive to data quality; poor or incomplete data will produce results of little value. It is important that environmental and species data are collected at the same place and at the same time (Shaw 2003). As outputs can be complex, they can be difficult to interpret. As with other ordination methods, it is not possible to statistically test the association between species composition and environmental variables.

Biplot

Often, a main objective of ordination is to explore the relations between floristic composition and environmental variables (such as soil characteristics, aspect or altitude). This can usefully be achieved by producing a biplot. Biplots can be produced by using different ordination techniques (such as PCA, DCA, and CCA), although the methods used differ according the ordination technique adopted. Biplots enable the properties of the variables measured (such as environmental variables) to be overlaid on top of the main ordination diagram. Usually the former are illustrated as arrows that run from the 0,0 point to the coordinate in question (Shaw 2003). This can help with data interpretation: for example, if an arrow points to a cluster of points, then this suggests a relation between the variable in question and the samples illustrated by the points in the cluster. For example, an arrow representing soil nitrogen concentration might be associated with a cluster of points representing the composition of a particular forest type. However, it should be remembered that a biplot provides no statistical analysis of the strength of this relation, and illustrates a correlative association rather than any causal relation.

3.10.4 Importance values

Importance values have been widely used as a measure of species composition that combines frequency, abundance and dominance importance values (Grieg-Smith 1957). This can be calculated according to the following equation (Husch *et al.* 2003):

$$I_j = 100 \left(\frac{n_j}{N} + \frac{d_j}{D} + \frac{x_j}{X} \right)$$

where I_j is the importance value of the *j*th species, n_j is the number of sampling units where the *j*th species is present, N is the total number of sampling units, d_j is the number of individuals of the *j*th species present in sample population, D is the total number of individuals in sample population ($D = \Sigma d_j$), x_j is the sum of size parameter (generally basal area or volume) for the *j*th species, and X is the total size parameter across all species $X = \Sigma x_j$.

3.11 Assessing the presence of threatened or endangered species

Conservation assessments typically depend on knowing not only which species are present at a particular location, but also the relative conservation importance of these species. At the global level, species at risk of extinction are listed on the IUCN Red List of Threatened Species; these species have been evaluated according to the IUCN Red List categories and criteria. Details of the listing process, including freely downloadable copies of the categories and criteria, are available at ⟨www.redlist.org/⟩. A list of taxa that are considered threatened is maintained in a searchable database, which can be accessed online from the same Internet site. The only taxonomic groups that have been comprehensively assessed are the amphibians, birds, mammals, conifers, and cycads. The vast majority of plant taxa are not yet included. To find out the conservation status of plant species, the

UNEP–WCMC Threatened Plants database should be consulted ⟨*www.wcmc. org.uk/species/plants/red_list.htm*⟩. For tree species, a separate database is available, the UNEP–WCMC Trees Database ⟨*www.unep-wcmc.org/*⟩, which contains information on threatened trees and other trees of conservation concern. This includes information published in the World List of Threatened Trees (Oldfield *et al.* 1998).

Two points are of particular concern when using the Red List and these databases.

- First, they are very incomplete. Application of the Red List criteria requires information on the current (and ideally historical) patterns of occurrence of individual species, which is often lacking. Most listing is undertaken on a voluntary basis by specialists familiar with a particular group of species. Although the Red List is continually being revised, it is certain that many species (particularly of plants) that are threatened with extinction are not currently listed. Indeed, assessing the conservation status of species is an important research endeavour in its own right, and readers are encouraged to become involved in this process.

- Second, it is important to note that the IUCN Red List includes only taxa that are globally threatened, in other words when the entire geographic range of the taxon is considered. In many cases, it is important to ascertain whether regionally or nationally scarce or threatened taxa are present at a particular site. For this purpose, it is necessary to refer to lists of threatened taxa developed at the national scale. Many national governments have developed such lists, some of which are available online. In some cases, such national lists have adopted IUCN national Red List criteria and categories (details of which are also available at ⟨*www.redlist.org/*⟩), and may be published in the form of national Red Lists. However, in many cases countries or individuals have developed their own categories and criteria by which species may be considered endangered or threatened. For example, a 'star' system for scoring different tree species in Ghana according to their rarity was developed by Hawthorne (1996) and Hawthorne and Abu-Juam (1995). It is therefore important to distinguish between globally and nationally threatened species when carrying out an assessment, and to consider the criteria by which species have been classified, as well as the definition of the classes adopted.

At the international level, another group of species that are considered as priorities for conservation are those listed on CITES ⟨*www.cites.org/*⟩. Species listed in the CITES Appendices are those that are considered threatened with extinction because of international trade. For a species to be listed, agreement must be reached between the Parties to the Convention. For this reason, species listed on CITES form only a small subset of those that are threatened with extinction. Their presence is useful to record, however, particularly in areas where forests are being subjected to some form of management, as CITES listing has significant implications for how resources of these species are traded and managed. Some 38 species of tree have now been listed in Appendices I and II of CITES, including economically important timber species such as mahogany (*Swietenia* spp.). The main research challenge for such species is to determine how they may be harvested sustainably in accordance with CITES requirements (see section 8.2.3).

At the national scale, species of conservation concern may also be listed in a variety of different forms of legislation or government policies, such as Biodiversity Action Plans or laws governing the protection of particular species. In some cases, species may be accorded conservation importance because of their high social, cultural, or economic value, rather than their rarity or extinction risk. Poorter *et al.* (2004), in their atlas of woody plant species of West African forests, provide an outstanding example of how ecological information on tree species, collated at the regional scale, can be analysed and presented in a way designed to support conservation efforts.

3.12 Vegetation classification

The classification of forests into different communities or types provides an important basis for basis for forest management and conservation planning. Numerous different classification systems have been developed and implemented in different parts of the world. Details of these different approaches are described by Mueller-Dombois and Ellenberg (1974) and Whittaker (1975). Recent examples of approaches to classifying forests are provided by McNab *et al.* (1999) and Carter *et al.* (1999). Kimmins (1997) presents a useful overview of the different approaches that have been used to classify forests in the past, which can be summarized as follows:

- *Climatic classifications* are based on the fact that climate is one of the major factors influencing vegetation distribution. Climate types can be defined in terms of temperature and precipitation, which are associated with particular types of vegetation—for example, tropical rainy and warm temperate climates are associated with tropical rain forest and temperate rain forest, respectively. It is important to note, however, that climatic classifications represent potential, rather than actual, vegetation distributions.
- *Landform or physiographic classifications* are based on soil characteristics and landform features such as topography and altitude. These are often employed when classifying vegetation using remote sensing data. Although they have a firm ecological basis, as with climatic classification, this approach is an indirect method of classifying vegetation as floristic composition is not taken into account.
- *Biophysical or ecosystemic approaches* are based on climate, soils, and landform together with information about vegetation composition. This approach is widely used to classify areas at the regional scale.
- *Physiognomic classifications* are based on the growth form of dominant plants and the environments in which they grow; for example, vegetation dominated

by coniferous trees, deciduous trees, or shrubs might be differentiated as different vegetation types.
- *Floristic composition* can also be used to define vegetation types. This is the main method used by ecologists, but different approaches have been adopted in different regions. For example, the Braun–Blanquet approach has been used widely in Europe, and the method developed by Daubenmire has been widely used in the USA.

This diversity of approaches greatly complicates the integration or comparison of maps of forest types produced in different areas, and inhibits the development of generalizations about the patterns of distribution of different forest types and their relative conservation status. If the objective is some form of conservation assessment, it is very important to note the classification methods used in producing any that are drawn upon. For example, a map representing potential forest distribution (as produced by climatic or physiographic approaches) would be of little use for assessing the actual current extent of a particular forest type.

The problems of using applying available forest classifications to conservation issues are illustrated by Miles *et al.* (2006) in their global assessment of the conservation status of tropical dry forests. A first glance at available maps of forest types, such as that presented by FAO (2001b), suggests that extensive areas of tropical dry forest remain. However, the FAO map of ecofloristic zones is based on a climatic classification of potential vegetation, which gives a very misleading impression of current vegetation cover. Miles *et al.* (2006) therefore used a global map of current forest cover derived from remote sensing imagery (MODIS) to examine the actual current extent of tropical dry forest. The next step was to overlay this map with ecological classifications of forest types. Blasco *et al.* (2000) provide a comparison of 10 regional classification schemes developed for tropical woody vegetation, and present a common framework based on 'bioclimatic types' to help make comparisons between them. However, this has proved to be of little use for mapping tropical dry forests, as forest types (or formations *sensu* Blasco *et al.*, 2000) labelled as 'dry forest' in available regional classifications are grouped under no fewer than four of the six bioclimatic types considered. This suggests either that tropical dry forests have a very broad edaphic tolerance range, or that the concept of what constitutes a tropical dry forest is interpreted variously by different authors.

To produce their assessment, Miles *et al.* (2006) used the global biogeographic classification presented by Olson *et al.* (2001). This has the great merit that it was explicitly developed to support global conservation assessments. Importantly, the classification was based primarily on biogeographic information rather than climate, and included information regarding distributions of animal species as well as plants, unlike most alternative classifications that are available. This classification is applicable for use at global and regional scales, but has also been used by WWF for conservation assessments at subregional (ecoregion) scales (Olson *et al.* 2000).

When choosing an appropriate classification scheme, it is necessary to decide whether compatibility with existing schemes is necessary or desirable. Such compatibility can be of great help in enabling communication between people working on a particular forest type. Selection of forest types used in a map will depend on the map's purpose, and whether the classes can be accurately and efficiently delimited. When a classification system has been selected, it is important to document the details of each class and apply these definitions in an objective and consistent manner.

Many vegetation classification systems use a hierarchical approach in which classes are nested; major classes are divided into subclasses, which themselves can be divided into further subclasses. Such systems can be easily adapted to different scales. Non-hierarchical approaches tend to be used at a specific scale, when they might be preferred because they can be more readily customized to the particular needs of the investigation in question.

4
Understanding forest dynamics

4.1 Introduction

Research into forest dynamics over the past three decades has transformed our understanding of forest communities and has had a major influence on the development of ecological science. Increasing recognition of the role of natural disturbance in shaping the forest composition and structure, through research such as that of Tom Veblen in south temperate forests (Veblen 1979, 1985, Veblen *et al.* 1981), has caused a paradigm shift in ecology that has major implications for forest conservation and management (Pickett and White 1985). Investigations of forest disturbance and its impacts on the population dynamics of trees have become widespread, and have generated a voluminous literature.

Forest dynamics can be studied in three main ways:

- by inferring dynamics from one-off assessments of stand structure
- by monitoring forests over time, for example through repeated surveys of *permanent sample plots* (PSP)
- through the use of models of forest dynamics.

Each of these approaches has its limitations. A key problem is that most trees are long-lived. Forests often change slowly, and such changes can be difficult to detect on a scale of a few years or decades, or even over a human lifetime. Inferences of dynamics from assessments of stand structure can be highly misleading. For example, some tree species only produce seed very infrequently. If a field visit does not coincide with a 'masting' year, when large amounts of seed are produced, an entirely erroneous impression might be gained about the reproductive capacity or regeneration potential of the species present. Modelling approaches have a great deal to offer the study of forest dynamics, enabling dynamics over long timescales to be simulated, but even here there are many problems to be overcome, such as the need for detailed parameterization and the high degree of uncertainty associated with ecological processes and their outcomes.

This chapter describes methods relevant to two of these three approaches, with a particular focus on field techniques. Modelling techniques are considered in Chapter 5. Remote sensing methods can also be used to monitor forest change, as described in Chapter 2. The impacts of human activities on forest dynamics are further considered in Chapter 8.

4.2 Characterizing forest disturbance regimes

Disturbances can be defined as relatively discrete events that disrupt ecosystem, community or population structure, and change the availability of resources or the physical environment (Pickett and White 1985). Analysis of the disturbance regime of a forest can be of great value for understanding patterns of structure and composition, as well as being important for defining appropriate management interventions. Detailed treatments of forest disturbance are provided by Attiwill (1994), Barnes *et al.* (1998), Oliver and Larson (1996), Peterken (1996), and Pickett and White (1985).

Characterizing the disturbance regime typically involves assessing the severity, timing, and spatial distribution of the different types of disturbance affecting the forest. It is useful to note the difference between the *intensity* and *severity* of disturbance. Intensity refers to its physical force, or the amount of energy released, whereas severity refers to the amount of living biomass that is either killed or removed as a result of the disturbance (Spies and Turner 1999). *Timing* refers to the seasonality (time of year), duration, frequency, and the return interval or rotation time of disturbance (Spies and Turner 1999).

Spatial distribution may include assessments of the extent and shape of disturbances, and can be characterized at different scales, for example at local, landscape, and regional scales (Barnes *et al.* 1998).

In order to characterize the disturbance regime of a particular forest, the following variables should therefore be measured (Gibson 2002):

- *extent* and spatial pattern of the disturbed area
- *intensity*, or the strength of the disturbance (for example, fire temperature or wind speed)
- *severity*, or the amount of damage that occurred to the forest (for example, number of individual trees killed or stems damaged)
- *timing*, including the frequency (the number of disturbances per unit time), the turnover rate or rotation period (the mean time taken for the entire forest area to be disturbed) and the turnover time or return interval (the mean time between disturbances)
- *interactions* between different types of disturbance (for example, drought increases fire intensity).

For each of these variables, the central tendency (mean, mode, or median) and variation (standard deviation or range, for example) should be calculated, and presented together with frequency distributions in order to fully characterize a disturbance regime (Gibson 2002).

The main types of natural disturbance affecting forests are fire, weather (including wind, ice, temperature, and precipitation), soil disturbance (erosion, deposition, flooding, and movement) and herbivory (Gibson 2002). Anthropogenic disturbances include logging, fuelwood cutting, road building, drainage, fire, forest clearing for agriculture and urban development, livestock husbandry, application of chemical fertilizers and pesticides, aerial pollution, and many others. Assessment of the impact of human activities on forests is of fundamental importance for conservation planning and management. Often, a key objective is to determine whether a given forest is able to withstand or tolerate a particular anthropogenic disturbance regime. Such analyses lie at the heart of defining approaches to sustainable forest management, and depend not only on characterizing the disturbance regime, but on understanding how the forest responds to different types and patterns of disturbance. An important principle is the idea of limiting human impacts to the frequency, size, and severity of disturbance to which species are adapted, leading to the development of forest management plans based on natural disturbance regimes (Spies and Turner 1999). However, this is often difficult because of a lack of understanding about past disturbance regimes, and the ecological impacts of current human activities.

The following sections describe methods for characterizing some of the most important types of forest disturbance. The methods described here are field techniques; remote sensing methods and GIS are also very widely used for assessing forest disturbance and are described in Chapter 2. Although the methods described here may be useful for characterizing the current disturbance regime of a particular forest, it can also be helpful to consider the *likelihood* (White 1979) or risk of disturbance occurring in the future. This is central to assessment of vulnerability, an important consideration in conservation planning, which is described in section 8.4. Types of disturbance not considered here in detail include aerial pollution and insect attack, both of which are of major concern to foresters, who have consequently developed a range of techniques for their assessment (see Horn 1988, Innes 1993, Knight and Heikkenen 1980).

4.2.1 Wind

It is helpful to differentiate the typical ('chronic') wind disturbance of a site from the effects of relatively rare wind events, such as hurricanes or storms (Ennos 1997). Rare events can have a major impact on forest ecology by destroying trees. The breakage or uprooting of trees by wind is referred to as *windthrow* or *blowdown*. This form of natural disturbance results from the interaction between climate, topography, stand structure, soil characteristics and the growth and form characteristics of individual trees. Damage can be in the form of stem failure, root failure or uprooting (Mergen 1954). Windthrow can be termed *endemic* or *catastrophic*. The latter results from winds with relatively long return periods and is influenced primarily by local wind speed and wind direction, whereas endemic windthrow results from peak winds with return intervals of less than 5 years and is influenced more strongly by site conditions (Lanquaye-Opoku and Mitchell 2005). Endemic windthrow is generally more predictable than catastrophic windthrow, and therefore most assessments of windthrow risk focus on the former rather than the latter (see section 8.4).

The impacts of wind disturbance on a forest can be assessed through a field survey, involving measures of the structure and species composition of the area affected (see Chapter 3) and the canopy gaps created (see section 4.4). Windthrow also results in production of woody debris (see section 7.2 for techniques for assessing deadwood volume). Treefalls not only create gaps in the canopy but also cause disturbance to the soil and influence the micro-environmental heterogeneity of the forest by creating pit-and-mound micro-relief with exposed root mats, bare mineral soil, and humus, as well as fallen logs. These features may provide suitable sites for seedling colonization in many forests, and are therefore often included in field surveys. For example, Ulanova (2000) mapped the distribution of pits and mounds caused by treefalls, as potential sites of seedling establishment. Some authors have used a decay scale for logs, with between five and nine divisions, to determine gap age or the time since a windthrow event (see, for example, Liu and Hytteborn 1991). However, the degree of decay of a log depends not only on the date of the windthrow but also on log size and species, its position (i.e. whether it rests directly on the ground or on broken branches), and its status at the time of the windthrow event (i.e. dead or alive) (Ulanova 2000).

Surveys may also be designed to assess tree condition, by recording different kinds of damage such as broken or bent stems, branch loss, snapped or uprooted trees, and root damage. The extent of canopy loss can be estimated visually by using methods for estimating canopy cover (see section 3.6.4) (Brommit *et al.* 2004). Wounding is often common in wind storms, caused by falling trees and branches, twisting, etc., and may also be recorded. Different tree species vary in the amount of crown damage caused by storms, reflecting differences in green wood strength and crown shape (Zimmermann *et al.* 1994). The capacity for recovery following wind damage, by resprouting of stumps or branches, also varies between species (Del Tredici 2001) and can usefully be recorded in a survey (Bellingham *et al.* 1994, Paciorek *et al.* 2000).

The amount of damage caused by wind is a function of wind speed, duration, and the direction from which the winds originate. Wind speed is usually measured by using a *cup anemometer*, consisting of three hemispherical or conical cups mounted on arms and attached to a spindle so that they can rotate in the wind (Coombs *et al.* 1985). An alternative design, the *vane anemometer*, consists of a number of light vanes radially mounted on a horizontal spindle. The main sources of error for such mechanical anemometers is that they have a threshold below which the friction of the system prevents rotation, and their inertia causes an overestimation of wind speed when the latter suddenly drops (Coombs *et al.* 1985). Continuous measurements from anemometers can be recorded by using data loggers (see section 4.6). If such instruments are not available, records may be obtained from national weather centres or meteorological surveys. Further details of devices for wind measurement are provided by Grace (1977).

Tatter flags offer a relatively simple alternative method of measuring exposure to wind. Described by Lines and Howell (1963) and Rutter (1966, 1968a, b), the method employs flags made out of cotton mounted on poles or wires. It is important that material is used that degrades with time—not the material usually used to make flags! The material should slowly degrade to lose 50–75% of its initial mass.

Tatter flags are also commercially available. The loss in mass or area of the flag measured over time can be attributed to attrition caused by wind. Studies of the tattering of flags in both controlled and field conditions have shown that the rate of tatter is closely correlated with wind exposure, but is also influenced by factors such as rainfall and atmospheric moisture (Jack and Saville 1973). Another possible source of error is freezing of flags in winter. As wind varies seasonally and from year to year, flags should be flown for 2–3 years and should be distributed throughout the area to be surveyed. Use of tatter flags in the UK has been described by Quine and White (1994).

Susceptibility of areas to wind damage can be assessed in several ways (Reed and Mroz 1997). It may be possible to identify soil types susceptible to wind damage because of shallow rooting depth. The incidence of wind damage in the past can potentially be ascertained from forest survey notes or weather records, and used to assess relative risks of different areas in the future. Climatic atlases often indicate severe storm frequencies. For example, Boose et al. (2001) describe methods for reconstructing hurricane disturbance in New England by using a combination of historical research and computer modelling. Wind disturbance history can also be inferred by means of dendrochronological techniques (see section 4.3). The presence of physical evidence such as broken trees, fallen trees, and tip-up mounds can also be used to infer the previous occurrence of wind damage.

Examples of studies that have assessed the damage caused by windstorms on forests include Bellingham et al. (1994, 1995), Brommit et al. (2004), Burslem et al. (2000), Peterson (2000), Peterson and Rebertus (1997), and Zimmermann et al. (1994). Further information on the effects of wind on forests is provided by Coutts and Grace (1995), Ennos (1997), and a special issue of the journal *Forest Ecology and Management* (2000), vol. 135. Approaches to modelling wind risk are briefly considered in section 8.4.

4.2.2 Fire

Fire temperatures can be measured by using thermocouples or temperature-sensitive paints. Thermocouples can be used to measure temperature by monitoring the voltage produced by the difference in temperature between two dissimilar metals, which are used in construction of the thermocouple. The temperature response of a thermocouple should be calibrated against a reference measurement. Ideally a thermocouple with a small bead (or tip) and small wire diameter should be used; sheathed thermocouples are also preferred because they protect the thermocouple junction from soot, which can influence the measurements made (Saito 2001). However, useful information can also be gained from thicker wire thermocouples (Iverson et al. 2004). Thermocouples need to be connected to a data logger for measurements to be recorded (see section 4.6). Arrays of thermocouples can be arranged in a grid and at different heights in a vegetation canopy, in order to characterize spatial variation in fire intensity (Jacoby et al. 1992).

Alternatively, *pyrometers* can be constructed by painting spots of temperature-indicating paint (such as Temiplaq), which melt at different temperatures, on to the unglazed side of ceramic tiles or metal tags, which are placed in the forest before the fire then collected afterwards for inspection (Gibson 2002). The minimum temperature attained in the fire around each tile is determined as the highest-temperature paint spot that melted in the fire (Hobbs *et al.* 1984). Maximum fire temperatures can also be estimated through the use of Tempil tablets, which melt at different temperatures (Grace and Platt 1995). The tablets should be wrapped in aluminium foil before they are placed in the forest, to facilitate relocating them after the fire, and their melting points should be calibrated in the laboratory (Gibson 2002).

Iverson *et al.* (2004) compared thermocouples and temperature-sensitive paints to measure fire intensity (Figure 4.1) and found that maximum temperatures recorded by the two measuring systems were highly correlated. These authors recommended the use of temperature-sensitive paints if only maximum temperature is required, because of their substantially lower cost involved. However, additional information can be collected by using thermocouples. For example, positioning of the thermocouples in a grid enables information on the rate of fire spread to be collected.

A further technique for measuring fire temperature involves use of an infrared camera and an image recording and analysis system, which is capable of obtaining a two-dimensional thermal image from a remote location (Saito 2001). This method has been used to produce temperature profiles in forest fires (Clark *et al.* 1999c).

Fire velocity can be measured by using a *pitot tube*, which can measure velocities in the range from a few metres per second to above 100 m s^{-1}, covering most of the wind velocity range in forest fires (Saito 2001). The technique is relatively simple and can measure a one-dimensional velocity component, although three-dimensional measurements can be obtained by changing the direction of the pitot tube head when the flow is at a steady state (Saito 2001). The method is described by Sabersky *et al.* (1989).

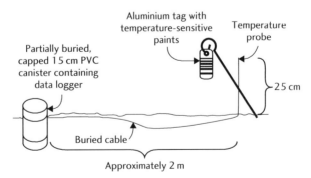

Fig. 4.1 Arrangement of thermocouple temperature probes and temperature-sensitive paints, used in a comparative trial to measure fire intensity. (From Iverson *et al.* 2004. International Journal of Wildland Fire, CSIRO Publishing. http://www.publish.csiro.au/nid/115/issue/871.htm.)

Trees may survive a fire but be damaged by it. Death of the cambium can lead to production of a fire scar, which is visible as a gap in the bark that is usually triangular in shape, becoming narrower with height, found on the leeward side of trees. Fire scars have been widely used to date fires, particularly in conifer forests, but also with some hardwood tree species. Such scars can be detected as blackened areas or damaged rings in increment cores (see following section), or in discs obtained from the trunks of felled trees (Schweingruber 1988). However, errors can be caused by missing, false, or indistinct rings, which should be corrected by cross-dating (Fritts 1976). This can be achieved by sampling trees that were undamaged by the fire, or sampling both sides of the trees affected. Further details of these methods are provided by Arno and Sneck (1977), Madany et al. (1982), and Schweingruber (1988).

Byram (1959) provided a simple index of fire intensity:

$$I = HwV$$

where H is the heat of combustion of fuel, w is the mass of fuel consumed per unit area, and V is the heading rate of spread of the fire. This index is thought to correlate closely with tissue necrosis and possible tree mortality. Iverson et al. (2004) indicate that measurements from thermocouples can be used to estimate I. Fire intensity is usually measured as the rate of heat energy released per unit length of fire line per unit time ($W\,m^{-1}\,s^{-1}$) or sometimes per unit area and time ($W\,m^{-2}\,s^{-1}$).

An ability to predict the spread of fires is something of great value to forest managers, and consequently this has received much attention from researchers. However, fire spread is influenced by a complex set of phenomena occurring over a range of scales, involving turbulent flow influenced by wind and topography, and the spatial distribution and amount of fuel available. A range of models have been developed in different parts of the world, which are reviewed by Weber (2001). Further information about the ecology of forest fires is provided by Agee (1993), Johnson (1992), Johnson and Myanishi (2001), and Whelan (1995). Methods for evaluating fire risk are considered further in Chapter 8.4.

4.2.3 Herbivory

In order to clearly demonstrate the effects of herbivory on plant populations, properly designed, replicated field experiments are required. Below-ground herbivory is generally assessed by applying some form of either physical or chemical exclusion as treatments in a field experiment. Chemical treatments range from the non-specific killing of most soil organisms through the use of chloroform fumigation (see, for example, Sarathchandra et al. 1995), to the use of biocides that are targeted to the control of specific organisms. Organophosphates such as isofenphos and ethoprop can be used to reduce populations of soil invertebrates, including coleopteran and lepidopteran larvae (Gibson 2002). Freezing of soil to -20 °C can be used to reduce the activity of soil nematodes (Sarathchandra et al. 1995). Experimental control of above-ground herbivory by invertebrates is often achieved through spraying insecticides such as carbaryl, chlorpyrifos, dimethoate, and malathion (Brown and Gange 1989, Gibson 2002). However, great care should be employed in the use of any chemical treatments; many can have unforeseen and potentially deleterious effects on components of the ecological communities other than those being investigated, and can have undesirable effects of the physical and chemical properties of soils. Some can alter plant growth, either positively or negatively, and this should be evaluated as part of the experiment (Gibson 2002). Insecticides that might be of use in ecological research are listed by appropriate guidebooks such as Page and Thompson (1997), which should be consulted carefully to ensure that the chemicals are selected, handled, and applied correctly (Gibson 2002) in a way that minimizes impact on the environment.

An important method of assessing herbivory impacts in forests is through the use of *fenced exclosures*, designed to exclude browsing mammals such as deer. Comparison of areas within and outside such exclosures can provide insights into the impacts that the animals are having on variables such as the extent and composition of ground flora and the seedling establishment of trees. The effectiveness of the exclosure in excluding the animals should be evaluated; the presence of a fence is no guarantee that it will be effective. Some species of deer are capable of jumping fences of at least 2 m in height, and can prove remarkably persistent in attempting to gain access to fenced areas. An appropriate design and size of fence should be used for the type of animal to be excluded, and animal populations should be monitored both within and outside the exclosures (see Sutherland (1996) for methods appropriate for censusing animal populations).

Fences should be regularly checked for damage and repaired, and fence supports need to be strong enough to withstand damage from any large animals that are present (Gibson 2002). For smaller mammals, an appropriate mesh size for the fencing material must be selected (i.e. 3–4 cm for rabbits, 0.5 cm for voles and mice; Gibson 2002). To exclude burrowing animals (such as rabbits), the fence must be partly buried in the soil. Care should also be taken to ensure that the fences do not have any negative effects on wildlife; for example, in Scotland, collisions with deer fences were found to be a significant cause of mortality in woodland grouse (Summers 1998).

The potential impact of herbivory on individual plants can be examined experimentally by introducing herbivores in appropriate cages, or by simulating the effects of herbivory by clipping leaves or other plant parts with scissors (Canham *et al.* 1994b). However, removal of plant material in this way may fail to mimic the effects of herbivory precisely (Gibson 2002). Generally, the effects of herbivory on individual plants are assessed by measuring damage to the plant parts consumed, although it should be noted that patterns of dry mass allocation within the plant may change in response to herbivory, and therefore this may need to be measured for a comprehensive assessment of impacts. Plant damage as a result of herbivory can be most simply assessed as the proportion of damaged leaves or shoots per plant (Gibson 2002). Alternatively, individual leaves can be assessed on a scoring system describing the proportion of leaf area removed. For example, Brown *et al.* (1987) used the following scoring system on a sample of 20–100 leaves per plant:

Damage rating	Estimated leaf area removed (%)
0	0
1	1–5
2	6–25
3	26–50
4	51–75
5	76–99
6	Total removal of leaf

Reimoser *et al.* (1999) provide a useful review of methods for measuring forest disturbance by ungulates. These authors note that impacts of ungulate herbivory are usually assessed by analysing variables such as density of tree regeneration, species composition, and growth rate of young trees. These variables can be measured in a field survey using quantitative methods (see Chapter 3 and elsewhere in this chapter). Surveys can also usefully assess the occurrence of variables such as:

- trampling (includes pawing, scraping, burrowing and rooting);
- browsing (includes 'unseen' browsing, i.e. feeding on seeds and seedlings, and 'visible' browsing, such as tree twig browsing)
- fraying
- peeling (of bark or surface roots, for example).

In situations where quantitative approaches are not possible, qualitative surveys can be considered. For example, Table 4.1 illustrates a classification scheme for forest stands developed in the UK, which can be used to characterize the intensity of grazing based on a series of indicators that can be rapidly assessed by observation.

A further technique widely used in studies of herbivory is the analysis of faeces and gut contents. The technique involves identifying plant fragments that have been ingested by herbivores, with the aim of identifying the dietary preferences of different species. The main advantage of faecal analysis is that the same population of herbivores can be continuously sampled, without direct interference. The method is described in detail by Bhadresa (1986), on which this account is based. The technique depends on development of a reference collection of the cuticles of plant species found in a study area. This can be achieved by peeling or scraping off the epidermis of a leaf of the chosen plant species, then mounting it in glycerol jelly on microscope slides. Permanent stained mounts can be produced by dehydrating slowly with alcohol and replacing by xylol, then mounting in Canada balsam or Euparal. The cuticle can then be stained with safranin, acid fuchsin, or gentian violet, conducted at the 70% alcohol stage.

Faeces can be collected from the field plots, after first clearing an area of existing faeces to ensure that any material sampled is fresh. The size of the plot varies with the animal concerned; Bhadresa (1986) recommends 4 × 4 m squares for rabbits, but larger areas are needed for larger mammals. Faeces may be collected at

Table 4.1 Indicators of different grazing or browsing pressures in north temperate forest (adapted from Reimoser et al. 1999).

Intensity	Indicator
Very heavy	No shrub layer; obvious browse line on mature trees; ground vegetation <3 cm tall with grasses, mosses or bracken (*Pteridium aquilinum*) predominating, and trampling down of ground flora; extensive patches of bare soil; surviving herb species usually dominated by unpalatable species; suppression of growth, and killing, of seedlings and saplings by browsing soon after germination and, therefore, virtually absent; very abundant dung from grazing animals; bark stripped from young and mature trees and from branches on the ground; mosses scarce or absent; possible invasion of weed species; the more palatable, grazing-sensitive shrubs and herbs confined to inaccessible areas or, at least, noticeably more abundant there.
Heavy	Shrubs absent or moribund; 'topiary' effects on remaining shrubs; a browse line on mature trees; ground vegetation <20 cm tall with grasses, mosses or bracken (*P. aquilinum*) dominating; few patches of bare soil; surviving herb species usually dominated by unpalatable species; tree seedlings not projecting above ground vegetation height; abundant dung from grazing animals; bark stripping occasionally occurring; bulky, common mosses favoured at the expense of the rarer species requiring deeper shade and cover; the more palatable, grazing-sensitive shrubs and herbs confined to inaccessible areas or, at least, noticeably more abundant there.
Moderate	Patchy shrubs showing evidence of pruning or a browse line; ground vegetation variable in height up to 30 cm, comprising a mixture of grasses, herbs, or dwarf-shrubs, including some of the more grazing-sensitive species of herb and showing direct evidence of browsing/grazing; localized close-cropped lawns where there is a concentration of grazing; patches of bare soil small and rare; tree saplings projecting above ground vegetation in a few areas; some dung from grazing animals; no bark stripping; wide range of moss species.
Light	Well-developed shrub layer, with no obvious browse line; a lush ground vegetation in places where the shrub layer covers not more than ~30–50% of the ground, dominated by grazing-sensitive species; tree saplings common in gaps; dung and tracks of grazing animals difficult to find; no bark stripping; browsed shoots scarce and localized, or totally absent; deep litter layer; ground mosses uncommon and consisting of few species.
None	As for Light but with no browse line on the shrub layer; no, or very few, saplings where there has been no grazing for many years; no herbivore dung or tracks present; no browsed shoots; no, or very few seedlings; extensive monospecific mats of vigorous ground-layer species may also occur on some sites.

regular intervals and stored after drying until required for analysis. Alternatively they may be stored in a mixture of formalin–acetic acid–alcohol–water in the ratio 10 : 5 : 50 : 35. To separate epidermal fragments for analysis, faeces may be dispersed in water, then after subsampling the epidermal fragments can be identified and counted under a microscope (Croker 1959). Alternatively the faeces may be soaked in 10% sodium hydroxide, or nitric acid and potassium hydroxide, to help disperse the epidermal fragments (see Box 4.1 below).

Once cuticular fragments have been extracted, they can be identified by comparison with a reference collection, following examination with a microscope under × 100 magnification. Estimates of the relative abundance of different species in the sample can be obtained by one of the following methods (Bhadresa 1986):

- counting the number of fragments in the entire subsample
- frequency counts in microscope fields at a particular magnification
- direct estimation of surface areas of fragments
- estimation of surface areas of fragments using the principle of point quadrats.

Similar methods may be used to analyse gut contents. For example, Erickson *et al.* (2003) analysed the feeding preferences of mangrove crabs based on analysis of leaf stomata in gut contents.

One problem to consider during studies of herbivory is the possibility that observations made by the researcher may affect the behaviour of animals and consequently the amount of herbivory taking place. This has been termed the 'herbivore uncertainty principle' by Cahill *et al.* (2001). Although Schnitzer *et al.* (2002) failed to find evidence of this effect, its possible occurrence should be borne in mind when designing studies of herbivory.

Crawley (1997) gives an overview of plant–herbivore interactions (see Figure 4.2), including analytical methods for analysing the effect of herbivores on plant

---

Box 4.1 Method for extracting epidermal fragments from faecal samples (from Bhadresa 1986).

The method refers to small (0.1 g) droppings; vary the amounts of reagents proportionately for larger or smaller samples.

Insert the dried faecal sample into a 50 ml conical flask and soak in 2 ml distilled water, followed after 5 minutes by 2 ml concentrated nitric acid. Warm the flask in a water bath at 50–60 °C for 5 minutes; tease with a glass rod. Add 10 ml of 1 mol/l potassium hydroxide. While shaking the flask on a flask shaker, take a 1 ml sample with a Pasteur pipette; place in a crucible and remove the supernatant with a Pasteur pipette after allowing to settle for about 5 minutes. The fragments can then be washed with distilled water, leaving 2–3 drops with the sample while pipetting off the rest. Scrape the fragments on to a labelled microscope slide and allow to dry. If the density of fragments is very high, it may be necessary to dilute the sample for analysis.

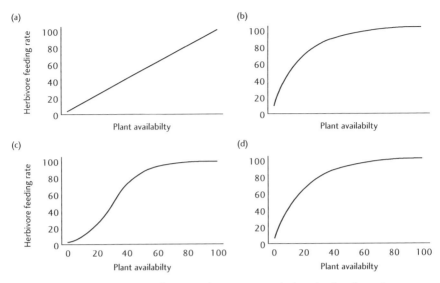

Fig. 4.2 Different herbivore functional responses relating the feeding of an individual to the amount of food available. (a) linear, (b) saturating, (c) sigmoid, (d) ratio-dependent. (After Crawley 1997.)

population dynamics. One important issue is the amount of plant material consumed per herbivore per unit time as a function of the availability of the plant material. Four different functional responses have been described (Crawley 1997):

- *Linear*, reflecting the fact that some plant species are eaten whenever they are encountered because they are so attractive; referred to as 'ice-cream plants'.
- *Saturating*, where the proportion of plant biomass eaten by herbivores declines as plant abundance increases, because of some upper limit at which the herbivores can depress plant abundance.
- *Sigmoid*, where herbivores only consume a particular plant species after it exceeds a particular threshold of abundance.
- *Ratio dependence*, where the functional response operates in terms of the amount of plant food available per herbivore, rather than the absolute density of plant food per unit area.

Changes in plant abundance in response to herbivory can usefully be modelled by using differential equations, such as the following formula (Noy-Meir 1975):

$$\frac{dV}{dt} = G(V) - c(V)H$$

where V is plant abundance, H is a constant herbivore density, and $c(V)$ is the herbivore functional response. $G(V) = rV(1 - (V/K))$, the logistic growth function where r is the intrinsic rate of increase, and K is the carrying capacity. Predictions depend on the herbivore's response to plant availability, $c(V)$ (Frelich 2002).

Crawley (1997) highlights the importance of both field observations and experiments for identifying not only which relationship best describes the situation being studied, but also the mechanisms underpinning the relationship observed. As the importance of herbivory in plant ecology is still the subject of active debate, Crawley (1997) highlights the need for more well-designed, replicated, long-term field studies in which herbivore species are excluded in combination with seed-sowing and soil-disturbance treatments. These observations certainly apply to forests, where discussion regarding the role of herbivores has been greatly stimulated by the recent monograph by Vera (2000). Further information about the impact of browsing by vertebrates on forests is presented by Gill (1992) and Putman (1986, 1996).

4.2.4 Harvesting

Timber harvesting, or logging, can be a major cause of anthropogenic disturbance to forests. However, it is important to note that harvesting of non-timber products, such as fruits, nuts, and extractives, can also result in significant ecological impacts (Peters 1996, Ticktin 2004). Generally the effects of harvesting are assessed by comparing field plots established in areas both within and outside areas subjected to harvesting, although some studies have compared a range of different forest management interventions varying in harvest intensity.

The impacts of harvesting are clearly most direct on the tree species being harvested. Harvesting for timber may be indiscriminate, but is more likely to target individuals above a particular stem diameter threshold. Harvesting of non-timber products might involve collection of some plant part other than the stem, such as roots, leaves, flowers, fruit, bark, or latex. In either case it is possible that the harvesting will not be lethal for the individual, as stumps may be able to resprout vegetatively, and other plant parts such as bark or roots may be able to recover after harvesting. A key objective is therefore to determine the impact of harvesting on the survival of individuals, requiring monitoring of individuals ideally before and after harvesting, and on the size or age structure of populations of the species of interest within the study area. Many of the techniques described in this chapter are relevant in this context. Population modelling using transition matrix models (see section 5.2.3) have been widely used to assess the impact of harvesting on individual species and to define whether or not current harvesting approaches are sustainable (Boot and Gullison 1995, Hall and Bawa 1993).

Harvesting may also result in impacts on other species. If trees are felled, this can cause significant damage to other trees and the understorey. Until recently, such impacts have not received much attention from researchers. An example is provided by Soehartono and Newton (2001), who assessed the impact of harvesting *Aquilaria* trees (*gaharu*) in Indonesia by accompanying teams of collectors on harvesting expeditions, and measuring not only the size of trees that were felled, but the number of seedlings, saplings, or adult trees of other species damaged or killed during the harvesting process (Figure 4.3).

A range of other environmental impacts can result from logging operations, the most significant of which is frequently the construction of access roads, which may be accompanied by clearing 20–30 m strips through the forest to assist the drying out of the road surface after rain (Johns 1992). Such roads, strips, timber loading

Fig. 4.3 Collection of *gaharu* (*Aquilaria* spp.) in Indonesia. This resinous wood is often collected by felling the tree, which can damage seedlings and mature trees located nearby. Such impacts should be considered when assessing the sustainability of harvesting forest products. (Photo by Tonny Soehartono.)

and landing areas, extraction tracks, and skidding trails all represent different forms of forest damage, which can best be estimated through field survey or through aerial photographs. Trees may also die some time after harvesting operations have taken place, as a result of insolation, drought, or increased risk of windthrow (Johns 1992), and therefore post-harvest surveys may need to be repeated for some time after harvesting to obtain a complete assessment of the damage that has occurred.

Harvesting can affect animal species living within the forest, and also biophysical variables such as soil fertility. In particular, use of heavy machinery can cause soil compaction, increasing surface run-off, erosion, and nutrient loss. Recent examples of research investigations that have assessed harvesting impacts include:

- Bertault and Sist (1997), who compared conventional and reduced impact logging techniques on the basis of pre- and post-harvesting stand inventories. This investigation is unusual in having collected data before harvesting, providing a design more powerful than those of many other studies.
- Feldpausch *et al.* (2005), who surveyed roads and log decks, canopy damage, and ground disturbance in skid trails and treefall gaps in southern Amazonia.
- Laffan *et al.* (2001), who investigated the impact of cable-logging on forest soils in Tasmania.

- Whitman *et al.* (1997), who mapped skid roads and tree-felling sites, and assessed soil compaction loss of canopy cover, damage to saplings and trees, and seedling survival after mahogany harvesting in northern Belize.
- Verissimo *et al.* (1995), who assessed damage caused by harvesting of mahogany in Amazonia by carrying out a field survey of recently harvested sites.

Further details of the environmental impacts of timber harvesting, with a particular focus on soil properties, are provided by Dyck *et al.* (1994).

4.3 Analysis of forest disturbance history

The material in this section is based largely on information provided by Frelich (2002), who notes that it is possible to infer historical patterns of forest disturbance by using information from a variety of sources, including:

- *Fossil evidence.* Fossil pollen and other durable plant parts such as seeds, needles, and wood fragments can be preserved in sediments, and provide a means of reconstructing forest composition in the past. Small forest hollows have proved to be particularly useful for assessing changes in forest composition at the local scale (Calcote 1995). The presence of charcoal in forest soil profiles can be used to infer historical fire frequency (see Clark and Royall 1995), and can be subjected to carbon-14 analysis to provide dating of fire events.
- *Historical records.* Forest managers often routinely collect information on specific disturbance events such as storms or fires, as well as records of timber extraction. Alternative historical sources include newspapers and other media, records made by surveyors or naturalists, and weather records.
- *Physical evidence.* The most common form of physical evidence is the presence of tip-up mounds, produced as a result of treefalls. The date of formation of the mound can sometimes be determined by excavating it, and ageing the trees growing nearby (Henry and Swan 1974). In northern temperate forests, tip-up mounds can remain visible for 200 years or more. Another form of physical evidence is the presence of cut stumps, which can be used to infer previous logging events.

Historical information can also be gathered from ancient documentary evidence, such as legal records, charters, and historical maps; archaeological artifacts; earthworks and surface features; iconography; and oral tradition (Rackham 2003). Use of such evidence has given rise to the discipline of *historical ecology*, which aims to identify the historical factors that have influenced the development of vegetation to its current state. The methods arguably lie closer to historical than to ecological research, and are therefore not considered in detail here. However, their potential value should not be underestimated. In the UK, the historical research of Oliver Rackham (Rackham 1986, 2003) has not only transformed our understanding of woodland ecology, but has led to a major shift in conservation policy. So-called 'ancient woods', which have a history of management stretching back over many centuries, are now rightly recognized as of exceptional conservation importance (Rackham 2003).

Most commonly, forest disturbance history is inferred from measurements of stand size and age structure. The frequency distribution of tree diameters can usefully be plotted to assess whether or not the stand is even-aged, indicated by a unimodal diameter distribution. However, without measurements of the age of the trees, interpretation of diameter-frequency distributions is subject to a great deal of potential error. To reduce such errors, diameter distributions should ideally be plotted separately for the dominant species in a single homogeneous stand, or for only those trees receiving direct sunlight on the top of the crown. Another structural measure that can be used is the diameter-exposed crown-area distribution, which is obtained by estimating the cross-sectional area of the portion of the crown of each tree exposed to the sun. These areas are then summed for all trees within each size class. The resulting measure represents the proportion of the total exposed crown area of the stand occupied by each size class, and has the benefit of equalizing for the different densities that can occur in various size classes.

Tree ages can be determined most readily by *increment corers*, at least in temperate forests (see section 3.6.1). Although the age distribution of trees can be very helpful for inferring stand disturbance history, the results may often be open to a variety of different interpretations. For example, the same type or intensity of disturbance can result in stands with different age structures, simply because of variation in the pattern of recolonization following the disturbance. Stands with even-aged structures can be produced by a series of disturbances, rather than just a single stand-levelling disturbance event, such as when seedling recruitment is interrupted by deer browsing or ground fires. Trees within a stand can display a wide variety of ages, reflecting the survival of individual trees following a range of different disturbance events.

Patterns of growth increment can also be obtained from increment cores and can provide a valuable source of information about disturbance history. If a large tree is blown down, then neighbouring trees that were formerly suppressed are likely to display a period of significantly higher radial growth. Analysis of such growth patterns can enable disturbance chronologies to be obtained for entire stands of trees, and enable those trees that were already in the canopy at the time of disturbance to be identified. However, care must be taken when inferring disturbance history from increment cores. A tree that has been released may subsequently be overtopped again by a competitor, making the original disturbance more difficult to detect. Trees some distance from a newly created gap can respond to increased light availability, and as a result, estimates of disturbance intensity may be inflated.

Details of the techniques of radial increment pattern analysis are presented by Frelich (2002), who provides the following guidance based on detailed analysis of northern temperate forests in the USA:

- Criteria of $> 100\%$ and $> 50\%$ increases in ring width after disturbance are typically used to indicate 'major release' and 'moderate release' for shade-tolerant

species, the former indicating transition from an understorey to a canopy position.
- For the release to be considered abrupt, indicating a sudden disturbance event, the > 50% or > 100% release in ring width should occur within a period of 1–5 years.
- Criteria of at least 15 years slow growth before release and at least 15 years of rapid growth after release are used to screen out growth patterns that are not related to the disturbance events of interest.

Use of this method, by applying these criteria, can enable disturbance chronologies to be produced that indicate the proportion of trees that entered the canopy in each decade.

Attention also needs to be paid to sampling approaches when taking cores for increment analysis. Typically, individual trees are selected by using a random or systematic method; for example, the trees closest to randomly located sample points may be selected for coring. Age structure may be represented either as the proportion of individuals in each age class, or as the proportion of area of a plot or landscape occupied by each age class, based on measurements of crown area. Area-based samples are commonly used in studies of disturbance, where disturbance rate is expressed as a percentage of forest area disturbed per unit time.

In some cases it is not possible to obtain complete increment cores, because the trees are hollow or are too large to core. The dbh and crown area of such trees should be measured, so that they can be taken into account in area-based calculations of disturbance. Small trees (<5 cm dbh) also tend not to be sampled by using increment borers; instead, a proportion of them may be felled to procure ring measurements.

How many trees should be sampled for increment analysis? The sample size determines the precision of the estimates obtained, and the chance of missing a cohort of trees in the sample. The probability of failure to detect an age class of trees can be calculated from the following equation (Frelich 2002):

$$Pf = (1 - Py)^x$$

where Pf is the probability of failure to detect age class y, Py is the proportional area occupied by age class y, and x is the number of independent sample points. Precision can be estimated by calculating the confidence limits for proportions (Frelich 2002):

$$\pm 1.96 \sqrt{(p(1-p)/N)}$$

where p is the sample estimate of the proportion of points belonging to a given cohort, and N is the number of sample points. As a rule of thumb, 5–10 cores may be enough to characterize the disturbance history of even-aged stands; 30 or more cores may be needed in complex multi-aged stands (Frelich 2002).

Fitting functions to stand age distributions (see section 3.7.1) can also provide insights into disturbance history. It is also useful to consider the hazard function, which expresses the chance of disturbance with stand age. For example, an equal hazard function (or equal probability of disturbance across all stand ages) results in the negative exponential age distribution. The formula for the negative exponential is:

$$A(t) = \exp(-(t/b))$$

where A is age, t is time in years, and b is the mean stand age. If the cumulative frequency of stand ages is plotted on a semi-log graph, a straight line is obtained if a negative exponential distribution fits (Frelich 2002). Changes in the slope of this graph indicate changes in the rate of disturbance ($1/b$).

In the case of the Weibull function (see section 3.7.1), when the shape parameter (c) is 1, the distribution is the same as the negative exponential, with a constant hazard function. When $c > 1$, disturbance hazard increases with age.

Rubino and McCarthy (2004) provide a recent review of the use of dendrochronological techniques to assess forest disturbance history, and highlight two limitations of radial-growth methods for the assessment of disturbance regimes:

- The length of time used for determining the mean growth rate must be long enough to take account of climatic anomalies (such as extended periods of increased or decreased precipitation), yet permit identification of short-term dynamics.
- Release identification methods may not be able to detect a disturbance event if two release events occur in rapid succession.

Despite such problems, the authors conclude that radial growth analysis is a useful technique for characterizing forest disturbance history, particularly where destructive and invasive sampling is undesirable or prohibited, such as in protected areas (Lorimer 1985).

4.4 Characterizing forest gaps

The concept of a *forest gap* or *gap phase* dates back to Watt (1947), who used the term to refer to a patch in a forest created by the death of a canopy tree. The composition and structure of many forest communities can be seen as the result of the responses of different species to the size, frequency, and distribution of such gaps. Investigations of forest gap dynamics became a central theme in ecological research in the 1980s (Brokaw 1985, Denslow 1980, Runkle 1981; see also the special issue of *Ecology* 70 (3), 1989), and despite criticisms (Brown and Jennings 1998), the concept remains important today (see, for example, Cumming *et al.* 2000, Wright *et al.* 2003). The concept is also used in some approaches to modelling forest dynamics (see Chapter 5).

The definition of what constitutes a gap has been variously interpreted in the literature, leading to some confusion. For example, a substantial proportion of tree mortality can fail to produce any break in the upper forest canopy; it is not always clear whether gaps include or exclude such tree deaths (Lieberman *et al.* 1985, Swaine *et al.* 1987). It has been argued that structural definitions are superficial and difficult to apply; rather, Swaine *et al.* (1987) suggest that a gap incorporates structural, microclimatic, edaphic, and biotic changes, and its size should be considered relative to the organism of interest. It is important to remember that although a gap in the canopy may often be most noticeable to human observers, a gap may also occur in the rooting zone.

One of the main problems of the gap concept is the difficulty of measuring a canopy gap. This is illustrated by one of the most beautifully titled forest ecology papers ever published: '*Forests are not just Swiss cheese*' (Lieberman *et al.* 1989). In other words, gaps can be difficult to map and measure because their boundaries are often imprecise. The dichotomy between 'gap' and 'understorey' can be a gross oversimplification, failing to recognize the pronounced spatial and temporal variation in light availability that is typically encountered in a forest. The issue of characterizing a gap or non-gap environment is therefore relevant to the broader problem of characterizing the light environment within a forest. Most commonly, gaps are characterized by measuring the light environment or degree of canopy closure (see following sections for appropriate methods). Alternatively, remote sensing methods may be used to produce maps of canopy gaps (see Chapter 2) (see, for example, Fujita *et al.* 2003a, b).

Gaps can be sampled by surveying transects and measuring length, width, date of formation, and composition of the gap-making trees and replacement trees in all gaps that are encountered (Runkle 1981, 1982). A protocol for characterizing gaps is presented by Runkle (1992), which is summarized in Box 4.2. This will have to be amended depending on the objectives of the particular study: it is important to measure only those variables that are relevant to the research question being addressed. The protocol will not work well in those forests, such as many old-growth forest stands, that are not characterized by the presence of discrete gaps or well-defined gap edges.

Box 4.2 A protocol for characterizing forest gaps, abridged from Runkle (1992).

- *Gap definition.* A gap is formed by the death (absence from the canopy) of at least one-half of a tree. The largest gap is created by the death of 10 canopy trees or has a ratio of canopy height to gap diameter equal to 1.0, whichever is larger for the forest studied. Gaps close when replacement stems reach a height indistinguishable from that of the surrounding closed forest. The edges of the gap are defined by a vertical projection of the canopy leaves of trees adjacent to the gap.
- *Sampling for gaps.* Line transects should be located randomly in a forest area with relatively homogeneous site conditions. The location on transects of the start and stop of each intersected gap are recorded. The start point and compass direction of the transect should be recorded. Location of

point-centred quarter-points can also be recorded, to characterize canopy composition and to aid in relocating the transect. Distance along the transect should be measured by using metre tapes. Gap orientation may be influenced by topography; it may be necessary to run transects both across and up and down slopes.
- *Recording gap makers.* Gap makers are those trees that formed the gap by their death. The species, dbh, original height, direction of fall if any (base to top), agent of death, and type of damage (uprooting, partial uprooting, breakage with stump height, standing dead, partial death) can be measured or estimated.
- *Gap size.* Many gaps can be approximated by ellipsoidal shapes. These gaps are measured by first locating a pair of perpendicular lines in the gap such that the first is the longest straight line that will fit in the gap, and the second is the longest straight line that will fit in the gap with the constraint that the line is perpendicular to the first. Gap area may often be calculated directly from the lengths of these two lines, fitted into the formula for an ellipse (area = $\pi LW/4$, where L is the length of the longer line and W is the length of the shorter line). Where the gap shape is more irregular, the length of each line segment from the intersection to the gap edge is recorded starting with the longest and moving in a clockwise direction. Gap size is calculated as the sum of the four quarter ellipses determined by the line segments. The orientation of each gap can be determined by recording the compass direction of the longest line segment. Where gap shape is too irregular to be characterized as an ellipse, distances from gap centre to edge are measured in at least the eight main compass directions, and the area of the resultant polygon is calculated. An optional addition is to determine the area of the 'expanded gap' (Runkle 1982). This is the area within a polygon constructed by drawing a line connecting the boles of all the trees whose crowns border the gap (canopy opening). Field procedure consists of selecting a point near the centre of the gap (preferably the intersection point previously used for estimating area) and measuring the distance and compass angle for a vector from the point to the bole of each marginal tree. Use of an optical rangefinder (see section 3.4.1) may facilitate this procedure.
- *Gap microhabitats.* Gaps consist of several identifiable microhabitats: pit, mound, log, branch pile, bark pile, and remainder. The length and width of representatives of these microhabitats can be recorded. Previously fallen logs can also be recorded.
- *Gap age.* The ability to age gaps helps determine which gaps are used in calculating gap-formation rates and forest turnover times. Several different features can be used to determine gap age: (1) the leaf and bud condition of fallen trees; (2) release dates of saplings from bud scar counts or radial increments; (3) ages of rings on scar regrowth, sprout age, and changes in branch growth direction; (4) release dates of adjacent trees; (5) dendrochronology of the dead tree; (6) ages of seedlings in the gap; and (7) decay state of the gap

maker. Some gaps are formed by several different episodes of tree mortality. In such cases, it may be necessary to subdivide the gap into sections of different ages.
- *Gap aperture.* This variable is measured from gap centre (the intersection of the major and minor axes) as the average of arcs of sky visible along both axes. It thus measures gap canopy openness scaling the size of the gap by the heights of the surrounding canopy trees. Measurements are taken at 1.5 m high.
- *Calculating gap properties.* The data collected by using the methods outlined above can be used to calculate a variety of different measures, including:

$$\text{percentage of total land area in gaps} = \frac{\text{transect distance in gaps}}{\text{total transect distance}} \times 100$$

percentage of total land area in gaps of specific size or age classes

$$= \frac{\text{transect distance in gaps in of specified class}}{\text{total transect distance}} \times 100$$

$$\text{gap formation rate} = \frac{\text{percentage of total land area in gaps} \leq n \text{ years old}}{n}$$

where n is the number of years for which a complete sample of gaps is obtained, and also

turnover time (turnover rate) = (gap formation rate)$^{-1}$

For gap size frequency distributions as a percentage of the number of gaps rather than the area occupied by gaps, correction factors may need to be applied, depending on the sampling procedure used.

4.5 Measuring light environments

4.5.1 Light sensors

A range of different types of sensor are available for measuring light. Quantum sensors can be constructed from photodiodes. Use of appropriate filters enables the sensor to measure *photosynthetically active radiation* (PAR); in other words, the component of solar radiation used in photosynthesis (400–700 nm) (Coombs *et al.* 1985). A white Perspex diffuser is typically used on top of the sensor, in order to obtain accurate measurements at all solar angles (cosine correction). The design of such a PAR quantum sensor is described by Woodward and Yaqub (1979). These sensors can be constructed reasonably cheaply, enabling large arrays to be produced, but each sensor must be carefully calibrated in order for accurate measurements to be made. They may also be obtained from a range of commercial suppliers.

PAR quantum sensors are probably the type of sensor most widely used by forest ecologists, providing measurements of the number of quanta incident on a particular area (in units of μmol m^{-2} s^{-1}), referred to as the *photosynthetic photon flux density* (PPFD). Alternatively, sensors may be used that measure solar energy, in units of watts per square metre (W m^{-2}). Such sensors (referred to as solarimeters or pyranometers) may either measure total solar energy, or energy within the photosynthetically active 400–700 nm waveband, depending on the use of filters. This type of sensor is preferred if information is needed on the energy relations of plants (relation between solar radiation and leaf temperature, for example). Note that it is not possible to convert from radiometric measures of solar energy to quantum measures without knowledge of the spectral composition of the vegetation (Jennings *et al.* 1999).

Other types of light sensor include:

- *Tube solarimeters*, which consist of a glass tube enclosing a strip thermopile of alternating black and white surfaces. Such solarimeters have been used very widely in crop science; they provide a measure of solar energy flux, and must be used with filters if measurements of the photosynthetically active part of the spectrum is required. As a result of their glass construction, solarimeters are relatively fragile.
- *Ceptometers*, such as the LP80 device manufactured by Decagon ⟨www.decagon.com/lp80/⟩, which consists of a linear probe 86.5 cm long that contains 80 quantum sensors sensitive to the PAR waveband. The instrument is particularly used for measurements of LAI (see section 3.7.3).
- *Plant canopy analysers* such as the LAI-2000 manufactured by LI-COR Inc. ⟨www.licor.com⟩, which is similarly used widely for measuring LAI (see section 3.7.3). The instrument consists of a near-hemispherical lens (148° field of view) held in front of five concentric silicon ring detectors, which enable canopy light interception at five angles to be measured. Radiation intercepted by the canopy is computed by dividing the above-canopy detector outputs by the below-canopy detector outputs (LI-COR Inc., 1992). The *diffuse non-interceptance* (DIFN) calculated by this instrument is similar to the instantaneous diffuse light transmission obtained on overcast days and has been found to be closely related to daily temporal variation of light (Hanan and Bégué 1995).

A further type of sensor is used to analyse the spectral composition of solar radiation. Options include spectral radiometers or the Red/Far-Red sensor marketed by Skye Instruments ⟨www.skyeinstruments.com⟩. This has a basic design similar to a quantum sensor, but with filters enabling ratio of red to far-red light (R : FR, 660: 730 nm) to be measured, a ratio that has a major influence on the development and growth of some plant species. It is important to calibrate such sensors regularly, as their performance can change over time.

Such sensors can be used with handheld meters for spot readings. Some are equipped with memory storage ('integrators') enabling measurements to be recorded over time. Alternatively, sensors may be attached to programmable data loggers, which enable readings from multiple sensors to be stored over prolonged periods. Note that luxmeters should not be used in ecological investigations, as the measurements that they provide (illuminance or brightness as perceived by the human eye, given in units of lumens, lux, or foot candles) are not relevant to tree growth and survival (Jennings et al. 1999).

Estimation of the light environment beneath a forest canopy canopies is very challenging because of the high spatial and temporal variability typically encountered. Underneath a forest canopy, a single point receives both direct and diffuse light. Direct light comes from the solar disc, and varies according to the time of day and year as the solar altitude changes. Diffuse light comes from all parts of the sky and is much more uniform, both spatially and temporally, than direct light under a forest canopy (Anderson 1964).

Although PAR sensors can provide accurate measurements, large numbers of sensors are needed to adequately characterize the light environment within a forest stand. For example, Baldocchi and Collineau (1994) estimated that sample plots in many tropical forests would require over 270 sensors for a representative description. It is important that the sensors are maintained in a level position, at a fixed height above ground level, and careful consideration is given to where they are located. Typically, if multiple sensors are available, they are located by using stratified random approaches or on points of a grid. Ideally, light should be measured continuously for several days in order to take account of temporal variation. The sensors should be kept clean and horizontal, or else significant errors can be introduced (Jennings et al. 1999).

Characterizing the light environment via instantaneous measurement of light transmission on clear, sunny days around noon has been very popular among researchers (Gendron et al. 1998). It is possible to take measurements at numerous locations in the understorey during such a period, assuming that irradiance above the canopy is similar for all measurements. Often, the amount of light recorded is expressed as a percentage of radiation incident at the top of the canopy (%PPFD, usually measured in an open area such as a large forest clearing) (Gendron et al. 1998). Messier and Puttonen (1995) proposed a new method to estimate light environments in the understorey, by measuring instantaneous diffuse light transmission on overcast days. This is based on the fact that under an overcast sky, %PPFD at any particular microsite tends to be very stable throughout the day (Messier and Puttonen 1995). Evidence presented by these researchers suggests that instantaneous percent above-canopy PPFD ((PPFD in understorey/PPFD above canopy) ×100) under completely overcast conditions, measured with PAR quantum sensors, can provide an accurate estimate of the mean daily percentage above-canopy PPFD over the course of a day and under all sky conditions (Messier and Puttonen 1995, Parent and Messier 1996). This result was supported by Gendron et al. (1998), who found that 10 minute averages taken on overcast days provided a more accurate assessment of %PPFD over a growing season than instantaneous measurements taken on sunny days around noon.

Given their relatively high cost and long cable lengths, quantum sensors are expensive and cumbersome for multiple-point sampling. For this reason, a number of simpler indirect methods are widely used, some of which are considered below.

4.5.2 Hemispherical photography

Hemispherical or 'fish-eye' photography has a long history of use in plant ecology, dating back to the pioneering efforts of Anderson (1964), Becker (1971), and Evans and Coombe (1959). As a result of the development of high-resolution digital cameras and advances in image-processing software, there has been a recent renewal of interest in this method (Bréda 2003). The technique is described in detail by Cannell and Grace (1993) and Rich (1989, 1990), and examples of its application are provided by Rich *et al.* (1993) and Whitmore *et al.* (1993).

The technique involves taking a photograph of a forest canopy by using a conventional camera with an unconventional lens, which has a very wide field of view (180°, hence the common name 'fish-eye'). It is important to note that a true 'fish-eye' lens should have a 180° field of view for accurate measurement of light environments; some hemispherical lenses do not have such a wide field of view and therefore do not capture the full range of incident light. The camera should be mounted so that it is level (Figure 4.4), and it is important

Fig. 4.4 The HemiView system used for taking hemispherical photographs of forest canopies. The system comprises a 180° fisheye lens with a high resolution digital camera, mounted in a self-levelling camera mount to ensure that it is held horizontally. (Photo courtesy of Delta-T Devices Ltd.)

Measuring light environments | 171

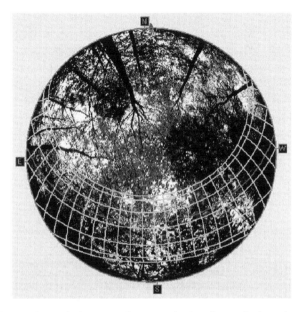

Fig. 4.5 A hemispherical photograph on to which solar paths have been superimposed, using HemiView software. This enables the occurrence of sunflecks and associated solar irradiance to be calculated on any day of the year. (Photo courtesy of Delta-T Devices Ltd.)

Fig. 4.6 An example of a hemispherical photograph of a forest canopy, taken using the Hemiview system. (Photo courtesy of Delta-T Devices Ltd.)

that orientation of the camera (due north, for example) is indicated on the image, to facilitate subsequent analysis. Photographs can be analysed manually or with computer software to determine the geometry and position of canopy openings and the path of the sun at various times, and to indirectly estimate the characteristics of light environments beneath plant canopies as well as properties of the canopies themselves (Roxburgh and Kelly 1995) (Figures 4.5, 4.6). This enables light transmission to be estimated for any specified period (daily, growing season etc.). Also, both diffuse and direct light components transmitted through the canopy can be estimated (often presented as 'site factors').

A number of commercial instruments are now available (Table 4.2) that enable hemispherical photographs to be analysed. In addition, a number of individual researchers have developed software for analysing hemispherical photographs, including Solarcalc (Chazdon and Field 1987), HEMIPHOT/WINPHOT (ter Steege 1994) ⟨www.bio.uu.nl/~boev/staff/personal/htsteege/htsteege.htm⟩ and Gap Light Analyser (GLA) (Frazer et al. 1999) ⟨www.rem.sfu.ca/forestry/index.htm⟩; ⟨www.ecostudies.org/⟩. Whichever analysis system is used, hemispherical photographs with both digital and film cameras must be taken under uniform sky conditions, such as those encountered just before sunrise or sunset or when the sky is evenly overcast.

During analysis, the different parts of the digitized image are classified as either black (completely blocked by foliage) or white (clear sky). The most critical step in image processing is determining the threshold between the sky and canopy elements (Bréda 2003). Small changes in the threshold value selected can result in relatively large changes in estimates of canopy closure, particularly beneath dense canopies; yet a consistent threshold value can be difficult to find (Jennings et al. 1999). Other shortcomings of the method are that the canopy is assumed to be a single layer of leaves; the presence of any leaves is assumed to completely block the passage of light. Furthermore, hemispherical analysis systems currently do not have the ability to assess reflection from leaves, or layers of leaves; reflection and transmission may be affected by leaf orientation relative to sun angle, which is not considered by the technique (Roxburgh and Kelly 1995). The method also assumes that there are no significant seasonal changes in the canopy throughout the growing season. However a number of authors have found close correspondence between direct measurements of PPFD using quantum sensors and estimates derived from hemispherical photography (Easter and Spies 1994, Rich et al. 1993), although the technique appears to be less reliable in shaded sites (Chazdon and Field 1987, Roxburgh and Kelly 1995).

One of the main drawbacks of hemispherical photography is the high cost, not only of complete analytical systems but also of the lens required. This has stimulated interest in using relatively low-cost, 'consumer' digital cameras, some of which offer the capacity to take 'fish eye' photographs. Following a comparison of such cameras with conventional systems, Frazer et al. (2001) caution against using consumer cameras for scientific applications, because of distortion detected in the

Table 4.2 Comparison of three commercial systems for analysis of hemispherical photographs, used for characterizing forest light environments (updated from Bréda 2003).

	Measurements	Field of view	Resolution	Company	Website
WinSCANOPY	LAI, leaf-angle distribution, and mean leaf angle, angular distribution of gap frequencies, sunfleck distribution, total radiation and site factors (direct, diffuse, and global)	180°	Depends on choice of camera; typically 6–12 megapixels	Regent Instruments Inc., Quebec, Canada	*www.regent.qc.ca*
HemiView	LAI, leaf-angle distribution and mean leaf angle, angular distribution of gap frequencies, sunfleck distribution, total radiation and site factors (direct, diffuse, and global)	180°	2592 × 1944 pixels	Delta-T Devices Ltd., Cambridge, UK	*www.delta-t.co.uk*
CI-110 Imager	LAI, sky view factor, mean foliage inclination angle, foliage distribution and extinction coefficient of the canopy	150°	768 × 494 pixels	CID Inc., Vancouver, USA	*www.cid-inc.com*

Note: The WinSCANOPY and HemiView devices are canopy analysis systems based on analysis of colour hemispherical images; the standard systems include a digital camera, a calibrated fish-eye lens, and a self-levelling system. Images are taken in the field and processed externally using specific software. Outputs are available by sky sector or aggregated into a single overall whole-sky value. The digital plant canopy imager CI-110 is different, because it is designed to capture and processes colour hemispherical images that can be analysed either in real-time in the field, or subsequently in the laboratory. The hemispherical lens is mounted on an auto-levelling design on the tip of a handle connected to a portable computer dedicated to the equipment. Note that the latter system has a relatively low field of view (150°).

imagery. However it is probable that this will become less of a problem as the quality of consumer cameras continues to improve, and acceptable results have been obtained by other authors (Englund *et al.* 2000).

4.5.3 Light-sensitive paper

Friend (1961) described a simple technique for measuring light based on the use of light-sensitive diazo paper, which was recently re-evaluated by Bardon *et al.* (1995), on which this account is based. Stacks of diazo (ozalid) paper are constructed into booklets of 20 sheets. These are then placed in Petri dishes for protection, maintained in position against the inner surface of the lid by a piece of sponge. A piece of black paper with a central hole of approx 0.95 cm^2 is placed inside the lid, allowing light to reach the surface of the booklet of paper. The hole is kept covered and the Petri dishes stored in the dark until the commencement of measurements. Once exposed, the amount of light received is estimated from the number of layers of paper that are bleached after dry development with ammonia vapour, with a development time of 20–25 min.

The assumption made by this technique is that the number of diazo sheets exposed is related to the total quantity of radiation received (i.e., duration × intensity). Bardon *et al.* (1995) tested whether PPFD measured with a quantum sensor correlates with the number of exposed sheets of diazo paper under a variety of conditions. A stronger linear relationship was found between the number of layers of exposed diazo paper and maximum instantaneous PPFD than the number of layers of exposed diazo paper and accumulated PPFD or $\log_{10}$ accumulated PPFD. Under field conditions, full sunlight resulted in exposure of no additional layers of diazo paper after about noon. The authors concluded that diazo paper seems to record irradiance at a low rate, giving the impression that it is recording accumulated PPFD, whereas in fact it is not. Bardon *et al.* (1995) therefore recommend that diazo paper should not be used to measure accumulated PPFD under field conditions, especially for periods that include a significant amount of time after noon or under conditions with light flecks or varying irradiance. However the method can provide an indication of the maximum intensity of solar radiation received during the period of observation. According to Jennings *et al.* (1999), however, this method can only provide an approximate estimate of PAR.

4.5.4 Measuring canopy closure

Canopy closure is the proportion of the sky hemisphere obscured by vegetation when viewed from a single point (Jennings *et al.* 1999). Note the difference between this term and *canopy cover*, which refers to the proportion of the forest floor covered by the vertical projection of the tree crowns (Figure 4.7) (Jennings *et al.* 1999). Methods for estimating the latter are presented in section 3.6.4. Canopy closure can be measured by using hemispherical photography, or by a number of other techniques detailed below.

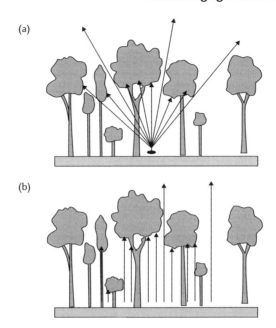

Fig. 4.7 The difference between canopy closure (a) and canopy cover (b). (From Jennings et al. 1999.)

Table 4.3 Crown position indices presented by Clark and Clark (1992).

Class	Description
1.0	No direct light (crown not lit directly either vertically or laterally)
1.5	Low lateral light (crown lit only from the site; no large or medium openings)
2.0	Medium lateral light (crown lit only from side: several small or one medium opening)
2.5	High lateral light (crown lit only from side: exposed to at least one major or several medium openings)
3.0	Some overhead light (10–90% of the vertical projection of the crown exposed to vertical illumination)
4.0	Full overhead light (≥90% of the vertical projection of the crown exposed to vertical light, lateral light blocked within some or all of the 90° inverted cone encompassing the crown)
5.0	Crown fully exposed to vertical and lateral illumination within the 90° inverted cone encompassing the crown)

Simple visual assessment

It is possible to produce a rapid, visual estimate of canopy closure by comparing the area of canopy with a standard scale. For example, Clark and Clark (1992) presented a simple index based on the crown illumination of individual trees (Table 4.3), which was found to be significantly correlated with measures derived from hemispherical photographs. The main limitation of this method is the potential for lack of repeatability, particularly between different observers. The magnitude of this error can be estimated by taking repeat measurements. Brown *et al.* (2000) found that repeatability could be improved by assessing the size of a hole in the canopy by comparing it with a series of ellipses of different size (ranging from 10.3 to 41.0 cm^2 in area) printed on a transparent Perspex screen, which is held at a fixed distance from the eye by attaching a cord 20 cm long. The score is determined by the size of the ellipse that fits entirely into the largest canopy opening visible in the canopy, whilst standing at the point of measurement (Brown *et al.* 2000).

An alternative approach was described by Lieberman *et al.* (1989), to provide a quantitative index of canopy closure based on the three-dimensional stereogeometry of trees at a specific point. The index is based on the following measures: (1) the horizontal distance between the focal tree and each taller neighbour within some given radius, (2) the height difference between the two trees, and (3) the distance from the top of the focal tree to the top of its neighbour, calculated from the height difference and the horizontal distance between the focal tree and the neighbour. The ratio of (2) to (3) is the sine of the included angle θ. The index of canopy closure, G, is defined as the sum of these ratios for all i taller neighbours within some specified radius:

$$G = \sum_{i=1}^{n} \sin \theta_i$$

The index is lowest for large trees with the fewest crowns above their own. The index can be calculated for any point within the forest volume, and therefore can be used to capture the three-dimensional characteristics of a forest canopy.

Spherical densiometer

A *spherical crown densiometer* is a relatively simple instrument for estimating crown closure. The instrument is described by Lemmon (1956), and consists of a convex or concave hemispherical mirror etched with a grid of 24 squares. The observer scores canopy cover by assessing whether sky or foliage is visible at four equally spaced points within each square. Strickler (1959) suggested that four readings be taken at each point, one for each of the cardinal directions. Potential problems with the technique are systematic differences between observers (Vales and Bunnell 1988), although this can potentially be estimated by taking repeat measurements by different observers. Bunnell and Vales (1990) and Cook *et al.* (1995) both reported that instruments that measure wide sky angles, such as the densiometer, underestimate canopy cover compared with methods that measure narrow angles such as the Moosehorn (Garrison 1949) considered below. Although the instrument is portable and robust, Jennings *et al.* (1999) conclude that spherical densiometers do not give a highly accurate measure of canopy closure; as the reflection of the canopy is small, they suffer from poor resolution. A further problem is that observations have to be made by viewing the instrument from the side rather than from above (to avoid an image of the person making the observations being reflected on the instrument), which interferes with making an accurate measurement. However, Englund *et al.* (2000) found in a comparison with hemispherical photography that spherical densiometers effectively characterized forest light environments, noting also that densiometer measurements significantly increased in consistency with user practice. These authors also note that the densiometer does not measure the entire sky angle; very little radiation enters beyond zenith angles greater than 58°.

Canopy-scope

Brown *et al.* (2000) describe a simple instrument that can be used for simple and rapid assessment of forest understorey light environments (Figure 4.8). The design is based on an earlier instrument, the Moosehorn (Garrison 1949), but is less cumbersome and less fragile. The *canopy-scope* consists of a transparent Perspex screen that has a 20 cm cord attached to one corner, which is used to maintain a consistent distance between the screen and the eye. The screen is engraved with 25 dots, approximately 1 mm in diameter, that are arranged in a 5 × 5 array, spaced 3 cm apart (centre to centre). The observer looks through the screen, placed 20 cm from the eye, and counts the number of dots that coincide with the sky rather than with the canopy. This ability to observe the canopy directly, through the instrument, represents a significant advantage over the densiometer. Whereas the original Moosehorn design required the observer to point the instrument vertically

Fig. 4.8 The canopy-scope, a simple instrument that can be used for rapid assessment of forest understorey light environments (Brown *et al.* 2000). (Photo by Nick Brown.)

and make a measurement of canopy openness centred on the zenith, Brown *et al.* (2000) suggest taking readings centred on the largest canopy gap above the point of measurement. These authors found that measurements of canopy closure made with the canopy-scope were significantly correlated with measurements derived from hemispherical photographs, for sites with canopy openness in the range 0–30%. Hale and Brown (2005) found that 8–10 canopy-scope measurements in 0.25 ha were sufficient to estimate canopy closure in a plantation forest site.

Which method for measuring light or canopy closure?

The choice of which method is most appropriate for light measurement depends on the objectives of the investigation, and the resources available. The preferred solution is to use an array of PAR quantum sensors, attached to a data logger. However, this is a time-consuming and expensive approach. Hemispherical photography has attracted a great deal of interest because obtaining and analysing an image is relatively rapid, and from a single photograph it is possible (with appropriate software) to estimate light transmission throughout the year—something difficult to achieve with quantum sensors. The main problem of hemispherical photography is the high cost involved. For this reason, a number attempts have been made to determine whether simpler and cheaper methods give acceptable results.

Machado and Reich (1999) found that in a deeply shaded conifer-dominated forest understorey, measurements made from hemispherical canopy photography and hemispherical sensors (LAI-2000) were positively and linearly related to the mean daily %PPFD measured with quantum sensors. However, the strength of the relation and closeness to a 1 : 1 fit was weaker for the hemispherical photograph technique (Gendron *et al.* 1998). Comeau *et al.* (1998), working in deciduous forests, found that similar results were obtained with hemispherical photographs, arrays of PAR quantum sensors and the LAI-2000. Ferment *et al.* (2001) compared five different methods in tropical forest, including photosensitive paper, a densiometer, hemispherical photographs, and light sensors. Results indicated that relatively simple, cheap, and portable methods, such as photographic papers, can provide similar information to more cumbersome or expensive methods such as quantum sensors or hemispherical photographs. Brown *et al.* (2000) noted that in situations where canopy openness exceeds 30%, such as in savannah woodland, methods such as the canopy-scope and crown index are unlikely to be of value. However, these authors suggested that for relatively dense forests, the canopy-scope is the best option available, taking its very low cost of construction into account. Jennings *et al.* (1999) recommended that details of the angle of view measured, the height to base of live crown, slope, and area of the forest assessed should be reported when canopy closure is measured.

4.6 Measuring other aspects of microclimate

Disturbance can influence aspects of forest microclimate other than light. Characterization of such changes can be of value in interpreting the potential impact of disturbance not only on the growth and survival of tree species, but on the availability of habitat for other species. In montane forests, for example, disturbance to the forest canopy may reduce air humidity in the understorey, which can have negative impacts on epiphytic mosses and lichens. The techniques and instrumentation required for measurements of microclimate in the field are described by Coombs *et al.* (1985), Hall *et al.* (1992), and Pearcy *et al.* (1990), on which this brief summary is based.

Temperature can be readily measured by using a mercury-in-glass thermometer, as used in meteorological stations (Figure 4.9). If recordings are required, either *thermocouples* or *thermistors* are generally used. Thermocouples are relatively cheap and easy to construct, typically out of copper–constantan or chromel–alumel, taking care to ensure a good junction between the two metals by soldering with either tin or silver. After soldering, the junctions should be cut with a blade under a binocular microscope to ensure that they are as small as possible (Coombs *et al.* 1985). All thermocouples should be individually calibrated, for example by immersing them in a water bath whose temperature can be controlled. Thermistors are semiconductors, composed of sintered mixtures of metallic oxides, and as for thermocouples, can either be constructed or purchased from a commercial supplier. Although more expensive than thermocouples, they are relatively robust. When measuring air temperature, the instrument should be shaded by a suitable screen. Surface temperatures (of leaves, for example) require contact between the sensor and the surface being measured, often achieved by using clips, springs, or tapes.

Humidity can be measured with a variety of different instruments, including psychrometers, dewpoint meters and electronic capacitance and resistance sensors. Although the latter are most convenient, they require frequent calibration. *Psychrometers* consist of a pair of thermometers, one of which is covered by a wet sleeve. Evaporation from the wetted sensor cools the thermometer, enabling vapour pressure to be calculated. Small ventilated psychrometers are available for use within or above plant canopies; for example, Delta-T devices (*www.delta-t.co.uk*). Available electronic sensors include *capacitance hygrometers*, which measure the change in electrical capacitance caused by water-absorption into a dielectric. *Infrared gas analysers* can also be used to measure water vapour concentration; although very expensive, these instruments are accurate and respond quickly to environmental changes. As for air temperature, instruments for measuring air humidity should be shaded when used in the field, and whichever instrument is used, it should be carefully calibrated.

Soil moisture can be described in two ways: in terms of the quantity of water present or in terms of the energy status of the water. The *gravimetric water content* is the mass of water in unit mass of dry soil (kg water/kg soil). Typically the wet mass of the soil sample is determined, then the sample is dried at 100–110 °C to constant mass and reweighed. The *volumetric water content* is expressed in terms of the volume of water per volume of soil (litres of water per litre of soil). This can be measured by multiplying the gravimetric water content by the soil bulk density (kg of dry soil/litre of soil). The energy status of water in soil can be expressed as the total soil water potential (MPa), which can be divided into the matric, solute and pressure potentials (Rundel and Jarrell 1989). Instruments used for measurement

Fig. 4.9 A micrometerological station for measuring light availability (using a PAR sensor) and wind speed (using an anemometer), to which a data-logger has been attached to record the measurements made. The instrument has been positioned in an experimental gap created at Harvard Forest, USA. (Photo by John Healey.)

of soil moisture include *tensiometers, gypsum resistance blocks, soil psychrometers*, and *neutron probes*. Of these, resistance blocks and tensiometers are the cheapest and easiest to use, the former being more suitable for drier soils and the latter for wetter soils (Rundel and Jarrell 1989).

Measurements of microclimate made over time can be recorded by using a *data logger*. A variety of different models are available from commercial suppliers such as Delta-T Devices, ⟨*www.delta-t.co.uk*⟩; Campbell Scientific, ⟨*www.campbellsci.com*⟩; or LI-COR, ⟨*www.licor.com*⟩. Data loggers can be programmed to collect and store information at a variety of different intervals, and for different lengths of time. Those designed for field use are available with weatherproof cabinets. Key features that differ between models include the number of inputs to which sensors can be attached, memory storage capacity, and battery life. Such data loggers are also used as the basis of automatic weather stations, which can be used to provide detailed measurements of meteorological variables such as wind speed, wind direction, and rainfall (Figure 4.9). Other sensors that can be attached to data loggers that are used in plant ecology include those used for measuring surface wetness and soil moisture content, as well as the PAR sensors described above.

4.7 Assessing the dynamics of tree populations

4.7.1 Permanent sample plots

Permanent sample plots (PSPs) are commonly employed to evaluate forest changes over time, enabling repeated measures to be made on the same individual trees. The locations of PSPs can be determined with a GPS (see section 3.4.2). Topographic maps and aerial photographs are also a useful aid to relocating sample plots. Information useful to relocating the plot should be recorded, including the distances and bearings of approach lines (determined using a compass) and reference points or landmarks. It is very important to collect and properly archive detailed information that will enable the plot to be relocated in the future, as it may be that attempts will be made to resurvey the plot many years hence.

Permanent markers should be positioned at the centre or corners of the plots, and referenced to a nearby permanent landmark. Plots may either be marked conspicuously, so that they can be relocated easily, or inconspicuously, to reduce the risk of interference from visitors. Small metal rods can be inserted as corner posts, inserted at depth into the soil, and projecting slightly above it (Avery and Burkhart 2002). Alternatively, galvanized metal posts may be used (Hill *et al.* 2005), which can be inserted entirely into the ground if they need to be entirely inconspicuous. Metal posts can potentially be relocated with a metal detector or a magnetic detection device (if iron posts are used).

Trees in PSPs are often marked to ensure that consistent measurements are taken during re-enumeration. For mature trees, the commonest method is to nail numbered metal tags into the trunks near ground level (Avery and Burkhart 2003). A variety of sizes and shapes of tag are available commercially, made from aluminium, copper, brass, plastic, or stainless steel. These can be purchased pre-numbered or blank. Alternatively, tags can readily be constructed from aluminium drink cans (steel cans tend to rust), by cutting off the top and bottom of the can to form a sheet, which can then be flattened and cut into the size of tag desired. Such tags can be labelled by pressing hard with a ballpoint pen, and can remain readable for many years (Kearns and Inouye 1993). Trees can also be marked with vinyl or polyethylene flagging, which can be wrapped around branches, stems, or trunks and labelled with a permanent marker (Kearns and Inouye 1993). However, such flags become brittle following exposure to sunlight, and may also be blown off in windy conditions, and as a result they tend not to last for more than a few months. In the case of younger trees, tags can be tied on with wire or twine, leaving enough space for the stem to grow. Plastic tags can also be used, on which numbers can be written on waterproof ink (Moore and Chapman 1986). As an alternative to tagging, individual trees can each be accurately mapped and numbered, although this is more labour-intensive than tagging.

A valuable, critical evaluation of the use of long-term PSP data to address a range of ecological questions is provided by the work of Douglas Sheil in Uganda (Sheil 1995, 1996, 1999, 2001a, b, Sheil *et al.* 2000, Sheil and Ducey 2002). Results of this detailed work have indicated the importance of taking into account site history when interpreting PSP data, the importance of large trees in determining forest dynamics and the value of collecting data over prolonged periods. A further useful illustration of the use of PSP data is provided by Oliver Phillips and co-workers, who have analysed a compilation of different data sets to examine changing patterns of turnover in tropical forests over time (Phillips and Gentry 1994, Phillips *et al.* 1994, 1998, 2002a, b). Details of methods for establishing PSPs in tropical forest are given by Alder and Synnott (1992) (see also Condit 1998).

4.7.2 Assessing natural regeneration

Surveys of natural regeneration are often carried out to support the development of forest management plans, and can also provide valuable insights into the ecology of individual tree species. Natural regeneration consists of seedlings, vegetative resprouts, and saplings. Definitions and size characteristics of these components varies between different studies, although generally seedlings are defined as small plants of tree species originating from seed, sprouts are stems that have originated from a dead or cut tree stem or from roots, and saplings are young trees that have not attained a given dbh or height (Husch *et al.* 2003). The origin of sprouts, whether from cut stumps or roots should be noted, but may be difficult to determine without excavation. Typically, adult trees are defined as ≥ 10 cm dbh, and saplings as <10 cm dbh but height >1.3 m, with any plant under this height threshold being referred to as a seedling (or young resprout). However, these definitions are arbitrary and may need to be adapted to the specific characteristics of the forest under investigation (for example in shrublands or thickets where the trees do not reach large girths). Some authors restrict use of the word 'seedling' to plants still bearing cotyledons or to plants <1 m in height (Turner 2001).

Typically, regeneration is assessed by using relatively small fixed-area plots, which are either circular or square in shape. Tree counts in each plot are made by species and type of regeneration, and often assessments of health or condition are also made. Distance-sampling methods can also be used to evaluate forest tree regeneration (see section 3.5.3).

The distribution of seedlings around parent plants is referred to as the *population recruitment curve* (PRC), which is plotted as seedling density against distance (Gibson 2002). Although seeds tend to be distributed away from the parent tree according to a negative exponential curve (see section 6.4.2), the same is not always true for seedlings, an issue that has received a great deal of attention from forest ecologists (see, for example, Augspurger 1983, and the 'Janzen–Connell escape hypothesis'; Janzen 1970, Connell 1971).

If the aim is to assess change in population numbers through time then repeated measurements are necessary, requiring individuals to be tagged (see above) or mapped (Gibson 2002). The frequency of sampling depends on the life-history of the species and the rate at which changes are occurring. Care should be taken when taking repeated measurements to reduce the risk of damaging the plant and affecting future measurements; interference with growing plants should be kept to a minimum. Techniques for recording the locations of plants in small field plots (<1 m^2) include the use of *pantographs, mapping tables, bar plotters,* and photographs (Hutchings 1986). For the larger plots typically used in forest investigations, a useful technique is to attach measuring tapes to adjacent corner posts of the sampling plot and to measure the distances from the corners to each plant in turn. The measurements can be converted to rectangular coordinates within the plot by using the cosine rule (Hutchings 1986) (Figure 4.10). Differential GPS units (see section 3.4.2) can also be used to map the location of trees to within a few centimetres, although this method is more useful in open areas than under a forest canopy, where obtaining a satellite fix can be difficult.

Rulers, tape measures, or calipers can be used to measure stem height, internode length, and leaf length and width. Leaf area can be measured by harvesting the leaves, then using a leaf area meter or planimeter. Care needs to be taken to keep the leaves fresh because if they dry out, leaf curl, roll, and shrinkage can occur, which will affect the accuracy of the measurement. Portable units suitable for field use are also available. Such machines can be accurate to within 1% if carefully maintained and calibrated (Norman and Campbell 1989). Alternatively, leaf area can be estimated non-destructively from measurements of leaf length and width, according to a simple formula (Norman and Campbell 1989):

$$\text{area} = k \,(\text{length} \times \text{maximum width})$$

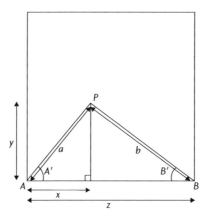

Fig. 4.10 Calculation of the rectangular coordinates, x and y, of plant P from measurements a and b, taken respectively from corners A and B of a study plot. (From Hutchings 1986): $\cos A' = (z^2 + a^2 - b^2/2za)$; $x = a \cos A'$; $y = \sqrt{a^2 - x^2}$.

where k is 0.5 for a triangle, 1 for a rectangle, and around 0.6 for many broadleaved trees. The value of k can be derived from a sample of leaves harvested for the purpose. Methods for measuring leaf area index are presented in section 3.7.3.

4.7.3 Measuring height and stem diameter growth

Height growth of a tree can be obtained by measuring the height at the beginning and end of a time interval, and subtracting the former measurement from the latter. Height measurements of mature trees can be made with hypsometers (section 3.6.3). Repeated measurements with mechanical handheld hypsometers generally do not provide sufficiently precise measurements for accurate increments to be determined. However, more recently developed electronic hypsometers do produce height estimates that are very precise and repeatable (Husch *et al.* 2003). Even these methods may underestimate height, if the top of the tree crown is not readily visible. Tree seedlings and saplings are measured by using measuring poles, tapes, or rulers. Seedlings should be measured at consistent points on the stem, typically from the root collar where the roots join the stem to the shoot tip. If there is a terminal bud at the shoot tip, then measurements will need to be made consistently to either the base or the tip of the bud.

Annual diameter increments of mature trees are generally small, and therefore instruments such as calipers and diameter tapes tend to be used only after intervals of several years when measurements of diameter growth are required. Precise measurement of very small-diameter increments can be obtained by using *dendrometer bands, dial-gauge micrometers, recording dendrographs,* and *transducers* (Husch *et al.* 2003). Dendrometer bands are probably the most widely used of these techniques, and consist of aluminium or zinc bands with vernier scales, enabling diameter measurements to be read directly from the instrument. The band is placed around the tree stem and held in place by a spring (Figure 4.11). Changes in diameter of less than 0.03 mm can be detected by this method (Bower and Blocker 1966). Sheil (2003) describes the use of dendrometer bands for growth assessment in tropical trees, and highlights the fact that stems contract and expand as stem water is depleted and replaced. Such daily changes are generally small (< 0.2 mm diameter) and are ignored in most growth measurements; however, Sheil (2003) notes that larger changes have sometimes been recorded (even >1 cm diameter), suggesting that significant measurement biases are possible. Comparison of dendrometer bands with multiple precision measurements found that the former instrument detected daily changes in stem diameter, but revealed less than a tenth of their magnitude (Sheil 2003).

Dial-gauge micrometers measure the distance of a hook screwed into the xylem to a metal contact glued to the bark; a degree of precision similar to that of a dendrometer band can be achieved. Precision dendrographs consist of a pen on an arm bearing on a fixed point on the tree stem; the pen records diameter changes (again to a precision of 0.03 mm) on a chart mounted on a drum. The method is described by Fritts and Fritts (1955). Growth of mature trees can also be assessed by using increment borers (see section 3.6.1). Diameter increment of seedlings and young saplings can best be measured with appropriate calipers, at a consistent

Fig. 4.11 Dendrometer band, affixed to a stem of the threatened conifer *Fitzroya cuppressoides*. Dendrometer bands can be used to obtain very precise measurements of stem diameter increment in trees. (Photo by Cristian Echeverría.)

point on the stem. Generally, measurements of tree seedlings are made at the root collar at the base of the stem.

4.7.4 Measuring survival and mortality

If information is required on rates of survival and mortality, then individual plants must be permanently tagged, as described above. Mortality can only be assessed by following a cohort of known individuals. Recruitment can only be assessed by re-enumerating the same area at different times (Hall *et al.* 1998). The finite survival rate is defined as the number of individuals alive at the end of the census period, divided by the number of individuals alive at the beginning. Finite mortality rate is defined as 1.0–finite survival rate. Values of both survival and mortality rate always relate to some specific time period (Krebs 1999).

The survival of plants in populations can be analysed by using either *depletion curves* or *survivorship curves*. In each case, the logarithm of the proportion of

186 | **Understanding forest dynamics**

individual plants surviving is plotted against time, on an arithmetic scale (Gibson 2002). Depletion curves are produced by plotting the survival of all the plants present on a given census date through time (Hutchings 1986). Such curves illustrate the survival of plants with a wide range of ages, and are therefore used when the population has an unknown age structure. Depletion curves can also be used to calculate half-life, or the time taken for a population to decline in size by 50% (Figure 4.12). Survivorship curves involve plotting the survival of a particular

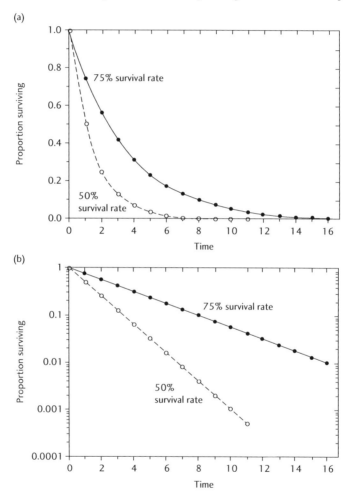

Fig. 4.12 Geometric population decline. Many survival analyses assume that a cohort of plants decreases at a constant survival rate. (a) Geometric declines at 50% and 75% per time period. (b) On a semi-logarithmic plot, in which the proportion surviving is expressed on a logarithmic scale, the same declines are linear. (From *Ecological Methodology*, 2nd ed. by Charles J. Krebs. Copyright © 1999 by Addison-Wesley Educational publishers, Inc. Reprinted by permission of Pearson Education, Inc.)

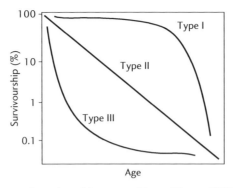

Fig. 4.13 Three types of survivorship curve. (From Gibson 2002.)

cohort of plants, which are uniform in terms of age. These curves may be used to examine age-specific mortality risks (Hutchings 1986), which can be of particular value for identifying which part of the lifespan conservation or management efforts should be focused on (Gibson 2002).

Three fundamental types of survivorship curve have been differentiated (Figure 4.13); much research has been devoted to determining which type best describes the survival of particular species. The types are (Gibson 2002, Hutchings 1986):

- *type I:* mortality increases as the maximum life span is approached.
- *type II:* mortality risk is constant throughout the life of the cohort.
- *type III:* mortality risk is highest for young plants and declines with age.

Most tree species display type III survivorship. Often the relation is defined by fitting a power function equation to the data (Hutchings 1986):

$$Y_t = Y_0 x^{-b}$$

or

$$\log_e Y_t = \log_e Y_0 - b \log_e x$$

where Y_t is the number of survivors at time t, Y_0 is the initial population size, and b represents mortality rate.

When it is possible to age individual trees, survival rates can be estimated directly from the ratio of numbers in each successive age group (Krebs 1999):

$$S_t = \frac{N_{t+1}}{N_t}$$

where S_t is the finite annual survival rate of individuals in age class t, N_{t+1} is the number of individuals in age class $t+1$, and N_t is the number of individuals in age class t. However, this simple approach is only applicable when the survival rate is constant for each age group, all year-classes are recruited at the same abundance, and all ages are sampled equally (Krebs 1999), assumptions that are rarely met in practice.

Sheil *et al.* (1995) critically examine different measures of mortality rate, and highlight some flaws in methods used previously. These authors recommend the following formula for estimating mortality per year, m:

$$N_1 = N_0(1-m)^t$$

which gives

$$m = 1-(N_1/N_0)^{1/t}$$

where N_0 and N_1 are population counts at the beginning and end of the measurement interval t. If counts of stems lost is more convenient to use, then the equation becomes:

$$m = 1-[1-(N_0-N_1)/N_0]^{1/t}$$

Analysing the survival of tree species presents many challenges. Sheil and May (1996) point out that estimated rates of mortality in heterogeneous forests are influenced by the length of the census period, emphasizing the need for care when comparing data collected with different census intervals. Survivorship of seedlings is difficult to measure accurately; the most intense mortality may occur with very small seedlings that are difficult to detect and identify (Turner 2001). Frequent observations are required early in the life cycle. Conversely, survivorship of mature trees may be difficult to measure accurately because of their longevity and the low rates of mortality occurring within any given census interval. Following the survival of cohorts throughout their entire lifespan is impossible for most tree species. Information on age-specific survival of mature trees therefore often has to be collected by using indirect means, such as annual growth rings, from which the age structure can be determined. A description of age structure can be used to estimate the probability of survival from one age class to the next, based on the assumption that age-specific survival rates and recruitment into the population have remained the same from year to year (Watkinson 1997). However, it is likely that these assumptions are rarely met, and therefore considerable care is needed when inferring survival from age structures. This is one of the reasons why plant population biologists frequently characterize tree populations by life cycle stage rather than by age (Watkinson 1997), enabling models to be produced describing their dynamics (see section 5.2.3).

Another issue is that estimation of survival or mortality rates is influenced by the length of time between assessments (census interval). This is because mortality rate estimates are often based on models that assume that a population is homogeneous. Sheil and May (1996) have shown that this can lead to an artefact in estimation of mortality rates, because higher-mortality stems die faster, leaving increasing proportions of the original cohort represented by lower-mortality stems. This has the effect that the lower-mortality stems dominate over time, leading to lower estimates of population mortality rates as the census interval increases. This hinders comparison between results of different investigations, which has led to suggestions that a standard census interval of 5 years be used for permanent plots established to assess trends in forest turnover (Lewis *et al.* 2004). However, as noted by these authors, the frequency of measurement and plot size needed for any field study will clearly depend upon the questions being asked and resources available. The most accurate stand level rates for comparisons with other plots always come from monitoring many trees for many years, and trends over time are probably most accurately elucidated by annual measurements (Lewis *et al.* 2004).

Causes of tree death in natural forests have rarely been investigated in detail. Uprooting or snapping of trees is generally attributed to the effects of wind. However, trees often die standing as a result of natural senescence, attack by fungal pathogens or insect pests, herbivory, drought, or fire, or a combination of these factors (Swaine *et al.* 1987). Attributing tree death to one or more of these causes can often be very difficult, and frequently requires close observation of the individual tree over a period of time. The spatial pattern of tree death merits attention; where groups of standing dead trees are encountered, pathogen attack is often assumed (Swaine *et al.* 1987).

A number of different statistical tests are available for comparing survivorship curves or differences in survival between populations. Examples include the log-rank chi-squared test (a non-parametric test) and the likelihood ratio test (a parametric equivalent). Further details of these and other tests are provided by Hutchings *et al.* (1991), Lee (1992), and Krebs (1999).

4.7.5 Plant growth analysis

A number of different metrics can be used to describe plant growth rates; full details are provided by Chiariello *et al.* (1989), Evans (1972), and Hunt (1978). The ecological importance of these measures is reviewed by Lambers and Poorter (1992), and examples of their application to tree species are provided by Poorter (1999, 2001).

- The *relative growth rate* (RGR) is the rate of relative growth in dry mass, and is expressed as:

$$RGR = \frac{\log_e M_2 - \log_e M_1}{T_2 - T_1}$$

where M_1 and M_2 are the total plant dry mass at the beginning (T_1) and the end (T_2) of the growth period, respectively. The units usually used are g kg^{-1} d^{-1} (or week^{-1}).

- *Leaf area ratio* (LAR) is an index of the leaf area (L_A) per unit dry mass of the whole plant (M), and may be calculated as an instantaneous value, LAR = L_A/M. Over time, LAR is calculated as:

$$LAR = \frac{L_{2A} - L_{1A}}{M_2 - M_1} \times \frac{\log_e M_2 - \log_e M_1}{\log_e L_{2A} - \log_e L_{1A}}$$

LAR is usually presented using the units m^2 kg^{-1}.

- *Specific leaf area* (SLA) is the average leaf area per unit dry mass of leaves (L_M) and is expressed in the units $m^2 \, kg^{-1}$. Instantaneously, SLA = L_A/L_M, and over time

$$\text{SLA} = [(L_{1A}/L_{1M}) + (L_{2A}/L_{2M})]/2$$

- *Leaf mass ratio* (LMR; $g\,g^{-1}$) is the average proportion of the total dry mass allocated to leaves. Instantaneously LMR = L_M/M, and over time it is calculated as:

$$\text{LMR} = [(L_{1M}/M_1) + (L_{2M}/M_2)]/2$$

A similar ratio can be calculated independently for roots and stems, to evaluate changes in the allocation of dry mass between different plant parts during growth.

Many of these plant growth variables require destructive harvests, and therefore they tend to be used most commonly in laboratory, glasshouse, or nursery experiments. However, it is perfectly possible to apply these measures to field experiments if planned appropriately; typically, a sample of seedlings is taken immediately before the application of experimental treatments, and again at the end of the experiment. Sample sizes may need to be larger in field experiments, because variation in plant growth tends to be higher. If non-destructive measurements of growth are required, often *relative growth rate of height* (RGRH) is calculated by substituting measurements of height for measurements of dry mass in the equation for RGR. Indirect estimates of plant biomass can be obtained by using linear dimensions (such as shoot height or leaf length) and relating them to biomass by using regression analysis with data obtained from calibration harvests. However, the height of tree seedlings is often poorly correlated with growth in dry mass, and therefore direct measures of dry mass increment are preferred as a measure of growth.

Fresh mass refers to biomass measured at time of sampling (Gibson 2002). To obtain fresh mass plants should either be weighted immediately after harvesting or stored at < 8 °C in watertight and sealed plastic bags. Storage time should be kept to a minimum to avoid decomposition and growth of fungi on plant tissues. Dry mass can be obtained after drying plant material to a constant mass (at least 24 h) at 80–105 °C in a forced draught oven. Often a two-phase drying process is used, with 60–90 min at 100 °C followed by drying to constant mass at 70 °C, to minimize respiratory losses (Chiariello *et al.* 1989). Samples should be weighed immediately after removal from the oven to avoid uptake of moisture from the atmosphere, or stored with a desiccant such as silica gel.

As an alternative to these 'classical' growth analysis techniques, growth parameters can be obtained by fitting functions to time trends of biomass and leaf area (Chiariello *et al.* 1989). A variety of different functions can be used, including exponential or logistic equations or polynomial expressions of different orders. The functions can be fitted to raw data or logarithmic transformations. Many of the functions that have been used are listed by Hunt and Parsons (1974).

4.7.6 Factors influencing tree growth and survival

In order to understand the population dynamics of tree species, it is necessary to examine the influence of environmental factors such as light availability, competition, herbivory, pathogen attack, weather, soil conditions, and different types of forest disturbance on the processes of tree growth and survival (Watkinson 1997). Although monitoring the performance of individuals growing in different locations can provide some insight into these processes, the amount of information that can be gained is always limited unless some form of experimental approach can be implemented. Experiments should be designed according to the principles mentioned in Chapter 1; in other words, experimental treatments should be applied randomly, and replicated, with adequate controls. Typically, the environmental factor of interest is manipulated as an experimental treatment and its impact on the growth and survival of individual trees is measured.

A large number of experiments investigating tree growth and survival have been conducted, employing a variety of different experimental approaches. Examples of environmental factors investigated experimentally include fungal pathogens (Augspurger and Kelly 1984), water availability (Fisher *et al.* 1991), root competition (Coomes and Grubb 1998), and herbivory (Molofsky and Fisher 1993). The majority of such experiments have investigated the effects of light availability on growth. As noted by Turner (2001), a range of different experimental methods have been adopted, including creation of artificial gaps above naturally occurring seedlings, and artificial establishment of seedlings under canopy gaps of different size, either by planting them in the soil or placing them in pots. Examples of these approaches are provided by Ashton *et al.* (1995), Kobe (1999), Osunkoya *et al.* (1994), Poorter (1999), and Whitmore and Brown (1996).

Field experiments pose a number of challenging problems. No two canopy gaps are exactly the same. Environmental heterogeneity can create differences between replicate treatments, and in addition many of the factors that can influence tree growth are impossible to control (Brown and Jennings 1998). Typically, attempts are made to measure as many of the factors influencing growth and survival as practicable, so that their influence may be taken into account during analysis, by using either covariance analysis or multivariate statistical techniques. However, the results of experiments where many factors are simultaneously explored can be very difficult to interpret.

As a consequence of such difficulties, many experiments have been conducted under more controlled conditions, for example in nurseries, shadehouses, glasshouses, or laboratory growth chambers. Care is needed in such investigations to avoid pseudoreplication (Hurlbert 1984): strictly, individual glasshouses, growth chambers, or shadehouses should be considered as single independent units that require replication. Different groups of plants grown in the same glasshouse or shadehouse should not be considered as independent replicates in any statistical analysis performed. This is a common error even in many published studies. Even in situations where the growth environment is controlled or regulated, variables such as light availability, temperature, and humidity should be regularly measured, so that the growth conditions can be accurately described and the experiment easily repeated by someone else.

Shadehouses are often used to examine the effects of different irradiance treatments on the growth and survival of tree seedlings (Turner 2001). Typically, shadehouses are constructed by covering a wooden frame with nylon or plastic netting; the sides of the frame should be covered as well as the roof (Figure 4.14). Variation in light availability can be achieved by using different thicknesses or mesh sizes of the shade material. Although a reduction in PAR may readily be achieved through the use of such materials, plastic netting generally has little effect on the spectral quality of the light. In natural forests, the ratio of red to far red light (R : FR) generally diminishes below a forest canopy, and this has a major influence on the development of some plant species, influencing growth variables such as leaf thickness and specific leaf area, internode elongation, and plant height. Investigations of the effects of shade should therefore seek to mimic natural regimes as far as possible, by varying the R : FR ratio along with total PAR. This can be achieved by using plastic filters or paints (sourced, for example, from theatrical lighting companies or horticultural suppliers). Alternatively, cut foliage, or even the shade provided by living plants grown in the nursery, can be used to create different light environments that mimic those found in natural forests. Ashton (1995) provides an example of an investigation where both the amount and the spectral quality of light were examined.

Fig. 4.14 A shadehouse constructed for examining the influence of light availability on seedling growth. Nylon netting has been used as shading material, attached to a wooden frame. Note that the sides of the shadehouse are also covered. Access is provided by a removable lid. (Photo by Adrian Newton.)

However, shadehouses often fail to reproduce the temporal pattern of light as experienced under a forest canopy, where brief sunflecks of relatively high intensity may be important for overall carbon balance. Also, use of fluctuating light by seedlings differs from that of uniform irradiance. The relative humidity and temperature (both of the soil and of the air) may be very different in a nursery or glasshouse than in a natural forest (Brown and Jennings 1998). As a result, the relevance of results obtained under controlled conditions to the field situation is always open to question. One potential solution is to do parallel experiments both in the nursery and in the field (see, for example, Newton and Pigott 1991), although this obviously involves substantial additional effort.

Growing seedlings in pots creates its own problems. As potted seedlings are free of root competition, the results may be substantially different from those obtained in the field, and may be difficult to interpret (Burslem *et al.* 1994, Newton and Pigott 1991). Plastic pots or containers are preferred to plastic bags, because of the potential for root spiralling in the latter. The base of the pot should be perforated so that water is able to drain freely. Pots should ideally be sunk in the substrate so that their surface is at ground level; otherwise, there is a tendency for them to reach higher temperatures than the soil, further affecting plant growth and development. Care must be taken to ensure that pots are neither overwatered or left in standing water, nor allowed to drain so freely that the plants are droughted (unless this is one of the experimental treatments being applied). Typically plants in controlled experiments are watered daily to field capacity (the point at which water ceases to drain from the soil surrounding the plant). Another criticism that is often levied at pot experiments is that the plants may not be colonized by the mycorrhizal fungi that are typically found in association with them in the wild. Incorporation of soil collected from around plants of the same species growing in the wild should help ensure that the appropriate mycorrhizal inoculum is present; disturbance to the soil should be minimized to ensure that mycelial systems are intact. The different growth media used in studies of plant mineral nutrition are described by Rorison and Robinson (1986).

Bhadresa (1986) describes the use of exclosures to protect plants from the effects of herbivory. Comparison of plants inside and outside fenced exclosures has long been used by plant ecologists as a technique to study the effects of herbivory on plant performance (see section 4.2.3). Such exclosures should be sited in areas that are homogeneous, in terms of vegetation, soil characteristics and topography, so that the differences detected can be attributed to the effects of herbivory. Fenced exclosures may also be required to protect field experiments from animals.

My own introduction to postgraduate research in forest ecology provides a powerful example of what can go wrong. The 3000 oak tree seedlings that I had carefully established in an experimental array to investigate mineral nutrition, situated within a fenced enclosure to prevent herbivory by deer, were all eaten by squirrels within a week. Such disasters are by no means rare in field-based research, and may be impossible to prevent—as in the case of an experiment established by a colleague of mine, which was de-replicated by a herd of elephants. As a result of this kind of experience, field researchers quickly learn the importance of contingency planning, and of maintaining a flexible approach to their research, so that setbacks can even be converted into opportunities. Long-term field experiments destroyed by hurricanes, for example, can be reinvented as investigations of storm damage. My personal oak tree disaster was eventually published as an investigation of squirrel predation—the observation of mass mortality turned out to be novel (Pigott *et al.* 1991).

A further problem with most experimental investigations of tree species is that they only examine responses in seedlings; mature trees have received very little attention from researchers by comparison. The relevance of seedling experiments to the ecological behaviour of mature trees is open to considerable doubt (Turner 2001). As noted by Clark and Clark (1992), studies are required on all size classes of trees, not just seedlings. Our current knowledge of the growth responses of mature trees, particularly with respect to variables such as stem density and soil characteristics, is primarily based on the forestry literature. Although experiments with mature trees present considerable logistical challenges, foresters have long adopted an experimental approach to silviculture, and a considerable body of valuable information has been collected for many economically important timber trees. Much of this information lies in the archives of national forest services, and can be difficult to locate. To cite just one example: a comprehensive review of the silviculture of mahogany (*Swietenia macrophylla*) identified a series of large-scale experimental trials investigating seedling regeneration and growth of mature trees, conducted in Belize in the 1920s and 1930s (Mayhew and Newton 1998). These trials produced useful information for understanding the ecology of the species, and have arguably never been bettered.

In some situations it is difficult to manipulate the environmental factor of interest, or to adequately control all of the many variables that can influence plant growth and survival. For example, the effects of major disturbance such as a hurricane or forest fire are difficult to simulate experimentally. A common solution is to use what has been termed the *comparative method* or a *natural experiment*; in other words, to compare situations occurring naturally that differ with respect to the variable of interest. An example is provided by the study of forest succession in Puerto Rico resulting from disturbance caused by a plane crash (Weaver 2000). As the forest in this area is currently protected, there was no possibility of creating experimental tree-fall gaps, and therefore the researcher made use of the only major disturbance event that was available (although suffering from the obvious problem of a lack of replication!).

The relative strengths and weaknesses of 'natural experiments' compared with randomized, controlled experiments are discussed in detail by Diamond (1986). The main limitation of the 'natural experiment' approach is that it is impossible to attribute any differences observed solely to the factor of interest; as a result, this method cannot be used to adequately test specific hypotheses (Underwood 1997). As pointed out by Diamond (1986), however, there may sometimes be no alternative. For this reason, purely observational studies that examine the distribution of individuals of a species in relation to patterns of environmental variation (often by means of multivariate statistics) are often carried out by forest ecologists (Clark and Clark 1992, Hubbell and Foster 1986, Newbery *et al*. 1988).

4.8 Seed bank studies

The soil seed bank refers to the reserve of persistent seeds in the soil and is usually assessed as the number of seed in a given volume of soil or for a given ground area (Gibson 2002). A review of soil seed banks is provided by Leck *et al*. (1989). Investigations of the seed bank occurring within forest soils are often undertaken in the context of understanding forest dynamics, for example when analysing the potential impact of disturbance on forest composition (see for example Alvarez-Aquino *et al*. 2005). The density of viable seeds in the seed bank often decreases with increasing age since disturbance (Hutchings 1986). Further details of the methods used for seed bank studies are provided by Hutchings (1986), Roberts (1981), and Thompson *et al*. (1997). Techniques for assessing seed dispersal and seed rain are considered in section 6.4.

The bank of seeds lying on the soil surface can be sampled by using vacuum pumps, although this method has the disadvantage of also collecting unwanted material (such as leaf litter) that subsequently has to be separated from the seeds (Hutchings 1986). Most investigations focus on the buried seed bank, which is usually sampled with a *soil corer* (or auger). Corers consist of a hollow cylinder that is twisted into the ground, enabling a cylindrical core of soil to be removed. Simple corers can be readily constructed from a length of metal pipe, the length and diameter of which determine the volume of soil sampled. A pair of holes drilled at one end of the pipe enable a handle to be fitted, which is helpful for twisting and extracting the corer (Figure 4.15). The rim at the other end should be sharpened, to assist penetration. Typically, the size of such corers lies in the range 2–20 cm in diameter and 5–20 cm in length (Bullock 1996). The size of the corer, and the volumes of soil extracted, should be reported with the results.

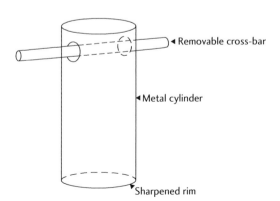

Fig. 4.15 A soil corer. (After Bullock 1996.)

Seed banks are spatially heterogeneous, both vertically and horizontally within the soil profile, and this should be taken into account when designing the sampling regime. It may be useful to separate the surface layers of the soil (for example 0–5 cm), where most seeds are generally found, from the deeper layers. Often, layers between 0–2 cm, 2–5 cm, and >5 cm are separated (Bullock 1996). The ratio of seed in the surface and deeper layers provides information about the longevity of seeds (Thompson et al. 1997). Roberts (1981) recommends that a large number of small samples should be taken, rather than the converse. Often, five or ten relatively small cores are pooled to form a single sample (Thompson et al. 1997). A preliminary sampling study can be of great help in determining the number of soil samples necessary to provide stable mean values of the variables being measured, with standard errors that are acceptably small (Hutchings 1986). It is important to remember that the larger the number of replicate samples taken on a particular site, the lower the variance is likely to be, increasing the chances of detecting differences between sites by statistical analysis. A minimum number of 50 samples should be taken in order to fully characterize the number of species present in a seed bank, although fewer samples may be sufficient to detect the presence of most species (Thompson 1993). The time of year at which the samples are taken also influences the composition of the seed bank recorded. Ideally, samples should be taken after the germination season is over and before new seeds are dispersed, if the objective is to estimate the persistent seed bank (Baskin and Baskin 1998). Alternatively, repeated samples may be taken at different times of year.

The total volume of soil that should be sampled varies with the objectives of the study and the characteristics of the community being sampled. Hutchings (1986) indicates that if the objective is to determine the species composition of the seed bank, the volume of soil needed to reveal the majority of species present increases as seed density declines. Suggested values are 0.8 l of soil for early successional vegetation, and 8–12 l for relatively undisturbed woodland. However, as noted by Thompson et al. (1997), these values would be inadequate for estimation of seed density, because of the high spatial heterogeneity of seed distribution. The volume of soil sampled has a major influence on the accuracy of values obtained, although precision is determined by the number of independent, replicate samples (Thompson et al. 1997). Accurate and precise estimates of the soil seed bank require a great deal of sampling effort; in the case of rare species, the number of samples required for such estimates may often be unfeasibly large (Thompson et al. 1997).

An estimate of the total number of all seeds in the soil (the total seed bank) can be obtained by extracting the seeds through sieving and flotation, then counting them. Soil samples can be washed on a coarse sieve to remove roots, pebbles, and stones, then on a fine sieve (i.e. 2 mm) to remove clay and silt. This is most easily done under running water (Bullock 1996). Seeds may be floated in water or in denser solutions, which allow more seeds to float. For example, Price and Reichman (1987) used a saturated solution of potassium carbonate for this purpose. Seeds may be divided into those that are dead or empty, versus those that are alive but dormant. Dormant seeds may be further divided into those that germinate under laboratory conditions, and those that do not. Whether or not seeds are alive can be determined by using the tetrazolium test (see below).

The germinable (or viable) seed bank, or the seed that can germinate if conditions are favourable, is generally assessed by thinly spreading a soil sample on a shallow tray kept in a laboratory or glasshouse environment, or sheltered conditions outdoors. The sample should be spread as thinly as possible (<5 mm), ideally on a sterilized medium (Thompson et al. 1997). The number of seedlings that subsequently emerge is then counted. The tray should be kept free of other dispersing seeds. Trays of sterilized growth medium should be included among the germination trials to act as controls, by indicating the presence of any contaminants. If any species appear in these controls, then their appearance in samples from forest soils should be viewed with caution (Gibson 2002). There may be a need to include a cold stratification period, if some species require this treatment in order to break seed dormancy. The trays should be kept moist through regular watering. Environmental conditions, particularly temperature and humidity, should be recorded regularly and reported. Once the rate of seedling appearance declines, the soil can be disturbed or stirred to stimulate germination of any buried seed that remains within the samples. Thompson (1993) recommend continuing to monitor seedling emergence for a period not longer than 6 months, and stirring the soil only once, although Baskin and Baskin (1998) recommend that a longer period be adopted.

The trays should be assessed regularly, and seedlings identified and removed as soon as possible thereafter. Identification can often prove difficult, particularly if the flora of the area is inadequately known. If some seedlings are difficult to identify, they can be preserved as voucher specimens, or transferred to individual pots to be grown on until they reach a large enough size to be identified more readily.

Problems encountered with the approaches described here include (Hutchings 1986):

- Viable seeds might die without germinating, and thus be omitted from the estimation of the viable seed bank.
- Some seedlings might germinate and die between survey dates.
- Some seeds may have very prolonged dormancy that is very difficult to break.

For these reasons, seed banks are always likely to be underestimated, and this should be borne in mind when interpreting results.

Germination test

In order to define the germination requirements of a particular species, some form of germination test is required. An example of an experimental protocol is described by Bradbeer (1988), and involves sowing air-dried samples of 100 seeds on moist filter paper placed in Petri dishes, which are then allowed to germinate for 28 days. Germination is indicated by the presence of a radicle, assessed every

24 hours, and results are expressed as the cumulative percentage of seed germinating. The following experimental treatments are suggested by Bradbeer (1988):

- Compare the effects of pretreatments such as storing air-dry seed in an unsealed container at laboratory temperature for three months, or chill at 5 °C in moist sand for 3 months.
- Examine germination under different combinations of illumination and temperature, such as 5, 10, 15, or 20 °C with or without illumination.
- Apply different physical or chemical treatments, such as removal of the fruit, scarification (rupturing the seed coat with a scalpel or concentrated sulfuric acid), leaching with water, application of gases such as carbon dioxide or ethylene, or chemicals such as 10^{-4} mol/l gallium arsenide, 10^{-4} mol/l kinetin, 0.1 mol/l thiourea, or 0.2% potassium nitrate.
- Such treatments should be applied in fully replicated, randomized designs. Results can be analysed by using chi-squared tests or ANOVA (if the data are arcsin transformed first) (Gibson 2002).

The effects of temperature on seed germination can be analysed in detail by using thermo-gradient bars. These provide gradients of temperature, along which seeds can be placed. Details of how to do such experiments are provided by Baskin and Baskin (1998), and further details of the method are provided by Larson (1971) and Thompson and Fox (1971).

Seed viability can also be tested by excising the embryo from fully imbibed seed, then placing it in a 0.1% aqueous solution of 2,3,5-triphenyl-tetrazolium chloride, pH 7 (Grabe 1970). If uncut seeds are tested, a 1.0% solution is recommended. After being placed in the dark for 24 h, living cells develop a red or purple colour, indicating that the seed was alive and germinable. Care is needed when interpreting the results of this test, because the presence of microorganisms can sometimes give a false-positive result (Gibson 2002). Also, the tissues of seeds of different species display different staining patterns, and in some cases, even non-viable seeds may exhibit tetrazolium staining (Hutchings 1986). Care should therefore be taken in interpreting the results of the test.

Alternatively, a 0.05% indigocarmine solution can be prepared in hot distilled water, then filtered and allowed to cool. Seed tissue is submerged in the stain for 2 h in the dark, then the stain is washed off with distilled water. A blue colour indicates dead tissue, whereas living tissue is colourless. As with tetrazolium, this stain is mildly poisonous, so should be handled with care; and the results may not always be completely reliable (Bullock 1996).

4.9 Defining functional groups of species

The classification of species into *functional groups* (or types) has received much attention from ecological researchers, partly as a way of simplifying the high taxonomic diversity encountered in many communities, and partly as a way of describing the ecological roles or 'functions' of different species. The concept is particularly important for modelling approaches that focus on understanding the behaviour of ecological systems, and many models of forest dynamics consider functional groups of species rather than all of the species present (see Chapter 5; Vanclay et al. 1997 provide an example).

There are three main methods of identifying functional groups (Gitay and Noble 1997):

- *Subjective approaches*, based simply on observations of ecosystems within which it is assumed that functional types exist. For example, a forest ecosystem might be divided subjectively into trees, shrubs and lianas.
- *Deductive approaches*, in which a functional classification is derived from an *a priori* statement or model of the importance of particular processes or properties in the functioning of an ecosystem. The concept of *keystone* species (Paine 1980) is an example of such an approach.
- *Data-defined approaches* use multivariate statistical techniques to identify clusters of species based on a set of characters. In most cases the analyses are based on morphological or growth characteristics. There are many examples of this approach in the scientific literature, with a wide variety of organisms. Leishman and Westoby (1992) provide an example for woodland plants in Australia.

The basis of any classification of functional groups depends on its purpose: in other words, different plant attributes are relevant depending on the purpose of the classification (Westoby and Leishman 1997). Functional classifications could potentially be used to address questions such as (Westoby and Leishman 1997, Bond 1997):

- Which species are most capable of dispersing in long jumps?
- Which species are most capable of establishing as seedlings?
- Which species are most capable of growth and competitive persistence as established plants?
- How are the diversity and distribution of vegetation influenced by climate change?

Characteristics of tree species that have been used for classification of functional types include life span, pollination, seed dispersal, tolerance of fire, tolerance of drought, tolerance of low nutrient availability, tolerance of shade, and dependence on a canopy gap for regeneration (Shugart 1997). For example, Condit et al. (1996) propose and test a classification of plant functional types for tropical trees based on demography, growth form, phenology, and moisture requirements, using data from a 50 ha forest dynamics plot in Panama. The classification presented by Noble and Slatyer (1980) has been widely used for predicting successional changes in vegetation, and is based on consideration of a set of three (sometimes four) 'vital' attributes that are critical in determining the continuing survival of a species on a site subjected to disturbance. The attributes are:

- *Method of persistence* on a site after disturbance, by dispersal from exterior populations, by persistence of seeds in a seed bank, by persistence of seeds

with protective measures in the canopy, or by vegetative regrowth following survival by some part of the individual.
- *Conditions for establishment*, based of division of species into *tolerant* species, which may regenerate at any time irrespective of whether other species already occupy the site; *intolerant species*, which are only able to regenerate after disturbance when competition is low; and species that require the presence of mature individuals of their own species or some other species to regenerate.
- *Life history*, based on the timing of critical events following disturbance, namely (1) the point at which propagules are plentiful enough to allow regeneration following disturbance, (2) the point at which individuals reach reproductive age, (3) the point at which the species is lost from the stand as reproducing individuals, (4) local extinction of the species when no viable propagules remain. If some measure of the relative abundance of species in the stand is required a fifth 'attribute' may be added, consisting of maximum size, growth rate, and mortality.

As an illustration of the application of this method to a conservation problem, Bradstock and Kenny (2003) used the 'vital attributes' approach to estimate optimal fire regimes for biodiversity planning in a national park. This was used to define groups of species that are likely to undergo a significant decline or extinction in response to particular fire regimes, by grouping species according to their juvenile periods, life span and the seed bank longevity.

Although characterization of functional groups has attracted increasing attention as a result of growing interest in the relationship between biodiversity and ecosystem function, it is not always obvious how such groups should be delineated or how species should be assigned to them (Magurran 2004). Some recently developed approaches include the following:

- Petchey and Gaston (2002a, b) have recently proposed a new method for quantifying functional diversity based on analysing a dendrogram constructed from species trait values. A trait matrix, consisting of a list of species and their traits, is assembled and then converted into a distance matrix. Only those traits linked to the ecosystem process of interest are used. Clustering methods are used to produce a dendrogram, then branch length of the dendrogram is used to define functional diversity.
- Gillison (2002) describes an alternative method for classifying plant functional types, based on assessment of a standard 'rule set' that can be scored in the field, and provides freely downloadable software (VegClass) to assist in the process.
- Pillar and Sosinski (2003) describe a new algorithm to numerically search for traits and identify optimal plant functional types. The algorithm uses three data matrices: describing populations by traits, communities by these populations and community sites by environmental factors or effects. Functional types are identified by cluster analysis, revealing types whose performance in communities is maximally associated to the specified

environmental variables. A free software program (SYNCSA Minor) can be downloaded from ⟨www.opuluspress.se/pub/archives/index.htm⟩ or ⟨http://ecoqua.ecologia.ufrgs.br⟩.
- Software tools are also freely available to assist in classifying species according to the widely used C–S–R functional classification (Hodgson *et al.* 1999), from ⟨www.people.ex.ac.uk/rh203/allocating_csr.html⟩.
- A practical handbook of methods for defining plant functional groups is provided by Cornelissen *et al.* (2003).

In the case of tropical trees, a number of different functional classifications have been proposed. For example, Swaine and Whitmore (1988) defined 'pioneer' and 'non-pioneer' groups of trees on the basis of seed germination requirements, the former species being found only in canopy gaps because their seed do not germinate elsewhere. Turner (2001) identified maximum height and regeneration class as factors that differentiate tropical tree species, the latter referring to the light requirements for regeneration, shade-intolerant ('pioneer') species being separated from species that are able to establish under canopy shade ('non-pioneers') (Swaine and Whitmore 1988). Baker *et al.* (2003) proposed that functional groups for tropical trees should be defined on the basis of species' associations with particular edaphic and climatic conditions, as well as regeneration requirements and maximum size. However, statistically robust comparative studies including large numbers of species are few in number. Sheil *et al.* (2006) examined the relations between relative crown exposure, ontogeny and phylogeny for 109 canopy species by using a generalized linear model (GLM) and found the interesting result that species achieving large mature sizes are generally shade-intolerant when small, suggesting a trade-off between these two variables.

The concept of shade tolerance is familiar to foresters as well as ecologists and can readily be inferred from observations of the survival and growth of seedlings or young trees under a forest canopy. For example, Augspurger (1984) used the slope of the logarithmic decay for seedling populations grown in deep shade as an index of shade tolerance. Alternatively, shade tolerance can be determined by comparing the growth of seedlings under different light availabilities in shadehouses or growth cabinets (see section 4.7.6) (see, for example, Agyeman *et al.* 1999, Veenendaal *et al.* 1996). However, it is important to consider that the degree of shade tolerance may vary with the age of the tree (Sheil *et al.* 2006). Hawthorne (1995, 1996) classified species according to their shade tolerance by simply observing crown exposure patterns of young trees (stems <5 cm dbh) and larger trees (>20 cm dbh), shade-intolerant species being defined as those that were consistently well exposed to light, shade-tolerant species being those consistently found primarily in shade. Similarly, Sheil *et al.* (2006) used crown exposure records (see section 4.5.4) collected during a forest inventory to classify species according to their apparent shade tolerance.

Kobe *et al.* (1995) characterized the juvenile survivorship of 10 dominant tree species in northern hardwood forests in the USA, and by using species-specific mathematical models were able to predict the probability of a sapling dying as a function of its recent growth history. Combined growth and mortality models were found to characterize a species' shade tolerance, by expressing a sapling's probability of mortality as a function of light availability. This was achieved by determining the numbers of live and recently dead saplings at each of a series of study sites, and randomly sampling a subset of them. Stem cross-sections were then removed at 10 cm height for every selected sapling, and the widths of at least the 10 most recent annual rings were measured. Mean radial growth over the five most recent years was used to predict mortality. The relations between growth and mortality were established by using maximum likelihood methods. Relationships with light availability were examined by using data from hemispherical photographs. This investigation provides a rare example of a quantitative analysis of shade tolerance under field conditions, and enabled shade tolerance to be specified along a continuum rather than as a few discrete categories. Species with relatively shade-tolerant saplings were able to better withstand periods of suppressed growth. An important finding from this research was that shade tolerance involves a trade-off between high-light growth and low-light survivorship. Therefore, both of these variables should be assessed when characterizing shade tolerance of tree species.

5
Modelling forest dynamics

5.1 Introduction

One of the biggest challenges to forest ecologists working on forest dynamics is the long timescale involved. Many significant disturbance events are very infrequent, but can influence ecological patterns for decades or centuries afterwards. Many individual trees are very long-lived, making observations of the entire life cycle impossible to achieve within a single human lifetime. For this reason, models are a very important tool for forest ecologists, providing a tool for predicting how forest structure and composition might be affected by disturbance events, potentially over very long timescales.

There are many different approaches to ecological modelling, so it is first helpful to consider what is meant by the term. Ford (2000) describes models as analogies representing important features of a system, with the principal aim being to explore ideas rather than to provide an exact description of how the system operates. Models provide a very useful tool for making predictions based on relevant theory, which can be tested through experimentation and observation. In this way, they can also provide a valuable framework for fieldwork, by focusing attention on those variables that need to be assessed.

If the objective is to understand the population dynamics of a single species, then *matrix modelling* approaches are often the method of choice. However, in situations where the management or conservation of entire forest communities is of interest, other modelling approaches are required that enable the interactions between species to be examined. The development of such models has been a central activity in forest ecology research over the past three decades, although application of these models to practical management situations has been fairly limited to date. At the same time, forestry researchers have devoted substantial efforts do developing models that enable forest growth and yield to be predicted, with the primary aim of exploring the impacts of different forest management options on timber production. Each of these approaches to forest modelling is considered in this chapter.

Recent reviews of forest modelling techniques are provided by Bugmann (2001), Johnsen *et al.* (2001), Landsberg (2003), Liu and Ashton (1995), Makela *et al.* (2000), Porte and Bartelink (2002), and Shugart and Smith (1996). Useful source texts include Amaro *et al.* (2003), Botkin (1993), Canham *et al.* (2003), Shugart (1984, 1998), Vanclay (1994), and West *et al.* (1981). General reference works about ecological modelling include Ford (2000) and Hilborn and Mangel

(1997), and useful introductions to ecological modelling techniques are provided by Jackson *et al.* (2000), Peck (2000), and Starfield *et al.* (1990).

5.2 Modelling population dynamics

Once data have been collected describing rates of growth and survival at different stages of the life cycle, as described in the previous chapter, they may be analysed in an integrated way that enables the population dynamics of individual species to be explored. Such analyses can be used to determine whether populations are stable or declining, an issue of fundamental importance to conservation. These methods can also be used to identify those life-history stages that are most limiting to population growth, which can help focus conservation management efforts and assist in the identification of extinction threats, as well as the understanding of invasive species (Gibson 2002). Sources of further information on techniques used for analysing and modelling plant population dynamics include Gibson (2002), Hutchings (1986), McCallum (2000), and Watkinson (1997). *Population viability analysis*, which builds on the methods described here, is presented in the following section.

5.2.1 The equation of population flux

The most basic equation describing population dynamics is the equation of population flux (Hutchings 1986):

$$N_t = N_{t-1} + B - D + I - E$$

where N_t is the number of individuals at a particular time, N_{t-1} is the number of individuals at some previous time, B is the number of 'births' or recruitment events, D is the number of deaths that have occurred, I is the number if immigrations to the population and E is the number of individuals that have emigrated from a population in the time interval being considered. Change in population size over this time interval is given as:

$$\Delta N = B - D + I - E$$

A negative value of ΔN indicates that the population is declining, whereas a positive value indicates population increase.

Another useful concept is λ (lambda), which is the net multiplication rate of a population. This is derived from the relation:

$$\frac{N_{t+1}}{N_t} = \lambda$$

In other words, λ provides a measure of population change; when $(B + I) > (D + E)$ then the population multiplies by λ each year in an exponential manner under unchanging conditions (Gibson 2002):

$$N_{t+x} = N_t e^{rx}$$

where x indicates some time interval (in years) after the initial time t, e is the base of natural logarithms, and r is the intrinsic rate of natural increase of the population. The latter is related to λ by the equation

$$\lambda = e^r$$

which can also be written as

$$r = \ln(\lambda)$$

In discrete time the value of r can be found from the Euler equation, as follows (Crawley 1997):

$$1 = \sum_{x=0} e^{-rx} l_x m_x$$

where l_x is survivorship and m_x is fecundity. The value of r can be found numerically by using a computer (i.e. an initial value is adjusted until the right-hand side is equal to 1; Crawley 1997). These equations can be used to develop a simple model of population growth from demographic data collected by using the techniques described in the previous chapter. However, such a model would be highly simplistic, failing to consider density-dependence, life-history components such as the presence of a seed bank, or changing survivorship or mortality over time. More elaborate models that consider such issues are presented by Watkinson (1997).

5.2.2 Life tables

A life table can be used to summarize information on mortality risks and reproduction associated with different categories of plant within a population. Categories can be defined in terms of age, size, or state. Although widely produced for animals, life tables are rarely produced for plants; most of the examples that are available refer to annual plants (Hutchings 1986). Life tables can be presented in the form of either tables or diagrams; details are provided by Begon and Mortimer (1981) and Ebert (1999). The standard form is the *cohort life table* in which a group of organisms born at the same time are followed throughout their life cycle. This is very difficult to achieve for long-lived plants such as trees. An alternative form derives information from the population at a particular time, and then determines the survival over one time period for each age or stage class in the population (McCallum 2000). However, such static life tables do not allow for age-specific changes in birth and death rates (Gibson 2002).

5.2.3 Transition matrix models

Matrix models provide a relatively simple means of modelling the population dynamics of individual species. The method has been used with a range of tree species, and has proved useful for exploring the impacts of different disturbance intensities, management interventions, or harvesting regimes on population size of selected species (and therefore for evaluating the sustainability of different harvesting approaches), and for examining the impact of potential threats on extinction risk (Fieberg and Ellner 2001). Matrix models have therefore become a standard method for researching the conservation biology and management of threatened plants (Silvertown *et al.* 1996). The method has been particularly widely used with palms (Freckleton *et al.* 2003, Olmsted and Alvarez-Buylla 1995, Pinard 1993); examples of other tree species investigated with this approach include *Araucaria araucana* (Bekessy *et al.* 2004), *Banksia* spp. (Drechsler *et al.* 1999), *Bertholletia excelsa* (Peres *et al.* 2003), *Fagus grandifolia* (Batista *et al.* 1998), *Nothofagus fusca* (Enright and Ogden 1979), and *Vochysia ferruginea* (Boucher and Mallona 1997).

In this approach, populations are divided into discrete classes based on age or life-history stage, and transitions from one stage to another are used to model survival, growth, and reproduction, enabling population dynamics to be forecast (Caswell 1989). Analysis is based on the use of matrix algebra. Matrix models can be used to assess the rate of population increase (λ), and through elasticity analysis the effects of different matrix elements on this parameter can also be evaluated (Caswell 1989). This approach has proved to be a particularly useful tool for examining demographic processes in plants, facilitating comparison between species and the development of generalizations regarding their population dynamics (Silvertown and Lovett Doust 1993, Silvertown *et al.* 1996). The following description of the method is based on that presented by Caswell (1989, 2001), Ebert (1999), and Gibson (2002).

The approach is based on multiplication of a matrix of demographic parameters with a column vector representing the age or stage structure of a population at a particular time. The column vector can be represented as follows:

$$n_t = \begin{bmatrix} n_1 \\ n_2 \\ n_3 \\ n_4 \\ \vdots \\ n_i \end{bmatrix}$$

where n_t is the age or stage structure of the population at time t, and $n_1, n_2, \ldots, n_i$ are the numbers of individuals in each age/stage class. The matrix of demographic parameters (referred to as the projection matrix) represents the probabilities (or transitions) of moving from one age/stage to another. There are two main types of projection matrix: a *Leslie matrix*, which is used for models structured by age, and a *Lefkovitch matrix*, which is used for models structure by life-cycle stage or plant size.

The *Leslie projection matrix* (A) takes the following form:

$$A = \begin{bmatrix} f_1 & f_2 & f_3 & f_{n-1} & f_n \\ p_1 & 0 & 0 & 0 & 0 \\ 0 & p_2 & 0 & 0 & 0 \\ 0 & 0 & p_3 & 0 & 0 \\ 0 & 0 & 0 & p_{n-1} & 0 \end{bmatrix}$$

where p_i is the probability that an individual will survive (values in the range 0 to 1), and f_i is the rate of reproduction (or fecundity) for an individual in age class i (values of any magnitude). This type of matrix only has non-zero elements in the top row (indicating the number of offspring produced), and in the subdiagonal (values representing annual survival; p_i). The other values are zero because plants cannot remain in the same age from one time step to the next.

The *Lefkovitch projection matrix* (*A*) takes the following form:

$$A = \begin{bmatrix} s_1 & f_2 & f_3 & \cdots & f_n \\ g_1 & s_2 & 0 & \cdots & 0 \\ 0 & g_2 & s_3 & \cdots & 0 \\ \vdots & \vdots & \vdots & \vdots & \vdots \\ 0 & 0 & \cdots & g_{n-1} & s_n \end{bmatrix}$$

where s_i is the probability that an individual will survive and remain in the same stage class (values in the range 0 to 1), g_i is the probability that an individual will survive and transfer to the next stage class (values in the range 0 to 1), and f_i is the rate of reproduction (or fecundity) for an individual in stage class i (values of any magnitude). In this type of matrix, elements in the leading diagonal (s_i) cannot be zeros, because there is always a possibility that a plant will remain in the same stage of the life cycle the following time step. The elements below the leading diagonal refer to the probability of growth to the next stage class, and elements above the leading diagonal refer to fecundity, with information presented relating to each size class.

The exact form of a Lefkovitch matrix differs between species, depending on the number of life-cycle stages that are defined. Multiple stages can be defined for life-cycle stages such as vegetative reproduction, the seed bank, and either juvenile or mature stages with different reproductive potential. It may be possible for plants to regress from one stage to another (for example if a tree is felled and subsequently resprouts vegetatively). An example of a Lefkovitch matrix for a threatened tree species is presented in Table 5.1.

To provide an estimate of population in the next time step, the projection matrix is multiplied by the column vector. Repeated iterations can be carried out to determine the age/stage structure in subsequent time steps. The same projection matrix (*A*) is generally used throughout such calculations, based on the assumption that demographic rates do not change over time. Multiplication of matrices can readily be done using spreadsheet software programs, such as the MMULT worksheet function in Microsoft Excel. Some mathematical or modelling software packages (such as Analytica; Lumina Decision Systems, ⟨www.lumina.com⟩) can also be used to do matrix calculations. Alternatively, dedicated software for modelling population viability can be used (see section 5.3).

If repeated iterations are performed, a point will be reached where the values in the vector cease to change with further multiplications by the projection matrix. This vector is referred to as the right eigenvector and represents a stable age/size structure. At this point, λ can be calculated as the *dominant eigenvalue* of *A*, which

Table 5.1 An example of (A) a life table and (B) a transition matrix for a population of the tropical tree *Aquilaria malaccensis* at a study site in West Kalimantan (from Soehartono and Newton 2001). All rates are expressed on an annual basis; growth rate units are cm year^{-1} for both height and diameter classes. Seedlings, saplings, and juvenile classes are classified on the basis of height. Poles and adult classes are based on diameter at breast height (dbh).

(A)

	Dimensions	Stage	N	Survival	Growth	Moving	Remaining	Fecundity
Seed		S0	4850	0.72	0	1.00	0.00	0
Seedling	<0.5 m height	S1	79	0.15	0.51	0.49	0.51	0
Sapling 1	0.5–<1 m height	S2	81	0.41	0.35	0.36	0.64	0
Sapling 2	1–<2 m height	S3	48	0.62	0.25	0.25	0.75	0
Juvenile	2–3 m height or <5 cm dbh	S4	7	0.66	0.11	0.12	0.88	0
Poles	5–10 cm dbh	S5	4	0.60	0.16	0.21	0.79	3544
Adult1	>10–<30 cm dbh	S6	6	0.70	0.03	0.03	0.97	9882
Adult2	30–50 cm dbh	S7	10	0.40	0.02	0.04	0.96	8146
Adult3	>50 cm dbh	S8	4	0.30	0.005	0	1	2622

(B)

Stages	S0	S1	S2	S3	S4	S5	S6	S7	S8
S0	0	0	0	0	0	3544	9882	8146	2662
S1	0.72	0.07	0	0	0	0	0	0	0
S2	0	0.07	0.26	0	0	0	0	0	0
S3	0	0	0.15	0.47	0	0	0	0	0
S4	0	0	0	0.16	0.51	0	0	0	0
S5	0	0	0	0	0.08	0.47	0	0	0
S6	0	0	0	0	0	0.12	0.67	0	0
S7	0	0	0	0	0	0	0.02	0.38	0
S8	0	0	0	0	0	0	0	0.01	0.3

is equivalent to the asymptotic population growth rate (Caswell 1989). This parameter is routinely calculated in studies employing transition matrix modelling techniques, and is often used to evaluate whether harvesting approaches are sustainable or not (indicated by whether λ is greater or less than 1, respectively). Other parameters often reported are elasticities, which indicate the sensitivity of λ to small changes in matrix elements. These values can be used to assess the importance of different components of the life cycle to λ, and also to fitness (De Kroon *et al.* 1986). The use of sensitivity analysis to explore the effects of different management strategies on population size is one of the most important practical applications of matrix modelling techniques to conservation. However, it should be noted that values of λ may sometimes be misleading as a measure of a population's short-term prospects for survival (Bierzychudek 1999).

One of the key decisions to be faced in any matrix modelling exercise is the definition of the age or stage categories to be included in the model. Categories may be defined on the basis of biological criteria, such as size, age, gender, reproductive state, development, or some combination of these. Alternatively, analytical methods are available for defining categories, which maximize within-class sample sizes and minimize error of estimates (Moloney 1986, Vandermeer 1978). However, these analytical methods also have their shortcomings (Caswell 2001) and for this reason many investigators choose categories that subdivide their plants into a number of well-represented groups, i.e. small, medium, and large reproductive plants (Gibson 2002). Many studies define categories that seem biologically reasonable based on regressions of survivorship and fecundity against size (Brigham and Thomson 2003). For a fixed number of trees for which data are collected, the sample size within each category decreases as the number of categories increases. The definition of categories therefore influences model output, and care should be taken to ensure that the sample size within each category is sufficiently large to ensure that estimates of transition probabilities are robust. However, if the categories are too large, the assumption that individuals falling within the same category have the same transition probabilities may be violated (Zuidema 2000).

Complete parameterization of a matrix model requires substantial effort. Ideally, data should be collected for a group of tagged individuals over a period of years. Data need to be collected on growth, mortality and fecundity for each of the age/stage categories defined. These data can be collected by the methods described elsewhere in this chapter (methods for assessing fecundity are described in section 6.4.1). The lack of sufficient data (either in terms of number of individuals included in the survey, or too few years of data collected) is one of the main shortcomings of investigations employing matrix modelling techniques. The fact that so many tree species are very long-lived presents a particular challenge, as accurate estimates of the very low rates of mortality typical among mature trees can be very difficult to obtain within the timescales generally available to researchers. As a result, mature individuals of long-lived tree species may essentially behave as if they are immortal (Drechsler *et al.* 1999). Often, in practice, transition values are guessed or a value of 1 is used, which may be unrealistic.

Two other important issues must also be considered when using this technique. First, the demographic parameters incorporated in the projection matrix may change over time in response to fluctuating environmental conditions. Such stochasticity can be incorporated in the model by estimating transition probabilities over several years or seasons. Monte Carlo simulations and resampling methods may also be used to provide a statistically rigorous method of estimating demographic rates in changing environments (Gibson 2002). Monte Carlo simulations assume particular statistical distributions (often normal) for vital rates, and the mean and variance for λ is calculated following several runs of the analysis by drawing vital rates from the assumed distribution. Resampling methods do not require such assumptions about the distribution, but resample individuals multiple times with replacement (bootstrapping) or random omission of inviduals (jackknifing) to obtain pseudovalues of λ (Gibson 2002). Estimates of the mean and variance of λ can then be derived from these pseudovalues (Alvarez-Buylla and Slatkin 1991, McPeek and Kalisz 1993). Methods for incorporating stochasticity in matrix models are reviewed by Fieberg and Ellner (2001).

A second key issue is density dependence. It is important to include density dependence because without it, population size in any model will tend rapidly towards either extinction or infinity, unless the parameters of the model are carefully adjusted so that net mortality exactly balances net fecundity. Failure to consider density dependence can lead to relatively high estimates of extinction risks (Ginzburg *et al.* 1990). Although methods are available for incorporating density dependence into matrix models (Caswell 2001, De Kroon *et al.* 2000), their use is relatively uncommon, reflecting the additional difficulties that this presents in terms of parameterization. One relatively simple approach is to define a population ceiling, limiting the maximum population size; this is provided as an option in some PVA software such as RAMAS *Metapop* (see next section). Alternatively, values of survival and growth may be varied in response to variation in density. An example of this approach is provided by Freckleton *et al.* (2003), in their investigation of the effects of harvesting of adults on population dynamics of the edible palm *Euterpe edulis*.

Assessing density dependence in field situations can be very challenging. Methods that have been used include analysis of time series data (an approach that is often criticized on statistical grounds), or experimental manipulation of population density, involving increasing or decreasing density (for example, by thinning a forest stand) and observing the consequences (see section 3.6.5). He and Duncan (2000) note that the importance of density dependence in tree population dynamics is most often inferred from field studies that investigate correlations or regressions between a measure of plant performance (such as growth or survival) and the density of neighbouring plants. A potential problem with such approaches is that patterns resulting from density dependence can be masked by variation in other environmental factors that influence the spatial distribution or performance of plants. Most field studies have failed to consider variation in such factors, although some have attempted to do so by selecting environmentally homogeneous sites or by controlling for such variation in statistical analyses. An example of the latter approach is provided by He and Duncan (2000) in a study of three conifer species in British Columbia, Canada. These authors tested for intra- and interspecific density-dependent effects on tree survival by analysing both the spatial patterning of trees in a field plot and the relations between neighbourhood density and tree survival, treating the effects of additional variables (such as elevation) by including them as covariates in the neighbourhood analyses. Furthermore, they were able to take advantage of the low decomposition rate of dead stems of these species to reconstruct the spatial patterning of live and dead individuals in the study plot.

Results from previous research indicate that the way in which the matrix model is constructed and parameterized has a major influence on the results obtained. Zuidema and Zagt (2000) provide a valuable review of the application of matrix modelling techniques to tree species and, on the basis of previous experience, make the following recommendations:

- The use of size-dependent relations to parameterize transition matrices yields more reliable model output than observed transition frequencies (Figure 5.1). In the latter case, the transition matrix depends strongly on the distribution of individuals among and within categories, especially when sample sizes are small. A number of different statistical methods can be used to analyse size-dependent relations in growth (linear regressions, non-linear regression) (see, for example, Zeide 1993) (Figure 5.2), survival (logistic or double-logistic regressions; Gompertz or Weibull distributions) and reproduction (logistic regression and/or linear regression). Data describing demographic rates can also be pooled between categories, if no significant differences are detected.
- It is important to ensure that the number of adults sampled in demographic field studies is sufficiently large. Producing accurate assessments of adult

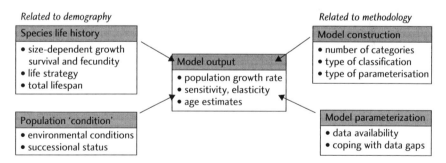

Fig. 5.1 Schematic diagram of factors influencing the output of size-structured population matrix models. Factors are grouped into four categories: the life history of the species, the environmental conditions experienced by the study population, factors related to the construction of the model, and those related to the parameterization of the model. Each of the four categories may influence the model output. (From Zuidema 2000.)

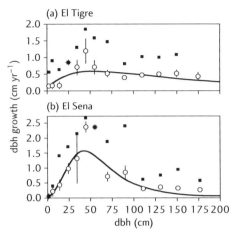

Fig. 5.2 Size-dependent pattern in dbh growth in *Bertholletia excelsa* in two sites in the Bolivian Amazon; El Tigre (a) and El Sena (b). Shown are fitted Hossfeld IV equations (lines), observed average (±1 SD) growth rate (open dots), and observed maximum growth rate (closed dots). (From Zuidema 2000 and Zuidema and Boot 2002, in *Journal of Tropical Ecology*.)

survival rates is one of the most significant challenges when modelling population dynamics of trees, and one of the main weaknesses of previous studies. Model outputs are particularly sensitive to adult survival, and therefore intense sampling, potentially over long periods of time, may be required. The population viability of many tree species appears to depend particularly strongly on the presence of long-lived individuals (Bond 1998, Kwit *et al.* 2004), further emphasizing the importance of evaluating the survival of such individuals.

- Care should be taken when using uncertain estimates or substituting missing values. The values of neighbouring categories can often be adopted. The impact of uncertain values on model output can be explored by varying the values and observing the outcome.
- It is often useful to pool demographic information from different sites or observation periods, when no statistically significant differences are found between such data sets. Differences in vital rates (survivorship and fecundity) can be determined by using standard statistical tests such as the *t*-test or the chi-squared test. When no significant differences are found, data from various years or populations can be pooled to produce a more reliable matrix model.
- Estimates of elasticities for vital rates provide a useful tool for demographic analysis, being relatively insensitive to matrix dimension, and providing a more direct indication of the dependency of λ on changes in measured parameters.
- It is important not to base conclusions (about the sustainability of harvesting regimes, for example) solely on the value of λ. As λ indicates the population growth rate when time goes to infinity (assuming conditions remain

unchanged), this may not always be the most appropriate parameter to assess population status with respect to population management. Estimates of λ are also often prone to a high degree of error.

Further advice regarding transition matrix modelling is provided in the following section.

5.3 Population viability analysis

Population viability analysis (PVA) is an important tool in conservation research and management that focuses on assessing the likelihood of population persistence over a given timespan (Brigham and Schwartz 2003, Menges 2000). PVA has been widely used to assess trends in population size, the risk of extinction, and the potential impacts of different management interventions on population viability, and has become a central component of many management and recovery plans for threatened plant species (Caswell 2000, Fieberg and Ellner 2001, Lindenmayer *et al.* 1995, Menges 1990, 2000, Silvertown *et al.* 1996). A useful overview of PVA is provided by Akçakaya and Sjögren-Gulve (2000) (Boxes 5.1 and 5.2), and Brigham and Schwartz (2003) provide an excellent recent source text for PVA specifically relating to plants.

In recent years there has been some debate about the usefulness and reliability of PVA (Brook *et al.* 2000, Coulson *et al.* 2001, Ellner *et al.* 2002). For example,

---

Box 5.1 Uses of PVA.

Akçakaya and Sjögren-Gulve (2000) provide a valuable overview of PVA, and highlight the fact that PVA can be used to address the following aspects of management for threatened species or other focal species:
- *Planning research and data collection*. PVA may reveal that population viability is especially sensitive to particular parameters, enabling research priorities to be identified, for example by targeting factors that have an important effect on probabilities of extinction or recovery.
- *Assessing vulnerability*. PVA may be used to estimate the relative vulnerability of populations to extinction. These results may be used to identify priorities for conservation action.
- *Impact assessment*. PVA may be used to assess the impact of human activities (such as timber harvesting or deforestation) by comparing results of models with and without the human activity.
- *Ranking management options*. PVA may be used to predict the likely responses of species to management interventions such as harvesting, reintroduction or habitat rehabilitation, or different designs for protected area networks.

> **Box 5.2** Approaches to PVA.
>
> The following should be considered when developing a PVA (from Akçakaya and Sjögren-Gulve 2000):
> - Model *structure* should be detailed enough to use all the relevant data, but no more detailed than this.
> - Model *results* should address the question being addressed.
> - The model should include *parameters* related directly to the question (for example, if the question involves the effect of timber harvest on population viability, then the model should include parameters that reflect such an effect realistically).
> - Model *assumptions* should be realistic with respect to the ecology of the species and the observed spatial structure (for example, if there is population subdivision, a metapopulation model should be considered).

Ludwig (1999) highlighted three potential limitations of PVA: the lack of precision regarding estimates of extinction probability, the sensitivity of such estimates to model assumptions, and the lack of attention to important factors influencing the extinction of populations. These findings were supported by further analyses based on diffusion approximations, indicating that reliable predictions of extinction probabilities can only be made for very short-term time horizons (Fieberg and Ellner 2000). In contrast, Brook *et al.* (2000) reported a high degree of correspondence between PVA predictions and field data from 21 long-term ecological studies, in the first such replicated evaluation of PVA. These conclusions were criticized by Ellner *et al.* (2002), who highlighted the lack of measures of statistical precision and power in this analysis (but see Brook *et al.* 2002 for a response). Despite the shortcomings of PVA, the method is described by Brook *et al.* (2002) as 'by far the best conservation management tool that we have'. Following a thorough review of the application of PVA specifically to plant species, Brigham and Schwartz (2003) concluded that PVA has been demonstrated to be most useful for assessing the relative extinction risks of different populations and the potential impacts of different management options, particularly for species with complex interactions between the factors influencing population dynamics.

Transition matrix models are most often used as the basis of PVA of plants (see previous section). Brigham and Thomson (2003) provide a valuable overview of PVA methods using this approach. A number of specialist software programs have been developed for PVA (Table 5.2), which incorporate a number of powerful features, such as the ability to explore the influence of stochasticity and density dependence, as well as different management regimes, on population viability. These programs differ in their approach; for example, VORTEX tracks the life of each individual in the population, whereas the RAMAS packages are matrix-based programs that track the number of individuals. The manuals and tutorials

Table 5.2 Selected specialist software programs available for population viability analysis (PVA).

Product	Reference	Website
ALEX	Possingham and Davies (1995)	www.rsbs.anu.edu.au/ResearchGroups/EDG/Products/Alex/index.asp
META-X	Frank et al. (2003), Grimm et al. (2004)	www.ufz.de/oesa/meta-x
RAMAS Metapop and RAMAS GIS	Akçakaya and Root (2002)	www.ramas.com/
VORTEX	Lacy (1993)	www.vortex9.org/vortex.html

VORTEX is probably the most widely used of these. It should be noted that all of these programs were developed primarily for animals, and have been relatively little used with plants. Whereas ALEX and VORTEX can be downloaded free of charge, META-X is provided with a book (Frank et al. 2003) and the various RAMAS programs are available as commercial products. In a rare comparative study, Brook et al. (2000) obtained similar results from different software programs, although other authors (e.g. Lindenmayer et al. 1995) have reported contrasting results from different programs. As noted by Lindenmayer et al. (1995), the choice of which program to use will depend on a range of criteria, including the objectives of the study, and the strengths, limitations, and assumptions that underpin the program and how these match the attributes, life-history parameters and available data for the target species.

available with these programs are another valuable source of guidance. Caution should always be exercised when using these tools, however. For example, Mills et al. (1999) demonstrated that different results can be obtained with the same data set by using different software programs (although Brook et al. (2002) reported a high degree of correspondence in results obtained for a range of species with different programs). An alternative approach to using an off-the-shelf software package is to develop a customized program written for a particular species (Frankham et al. 2002). A valuable online resource is available at ⟨www.ramas.com/⟩, designed to help avoid mistakes when population modelling (Box 5.3). Methods for testing the accuracy of PVA are reviewed by McCarthy et al. (2001). An example of a PVA developed for a threatened tree species is given in Box 5.4.

One of the main ways in which PVA programs differ from each other relates to how they model density dependence. This can have a major influence on model output. As noted in the previous section, density dependence can be difficult to evaluate under field conditions, and this lack of field data is one of the most important shortcomings of many PVA studies. The different methods that have been used to incorporate density dependence in PVA models are reviewed by Henle et al. (2004), who suggest that spatially explicit models hold particular promise for analysing the effects of density dependence on population viability. However, such approaches are labour-intensive and require a thorough knowledge of the biology of the species under consideration.

Increasingly, spatially structured models are being used for PVA. For example, matrices can be constructed for different populations and linked by using a

Box 5.3 Avoiding mistakes when population modelling.

A very useful online information resource is provided by H. Resit Akçakaya, one of the designers of the RAMAS *Metapop* software, at the RAMAS website ⟨www.ramas.com/mistakes.htm⟩. The resource is designed to help identify and avoid common mistakes when modelling population viability. The list of potential mistakes includes the following:

- invalid model assumptions
- model too complex; or conversely model too simple
- internal inconsistency in the model
- bias in estimation of fecundity or survival, and uncertainty in estimates of these parameters
- too many (or too few) age classes or stages
- using the wrong type of density dependence, or failing to consider density dependence
- over- or underestimating maximum growth rate of the population
- not considering demographic stochasticity
- duration (simulation time horizon) too long or too short
- ignoring spatial structure
- considering too many (or too few) populations
- dispersal rates incorrect or uncertain
- estimating risk of extinction rather than decline.

Box 5.4 An example of a population viability model developed for a threatened tree species: the case of *Araucaria araucana*.

Araucaria araucana (monkey puzzle, *pehuén*) is a large and long-lived conifer, endemic to the southern Andes and a restricted area of the coastal region of southern Chile (Figure 5.3). The historical range of the species has been much reduced as a result of logging, human-set fires, and clearance of land for agriculture following European colonization of the region. The species is of exceptional evolutionary as well as conservation interest, with similar fossils known from 200 Ma ago. The species is listed in Appendix I of CITES.

Bekessy *et al.* (2004) describe a PVA developed for this species by using RAMAS *Metapop* software, based on extensive field observations and glasshouse experiments. Development of the model presented a number of significant challenges. First, adult trees can be very long-lived (>1000 years), and growth rates are often very low (<1 mm radial increment per year). This made it difficult to create a model that is meaningful in terms of the typical timescales used in conservation management. Second, modelling density dependence was difficult to incorporate, although field observations suggested that self-thinning is common in pole-stage stands. Third, the species occurs in different forest associations and

Population viability analysis | 217

Fig. 5.3 Wild population of *Araucaria araucana* in the southern Andes. Many populations of this species are located on active volcanoes. (Photo by Cristian Echeverría.)

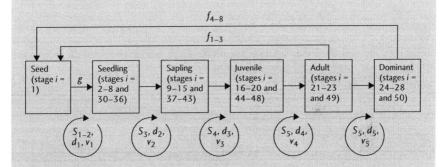

Fig. 5.4 Life cycle of *Araucaria araucana* and its population parameters: g, germination rate; s, survival; d, fire survival; v, survival following volcanic activity; t, transition; f, fecundity. (From Bekessy et al. 2004.)

displays different ecological characteristics in each. This may be related to the high degree of genetic differentiation recorded between populations and the broad edaphic range of the species, spanning a rainfall gradient (Bekessy *et al.* 2002a).

A model was constructed with 50 stages, including seeds, seedlings, saplings, juveniles, adults, and dominants (Figure 5.4). These were considered to reflect important biological stages in the life history of the species, with distinct growth rates, survivorship, fecundity, and sensitivity to disturbance. It was necessary to include many stages in the model to delay transitions, which otherwise would have led to unrealistic times for individuals to move through the various stages. Because of their extreme longevity, adult and dominant stages were assumed to be immortal. The model included both environmental and demographic stochasticity.

Simulations examined the impact of fire, volcanic activity, seed harvest, and timber harvest on population viability (Figure 5.5). Results indicated that the species has very limited ability to recover after disturbance, although seed harvesting appeared to be having relatively little effect on population viability compared with the other forms of disturbance assessed. Most importantly, the viability of the entire species appeared to be particularly sensitive to the fate of an individual population, which is located in an area of active volcanism.

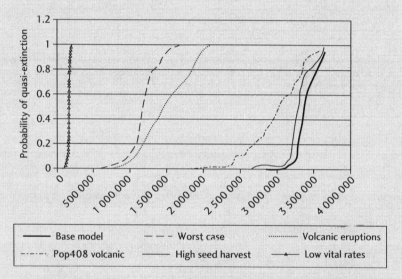

Fig. 5.5 A set of extinction risk curves for *Araucaria araucana*, derived from a meta-population model run for 100 years. (From Bekessy *et al.* 2004.)

dispersal function to produce a form of metapopulation model (for example RAMAS *Metapop*; see Table 5.2). Other approaches to modelling metapopulations predict the occupancy of discrete habitat patches through time by using a variety of analytical methods, such as the incidence function model and logistic regression models (for details see Hanski 1994, 1999; Sjögren-Gulve and Hanski 2000). A range of software tools to help perform such analyses can be freely downloaded from the University of Helsinki website; an example is SPOMSIM 1.0, which is a stochastic patch occupancy model ⟨*www.helsinki.fi/science/metapop/english/Software.htm*⟩.

Box 5.5 An example score sheet for a relative population viability assessment (RPVA).

This qualitative approach, based on expert judgement, may be used to provide a preliminary assessment of the vulnerability of species to extinction, where more formal quantitative information is lacking. Assessors evaluate each criterion for each species and score them from no degradation (0 points) to degradation sufficient to render the population inviable (20 points). These points are then summed to provide an assessment of relative risk of extinction for different species (after Schwartz 2003). The scoring system illustrated in Table 5.3 could potentially be adapted or further elaborated for any particular group of tree species under consideration.

Table 5.3 Criteria and point-scoring system.

Criteria	Examples	Score
Habitat loss	Multiple protected populations	0 points
	Few protected populations	5 points
	Moderate loss, little protected habitat	10 points
	Severe loss, no protected habitat	15 points
	Nearly complete habitat loss	20 points
Disturbance regime disruption	Natural disturbance regime maintained	0 points
	Severe disruption of disturbance regime (e.g. fire suppression with obvious indication of community change)	20 points
Habitat degradation	Habitat intact	0 points
	Severe habitat degradation (e.g. invasive weed colonizing habitat)	20 points
Population performance	Apparently normal	0 points
	Poor performance (e.g. disease killing all seedlings)	20 points

The recent development of metapopulation theory, by Ilkka Hanski and others, undoubtedly represents one of the most important theoretical contributions to conservation science, and is beginning to have a significant influence on conservation policy and management practice. However, the relevance of the theory to practical conservation problems is open to debate (see, for example, Baguette 2004, and Hanski 2004 for a response). The application of metapopulation models to plants has been the subject of particular controversy; it has been argued that single population approaches rather than metapopulation approaches may be appropriate for most plant species (Harrison and Ray 2002).

Other spatially explicit modelling approaches examine the locations and densities of individuals across a landscape (see section 5.5.3), rather than the occupancy of habitat patches. One of the main problems with spatial modelling approaches is that they require estimates of dispersal distances and probabilities that can be difficult to acquire (see section 6.4.2), yet are of critical importance to the predictions of the model. However, the incorporation of spatial pattern into population models can undoubtedly provide valuable insights into the processes influencing forest dynamics (Wiegand *et al.* 2003).

Although the vast majority of PVAs performed with plant species have employed stage- or age-structured models, there are alternative approaches. Schwartz (2003) notes that community-level models of forest dynamics (see section 5.5.1) could be used to assess population viability of tree species, but this has apparently not an approach that has been used to date. Analysis of population age or size structure (see section 3.7.1) can be used to identify recruitment failure, which may indicate that the population is not viable (although it should be noted that many tree species are naturally characterized by episodic recruitment) (Schwartz 2003). Another approach uses a time series of population counts to estimate the mean and variance of the stochastic population growth rate, which can then be used to predict extinction probabilities and time to extinction (Brigham and Thomson 2003). The method requires far fewer data than structured approaches to PVA, and is analytically relatively simple. However, observations made over a large number of years (perhaps at least 10) may be required to estimate extinction probabilities accurately. Although the technique has rarely been applied to plants, it clearly has potential (Elderd *et al.* 2003). Alternatively, relative extinction vulnerability can be qualitatively assessed (Box 5.5) by surveying expert opinion (Schwartz 2003).

5.4 Growth and yield models

Mathematical approaches to predicting growth and yield of timber trees have a long tradition in forestry, and are widely used as a tool for forest management. A very large number of simulation models have now been developed for forests in many parts of the world (see, for example, Vanclay 1995). Most yield models share the same basic approach, aiming to predict timber yield over time for a specific site. Models are generally species-specific and aim to explore the relation between stand density and tree growth, measured in terms of both stem diameter (dbh) and tree height. In some models, growth is also related to measures of crown size. Growth and yield models are generally produced by empirically deriving equations that describe relations between these growth variables by using standard statistical procedures such as regression.

Typically, regressions are carried out on multiple measurements of individual trees, although growth–yield models differ with respect to the regression functions used and the variables included. Analyses are generally performed on forest inventory data, perhaps employing measurements of forest stands growing on different sites that have been subjected to different silvicultural treatments or management regimes (Vanclay *et al.* 1995). The data are usually derived from individual trees that have been marked within permanent sample plots. These data may be used to produce *yield tables*, which give estimates of the changes in stand variables (such as height, basal area, and volume) with age. In general, models are produce for a range of different *yield classes*, reflecting the variation in growth rate on sites with different environmental characteristics. Often, a large number of plots (hundreds or even thousands) are used as a source of data. Details of the methods are provided by Munro (1974) and Ek *et al.* (1988); uneven-aged stands are considered by Peng (2000). Some examples are provided by Amaro *et al.* (2003) and Vanclay (1994).

Burkhart (2003) provides some suggestions for using this approach, including:

- Keep the model as simple as possible
- Ensure that the model is as accurate as possible, or by increasing the size of the data set, improving the quality of the data obtained
- When planning data collection, include a wide range of site and stand conditions in the sample

The main value of forest yield models is their ability to make detailed predictions of tree and stand dynamics, particularly stem-size distribution predictions that aggregate accurately to the stand level (Monserud 2003). In addition, tree mortality is often modelled in some detail, incorporating both competitive interactions between individual trees and stand-level responses such as the $-3/2$ power law (section 3.7.5). The main disadvantage of this type of forest model is that it does not explicitly consider the underlying ecological processes responsible for variation in growth and yield, such as uptake of moisture and mineral nutrients and the process of carbon fixation. These models therefore have limited value for understanding how forest ecosystems function. Most growth–yield models are designed for a certain forest type or region and the range of species considered is usually restricted to a few that are commercially important. As ecological processes are not incorporated in the models, they are restricted to the range of environmental conditions under which the data were originally collected, and have limited value for predicting how a forest might respond to changing environmental conditions. They therefore have limited applicability to forest conservation. However, they offer a useful technique for assessing forest biomass, which is important for consideration of environmental services such as carbon sequestration and nutrient cycling.

5.5 Ecological models

The term *ecological models* is used here to refer to all forest dynamics models originating from an ecological perspective or simulating ecological processes or characteristics of forests (following Hope 2003). These may be classified in a variety of different ways; see, for example, Liu and Ashton (1995), Porte and Bartelink (2002), and Shugart (1998). Here, following Hope (2003), two main types of ecological model are differentiated.

- *Process models* focus on modelling physiological processes such as photosynthesis and respiration (Vanclay 1994). Process models generally do not depict changes in species composition or stand structure, and therefore have relatively little relevance to forest conservation or ecological dynamics. However, some models of this type are of value for exploring the provision of ecological services.
- The other main group of models may be described as *succession models* (Hope 2003), as they focus on simulating the ecological dynamics of forest communities; it is these models that are considered below. This type of model may be further subdivided into *gap* and *transition* models (Figure 5.6).

5.5.1 Gap models

Forest ecologists have devoted substantial efforts to the development of models that simulate the process of forest succession that follows the creation of a canopy gap. During the past 30 years, a large number of such models have been developed, all of which are linked to the first such model, JABOWA (Botkin *et al.* 1972a, b), and its direct descendent FORET (Shugart and West 1977). Gap models simulate the establishment, growth and mortality of each individual tree within a specified area (often about 0.1 ha, or the size of a typical canopy gap). Typical model outputs include species composition, age structure, size distribution of trees and vertical stratification. Gap models have also been applied to other components of forest communities, such as the ground vegetation (Kellomäki and Väisänen 1991).

It is useful to differentiate between *deterministic* models, for which the predicted values may be computed exactly, and *stochastic* models, for which the predicted values depend on probability distributions. Many gap models are composed of a mixture of deterministic and stochastic sub-models; they are generally described as stochastic if they contain any stochastic elements (see, for example, Liu and Ashton

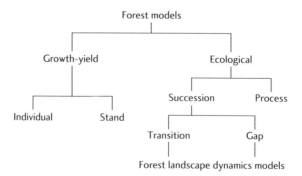

Fig. 5.6 A simple hierarchical classification of forest dynamics models. (From Hope 2003). Whereas growth–yield models have primarily been developed by foresters for estimating timber yield, those developed by ecologists tend to focus on forest succession.

1998, He and Mladenoff 1999). Use of stochastic models raises the need to determine the appropriate probability distributions of the stochastic elements, and the models may need to be run multiple times in order to yield meaningful results (Hope 2003).

Gap models generally have six components (Hope 2003):

- *site variables* (including gap size, soil fertility, soil moisture, accumulated temperature)
- *species variables* (such as maximum age, maximum diameter, maximum height, growth rate, shade tolerance, tolerance of low nitrogen availability, tolerance of extreme wet and drought conditions, etc.)
- *growth sub-model* (deterministic, in which growth of individual trees is often expressed in terms of dbh, height, leaf area, and stem volume)
- *resource sub-model* (deterministic, in which trees interact with each other by influencing the gap neighbourhood)
- *recruitment sub-model* (stochastic; often the individuals that colonize a gap are defined by selecting of individuals from the species list at random)
- *mortality sub-model* (stochastic, often modelled as a function of age and growth environment, although harvesting regimes may be included as a cause of mortality).

Gap models calculate annual change within the specific area by calculating the growth increment of each tree, the addition of new young trees (both from seeds and by sprouting) and mortality (Shugart 1984). Growth increment is described by equations that relate growth to current diameter and the maximum diameter observed for the species, based on the assumption that tree growth is directly proportional to the abundance of leaves and inversely proportional to the amount of respiring, non-photosynthetic tissue (Hinckley *et al.* 1996). The variables leaf area, tree height, and species maximum height are defined in the basic diameter growth equation as a function of current stem diameter. Species-specific regeneration within the gap is generally defined as a function of light reaching the forest floor. Interactions between individual trees are described by relatively simple rules, mostly relating to competition for resources, especially light.

Different gap models have been developed for forests with different characteristics, and vary with respect to how many species are considered, and how different forest structures are treated. Some models consider the explicit location of each individual tree whereas others do not (referred to as *spatial* or *non-spatial* models, respectively). Whereas the processes of recruitment, growth, competition, and mortality are considered by all gap models, the relative emphasis on these different processes varies between models. Parameterization of gap models is often achieved by referring to published information on the characteristics of the tree species of interest, for example their establishment requirements, growth rates, and height–diameter relations. If these data are not available from previous work, then they will need to be collected via an appropriate field survey. Data describing environmental conditions (for example soil moisture, light, temperature, frost, elevation, elevation, aspects, altitude, nutrients) should be collected for the sites to be modelled (Shugart 1984).

It is important to note that the aim of gap models, to explore and test ecological theory, is very different from the objective of the growth and yield models described in the previous section. These two modelling approaches have largely developed in isolation from each other. Liu and Ashton (1995) compare growth–yield and gap models, and highlight the following differences:

- Although the two types of model share some similar features, they differ in model structure and data requirements.
- Growth–yield models tend to be used by foresters to assist timber production and evaluate growth and yield of timber species in managed forests, whereas gap models are generally used by ecologists to explore ecological processes and dynamics in natural forest ecosystems.
- Site-specific environmental and species information is necessary for constructing growth–yield models, whereas gap models require species-specific biological information on individual trees and site-specific environmental data.
- Growth–yield models are more diverse in terms of model structure, whereas most gap models are derived from the original antecedent (JABOWA), although they differ in detail.

One of the attributes of gap models is that gap size strongly influences the predicted population dynamics (Shugart 1984). In contrast, forest yield models are generally unaffected by a change in plot size (Monserud 2003). Typically, gap models are not designed to make direct use of forest inventory data, and do not provide estimates of timber yield required by foresters. However, some researchers have sought to achieve greater realism by incorporating physiological processes within the models (Friend *et al.* 1993), as in the case of HYBRID (Friend *et al.* 1997).

Gap models have recently been criticized by a number of researchers. For example, Lindner *et al.* (1997) compared output of a second-generation gap model, FORSKA, with long-term data from Bavaria, and found a poor correspondence with individual tree dimensions and stand structures. Changing stand density over time influenced the height–diameter relations of trees and rates of mortality, and these were found to be inadequately represented in the model. Yaussy (2000) similarly found that projections made by using the gap model ZELIG failed to produce accurate estimates of biomass or volume, when compared with 30 years' inventory data from Kentucky. These experiences highlight the importance of carefully calibrating the underlying growth model if this type of modelling approach is to be used for guiding management decisions.

Despite these problems, gap models continue to be the focus of active research, and new versions continue to be developed. Details of some of the models currently used are listed in Table 5.4. It is notable that few attempts have been made to model dynamics of relatively species-rich forests, such as lowland tropical forests. Also, although gap models have undoubtedly proved to be of value for

Table 5.4 Selected models used for exploration of forest dynamics.

Name	Comments	Reference	URL
JABOWA	The ancestor of most forest gap models	Botkin et al. (1972a, b); Botkin (1993)	www.naturestudy.org/services/jabowa.htm
FORET	An early descendent of JABOWA	Shugart (1984); Shugart and West (1977)	
SORTIE SORTIE-ND	Individually based and spatially explicit model developed for mixed temperate forests	Pacala et al. (1993, 1996), Canham et al. (1994a)	www.sortie-nd.org/
LANDIS LANDIS II	Uses an object-oriented modelling approach operating on raster GIS maps, enabling modelling of forest dynamics at the landscape scale	He and Mladenoff (1999), Mladenoff and He (1999), Mladenoff (2004)	www.snr.missouri.edu/LANDIS/landis Also available commercially from Applied Biomathematics ((www.ramas.com))
ZELIG	Based on FORET, but unlike earlier gap models, incorporated spatial interactions	Urban (1990), Urban et al. (1991)	
MOSAIC	A semi-Markov derivative of the ZELIG gap model	Acevedo et al. (1995)	http://emod.unt.edu/
FORMIX FORMINd	Simulates the dynamics of tropical rainforests based on a few plant functional types, but with a relatively detailed treatment of physiological processes	Huth et al. (1998), Huth and Ditzer (2000), Köhler and Huth (1998)	

exploring hypotheses about forest dynamics, they have rarely been used to address forest conservation issues directly. On the other hand, as noted by Bugmann (2001) in his review of forest gap models, no alternative approach offers such an intuitive and elegant means of exploring the effects of competitive interactions on tree population dynamics. He also notes that gap models have evolved considerably from the original versions and contain far greater detail than is sometimes recognized; there is currently a wide variety of approaches that are being used to incorporate the various ecological processes relevant to tree population dynamics. A further advantage of gap models is that the output gives detailed representations of species composition and physical structure, which could potentially be used to develop models of habitat attributes (Hope 2003) (see Chapter 7).

5.5.2 Transition models

Other modelling approaches that can be used to explore forest dynamics at the landscape scale include Markov and semi-Markov models. These are constructed by determining the probability that the vegetation in a specific area will have developed or been converted into some other vegetation type within a given time interval. The probabilities for this change or conversion are referred to as *transition probabilities*. The vegetation must therefore be classified into identifiable categories in order to apply this approach. Markov models have two important characteristics (Shugart 1998):

- The transition probabilities at time t_n depend only on the immediate past value at time t_{n-1} and are independent of the state of the system at any time earlier than t_{n-1}. Sometimes this condition is referred to as a first-order Markov process.
- The system is stable if the transition processes do not change over time.

Vegetation categories may be defined in a number of different ways, in terms of species of canopy tree present (Horn 1975), the most abundant species of tree in the forest canopy, or the number of individuals of different species present. Analysis of Markov models is very similar to the transition matrix models described earlier (section 5.2.3), depending upon the multiplication of the transition probabilities by a vector representing the proportions of each vegetation category present within an area at a particular time. Markov modelling can therefore be carried out with any appropriate software that supports matrix algebra (see section 5.2.3). Model outputs describe the projected composition of the vegetation, in terms of the proportional cover of different vegetation categories, at different times into the future.

Markov models have the attraction of being analytically relatively simple and easy to interpret. The main challenge in using them relates to deriving accurate transition probabilities. These can be obtained either by monitoring vegetation over prolonged periods in permanent sample plots (see section 4.7.1), or by using remote sensing imagery to identify vegetation changes that have occurred in the past (see Chapter 2). The probability of transition from any vegetation category to any other category (or state) must be determined; as a consequence, the number of model parameters is a function of the square of the number of categories in the model (Shugart 1998). There is therefore a trade-off between the increased resolution offered by incorporating a larger number of vegetation categories (or states), and the increased difficulty of accurately parameterizing a model with a larger number of transition probabilities. This is particularly problematic given that relatively rare transitions need to be estimated with equivalent precision to relatively common transitions (Shugart 1998). One potential way round this problem is to estimate model parameters on the basis of some theory (Horn 1975). However, Markov models tend to be highly specific to the forest type (and even the particular study area) for which they were created, limiting their practical value (Hope 2003).

Examples of applying Markov models to forest dynamics are provided by Horn (1975, 1976) and Waggoner and Stephens (1970). One of the main drawbacks of this type of model is that the transition probabilities remain the same through time (Moore and Noble 1990). As an alternative, semi-Markov models have been developed (Acevedo *et al.* 1996a,b) in which the transition probabilities are not fixed but instead vary depending on the time that the vegetation unit entered the current state. Transitions representing successional change are considered to have *holding times* associated with them, whereby the transition cannot occur until the vegetation unit has occupied a successional stage for a fixed time period (Hope 2003). Various hybrid approaches have also been developed. For example, Acevedo *et al.* (1995) linked a semi-Markov model to output from the gap model ZELIG, enabling the behaviour of the gap model to be explored at a much larger spatial scale than would have been possible with the gap model alone. This semi-Markov approach can be used to assess forest dynamics at the landscape level by making the parameters (probabilities and holding times) depend on environmental variables such as elevation, slope, aspect, and soils, and visualizing the output in GIS (Acevedo *et al.* 1995, 1996a).

Another related approach is the use of *cellular automata*, in which the area of interest is divided into cells, the states of which are defined by rules regarding the states of neighbouring cells. Interactions between cells are more likely to occur between close neighbours than cells far apart, and can relate to spatial processes such as dispersal or the spread of fire, enabling vegetation change to be modelled at the landscape scale. Cells are considered to be identical or of a relatively small number of types, so that even though the number of cells is large the number of parameters needed may be relatively low. Analysis is done by multiplying each cell by the probabilities of plants going from one state to another, for example by using a transition matrix. The results of the multiplications, however, are weighted by a set of constraints, reflecting interactions with neighbours (Gibson 2002). Environmental constraints and disturbances can be included; for example, Hochberg *et al.* (1994) included the constraint that tree seedlings were susceptible to fire-induced mortality following burning, if not surrounded by and protected by a certain number of adult trees. Cellular automata are particularly useful for making qualitative predictions regarding spatial pattern formation and have been applied to the study of tropical rain forests (Alonso and Sole 2000), as well as for simulating the long-term population dynamics of threatened tree species, such as *Fitzroya cupressoides* (Cannas *et al*. 1999). A review of the use of cellular automaton models in ecology is provided by Balzter *et al.* (1998), and further guidance is provided by Durrett and Levin (1994).

5.5.3 Other modelling approaches

Attempts have been made to address the limitations of traditional gap models through the development of alternative approaches. SORTIE is an empirically based model of forest dynamics, developed for mixed temperate forests, that realistically simulates the spatial interactions between individual trees (Pacala *et al*. 1996). SORTIE tracks the exact location of each tree and determines its performance based on its local neighbourhood. The model incorporates a sophisticated means of calculating the transmission of light through the forest canopy, and simulates the responses of individual trees to the light conditions occurring locally. Another key feature of this model is the much larger areas of land that can be considered compared with conventional gap models. SORTIE is a powerful model that has been used successfully to investigate forest dynamics over a variety of scales. The main obstacle to using SORTIE is the substantial amount of field data required to parameterize it. It is also demanding in terms of processing power (Bugmann 2001).

Kohyama (1993) and Kohyama and Shigesada (1995) describe a fundamentally different approach to modelling forest dynamics, based on the use of partial differential equations to approximate the shifting-patch mosaic of forest landscapes according to the age distribution of patches with different tree-size structures. The creation of a treefall gap corresponds with the 'death' of a patch of particular age, resetting patch age to zero. Tree size structure is developed in relation to patch age, and the demography of trees in each patch is regulated by patch-scale tree-size structure (Kohyama 2005). This approach has recently been extended to larger spatial scales and longer timescales (Kohyama 2005).

As noted earlier, process-based models that simulate physiological processes under changing environmental conditions offer another approach to forest modelling (Landsberg and Gower 1997). Although not considered in detail here, increasingly such models are being used to address questions relating to forest management. An example is provided by FORMIX and its successor FORMIND, which includes more information relating to forest dynamics. The latter has been used to analyse the growth and yield of logged-over forest in Venezuela under different logging scenarios (Kammesheidt *et al*. 2001).

One of the main developments in recent years is the growth in interest in modelling forest dynamics at the landscape scale. Hope (2003) uses the term *forest landscape dynamics model* (FLDM) to refer to a spatially explicit forest dynamics model acting at landscape scales (see Figure 5.6). This approach to modelling has been greatly supported by recent developments in GIS technologies (see section 2.6), which now enable GIS to be linked to models of forest dynamics, either for preprocessing data for use in a non-spatial modelling, or for displaying model output. Closer linkages between the model and GIS can be achieved if they share the same data structures. Some models are now implemented entirely within a GIS environment, sometimes using custom-designed GIS modelling packages such as PCRASTER (Wesseling *et al.* 1996). Examples of models that have been developed specifically to operate at landscape scales are provided by Frelich *et al.* (1998), Frelich and Lorimer (1991), and Liu and Ashton (1998).

An example of this approach is provided by the model LANDIS (Figure 5.7), which is based on an object-oriented modelling approach operating on raster GIS maps (He *et al.* 1999, Mladenoff and He 1999). The principal modules of LANDIS relate to forest succession, seed dispersal, wind disturbance, fire, and timber

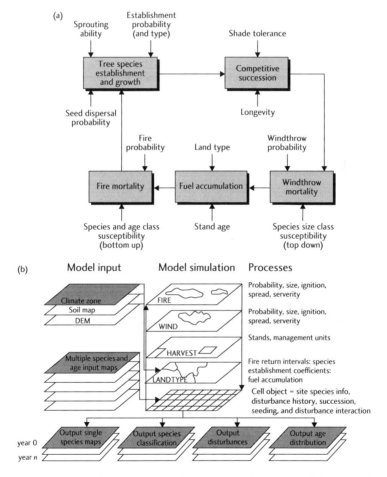

Fig. 5.7 Structure of the LANDIS model: (a) original conceptual diagram of the model, and (b) computer operational design—note the use of raster data organized in layers. (Reprinted from *Ecological Modelling*, 180, Mladenoff, LANDIS and Forest Landscape Modelling, pp 7–19, (2004), with permission from Elsevier.)

harvesting. In LANDIS, each cell is a spatial object containing species, environment, disturbance, and harvesting information (Mladenoff and He 1999). Tree species are simulated as the presence or absence of 10 year age cohorts in each cell, rather than as individual trees, greatly reducing the processing power required to carry out simulations over large areas. The integration of LANDIS with GIS provides a powerful set of tools with which to explore the potential impacts of management interventions at the landscape scale, in relation to the principles of landscape ecology (Mladenoff 2004). The model is considered further in section 5.5.4 below.

One of the main limitations of LANDIS, and other landscape-level models such as LANDSIM (Roberts 1996), is that they cannot be used to address detailed spatial dynamics of forest structure within stands. Recent progress in incorporating spatial processes within individual-based models, such as gap models and SORTIE, is reviewed by Busing and Mailly (2004). Further information regarding spatial modelling of forest dynamics using statistical approaches and GIS is provided in sections 7.8 and 8.4.

5.5.4 Using models in practice

How can forest models be used in practice? It is important to note that most available models of forest dynamics were developed as research tools, rather than as practical tools to support decision-making. They may therefore be poorly designed to address the specific question of interest, and have rarely been used to address issues relating to practical forest management or conservation. Many models require a great deal of data in order to provide useful outputs. For this reason, Gibson (2002) suggests that models of forest dynamics are of value only to a small minority of specialist researchers. However, there is no doubt that modelling approaches have proved to be of great value in understanding the processes of forest dynamics, and their popularity is likely to increase in future as increased computing power becomes more widely available. Forest models offer a uniquely powerful method for exploring how forest structure and composition might change in response to human activities. They also provide a very useful analytical framework for underpinning field-based investigations. For these reasons, increased use of forest models in both forest research and management is something to be encouraged.

How, then, can forest modelling methods best be employed? A starting point is to use one of the forest models that already been developed (see Table 5.4). Many of these have been adapted by individual researchers to their own individual circumstances, and as a result, there is now a wide variety of different models available. In particular, the JABOWA/FORET/ZELIG family of 'gap' models have been widely used by researchers, and as a result a wide variety of different versions have been developed for specific forests and applications. A useful compendium of available models, and associated information, is provided by the WWW-Server for Ecological Modelling ⟨*http://eco.wiz.uni-kassel.de/ecobas.html*⟩. It may be that a version of one of these models is available that can be applied directly to the problem of interest. More typically, researchers use them as a starting point for developing their own models. This can be a significant undertaking, and generally requires skills in both computer programming and mathematics. Many individual PhD projects, each of several years' duration, have been devoted to this form of model development and application. On the other hand, some models (such as LANDIS and SORTIE-ND) are now used by communities of researchers, who may welcome new collaborators and provide a valuable source of support. Some textbooks are now available, such as Botkin (1993) and Shugart (1998), which provide a helpful introduction to forest modelling methods.

Which type of model is most likely to be of value to forest conservation? This will depend upon the objectives of the investigation. It is therefore very important to be clear at the outset precisely what the objectives are. To date, relatively little forest modelling research has been undertaken that is directly relevant to forest conservation. LANDIS is one of the few models that has been designed with this kind of objective in mind, and has much to recommend it. First, its close linkage with GIS and its ability to both input and output spatial data, as well as to explore spatial processes such as dispersal, make it particularly suitable for exploration of dynamics at the landscape scale. The model has partly been inspired by the emerging discipline of landscape ecology, and is consistent with the widespread growth of interest in developing conservation management approaches at the landscape scale. Second, LANDIS has also been developed as a commercial product (as RAMAS Landscape), available from Applied Biomathematics, ⟨www.ramas.com⟩, as part of an integrated system that combines LANDIS with one of the leading software programs for examining metapopulation dynamics of individual species, RAMAS GIS. This system provides the first integrated tool that enables explorations of forest dynamics to be linked directly with a species metapopulation model (Akçakaya *et al.*, 2004). The LANDIS model itself is freely available, and has been used in an increasing number of forest communities in different parts of the world (Mladenoff 2004). For example, Pennanen and Kuuluvainen (2002) present a modification of LANDIS designed to allow simulation of fire-prone landscapes in Fennoscandinavia. The main limitation of LANDIS is perhaps its relatively simplistic representation of cohorts (Hope 2003).

Whether an existing model is used or adapted, or a new model constructed, a number of key decisions have to be made (following Hope 2003), regarding:

- *Organizational resolution*, for example whether the model needs to be *individual-based* (where each tree is modelled as an individual entity), or *stand-based* (where tree attributes are aggregated over the whole stand).
- *Operating scale*, referring to whether the model should operate at the *stand* or *landscape scale*; in general a landscape is heterogeneous in terms of climate, soil type and vegetation cover, whereas stands are generally considered to be homogeneous in terms of these variables. Typically, stand scales refer to areas less than a few hectares whereas landscape scales range from hundreds to million of hectares.
- *Spatiality*; typically, a spatial model is composed of non-spatial sub-models linked by spatial processes. Alternatively (or additionally), it may be

parameterized by spatial data. Recently, models have tended to become more spatially explicit, reflecting developments in GIS technologies and the increasing availability of computing power.
- *Management objectives*: where models are aimed at informing management decisions, a distinction may be made between *strategic* and *tactical* decision-making processes, the former involving the development of broad strategies for dealing with problems whereas tactical management focuses on actions at the site level.

Similarly, Botkin (1993) suggests the following stepwise approach when developing a model (see also Table 5.5):

- Choose *variables of interest*. (What are the variables that you want to project for the forest—biomass, timber production, diversity, stand structure or some other variable?)
- Determine the *conceptual level* at which the model should operate. It can be useful to consider the hierarchical level of the desired output, to help define the level of detail required. For example, in many applications the desired level of output refers to populations of trees; to produce output at this level, it may be necessary to model at the individual level.
- Determine the *range of phenomena* that the model is required to reproduce.

Table 5.5 Key stages in model development, testing, and application (adapted from Gardner and Urban 2003).

Stage of development	Tools for analysis	Information derived
1. Conceptualization and selection.	Mathematical and graphical analysis	Class of dynamics defined
2. Parameter estimation	Calibration	Adequacy of model representation quantified
3. Parameter refinement	Sensitivity analysis, uncertainty analysis	Important parameters and processes identified
4. Model evaluation	Preliminary exploration of model ('experiments'), evaluation of uncertainties	Alternative hypotheses, scenarios may be tested
5. Validation	Statistical comparison of model projections with independent data	Reliability of model projections established
6. Application	Simulation of relevant management and/or policy scenarios, synthesis of results	Understanding of system dynamics

Note that these stages may be performed iteratively rather than sequentially. Calibration refers to curve-fitting exercises or other statistical procedures designed to help identify the most appropriate model formulations and parameter values; sensitivity analysis refers to the examination of model response to changes in parameter values.

- Determine whether the goal is *realism* (for example, the qualitative shape of an output curve) or *accuracy* (the quantitative difference between observed and projected values). If the goal is realism, then some rules should be adopted regarding what level of correspondence between an output curve and actual observations is acceptable. It is important to remember that a model is an abstraction of nature, and will therefore always represent a simplification; a key decision therefore is to decide the degree of simplification that is acceptable. There are many reasons to keep a model as simple as possible (Canham *et al.* 2003).
- Define the *level of accuracy* required. Botkin (1993) suggests that for forest stand models 10–20% accuracy is reasonable.

Whatever model is used, there is value in conducting a *sensitivity analysis* in which the model is run with a range of parameter estimates, and the effect of changing these values on the model outputs is observed to evaluate how robust the conclusions are (Gibson 2002). Often it is impossible to measure ecological parameters accurately, and therefore there will be a degree of uncertainty about many of the values included in the model. The potential influence of this uncertainty can be explored through sensitivity analysis, typically by varying each of the uncertain parameters in turn, recording the response of the model, while holding all other parameters constant at their most likely values. Further details of sensitivity analysis methods are presented by Swartzman and Kaluzny (1987).

As noted by Shugart (1984), whichever modelling approach is adopted, a key question is: 'how well does it work?'. Models can be viewed as hypotheses, which need to be tested. This may involve two main types of procedure (Shugart and West 1980):

- *Verification procedures*, in which a model is tested to determine whether it can be made consistent with some set of observations. Usually, forest models are verified by comparing model structures and parameters with what is known about the ecological system being modelled.
- *Validation procedures*, in which a model is tested for its agreement with a set of observations that are independent of those observations used to structure the model and to estimate its parameters. It is important that the data used to test a model in this way are genuinely independent from the data used to parameterize it.

One of the challenges of modelling forest dynamics is that validation can often be difficult, because of the lack of long-term data describing the ecological behaviour of forests. As long-term data sets describing forest dynamics are so scarce, most efforts at model evaluation have focused on comparing model simulations against measured data that refer to a single point in time (Bugmann 2001). Statistical procedures used to test ecological models, such as the kappa statistic, contingency tables (or the 'confusion' matrix), and receiver–operator characteristic (ROC) curves, are described by Gardner and Urban (2003). Although predictions may be difficult to test, every effort should be made to validate models as thoroughly as

Table 5.6 Definition of terms used in modelling forest dynamics (adapted from Bugmann 2003).

Term	Definition
Prediction	Commonly denotes inference from facts or accepted laws of nature, implying *certainty*.
Forecast	Differs from prediction in being concerned with *probabilities*; implies the anticipation of outcomes.
Projection	An estimate of future *possibilities*.
Scenario	An account or synopsis of a *possible* course of action or events.

possible. Bugmann (2001) suggests that model behaviour should be evaluated through a combination of sensitivity analyses, qualitative examinations of process formulations, and quantitative tests of model outputs against various kinds of empirical data. However, models may also be judged on their usefulness for providing insights into theoretical or practical problems (Shugart 1984). According to Bugmann (2001), the value of forest models does not lie in their ability to 'predict' the future, but rather in their ability to help understanding of processes and patterns by allowing exploration of the consequences of a set of explicitly stated assumptions that are too complex to explore by other methods.

A further important point relates to the outputs of a model, and how these are described. Many of the terms used by modellers are interpreted variously by different authors. Often, model outputs are described as *predictions*, a word that implies a high degree of certainty. As uncertainties are inevitably involved in any attempt at ecological modelling, Bugmann (2003) argues that the word 'prediction' is not appropriate in this context, and suggests use of the words '*forecast*' or '*projection*' instead (see Table 5.6). Models might also be used to support development of *scenarios*, a technique that is considered in section 8.7. The word 'prediction' is perhaps best used when referring to formal derivations from the logical structure of a theory, which serve as a means of testing and evaluating that theory (Pickett *et al.* 1994).

Finally it is pertinent to remember one of the axioms of modelling: GIGO; or 'garbage in, garbage out'. In other words, the quality and value of any model output ultimately depend on the quality of the data that are used as input. Careful attention to ensuring that accurate and precise measurements are obtained during field surveys is therefore of paramount importance. Botkin (1993) highlights the importance of understanding the assumptions on which a model is built, and of questioning and testing these assumptions rather than simply applying existing models uncritically. It can also be helpful to remember another axiom: all models are wrong, but some models are useful (G. E. P. Box, cited in Ryan 1997). A useful outcome of modelling might be an increased understanding of the consequences of the assumptions made, or a demonstration of areas of ignorance (Botkin 1993).

6
Reproductive ecology and genetic variation

6.1 Introduction

Reproductive ecology tends to receive relatively little attention from researchers; as a consequence, remarkably little is known about even many relatively widespread tree species. Yet a firm understanding of the processes underpinning reproductive success is of paramount importance to forest conservation and management. The viability of a population depends critically on the process of reproduction, and may be influenced by factors such as the availability and behaviour of pollinators, the breeding system of the plant, and the processes of fruit development and dispersal. The Brazil nut (*Bertholletia excelsa*, Lecythidaceae) provides a striking example. Brazil nuts are one of the most economically important forest products of the Amazon, but the nuts are not produced without successful cross-pollination. Initial attempts at producing Brazil nuts in cultivation failed because the main pollinators, euglossine bees, also need epiphytic orchids to complete their life cycle (Smith *et al.* 1992).

This chapter first considers the process of flowering phenology and pollination, and techniques for measuring fruit production, dispersal, and predation. Methods for analysing the mating systems of plants and the genetic structure of populations are then presented. The recent development of molecular markers has revolutionized our understanding of the processes influencing genetic variation, and these techniques are now being widely applied to address conservation problems, such as estimating rates of gene flow and identifying conservation units. However, the more traditional methods of assessing quantitative genetic variation still have a valuable contribution to make, and are therefore also considered here. Overviews of plant reproductive ecology are provided by Willson (1983) and Bawa and Hadley (1990).

6.2 Pollination ecology

Techniques for pollination ecology are described in detail by Dafni (1992) and Kearns and Inouye (1993), on which this account is based. Pollination biology is a broad discipline embracing evolutionary ecology and taxonomy, as well as animal behaviour. Here, only a small selection of methods is presented, focusing on those field-based techniques relevant to *in situ* forest conservation. The significance of pollination ecology for conservation lies in understanding the processes influencing gene flow and reproductive success, and therefore the evolutionary viability of populations. Practical conservation challenges include low production of viable seed by populations of threatened species, perhaps caused by loss of native pollinators, and the problem of reestablishing populations of species in areas where specialist pollinators are absent. Where threatened species are dependent on specific pollinators for reproduction, conservation action must obviously address the pollinators as well as the plant species of concern.

6.2.1 Tagging or marking flowers

In many pollination studies, flowers need to be marked or tagged. It is important not to use a method that might alter the attractiveness of the flower to the pollinator (for example, the use of red tags may attract hummingbirds). Inconspicuous tags can be made by tying sewing or embroidery thread on to individual flowers; it is also possible to write on the flowers themselves with felt-tip markers, indelible ink, or paint. Jewellery tags (made out of small rectangles of stiff paper), or gummed labels wrapped around the stem or pedicel, can also be used to label flowers. Alternative methods of marking flowers include thin copper wire, lengths of plastic drinking straws slit lengthwise, and coloured waterproof tape. Techniques for marking entire plants are described in section 4.7.1.

6.2.2 Pollen viability

One potential cause of reproductive failure is that the pollen is not viable. Direct tests of pollen viability involve depositing pollen on receptive stigmas and observing whether seeds are produced. Although it gives an unequivocal answer, the method is time-consuming. Indirect methods focus on correlating the ability to fertilize an ovule with some physiological or physical characteristic that can be determined relatively rapidly.

The most widely used and reliable indirect method for assessing pollen viability is the *fluorochromatic reaction (FCR) test* (Dafni 1992). The reagent is prepared by placing 10 ml of freshly made 15% sucrose solution in a transparent vial. A solution of 20 mg fluorescein diacetate in 10 ml acetone is prepared, and added drop by drop (1–3 drops in total) to the sucrose solution until it turns a light milky or greyish colour. Dehydrated pollen grains should be stored for 10–30 min under high relative humidity before the test, to enable membrane recovery. The pollen sample is dispersed in a drop of the fluorescein diacetate solution; the microscope slide is placed in a Petri dish lined with wet filter paper for 10 min and then the drop is covered with a coverslip. The drop can then be examined under a fluorescent microscope, through a violet exciter filter. Pollen grains with bright golden-yellow fluorescence can be scored as viable; undeveloped or empty grains will not fluoresce. To record percentage viability, first count the grains under white light then assess how many grains remain visible after switching to fluorescent light. Observations should be completed within 10 min after placing the coverslip over the sample.

Alternatively, pollen grains can be stained with a vital dye such as methylene blue (1%), neutral red (1%), or aniline blue (1%). A sample of the pollen grains is placed in a droplet of the dye and covered with a coverslip; the dye is replaced with water or glycerol after 5 min. The percentage of dyed pollen grains can be counted. This method should be used with caution or only as a preliminary assessment, because values obtained may depart significantly from the actual value of pollen viability.

Pollen germination ability for many species can be examined in aqueous solutions of sucrose. Germination can be compared in solutions of different sucrose concentration (0–60%) as percentage by mass (g sucrose/100 g solution) to determine the optimum concentration for pollen germination, and the maximal germination rate as an indicator of pollen viability. Pollen grains are left in the sucrose solution for 24 h at room temperature, then examined under a microscope with a small drop of methylene blue. Germination of some species will be low unless they are kept under high humidity (95%) for a period beforehand (for example 30 min).

Pollen viability can also be assessed by examining the germination of pollen grains on the stigma, which also provides an indication of the stigma's receptivity. Pollen tubes can be detected in the style by using the following method (Dafni 1992). A solution of FPA is prepared (formalin 40%, concentrated propionic acid, 50% ethanol 5 : 5 : 90 by volume). The excised stigma and style is washed in FPA for 24 h, then stored in 70% ethanol. The style is then washed in tap water, softened for approximately 5 h in sodium hydroxide, then rinsed again in tap water for 1–3 h. Stain with 0.1% aniline blue in potassium acetate for 4 h, then squash the stained style under a coverslip and observe under a fluorescent microscope equipped with a filter set (maximum transmission 365 nm). Pollen tube walls and callose plugs should display bright yellow to yellow-green fluorescence. Alternatively, germinating pollen and pollen tubes can be stained with a hot solution of dye, composed of 150 mg of safranin O and 20 mg aniline blue in 25 ml hot (60 °C) glacial acetic acid, which should be filtered before use. The styles are first hydrolysed in 45% acetic acid at 60 °C for 10–60 min (until the style becomes soft enough to be squashed), then split longitudinally before staining. Pollen grains are stained blue, but the ends of pollen tubes stain red.

6.2.3 Pollen dispersal

Understanding patterns of pollen dispersal is important for studies of gene flow. However, direct measurements of pollen movement are difficult to obtain. One option is to measure foraging distances travelled by pollinators, although it should be noted that not all pollen is transferred to the next flower visited, and carryover to a sequence of flowers can be extensive. Studies of the direct measurement of pollinator movements have recently been transformed by the development of radio tracking methods suitable for use with small birds and insects (Naef-Daenzer *et al.* 2005), providing some remarkable new insights into pollination ecology. Techniques include the use of *miniaturized VHF radio-transmitters, harmonic*

238 | Reproductive ecology and genetic variation

radar transponders (Cant *et al.* 2005), and *passive integrated transponders* (PIT tags). Further technological developments are likely in this area with continuing progress in the miniaturization of transmitters.

Indirect methods of measuring pollen dispersal include the following (Dafni 1992, Kearns and Inouye 1993):

- *Pollen stains.* Histochemical stains can be used to label pollen, including brilliant green (1% w/v), Bismarck brown (1%), methylene aniline blue (1%), orange G (10%), rhodamine (0.2%), and trypan red (2%). The stain may be injected into anthers with a 10 μl syringe, or applied to the pollen exposure surface of a freshly dehisced anther by means of a toothpick or fine brush. Dispersal can be assessed by examining flowers in the target population to assess whether stained pollen are present. Collected stigmas can be stored in 70% ethanol before analysis. The stigmas are examined by squashing and examining through a microscope. The ratio of stained to non-stained pollen grains can be calculated as a function of the distance from the dyed pollen source and the time since marking.
- *Fluorescent powdered dyes.* Such dyes are used to make fluorescent paints and are commercially available from companies supplying these paints. Dyes can be applied with an atomizer or to individual anthers by using toothpicks. Dye movement is tracked on the assumption that the dye mimics pollen. An insect that visits a marked flower is followed to successive flowers; these flowers are tagged, and then their stigmas are examined for dye particles (Figure 6.1). The fluorescent dyes can readily be detected under ultraviolet light, as they glow brightly; an ultraviolet light source is therefore required to assess the flowers.

Fig. 6.1 Seed of the tropical tree *Cordia alliodora* covered with fluorescent dye, photographed under ultraviolet light. The technique can be used to measure seed dispersal. (Photo by David Boshier.)

Portable ultraviolet light sources enable this to be done in the field at night, but recipient flowers can also be harvested and examined under a dissecting microscope in the laboratory (using an epifluorescent microscope, for example). It should be noted that, although such dyes have often been found to mimic pollen effectively, this is not always the case. An example of the technique is provided by Campbell and Waser (1989).
- *Molecular markers.* See section 6.5.1.

Pollen can also be extracted from pollinators, to provide measures of the quantity of pollen transported or deposited during visits. The most widely used method for collecting pollen from insects consists of using glycerine jelly containing a stain (pararosaniline or fuchsin) to make a semi-permanent microscope slide. A small cube of hardened jelly is placed on the point of a dissecting needle or held within forceps, and used to pick the pollen off a captured insect. The pollen can then be transferred to a clean microscope slide for examination and counting. In the case of birds and other vertebrates, pollen can be brushed off captured animals and collected in a folded piece of paper, or lengths of adhesive tape can be pressed on to the surface of the animal and then removed and examined microscopically for the presence of pollen (Kearns and Inouye 1993).

6.2.4 Mating system

Plant mating systems have a major influence on patterns of genetic variation within populations, by affecting the extent of gene flow. There are five main types of mating system (Brown 1990), although these are best considered as points along a continuum rather than discrete categories:

- predominantly selfing
- predominantly outcrossing
- mixed mating
- partial apomixis
- partial selfing of gametophytes (in ferns).

The mating system can vary between populations of the same species, and even within a population there may be differences between individuals in the amount of selfing or outcrossing that occurs. Within individual trees, variation in the amount of selfing has been recorded in different parts of the crown. These sources of variation should be borne in mind when sampling populations or individuals for analysis.

Excluding flower visitors is a useful method of determining whether the plant is self-pollinating or not. This is usually achieved by enclosing the flower in bags made from fine-mesh cloth, gauze, cheesecloth, or nylon mesh (such as mosquito netting). Synthetic fabrics such as nylon are often preferred because they shed water more easily. Paper and cellophane have also been used to construct pollination bags, but are not very durable. The bags can be tied on to the plant with string or wire ties. Birds or bats can be excluded from flowers by using chicken-wire cages.

It is important to be aware that the use of bags or cages can alter the microclimate around the flower, which may alter patterns of flowering phenology. If there is a possibility that thrips are acting as pollinators, and there is a need to remove them, they can potentially be treated by regular applications of an insecticide such as malathion. Toothpaste has also been successfully used to exclude insects but not hummingbirds from tropical passionflowers (Gill *et al.* 1982).

To test for wind pollination (anemophily), airborne pollen can be excluded by using bags with a very fine mesh, made out of nylon stockings or cotton fabric, or even plastic bottles (Figures 6.2, 6.3). To determine whether pollen is being carried by the wind, microscope slides coated with silicone grease or glycerine jelly can be used as pollen traps. These may be placed both inside and outside bags placed around flowers, to test whether the bag is effective at excluding wind-borne pollen. If there is a possibility of both wind-borne and animal-borne pollen entering a flower, the relative importance of each pollination mechanism can be observed by comparing fertilization and seed-set with a variety of different bags designed to exclude different pollen sources. Bags with a mesh of 0.25×0.25 mm up to 1×1 mm are regarded as generally suitable for enabling airborne pollen to penetrate, but dense enough to exclude pollinators (Dafni 1992)—with the possible exception of thrips.

Exclusion of pollinators before anthesis, by using bagging or caging techniques, can be used in combination with hand pollination methods (see below) to identify whether flowers are able to self-pollinate and self-fertilize. Open-pollinated emasculation treatments (involving removing the anthers before anthesis) should

Fig. 6.2 Pollination bag being used to exclude pollinators from flowers of the tropical tree *Cordia alliodora*. (Photo by David Boshier.)

be included in such investigations to differentiate between facultatively outcrossing and facultatively selfing plants (Kearns and Inouye 1993). Table 6.1 summarizes how the results of such experiments should be interpreted, and Table 6.2 presents additional procedures that can be used in this type of experiment to measure reproductive success.

Results from bagging and hand-pollination experiments provide information on the potential methods by which seed set can occur in a particular species, which can be analysed in a variety of ways. Zapata and Arroyo (1978) suggested the following index to measure self-incompatibility (ISI):

$$\text{ISI} = \frac{\text{fruit set from self-pollination}}{\text{fruit set from cross-pollination}}$$

Fig. 6.3 Crane being used to attach pollination bags to mature ash trees (*Fraxinus excelsior*). (Photo by David Boshier.)

Table 6.1 Field tests for the determining the ability to self-pollinate (adapted from Kearns and Inouye 1993).

Treatment	Breeding system selfing	Facultatively outcrossing	Outcrossing
Bagged/caged	±	+	−
Bagged/caged and emasculated	−	−	−
Bagged/aged and self-pollinated	+	+	−
Emasculated and open-pollinated	−	+	+
Open-pollinated	+	+	+

+, seed set recorded; −, no seed set.

Table 6.2 Variables used to measure reproductive success (after Dafni 1992).

Variable	Procedure
Pollen germination on the stigma	Examined (or fixed) several hours after natural or artificial pollination
Growth of pollen tubes	Examined (or fixed) several hours after pollination
Fruit set	Counting fruit number as a result of the treatment (% of the test flowers) at the end of the reproduction cycle
Number or mass of seeds per fruit	Weighing and counting seeds per fruit
Seed viability	Germination or tetrazolium test (see Section 4.8)

Values of ISI can be interpreted as follows:

- >1 = self-compatible
- $>0.2<1$ = partly self-incompatible
- <0.2 = mostly self-incompatible
- 0 = completely self-incompatible.

The frequency of self-pollination (S) can be estimated by comparison of seeds from naturally pollinated flowers (P_0) with those of hand self-pollination (P_s) and hand cross-pollination (P_x), as follows (Dafni 1992):

$$S = \frac{P_x - P_0}{P_x - P_s}$$

The selfing rate (S) can also be defined as $1-t$, where t is the outcrossing rate. Charlesworth (1988) presented a mathematical method of estimating outcrossing rates in populations of self-compatible plants, based on differences in the viability of zygotes produced by open pollination, hand-outcrossing and hand-selfing. Alternatively, appropriate molecular markers can be used to determine outcrossing rate (see section 6.5.1).

6.2.5 Hand pollination

Flowers can be artificially pollinated by hand (Figure 6.4), which is a useful method to test the viability of pollen (see next section), or to test whether the absence of pollinators is a cause of reproductive failure. In the latter case, hand-pollination offers a practical conservation technique, enabling plants with no pollinators to reproduce. It is important to remember that more than one pollen grain is usually required to initiate seed production. Methods of collecting and transferring pollen include the following (Kearns and Inouye 1993):

- Collect pollen from dehiscing anthers with toothpicks, needles, small paintbrushes, or forceps, or on small pieces of tissue paper wrapped around forceps. Sometimes fabric has been used attached to wooden sticks, to simulate bee hairs. Pencil tips have also been used to scoop anthers out of a flower and transfer them to a recipient stigma.
- Collect pollen by tapping dehiscing anthers over a Petri dish.
- To transport pollen collected on toothpicks or needles, poke the clean end of a toothpick through the lid of a plastic vial and then close the vial around the end with the pollen (Price and Waser 1979).
- Rub an entire male-phase flower or anther over the recipient stigma.
- Mimic pollen application by birds or insects by using dead animals to transfer pollen, for example introduce the beak of a bird or the head of a bee into the

Fig. 6.4 Artificial pollination of the night-flowering, bat-pollinated tropical tree *Bombacopsis quinata*. A small brush is being used to transfer pollen to a receptive stigma. (Photo by Mark Sandiford.)

flower of a pollen donor and then the receptive flower. Construct 'bee sticks' from dead bees and cocktail sticks, by using glue to attach the bee thorax to the stick.
- Larger amounts of pollen can be transferred by using hypodermic syringes equipped with rubber bulbs.

6.2.6 Pollinator foraging behaviour and visitation rates

In order to determine which species are responsible for pollination, it is not sufficient simply to make observations of animals visiting a flower (flower visitors), as some flower visitors may rob the nectar without transferring pollen. To identify the pollinator of a species, it should be determined whether:

- pollen is transferred from the visitor to the stigma
- pollen is transferred between flowers on a plant or among plants
- fertilization or seed production has occurred as a result of the pollination.

Methods described in the previous sections can be used to make this determination.

The behaviour of pollinators can affect the success of pollination and patterns of gene flow. Foraging behaviour can be examined experimentally, for example by using techniques such as artificial flowers or by manipulating flowers. It is even sometimes possible to train pollinators as part of experimental investigations, as in the case of hummingbirds studied by Feinsinger and Busby (1987).

When assessing animal foraging behaviour, careful consideration should be given to sampling design. Ideally, a sampling unit should be reproducible to enable comparison of the activities of foragers in different times or places. Transect methods are widely used for this purpose, in which an observer walks at a constant pace along a line while recording the presence of visitors on the flowers in the sample. Typical lengths of transects are 100–200 m $\pm$ 3 m broad (Dafni 1992). Remember that points along the transect are not statistically independent; it is the transect itself that is the sample unit, and the entire transect that must be replicated. Alternatively observations may be carried out from fixed points, for the same number of flowers or inflorescences, over a standard period of time (10–30 min is typical). Video recording offers a method for recording animal behaviour that can then be analysed in detail back in the laboratory. Some of the variables commonly measured in field studies are listed in Table 6.3.

Visitation rates (number of visits to a flower per unit time) can be measured from the perspective of either the flower or the visitor, by timed field observations (timed with stopwatches). From the perspective of the visitor, it may be helpful to record the time expended to visit a single flower (which can be further subdivided into the time taken to land on and manipulate the flower, and the time taken to extract nectar), the spatial distribution of flowers, and how rapidly the visitor can move within and between inflorescences. Visitation can sometimes be inferred indirectly; for example, some nectar and pollen robbers leave a hole in the flower that can be recorded. Some flowers may be scarred or otherwise damaged by

Table 6.3 Variables commonly measured in assessments of forager activity in pollination studies (after Dafni 1992).

Variable	Procedure
Index of visitation rate	No. of total visits at the observation period / no. of available flowers at this period.
Visitation rate	No of visits / flower×hours.
Effective visitation rate per flower per time unit (V)	$V = (A \times N)/C$, where A is the no. of visitors per plant per time unit, N is the no. of visited flowers per visitor, and C is the no. of flowers per plant.
Visitation rate (VR)	$VR = FT/(HT \times FN)$, where FT is the foraging time per hour, HT is the species-specific handling time of the flower, and FN is the no. of flowers observed.
Attractiveness index	No. of visitors/available flowers per time unit.
Foraging rate	No. of flowers visited/time unit.

visitors, or may display a morphological response to pollination (such as changes in petal colour, wilting, or flower abscission). However, in most cases visitation can only be quantified by direct observation, which can be very time-consuming. Automated photography techniques have sometimes been used to reduce the labour required for such studies (see, for example, Goldingay *et al.* 1991). Visitation data can be summarized as the number of visits per flower per minute.

Patterns of movement of pollinators can be described by measuring flight distance, or the linear distance between two flowers visited in succession, and change in direction, or the difference in angular direction from one flower to the next in relation to the direction of approach to the first flower. Such patterns can vary with the density and spatial distribution of flower resources, and can be an important determinant of pollen flow (Kearns and Inouye 1993). Movement patterns may be determined by following an individual flower visitor as it forages, with a compass to measure the direction of each flight and a voice recorder to measure the direction, time, and distance of each flight. Alternatively, all of the plants within a study area can be mapped and the movements of animals as they visit the plants recorded. Such observations can be made in three dimensions as well as two, for example by labelling branches or flowers in a volume of forest canopy.

6.3 Flowering and fruiting phenology

Phenology refers to the seasonal timing of flowering and fruiting. Describing patterns of phenology can be important for understanding the relative potential for gene flow at different times of the year, and the reproductive isolation of individuals or populations. Seasonal variation in flower and fruit availability will also affect populations of animals dependent on these structures as a source of food.

Phenology may be studied at the level of individual flowers, plants, populations, or communities.

At the simplest level, records of the first date of flowering and fruiting can be obtained from regular field surveys. Quantitative studies of phenology can be conducted by counting the numbers of flowers or fruit, from which a number of different variables can be derived (see Table 6.4). Such data can be used to produce *flowering curves* (the number of flowers or inflorescences open plotted against census date) or *fruiting curves* (similarly, the number of fruits available over time). Decisions must be made regarding how frequently counts should be done: some flowers may last only days or even hours, and therefore observations may need to be made hourly, whereas at the other extreme weekly or even monthly observations

Table 6.4 Selected variables commonly recorded in phenological studies (following Dafni 1992). Although the variables described here refer to flowering, similar approaches can be applied to assessment of fruiting phenology.

Variable	Definition
Flowering commencement	The date (or day number in the year) of the first flowering.
Rate of flowering	The cumulative numbers of flowers versus time.
Course of flowering	The number of flowering units versus time.
Peak of flowering	The date of maximum number of flowers, flowering plants, flowering species.
Relative flowering intensity	The number of flowers at the individual plant peak as a percentage of the highest number of flowers of its conspecific individual.
Index of flowering magnitude	The number of simultaneous open flowers in a given time as the percentage of the total number of flowers.
Mean flowering duration	The mean flowering duration of the sample in days.
Midpoint flowering time	The midpoint of extreme record dates of flowering.
Dispersion of the flowering curves during the flowering season	The observed variance of the distance between mean flowering dates is compared with the expected variance based on randomly dispersed means and a uniform distribution. If the ratio of the expected variance is much greater than 1, then the dispersion is aggregated; if much less than 1, then the dispersion is even.
Flowering overlaps	At the community level, the observed overlaps in flowering between each two species may be compared with those overlaps generated by a series of randomizations.
Flowering termination	Last date of flowering.

may be more appropriate, depending on the characteristics of the species being studied. Attention must also be paid to the sampling design adopted. Many studies involve sampling individual plants within (randomly located) permanent plots, which are then repeatedly visited over time. Alternatively, individual trees or branches may be selected for study.

Phenological studies of mature trees are complicated by the difficulty of observing reproductive structures in the forest canopy, and of producing reliable estimates for large trees. Binoculars can be used to count inflorescences or fruits. Rather than attempting to sample the entire canopy, a sample of branches (typically 10–20) can be marked at different locations in the canopy, and subsequently used to make extrapolations to the whole plant. Another approach is to collect flowers or fruits in traps (see next section). For example, House (1989) measured the rate and duration of flower production in rain forest trees by collecting flowers in traps as they fell from the canopy, in units of numbers of flowers trapped per square metre of crown shadow per unit time. From such data, a number of statistics can be readily calculated, such as length of flowering or fruiting period, maximum number of flowers in bloom, number of species flowering at a particular date, etc. Long-term data can be of great value in determining annual patterns of variation in flower or fruit production, and are essential to detect the masting behaviour demonstrated by many tree species.

D'eca-Neves and Morellato (2004) highlight the fact that many different methods of sampling and describing plant phenology have been used during the last 30 years, making it difficult to compare the results of different studies. Among the 60 studies analysed, these authors recorded the following distribution of sampling methods: trails (20%), transects (18%), field plots (15%), and traps (10%). Furthermore, Hemingway and Overdorff (1999) found that the method used to collect phenology data can affect the results obtained, transect methods detecting a higher number of food resources used by primates than selected tree methods.

A variety of different methods have been used to analyse and present phenological information. Simple graphs or calendars can be produced illustrating the change in phenological state of an individual plant or community throughout a growing season. However, as pointed out by Newstrom *et al.* (1994), the choice of graphical style has a major influence on the detection of phenological patterns and the interpretation of results. Newstrom *et al.* (1994) present three different types of graph: *time series graphs*, which show the frequency and regularity of phenological cycles; *matrix graphs*, showing the duration and date; and *bar graphs*, to illustrate seasonality of flowering frequency and amplitude (Figures 6.5 and 6.6).

Patterns of staggered flowering described for tropical forests illustrate the problem of presenting and analysing phenological data. For example, Stiles (1977, 1978) described sequential flowering in a group of 10 different plant species, which were pollinated by different hummingbird species. Similar results have been obtained with pollination by insects (Figure 6.7). This pattern could be interpreted as evidence for competitive interactions or co-evolution between species. However, differences in the seasonal overlap of flowering are difficult to test statistically.

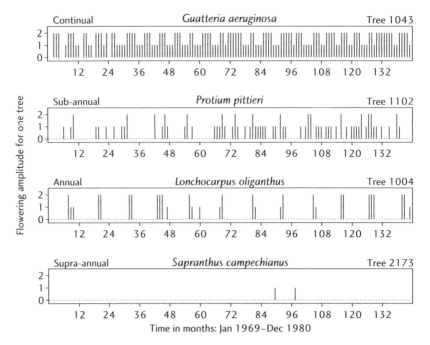

Fig. 6.5 Time series graphs showing the frequency and regularity in four basic flowering patterns. Each graph shows monthly flowering from 1969 to 1980 for one tree of each of four species. Amplitude categories are 1, light flowering; 2, heavy flowering; dots on the x-axis, no flowering. (From Newstrom et al. 1994).

Estabrook et al. (1982) describe a simple non-parametric statistical procedure to address this issue. Phenological observations, such as the day on which a particular flower bud opened, can be compared between two groups of plants by testing whether the probability that a bud will open on any given day is the same for buds in both groups of plants. This can be achieved by using the formula:

$$P(D) > 1.36\sqrt{[(m+n)/mn]} = 0.05$$

where m is the number of buds that eventually flowered in one group, n is the number of buds that eventually flowered in the other group, and D is the difference in cumulative frequency. The larger the value of D, the stronger is the evidence that the two phenological patterns are different. The formula enables the threshold value of D to be determined, at $P = 0.05$. The value of D can be determined directly from the data, as the difference in cumulative frequency (for example in bud break) between the two groups, and if this is larger than the threshold value, then the two groups of plants are statistically different at $P < 0.05$.

A variety of other methods have been used to analyse phenological data. For example, Osawa et al. (1983) developed a stochastic model of bud phenology by using a maximum likelihood technique for parameter estimation, which was fitted

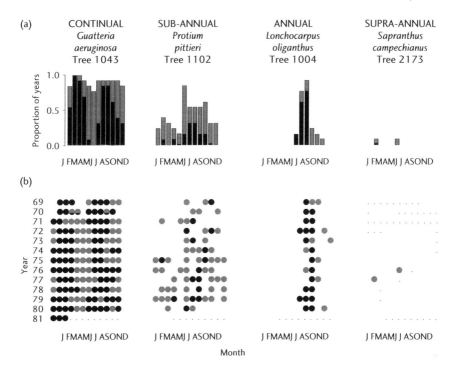

Fig. 6.6 (a) Bar graphs showing seasonality in four basic flowering patterns as proportion of years in which heavy (dark bars) or light flowering (light bars) occurred for each month of the year. (b) Matrix graphs showing duration and date in four basic flowering patterns. Dark shaded circles, heavy flowering; light shaded circles, light flowering; blanks, no flowering. (From Newstrom *et al.* 1994.)

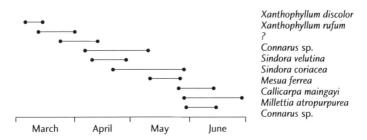

Fig. 6.7 Sequential flowering of 10 species of tropical plant all pollinated by the same carpenter bees (*Xylocopa* spp.). (From Whitmore 1990.)

to balsam fir (*Abies balsamea*) data collected in Quebec. The model was further extended by Normand *et al.* (2002), then tested by using experimental data for populations of strawberry guava (*Psidium cattleianum*) on the island of Réunion. Schirone *et al.* (1990) describe a different approach, based on definition of different phenological stages for leaf development and abscission. A quantitative estimation of each of these '*phenophases*' is obtained by using a seven-point scale for percentage representation, with the unit of assessment being an individual tree or all the branches in a tree crown. Graphs describing the change in phenophase over time, either for individual trees or for a forest stand, can be produced by fitting curves to the data, by using logistic, Gompertz, or Richards functions (Schirone *et al*. 1990). In contrast, Chapman *et al*. (1999) used *spectral analysis* (Fourier analysis) to analyse phenological patterns. This is a type of analysis of variance used to detect cycles of various frequencies in time series data, and involves comparing the variation in the time series about the mean to sine functions of different frequencies. The result is a periodogram that displays the least-squares fit of each frequency to the time series, and was used by Chapman *et al*. (1999) to identify peaks in flowering and fruiting.

6.4 Seed ecology

Units of dispersal can include fruit and vegetative reproductive fragments, but these are referred to here collectively as seed. Further information about seed ecology is provided by Baskin and Baskin (1998), Bradbeer (1988), and Fenner (2000).

6.4.1 Seed production

Measures of seed production are required in order to estimate the fecundity of trees, an important variable when modelling population dynamics and viability (see Chapter 5). Seed production is also often of relevance to studies of those animals that use them as a food source. The simplest method of estimating seed production is by direct observation, for example by counting fruit or cones on an individual tree. For example, Koenig *et al*. (1994a) described a method whereby observers count acorns in the tree canopy within a given period of time (30 s). Other visual surveys have employed simple categorical measures to evaluate production, for example by estimating the percentage of a tree's canopy containing seed, the percentage of twigs containing seed, and the average number of seeds per twig to derive an overall score (Whitehead 1969).

The main problem with observational approaches is that fruits may be difficult to count because they are small, inconspicuous, or obscured by foliage. Such measures are therefore likely to be underestimates, particularly in closed forests. Counts can sometimes obtained from open-grown trees (see, for example, Koenig *et al*. 1994b), but such values are often likely to be biased, because seed production by many tree species is higher in open-grown conditions than in a forest stand. Alternatively, observations can be made from above the forest canopy by using cranes or towers (LaDeau and Clark 2001).

More commonly, some form of seed trap is used (see also section 6.4.2), a technique that is most appropriate for use when trees are isolated from conspecifics and dispersal distances are low. Traps used for seed production studies are usually constructed out of a square wooden frame (typically of 50×50 cm) supported on short legs, with a base of nylon cloth of small mesh size (1 mm). A large number of traps are typically needed for accurate estimates of seed production (Zhang and Wang 1995); published studies have surveyed 0.000 03–0.017% of the forest area being investigated (Chapman et al. 1992). Traps should be visited every few days to avoid decomposition, removal, and damage of fruits and seeds by insects and terrestrial vertebrates. Seed numbers can be counted or their mass measured, and values can be extrapolated to provide estimates per unit forest area (Parrado-Rosselli et al. 2006).

Relatively few studies have compared different methods for assessing seed production (Chapman et al. 1992, Stevenson et al. 1998, Zhang and Wang 1995). In French Guiana, Zhang and Wang (1995) compared three fruit census methods: fruit-traps, observation from platforms situated in the canopy, and a raked-ground survey. The last method involved surveying fallen fruits along a fixed route at regular time intervals, removing the checked fruit after each census. While recognizing the main limitation of fruit-traps (that some of the fruit may be consumed before fruit fall), these authors note one of the key advantages of this method, in being able to sample a relatively large area of forest. However, the number of traps needed may be high, with at least 80 traps needed for accurate assessments of fruiting species richness in diverse forests such as this. The main cause of loss from fruit traps was decomposition, although the risk of terrestrial mammals or insects eating the fruits while in the trap was also noted. Although the raked-ground survey was found to be simple to do, it was found to be inaccurate for determining the quantity of fruit falling from tree crowns, because of rapid consumption by terrestrial animals. Platform observations were found to give the most accurate measurements of the quantity of fruit in tree crowns, the main problem being the logistical difficulty of creating and accessing the platforms, and the limited proportion of the forest sampled.

Parrado-Rosselli et al. (2006) recently compared fruiting data derived from fruit-traps placed on the ground with data from canopy-surveyed plots in a terra firme rain forest, in Colombian Amazonia. Results indicated that estimates obtained by using the two methods were not correlated: values derived from the canopy-surveyed plots tended to be higher than fruit-trap estimates, suggesting that the latter method tends to underestimate seed production. Consequently, these authors suggest that the use of traps should be restricted to particular types of study such as estimates of fruit available for terrestrial frugivores, scatter hoarding rates, and for long-distance dispersal estimates. In contrast, traps should especially be avoided in studies aimed at measuring fruit availability for arboreal and flying frugivores, because a residual quantity of fruits is sampled.

Chapman et al. (1992) compared three methods to estimate fruit abundance of tropical trees: visual estimation, and relations with tree stem diameter (dbh) and crown volume. Estimations from measures of dbh were found to be consistently the most accurate and precise. Allometric models are often used to estimate fecundity of trees, usually as a function of stem diameter (dbh), by using equations such as:

$$y_i = \mu_i + \varepsilon_i$$

where y_i is the annual log seed production by the ith tree having log diameter d_i, and

$$\mu_i = \alpha_0 + \alpha_1 d_i$$

where α_0 and α_1 are regression parameters, and ε_i is a zero-mean error process, $\varepsilon_i \sim N(0, \sigma^2)$ (Clark *et al.* 2004). If fecundity is proportional to basal area (diameter squared), then $\alpha_1 = 2$. Clark *et al.* (2004) provide a more sophisticated model, employing a hierarchical Bayes modelling structure and Markov-chain Monte Carlo techniques, to estimate fecundity from the two types of data that ecologists typically collect, including seed-trap counts and observations of trees. The relations between tree size, seed mass, and seed production are reviewed by Greene and Johnson (1994), who showed that tree size (basal area) is related to seed production, and that seed production is highly negatively correlated with mean seed mass according to a power law relation.

6.4.2 Seed dispersal and predation

It is useful to differentiate between primary dispersal, which is the movement of seeds from the inflorescence on the parent plant to its first settling point, and secondary dispersal, which is any subsequent movement prior to germination (Gibson 2002). Zoochory refers to dispersal by animal vectors, anemochory by wind, and hydrochory by water. A complete assessment of seed dispersal in a particular species should involve consideration of all possible vectors. A detailed review of different dispersal mechanisms is provided by Van der Pijl (1982). Methods for measuring and modelling seed dispersal are reviewed by Greene and Calogeropoulos (2002).

Seed traps

Measurements of the density of seed falling on a particular area, or the *seed rain*, are generally made by using a seed trap (Figures 6.8 and 6.9). The following attributes of a seed trap are important (Kollmann and Goetze 1998):

- It should be designed to allow ready separation of seed from litter, soil particles and insects.
- There should be some protection against seed predation.
- The seeds must not be allowed to rot or decay before examination.

The seed rain is likely to display pronounced spatial variation, and this needs to be taken into account in designing an appropriate sampling approach. The most difficult part to measure of the seed dispersal curve is the tail of the distribution. Ideally, trap area should be increased so that the same proportional area is sampled

Seed ecology | 253

Fig. 6.8 Seed traps manufactured from wire and plastic netting, used in seasonal tropical montane forest in northern Thailand. (Photo by Kate Hardwick.)

at increasing distances from the parent plants (Bullock and Clarke 2000). All methods for assessing seed rain are labour-intensive; a large sample size is needed and it may be necessary to sample in different seasons or even all year round (Bullock 1996).

The following types of seed trap have been used (Gibson 2002, Kollmann and Goetze 1998):

- *Flat traps*. This involves spreading tissue paper or aluminium foil on the ground in order to catch seed. This method is generally very inaccurate because of high seed losses due to the effects of rain, wind, or predation. However, the method was successfully applied by Greene and Johnson (1997) in their assessment of birch seed dispersal across snow.
- *Soil traps*. Pots or trays of sterilized soil or other growth medium are placed in the field. The trays can then be taken to the laboratory or greenhouse for

Fig. 6.9 Another design of trap used to collect fruit, to measure production. In this case fruit production by a single tree of *Aquilaria malaccensis* is being measured. The trap is constructed from nylon netting on a wooden frame, which extends beyond the limit of the tree canopy. The man pictured is the proud creator of the trap, Tonny Soehartono. (Photo by Adrian Newton.)

germination trials; the number of seedlings that subsequently appear are taken to indicate the density of seeds deposited on the site. The accuracy of this method is not high because of variation in germination rates, mortality of seed and seedlings, seed predation, and secondary dispersal.
- *Sticky traps.* Trays, boards, or filter paper 10–30 cm in diameter can be covered with non-drying glue or grease, then exposed either vertically or horizontally. Once seeds land on the glue, they become fixed, enabling the trap to be examined in detail in the laboratory. An example is provided by Werner (1975), who used 'Tanglefoot' glue, a bird repellent available in many countries.

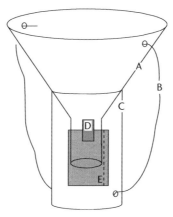

Fig. 6.10 Example of a design for a funnel seed trap. (From Cottrell 2004). A polyethylene funnel (A) is held by wire (B) connected to a PVC pipe (C). A small piece of duct tape (D) is used to attach narrow mesh (100 μm) cloth (E) to the funnel. The funnel is supported at a desired height by the PVC pipe, and should always be at least 2 cm above soil level to limit insect entry.

  The sticky cards used to trap insect pests can also be used (Bullock 1996). The design is highly efficient for wind-dispersed species, but examination of the traps can be time-consuming because they collect other debris, such as insects, dust, and leaf litter. If the traps are covered by snow or overhanging vegetation, their performance is impaired.
- *Wet traps.* Seeds can be caught in shallow containers filled with water or kerosene; if water is used, evaporation can be reduced by using a thin layer of paraffin oil. Kerosene is used to prevent freezing during cold weather (Matlack 1989).
- *Bucket traps.* Deeper containers, such as plastic buckets, can be inserted into the soil or fixed to the ground. A hole in the bottom of the container permits drainage of water. A screen can be placed on top to prevent granivory. This form of trap can work well for assessing the seed rain of zoochorous species.
- *Nets.* Traps can be constructed as bags made out of narrow mesh nylon netting and suspended either below tree branches, or on wooden frames at ground level. This method can provide useful measurements of seed production by trees, but traps must be emptied regularly to prevent loss of seed as a result of wind or the actions of seed predators. An example is described by Hughes *et al.* (1987).
- *Funnel-shaped or cylindric dry traps.* This method has been widely used and is considered to be the most efficient type of seed trap (Kollmann and Goetze 1998). The most important aspect of the design is to ensure that seeds remain in the trap after being caught, particularly during windy conditions. A simple version of this trap can be constructed by cutting the top section off a plastic bottle, then inverting it over the base section to provide a funnel (Figure 6.10). Wire gauze placed over the top of the trap can be used to

exclude larger seed predators. The traps can be inserted in the soil, or placed on the soil surface, and should be inspected regularly to minimize seed losses (further details are provided by Cottrell 2004). Hydrochorous seed can be captured by using floating funnel-traps that remain on the water surface during fluctuating water levels (Middleton 1995).
- *Natural traps.* Zoochorous seed can be sampled from animal faeces, hair, fur, and feathers. Examples are provided by Izhaki *et al.* (1991), who counted the density of seeds in 'droppings traps' constructed out of nylon sheets that were placed under trees; Campbell and Gibson (2001), who counted seedlings emerging from samples of horse dung; and Voysey *et al.* (1999), who assessed tree seed dispersal by gorillas by using this method.

Once seed have been trapped, their viability can be tested by using the methods described in section 4.8.

Trapless methods: examining seeds, germinants, and seedlings

In some cases it is possible to sample seeds directly on the ground surface, for example by counting the number of seeds present in quadrats located randomly or along transects. This approach is particularly useful with large seeds (see, for example, Augspurger and Hogan 1983), but is very time-consuming with small seeds, which may be difficult to detect among leaf litter. Examination of germinated seeds (germinants) and seedlings can also be used to infer dispersal patterns. For example, Ribbens *et al.* (1994) used the relationship between seedlings and conspecific trees to estimate seedling production and dispersal. These authors present a model that predicts seedling density as the summed contribution of seedlings from all trees in a sample plot, estimating fecundity and dispersal distance of seedlings based on the summed contributions of potential parent trees. The main problem with inferring dispersal from seedling distribution is that other factors than purely dispersal may account for the distribution of seedlings, including the availability of seedbeds or microsites suitable for seed germination, herbivore density, the availability of light, or competition with other plants. Such factors may vary with increasing distance from the parent plant.

Both traps and examination of seeds or seedlings share the problem that the source cannot be known with certainty except in situations where a single plant or group of plants is isolated from other conspecifics. Therefore, when using these methods, isolated trees should be selected for analysis. The following methods provide an alternative way of overcoming this problem.

Observations of seed movement

In the field, the usual approach is to place or release seeds on or close to the maternal plant, or to mark seeds still attached to the plant before their natural release, and then to track their movement (Gibson 2002). Often, isolated individuals are used to facilitate tracking. Although it is possible to follow seed from the source to the landing site by visual observation, this method is time-consuming and often difficult. A more efficient approach is to mark the seed in a way that allows subsequent identification. Care is needed to ensure that the method of marking does not in itself alter the pattern of dispersal, particularly in the case of small seeds (Gibson 2002).

One of the most commonly used methods of marking seed is spray painting (Greene and Johnson 1997), including the use of fluorescent paint, which enables seeds to be identified during subsequent searching by using a ultraviolet lamp (Bossard 1990). Other methods include the use of radioactive tagging (Winn 1989), which involves injecting the parent plants with a radionuclide (often scandium-46) in solution while the seeds are developing, then using a counter to detect the seeds after dispersal. In the case of large seeds, small pieces of metal can be inserted in the seed, which can then be relocated by using a metal detector (Sork 1984, Mack 1995). This method was recently tested by López-Barrera *et al.* (2006a) with oaks in Mexico. The acorns were drilled with a hand drill and a small nail (15×3 mm) was inserted inside each acorn (following Sork 1984). The head of the nail in each acorn was exposed above the surface and painted with fluorescent spray paint to facilitate relocation. A number of different models of metal detector were tried before finding an instrument that successfully detected the tags under field conditions. However, in a trial of more than 4000 tagged acorns, only 2.2% of the dispersed acorns were recovered, the others presumably having been cached by the small mammals responsible for dispersing them. This highlights one of the key problems with tagging seeds: often the proportion recovered is very low, and it is impossible to know whether this results from insufficient sampling or that a fraction of the seeds has passed beyond the maximum sampling distance (Greene and Calogeropoulos 2002).

Alternatively, individual seeds may be followed during transit (the Lagrangian method), an approach that has been used with wind-dispersed seeds as well as with those dispersed by birds, insects, and mammals. For example, an animal can be followed, and its caches or dung piles examined. However, this approach only works for those seeds travelling short distances, because those travelling further can be difficult to follow (Greene and Calogeropoulos 2002).

Forget and Wenny (2005) review the methods used to study seed removal and secondary seed dispersal, and make the following points:

- Although seed dispersal behaviour can be observed directly in open habitats, with small animals that travel short distances, few observational studies include data on final post-dispersal seed fate. In closed forests, visual observation of animals removing seeds is more difficult, although remote cameras and video surveillance systems have successfully been used. Marking seeds is essential in the wild when seeds are taken by vertebrates and transported out of sight.
- As an alternative to direct observation of animals moving seeds, indirect methods of recording animal movements can be used. For example, trapped rodents can be marked with a fluorescent powder and their travel route

retraced with an ultraviolet lamp. The method works best under dry conditions and is therefore not suitable in rain forests.
- Attaching seeds with a nylon thread to a fixed point such as a small tree or twig is useful for distinguishing between biotic and abiotic causes of seed movement, and for determining which types of animals remove seeds. The method also enables seed fate to be determined. The seed can be glued to the line or passed through the seed after drilling a small hole. Line lengths of 30–200 cm have often been used; shorter lengths are less likely to get tangled. Often, in such experiments, the location of seed is marked by using coloured stakes, flagging tape, or toothpicks. If so, care should be taken to ensure that the method of labelling the locations does not attract seed predators and therefore bias the results.
- The spool-and-line protocol consists of a thread-filled bobbin from which the line is dispensed as the seed attached to the end of the line is carried away. The travel route and final location of the seed can be determined by following the line. The main problem is that the line can often be broken, and tracing the line can be difficult. The method works best when studying ground-dwelling animals, but preparing spools can be time-consuming.
- Most studies of seed fate use a free marking method to relocate the removed seeds, and to discriminate between seed predation and hoarding. In addition to radioisotopes or metal objects mentioned above, miniature radio transmitters or magnets can be used. If the seeds are large enough it is useful to number and mark them individually, allowing individual fates to be determined. This can be achieved by using indelible ink.

Overall, Forget and Wenny (2005) suggested that, although the use of magnets or radiotransmitters is efficient in retrieving seeds, especially over long distances, they are relatively expensive, and therefore free-line methods tend to be more commonly used in practice. Although radioisotopes have proved very effective, concerns about the environmental impacts of radioactivity have deterred researchers from using them. Most methods work best with larger seeds, and therefore as a result relatively little is known about smaller seeds. Most studies focus on common, short-distance events, even though rare, long-distance events may well be more important to plant community dynamics. Studies that combine techniques are therefore recommended.

Seed predation

Many studies of post-dispersal seed fate use seed removal as an indicator of seed predation. Commonly, a population of seeds is placed on the ground, either singly or in clumps, and their removal is monitored over time. From such observations, rates of seed loss can be calculated. However, it may be more appropriate to analyse such seed loss data as a categorical variable. For example, in a seed predation experiment in Mexico, López-Barrera *et al.* (2005) transformed the proportion of seeds removed into a categorical variable with three levels (low: 0–33%, medium: 33–66% and high removal: 66–100%). In a second experiment, data appeared to be bimodal (all or most of the seeds were either present or missing), therefore acorn removal was assigned to two removal categories (low: 0–50% and high: 50–100%). Such data can be analysed by using categorical modelling (such as Proc CATMOD in SAS; see SAS 2002), a procedure analogous to ANOVA but with categorical data represented in contingency tables (López-Barrera et al. 2005).

Investigations that interpret removal of seed purely as seed predation overlook the possible occurrence of secondary dispersal, and therefore overestimate the magnitude of predation. Vander Wall et al. (2005) examine the recent literature on seed removal studies, and indicate how the results of some of these investigations may have been misinterpreted, highlighting the importance of detailed studies of seed fates in order to assess predation. Methods used to assess seed fates by tracking are described in the previous section. Following seeds can be very challenging but is the only sure way of gaining knowledge about seed fates (Vander Wall et al. 2005).

Fitting a dispersal curve

Measurements of seed density or the seed rain can be used to produce a *seed dispersal curve*, which describes the frequency distribution of dispersal distances, and the *seed shadow*, which is the post-dispersal spatial distribution of seeds around the maternal plant (Gibson 2002). The seed dispersal curve for wind-dispersed seed is usually described by fitting a negative exponential curve to the data, of the form:

$$S_D = a_1 \exp(-b_1 D)$$

where S_D is the density of seeds at distance D from the source, and a_1 and b_1 are constants indicating the density of seeds falling at the source and the slope of the decline in seed density with distance, respectively (Bullock and Clarke 2000). This gives a linear relation between $\ln(S_D)$ and D; the slope of the line provides a measure of the rate of decline in seed density with distance (Gibson 2002). Some authors (for example Willson 1993) have suggested that logging both axes gives a more realistic relation, which accords with the inverse power model (Bullock and Clarke 2000):

$$S_D = a_2 D^{-b_2}$$

These authors found that an empirical mixed model based on the negative exponential and power models provided a better fit than either individual function. This is described by:

$$S_N = T_A \left(a_3 \exp(-b_3 D) + c_3 D^{-p_3} \right)$$

where S_N and T_A are the total number of seeds trapped and the total area of traps as distance D from the centre of the plant; a, b, and c are constants; and p is the inverse power parameter (Bullock and Clarke 2000). The length of the tail of the fitted distribution can be reported as the distance within which 84% of the seeds fall (equivalent to one standard deviation of the mean trajectory) (Greene and Johnson

1989). The range and mean of the measured dispersal distances should also be reported (Gibson 2002). Bullock and Clarke (2000) suggest that researchers should measure the tail of dispersal curves and examine carefully whether the widely used exponential and power models are, in fact, valid.

Recruitment limitation

As a result of the difficulties of measuring seed production and dispersal in forests, the influence of these processes on the population dynamics of tree species has been largely overlooked. Recently, there has been growing awareness that seed may often be in short supply and can limit seedling recruitment. The phenomenon of recruitment limitation (Figure 6.11) has attracted increasing attention from researchers working in a variety of different forests (Hubbell *et al.* 1999). Clark *et al.* (1999a) review the concept, and highlight the fact that most studies of seed rain, seed banks, and seedlings are undertaken for relatively short periods (usually 1 year, and very rarely as much as 5 years). These authors highlight the inadequacy of such sampling approaches, and indicate the need for data to be collected across multiple years and multiple forest stands, something rarely achieved to date. In addition, most efforts to quantify seed production and dispersal (such as observations of seedling distribution) are highly indirect. Clark *et al.* (1998b) illustrate how recruitment limitation can be analysed by using a combination of modelling approaches and in intensive analysis of fecundity, seed dispersal, and establishment, by using data from an array

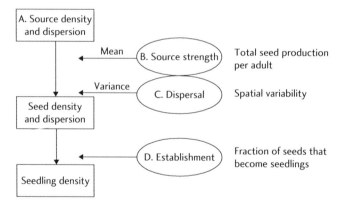

Fig. 6.11 Processes responsible for recruitment limitation in forests. (From Clark *et al.* 1999b). Seed arrival depends on: (A) density and dispersion of adults, (B) adult fecundities, and (C) dispersal distances of seed. Seed arrival can be constrained by source-density (A), source-strength (B), and dispersal limitation (C). A fourth limitation on recruitment, establishment limitation (D), depends on seed survival and germination and seedling survival, which depends on physical (e.g. light, water and nutrients) and biotic (e.g. litter depth, seed and seedling predators, pathogens, and competitors) factors that vary at several spatial scales, both within and among forest stands.

of 100 seed traps, and surveys of several thousand mapped trees and seedlings in five southern Appalachian forest stands undertaken over 5 years. Another useful technique for studying recruitment limitation is to perform seed-sowing experiments, an approach receiving increasing interest from researchers, although relatively few such studies have been carried out with woody plants to date (Flinn and Vellend 2005, McEuen and Curran 2005).

Long-distance dispersal

Although it is widely recognized that dispersal over long distances is ecologically very important, it is very difficult to measure. Nathan *et al.* (2003) review methods for estimating long-distance dispersal and identify a variety of alternatives:

- Drawing inference from *biogeographical distribution patterns*; however, dispersal rate and method of dispersal cannot readily be estimated by using this approach.
- Observation of *movement* (for example by the Lagrangian method, see above); this can give accurate estimates (by feeding labelled seeds to animals, for example; Yumoto 1999), but is difficult over long distances.
- Short-term and long-term *genetic analyses* (see next section).
- *Modelling*, including both empirical and mechanistic models.

Greene and Calogeropoulos (2002) further review methods for assessing long-distance seed dispersal, and conclude that there is no single preferred method. Trapping methods are particularly inefficient when used over the large areas needed to estimated long-distance dispersal. These authors also highlight the need to test models with empirical data describing the far tail of the distribution, where seed densities are relatively low—something rarely achieved to date.

A range of complex mechanistic models are available, taking into account windspeed and height of release (Andersen 1991, Greene and Johnson 1989, Katul *et al.* 2005, Okubo and Levin 1989). However, the applicability of these models has rarely been tested (Bullock and Clarke 2000). Horn *et al.* (2001) highlight the fact that dispersal of forest tree seeds by wind is biphasic: seeds that fall within the canopy have no chance of long-distance dispersal, but seeds that rise above the canopy do. For this reason, mixed models such as those presented by Bullock and Clarke (2000) and Clark *et al.* (1999b) are appropriate, as they provide an opportunity to estimate separately the partitioning of seeds between those that are dispersed over long distances and those that fall locally. However, zoochorous species can display dispersal curves very different from those of wind-dispersed species, and tend to be highly clumped (for example, high densities of bird-dispersed species may be found under perching sites). The ecology of seed dispersal is reviewed in detail by Levin *et al.* (2003).

Trakhtenbrot *et al.* (2005) review the importance of estimating long-distance dispersal for conservation planning, and highlight two separate issues: the dispersal of alien or exotic species, which may invade ecological communities, and reduction in dispersal of native species through processes such as habitat fragmentation. The authors highlight the fact that quantitative assessment of long-distance dispersal is highly informative but also very costly in terms of data collection, and therefore recommend that such analyses be restricted to addressing the most crucial threats (such as invasive species). Qualitative assessments can provide a useful first step, for example by enabling potentially invasive species to be identified through a consideration of dispersal vectors. Where quantitative analysis is needed, the use of mechanistic models is recommended, ideally supported by empirical tests.

Molecular methods are useful for estimating both short- and long-distance dispersal, especially when used in conjunction with studies of germinants (Cain *et al.* 2000) or trapping methods. For example, Jones *et al.* (2005) used microsatellites to identify the parent trees of seed collected in a large array of traps, for the neotropical tree species *Jacaranda copaia* (Bignoniaceae). Use of molecular marker variation to analyse dispersal is considered below (see section 6.5.1).

6.5 Assessment of genetic variation

Information about the extent of genetic variation within species is usually obtained either by using molecular markers, or by assessing variation in quantitative traits. Both of these approaches are considered below.

6.5.1 Molecular markers

Most often, molecular markers are used to assess patterns of genetic variation within and between populations, but they can also be used to provide information about plant breeding systems, clonal structure, the evolutionary history of populations, and estimates of gene flow. Such measures have direct relevance to forest conservation and management (Haig 1998). Molecular methods are also widely used for addressing questions relating to evolutionary relationships and taxonomy, which are not considered further here. Further details regarding application of molecular methods in ecology are provided by Baker (2000), McRoberts *et al.* (1999), Ouborg *et al.* (1999), and Parker *et al.* (1998). Two recent introductory texts on ecological and conservation genetics are by Frankham *et al.* (2002) and Lowe *et al.* (2004). Young *et al.* (2000) present a valuable compilation of methods relating explicitly to forest genetics, and an overview of approaches to managing and conserving forest genetic resources is provided by the National Research Council (1991). The application of molecular methods to conservation of tree species is considered by Newton *et al.* (1999a).

Practical details of molecular methods are not presented here in depth; the reader is referred to specific texts such as Lowe *et al.* (2004) and Young *et al.* (2000). The subject will also be addressed by another book in this series (P. Taberlet, in preparation). Here, the relative strengths and weaknesses of the main different types of marker are considered in relation to different questions of interest, to help identify which method is most appropriate for a particular situation. Some guidance on use of molecular marker methods for addressing different questions is also provided.

Collection of material

Careful attention should be paid to sampling when collecting material for genetic analysis. Typically, samples are collected from populations in the absence of any information about the patterns of genetic variation that exist, or about the size of the breeding populations of the species. As a result, there is a high risk that any sampling design adopted will provide an inaccurate assessment of patterns of variation within the species. Sampling decisions will obviously depend on the objectives of the investigation, which should be stated clearly at the outset. For example, for an assessment of genetic variation across the entire geographic range of a species, populations must be sampled throughout the entire range, perhaps sampling a relatively small number of individuals (typically in the range 5–30) within each individual population.

Ideally, for this kind of investigation, individuals should be selected for sampling by using random or stratified random procedures. In practice, sampling for genetic studies is often rather haphazard or opportunistic, being governed by which populations and individuals are relatively accessible (Lowe *et al.* 2004). It is important to be aware that any departure from random sampling may introduce bias into the results obtained, which should be borne in mind when interpreting results. If an assessment of genetic variation within a population is required, it is also important to ensure that clonal individuals are not being repeatedly sampled. This can be achieved by stipulating a minimum distance (typically > 100 m) between individuals to be sampled. Investigations at a more local scale, such as studies of gene flow across landscapes, may require all individuals within a given area to be sampled and mapped. In this case, such studies should ideally be replicated at different sites. It may also be necessary to sample a large number of progeny (> 200) for accurate assessments of gene flow (Lowe *et al.* 2004) (see section on p. 270).

Samples of leaf tissue are generally used for DNA analysis, although other plant parts such as roots, seeds, or the cambium layer can also be used (see, for example, Colpaert *et al.* 2005). Leaf samples collected in the field for DNA analysis can be dried by using silica gel following the method described by Chase and Hills (1991), on which the following account is based:

- Fresh young leaf material should be selected, free from any fungal attack or dirt. Remove any surface water from the leaves before sampling. Torn leaf pieces ($<2\,cm^2$) should be removed and placed into a small ($12\times8\,cm$) sealable plastic bags, or 50 ml tubes with screw caps (for example plastic centrifuge tubes).
- Add 50–60 g of 20–200 mesh size, grade 12 silica gel (if the gel is to be reused, a smaller mesh grade should be used: for example, 6–16 mesh size, grade 42). Generous amounts of silica gel should be used; a minimum ratio of

10 : 1 silica gel : leaf mass is required for effective preservation of leaf samples. A small amount (5%) of indicator silica gel can usefully be added. Never reuse gel without first making sure that no contamination from previous specimens is present (by baking or using ultraviolet light to sterilize it, for example). Shake the bag to distribute the gel between the layers of leaves.
- The leaf tissue should be dry after 12–24 hours, by which point the leaf fragments should snap and break cleanly. The sample bags should be kept sealed and ideally stored in a refrigerator before analysis.

For isozyme studies either fresh or rapidly frozen material is required. Seed tissues, or seedlings that have recently germinated, are often preferred for isozyme studies because of the relatively high number of loci that can generally be resolved when using such tissues. However, other plant parts such as buds or root tips can also be used (Soltis and Soltis 1990).

Types of molecular marker

A wide range of different molecular markers are available. Some of those most widely used to assess genetic variation in tree species are briefly described below:

- *Isozymes.* This method involves the separation of different molecular forms of enzymes by using gel electrophoresis. A variety of specific stains are used to distinguish particular enzymes in a tissue extract. When the polypeptide constituents of enzymes are coded by more than one gene, then they are referred to as isozymes, but when coded for by a single gene, they are referred to as *allozymes* (Gibson 2002). Different banding patterns between individuals obtained on a starch or polyacrylamide gel are interpreted as indicating the presence of alternate alleles at a given locus. Isozymes can underestimate genetic variation because only a very small part of the plant genome is being considered, even if a large number of loci are screened. Another problem is that the amount of variation detected may sometimes be small; more than half of all loci may often be monomorphic (Parker *et al.* 1998). Details of the method are described by Soltis and Soltis (1990).
- *Restriction fragment length polymorphism analysis (RFLP).* DNA is digested with restriction enzymes (usually 4–6 bp cutters), then the fragments produced are separated by gel electrophoresis and blotted on to a filter (Gibson 2002). The Southern blot procedure is used to hybridize labelled probes to the bound DNA, to allow discrimination of target fragments homologous to the probe. Fragments that have the same restriction sites (i.e. similar DNA sequence variation) migrate to the same location on the gel. A modification of the method (PCR–RFLP) is widely used, based on the restriction digestion of PCR-amplified products, with the use of primers designed from universal cpDNA, mtDNA, or nuclear DNA sequences.
- *Random amplified polymorphic DNA (RAPD).* A PCR-based method based on the amplification of arbitrarily derived ('random') DNA segments. PCR-amplified products are separated on agarose gels in the presence of

ethidium bromide and visualized under ultraviolet light. Co-migrating bands are assumed to represent identical genome segments for that particular primer pair.
- *Amplified fragment length polymorphisms (AFLP)*. Genomic DNA is digested, then the cleaved fragments undergo several rounds of selective amplification. The amplified products are radioactively or fluorescently labelled and separated on sequencing gels.
- *Microsatellites (SSR)* are short (10–15 copies) tandem repeats that are assumed to be randomly distributed throughout the genome. Primers are designed for the conserved regions flanking the variable SSR. Polymorphism of the SSR is detected by using PCR and separation of products on agarose, polyacrylamide, or DNA sequencing gels.

Choice of marker system

Marker systems differ in a range of characteristics, such as the amount of variation detected, their ease of use, and the costs of development and implementation (see Table 6.5). The choice of a marker is therefore based on the objectives of the investigation, the properties of the marker system, and the resources available (Lowe *et al.* 2004). To some extent, there are trade-offs to be made between cost, ease of use, and information obtained. For example, SSRs are preferred in many investigations because of the high level of polymorphism detected and the high reproducibility. However, this method is relatively expensive, and requires DNA sequence information for the species of interest. As a result, relatively cheap and easy methods such as RAPD and AFLP, which do not require any DNA sequence information, have been widely used with tree species. These methods are particularly useful for an initial examination of the partitioning of genetic variation within a species or for locating centres of genetic diversity. However, the reliability of RAPD has been called into question, and investigations by this method are becoming increasingly difficult to publish in the scientific literature. The technique is effectively being replaced by AFLP, which is a somewhat less straightforward technique.

Marker systems are classified as either *dominant* or *co-dominant* according to their mode of inheritance. This difference has major implications for the type of information obtained, how the data can be analysed, and the kind of question that can be addressed. Dominant markers provide much less genetic information than co-dominant markers, resulting in estimates of population genetic statistics that are much less precise than those obtained from co-dominant markers. The dominance of RAPD and AFLP markers is one of their main limitations; as a result, many more loci need to be assayed to obtain sufficient statistical power to answer a particular question (Glaubitz and Moran 2000).

As isozymes are relatively cheap and easy to do, and are co-dominant, they are a useful starting point for any ecological investigation. They also have the advantage that a substantial body of isozyme data has been collected previously for a wide range of tree species, providing a basis for comparison (Hamrick and Godt 1990,

Table 6.5 Comparison of some of the molecular techniques most commonly used in forest ecology (adapted from Lowe et al. 2004, Newton et al. 1999, and Young et al. 2000). Note that this is not a comprehensive list.

	Isozymes	RFLP	SSR	RAPD	AFLP
Details of method	Gel electrophoresis and histochemical staining of cellular enzymes and proteins	Total genomic DNA digested with restriction endonucleases then probed with specific DNA fragments by Southern blotting and hybridization	Specific PCR primers used to amplify previously characterized hypervariable repeat motifs in nuclear or organelle genomes	Short sequence primers (usually 10-mers) used to PCR amplify random loci throughout the entire genome	Total genomic DNA digested with two restriction enzymes, DNA adaptors fitted to cut sites, and products selectively amplified by using PCR primers
Advantages/ disadvantages	Well-documented enzyme systems can provide unequivocal measures of allele frequencies. Fresh and often specific tissues required, e.g. buds, germinated seeds	Same probes / methods applicable to different taxa. Most commonly requires use of radiolabelled probes. Requires relatively large amounts of sample DNA	Unequivocal single locus alleles can be scored. Microsatellite-containing regions differ between taxa therefore expensive and laborious development required for each new species	Coding and non-coding DNA of potentially all three plant genomes, randomly analysed. Can give low reproducibility and artefactual markers (owing to competition and/or heteroduplexes). Genomic location unknown without controlled crosses	AFLPs are more reproducible than RAPDs, but more expensive. Automation of marker scoring available. Radioactive labels may be required

Number of loci typically obtained	30–50	100s	10s	1000s	1000s
Degree of polymorphism	Low	Moderate	High	Moderate	Moderate
Dominance	Codominant (usually)	Codominant	Codominant	Dominant	Dominant (usually)
Reliability (reproducibility)	High	Very high	High	Low to medium	Medium to high
Amount of sample required per assay	Milligrams of tissue	2–10 µg DNA per lane	25–50 ng DNA	5–10 ng DNA	25 ng DNA
Ease of assay	Easy	Difficult	Easy to moderately difficult	Easy	Moderately difficult
DNA sequence information needed	No	No	Yes	No	No
Cost of equipment	Low	Moderate/high	High to very high	Low / moderate	Moderate / high
Cost of development	Low	Moderate/high	Very high	Low / moderate	Moderate
Cost of assay	Low, although in some countries, the reagents can be difficult or expensive to obtain	High	Moderate to high	Moderate	Moderate

Hamrick *et al.* 1991, 1992, Loveless 1992). (A similar review of use of RAPD with plant species was presented by Nybom and Bartish 2000.) The main limitations of isozymes are the need for fresh or rapidly frozen material for analysis, which can be difficult to obtain when sampling in remote areas, and the relatively low number of loci often obtained. DNA methods might be preferred in such situations, or used as a complement to studies of isozyme variation.

Glaubitz and Moran (2000) suggest that for applications requiring a large number of loci, AFLPs or RFLPs might be the DNA marker of choice. Such applications include:

- measuring genetic variation and differentiation
- estimating rates of gene flow or migration between populations
- genetic linkage mapping, or localization of quantitative trait loci (see following section).

As noted above, for applications requiring a high discrimination power, microsatellites are the preferred marker, because of the very high degree of polymorphism obtained. As few as 5–6 microsatellite markers can often answer conservation genetic questions (such as paternity and pollen flow) that cannot be answered with 30 or more isozyme loci (Glaubitz and Moran 2000). Relevant applications for microsatellites include:

- characterizing mating systems
- analysing paternity or parentage
- characterizing patterns of gene flow or migration within populations.

Some of these various applications are considered in more detail in the following sections.

Measuring genetic diversity and differentiation

Ideally, for any investigation employing molecular markers, information would be obtained on how the putative loci being investigated are inherited. This should be determined by crossing experiments, but this is very difficult to achieve with most tree species because of the long time taken to reach reproductive maturity. Gillet and Gregorius (2000) describe a method for identifying marker inheritance from the analysis of progeny arrays. However, in most studies, such detailed genetic analyses are not carried out and loci and alleles are interpreted based on expected patterns of segregation (Lowe *et al.* 2004).

Many statistics describing genetic diversity and differentiation have been developed for use with isozyme data. The use of such statistics with dominant markers, such as RAPD and AFLP, is problematic because of the unknown proportion of heterozygotes within the population, although analytical methods have been proposed to circumvent this problem (Lynch and Milligan 1994). Kremer *et al.* (2005) provide guidance for estimating diversity by using dominant marker data, emphasizing the importance of using a large number of loci. Simple measures of genetic diversity include estimates of allelic diversity (the number of alleles per locus or the number of polymorphic loci), the percentage of polymorphic loci, and the mean observed heterozygosity (the mean number of heterozygotes recorded at a particular locus expressed as a proportion of the total number of loci surveyed) (Lowe *et al.* 2004).

Nei's (1973) measure of gene diversity is a very widely used measure, and uses the expected heterozygosity across the total species (H_T) calculated as:

$$H_T = 1 - \sum_{i=1}^{i=K} p_i^2$$

where p is the mean frequency of the ith of K alleles across all populations surveyed.

Shannon's index of diversity (H) is another diversity measure widely used in ecological genetics (Lewontin 1972):

$$H = -\sum p_i \ln(p_i)$$

where p_i is the proportion of the ith allele in the population. This method has the advantage of being appropriate for use with dominant markers such as RAPD (see, for example, Allnutt *et al.* 1999).

Genetic differentiation is often assessed by using Nei's G_{ST} (Nei 1973), which is the gene diversity between populations relative to the combined populations. G_{ST} is calculated by using the equation:

$$G_{ST} = D_{ST}/H_T$$

where D_{ST} is the proportion of gene diversity of the species that is present within populations, and H_T is total gene diversity, measured in terms of the total expected heterozygosity. D_{ST} can be calculated as $H_T - H_S$, where H_S is the mean of expected heterozygosities within each population (assuming Hardy–Weinberg equilibrium). G_{ST} varies between 0 and 1, with 0 indicating no differentiation between populations, and 1 indicating that all of the variation detected is attributable to differences between populations.

Wright's F-statistics are also very widely used in genetic studies (Wright 1951). Wright described H_T and H_S as the total expected heterozygosity in the total population and the mean expected heterozygosity within populations, respectively (assuming Hardy–Weinberg equilibrium). Note that these definitions differ from those of Nei (1973), although they have the same mathematical basis (Lowe *et al.* 2004). Wright (1951) also defined H_I as the mean observed heterozygosity per individual. Three main statistics are commonly used (Lowe *et al.* 2004):

$$F_{IS} = (H_S - H_I)/H_S$$

where F_{IS} is the inbreeding coefficient (describing the divergence of observed heterozygosity from expected heterozygosity within populations assuming panmixia);

$$F_{ST} = (H_T - H_S)/H_T$$

where F_{ST} is the fixation index (describing the reduction in heterozygosity within populations relative to the total population owing to selection or genetic drift); and

$$F_{IT} = (H_T - H_I)/H_T$$

where F_{IT} is the overall inbreeding coefficient (describing the reduction of heterozygosity within individuals relative to the total population owing to non-random mating within subpopulations (F_{IS}) and population subdivision (F_{ST})).

These statistics are usually produced by using specialist software programs such as:

- FSTAT (developed by J. Goudet, freely available from ⟨www.unil.ch/izea/softwares/fstat.html⟩)
- POPGENE ⟨www.ualberta.ca/⟩; also provided by Young *et al.* (2000)
- GENEPOP ⟨http://wbiomed.curtin.edu.au/genepop/⟩.

Analysing gene flow and mating system

Gene flow refers to the movement of genes between populations and is a key process influencing the pattern of genetic variation within species. As a result of widespread concern about the occurrence of habitat fragmentation, molecular markers are increasingly being used to provide estimates of gene flow, so that the genetic impacts of fragmentation can be elucidated (Young and Clarke 2000). Gene flow estimates can be made by using molecular markers and either *indirect* methods (in which gene flow is inferred from analysis of the genetic structure of populations) or *direct* methods (in which gametes or progeny arrays are analysed).

Indirect estimates of gene flow can be obtained from *F*-statistics (see above), which provide an indication of historical dispersal between populations. For example, the effective number of immigrants to a population ($N_e m$) can be estimated as described by Hamrick and Nason (2000):

$$N_e m = \frac{1 - F_{ST}}{4 F_{ST}}$$

By using markers with different inheritance patterns it is possible to compare rates of gene flow by seed and by pollen (see, for example, Bacles *et al.* 2004, 2006). Ennos (1994) describes a method for achieving this based on comparison of differentiation in the maternally inherited chloroplast genome and the biparentally inherited nuclear genome. However, such indirect methods have a number of limitations: they can only provide relative estimates, rather than absolute distances and frequencies of gene flow; and they illustrate cumulative gene flow over time, rather than current patterns. Indirect methods are also often based on a number of assumptions (such as an equilibrium having been met between gene flow and drift, and random mating) that are often not met in reality (Lowe *et al.* 2004). As a consequence, direct methods are generally preferred.

Direct estimates of gene flow can be obtained by genetically identifying the parents of gametes and/or progeny (*parentage analysis*; see Box 6.1). The method

Box 6.1 Forest genetics and restoration: the Carrifran Wildwood project.

Few areas in the world are as much in need of ecological restoration as southern Scotland, where less than 0.1% of the original forest area remains. Much of the original forest was cleared for agriculture, a process which began with the colonization of Scotland by Neolithic people some 6000 years ago. Today, native woodlands are reduced to small, scattered remnants on inaccessible cliffs and valley sides. On 1 January 2000, the first trees were planted by the Carrifran Wildwood Project, an ambitious attempt to restore native forest over an area of some 600 ha in the Southern Uplands of Scotland (Newton 1998) (Figures 6.12 and 6.13). The project faced substantial obstacles, some of which were genetic in nature. As very few trees remained on the site, planting stock had to be obtained from elsewhere. It was recognized that planting material should be well adapted to the site, and therefore sourced locally from genuinely native populations. However, there were serious concerns that the extent of genetic variation within such planting stock may be very low, because native woodland fragments in the region are small and geographically isolated, increasing the risk of genetic drift and inbreeding (Newton and Ashmole 1998).

To address these questions, a research programme was undertaken employing a range of molecular markers (isozyme and PCR–RFLP of cpDNA) (Bacles *et al.* 2004). Severely fragmented populations of *Sorbus aucuparia* (Rosaceae) displayed surprisingly high levels of gene diversity (Figure 6.14), similar to values

Fig. 6.12 Carrifran valley, southern Scotland, scene of a major community-based project aiming to restore native woodland to a severely deforested landscape, being undertaken by the Carrifran Wildwood Project. (Photo by Adrian Newton.)

Fig. 6.13 The first tree planting at Carrifran, on 1 January 2000. The author and his young son are pictured at the centre of the photograph, planting a hazel (*Corylus avellana*) seedling. (Photo by Lynn Davy.)

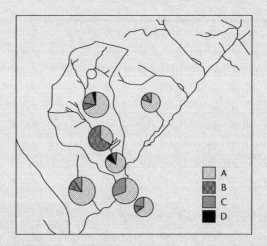

Fig. 6.14 Map of the distribution of four cpDNA haplotypes in sampled populations of *Sorbus aucuparia* in the vicinity of the Carrifran Wildwood Project. Pie chart diameter reflects relative sampling effort in each population. (From Bacles *et al.* (2004). Genetic effects of chronic habitat fragmentation on tree species. *Molecular Ecology*, 13, 573–584, Blackwell Publishing.)

from non-fragmented populations in continental Europe, even though the latter were sampled over a much larger spatial scale. No genetic bottleneck or departures from random mating were detected. The ratio of pollen flow to seed flow between fragments was estimated by using these markers, and was found to be close to 1. Results indicated that reduced gene flow by pollen movement is a likely consequence of habitat fragmentation, but effective seed dispersal by birds appears to be maintaining high levels of genetic diversity within forest fragments. In a further investigation with the wind-dispersed tree *Fraxinus excelsior* (Oleaceae), seed dispersal was found to be up to six times more effective than pollen dispersal at maintaining genetic connectivity among forest remnants (Bacles *et al.* 2006) (Figure 6.15).

A second concern related to how far away from a site seed may be safely collected, to avoid the risk of poorly adapted planting stock. To answer this question, provenance trials would need to be established for native species, to identify patterns of adaptive variation. Few such trials have been established to date. Although seed transfer zones have been proposed, these may be difficult to implement in practice and are based on very inadequate information (Ennos *et al.* 1998). As an alternative, Ennos (1998) suggested that comparisons should be made between the ecological characteristics of the planting site and those of the planting sources. The planting source most nearly matching the planting site should then be chosen.

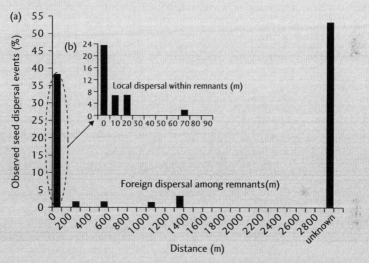

Fig. 6.15 Seed dispersal within and among *Fraxinus excelsior* remnants in the area surrounding Carrifran, southern Scotland. (a) Frequency distribution of effective seed dispersal events within (<3000 m) and outside of the study area. (b) Close-up on local seed dispersal (<100 m). Effective seed dispersal events were estimated by means of maximum likelihood parentage analysis. When a single parent was identified, it was assumed to be the maternal parent. When a parent pair was identified, the nearest parent was assumed to be the maternal parent. (From Bacles, C. F. E., Lowe, A. J., and Ennos, R. A. (2006). Effective seed dispersal across a fragmented landscape. *Science*, 311, 628. Reprinted with permission from AAAS.)

requires a co-dominant, highly polymorphic marker, such as microsatellites. In the case of tree species, seed can be collected from a mother tree and analysed, along with the mother tree itself. Once the genotype of both the mother and progeny have been established, the identity of the father can be determined by comparing the genotype of the progeny with potential males in the population, once the maternal genotype has been subtracted. This can be most simply be achieved by locating an individual or cluster of individuals with a unique allele, then examining the distribution of this allele within progeny arrays from other individuals or in seedling cohorts within the population (Hamrick and Nason 2000). Alternatively information from all available loci may be used to exclude particular males (Lowe et al. 2004).

If tree seedlings are analysed, rather than seed, then the problem is more difficult, as the identity of both the mother and the father needs to be determined. Organelle markers (such as cpDNA) that are maternally inherited, or tissues derived from only one parent (such as the pericarp or megagametophyte of plant seeds) can be used to differentiate the genetic contribution of different parents (Lowe et al. 2004). Alternatively, assumptions can be made regarding the spatial position of the seedling in relation to its putative parents. For example Dow and Ashley (1996) assumed that the closest potential parent to an oak seedling must be the mother, as acorns disperse less readily than oak pollen.

The advantage of direct methods is that information of great value to conservation can be derived, including the fertility or mating success of an individual or population, propagule dispersal frequency between populations or among groups of trees within a population, and the distance and frequency of propagule dispersal (Lowe et al. 2004). The main problem is the amount of work involved. Identification of all potential parents of a seed or seedling may require a large number of adults to be sampled and included in the analysis. The spatial location of putative parents also needs to be recorded. Problems arise when parents fail to be identified (for example as a result of long-distance pollen or seed dispersal), or when several potential parents are identified (when the marker system fails to differentiate between the actual parent and other individuals).

Molecular markers may also be used to provide estimates of the mating system of plants. Wright's fixation index F_{IS} (see above) can be used as indicator of inbreeding within populations. More frequently, estimates are derived by assessing outcrossing rates by using co-dominant markers to assess variation within progeny arrays. If only the maternal alleles are present in the progeny, then selfing can be assumed, whereas outcrossing can be inferred if each progeny has a single maternal allele and a second different allele from the father. The total number of alleles present within a progeny array can itself provide an indication of the extent of outcrossing (with more than two alleles present indicating the occurrence of outcrossing). In addition, a chi-squared test can be used to compare observed versus expected segregation ratios of alleles within a progeny array (no deviation from expected values indicating selfed progeny) (Lowe et al. 2004).

Ritland (1986, 2002) has developed a series of additional measures for describing the mating system of a species, based on calculating two estimates of outcrossing: a single-locus estimate based on the number of non-maternal alleles, the total number of alleles, and their segretation ratios; and a multi-locus estimate based on multiple or all loci. Computer programs for estimating outcrossing rates and other mating parameters are available (Ritland 1990), for example MLTR (*http://genetics.forestry.ubc.ca/ritl&/programs.html*). An example of a study that estimated the mating system of a tropical tree (*Cordia alliodora*) is provided by Boshier *et al.* (1995). A recent review of research into outcrossing rate and gene flow in neotropical trees is provided by Ward *et al.* (2005).

Assessing the impacts of anthropogenic disturbance

It is increasingly being recognized that deforestation and forest fragmentation can have substantial effects on processes influencing genetic variation within tree species, such as random genetic drift, inbreeding, and gene flow. Recent progress in assessing these effects by using molecular markers is reviewed by Lowe *et al.* (2005), with respect to neotropical trees. These authors make the following recommendations for such studies:

- Studies over small spatial scales should aim to examine genetic variation both before and after disturbance, include a large number of replications, and be conducted over relatively long timescales, to incorporate descriptions of annual or seasonal variation.
- Studies should aim to compare a variety of species with contrasting life history characteristics within the same landscape that contains populations that vary in size, isolation, and duration of impact. Such studies would highlight which life history traits are most important in mitigating genetic resource impacts of habitat degradation.
- Sampling should target specific questions, rather than relying on opportunistic availability of material, which may compromise experimental design. Large-scale studies, especially for the purpose of developing species-specific conservation recommendations, should sample across the geographical range of a species (as generally implemented when undertaking provenance trials for forestry), rather than just a few populations from a restricted area.
- Simulation modelling can enable comparisons to be made between species with very different life history traits and between landscapes that have experienced very different levels of disturbance. Modelling approaches should integrate population genetics and demographic processes to simulate the impact of different management scenarios on the genetic make-up of species being modelled (for example ECO-GENE, Degen *et al.* 1996).

Further recommendations are provided by Cavers *et al.* (2005), who consider the potential impacts of anthropogenic disturbance on the genetic structure of tree populations. Such structure (defined as the non-random distribution of genotypes) may occur at a variety of spatial scales, as a result of ecological processes such as dispersal, competition, and succession. Anthropogenic disturbance may affect such structure and influence the regeneration potential and evolutionary viability of populations.

Use of highly variable molecular markers (such as microsatellites) has enabled a number of recent investigation of such structure, most commonly by using spatial autocorrelation methods (Sokal and Oden 1978), comparing patterns of genetic variation with patterns of spatial distribution. Computer programs are available for such analyses (for example Degen *et al.* 2001). In contrast to population genetic estimators (such as *F*-statistics, described above), which involve averaging across populations, spatial autocorrelation analysis uses data from pairs of individual locations across the sampled area and therefore accesses more of the available information at the population scale (Cavers *et al.* 2005). To carry out such analyses, the spatial location of the individual trees must be recorded. The statistical power of the technique depends strongly on the sample size and how sampling is done in relation to the spatial structure of the population. Cavers *et al.* (2005) make the following recommendations for such studies:

- The characteristics of the target species (mating system, seed and pollen dispersal mechanisms) should be considered when planning the sampling strategy and selecting the molecular marker. For example, species with long-distance dispersal mechanisms (eg wind-dispersed pollen or animal-dispersed seed) are expected to show little genetic structure within populations; in such cases it is likely to be more efficient to devote greater effort into sampling larger numbers of individuals than increasing the number of markers used. The age structure of the population should also be considered and recorded where possible.
- A mixed sampling strategy is recommended, balancing high density of local sampling with wider scale coverage. At the same time, the sampling strategy should ensure that sufficient numbers of pairwise comparisons are produced in each distance class to achieve statistical significance (a minimum of 30 pairs per class is recommended; Cavers *et al.* 2005).
- The sampling effort required (for both individuals and of loci) is much greater for AFLP markers than for microsatellite markers. When using microsatellites it is more effective to increase individual sample numbers than increase numbers of loci, once 5 loci are available. With 5 loci, 100 individuals should be sufficient. When using dominant markers, the number of both loci and individuals required is much higher: at least 100 loci and 150 individuals. However, the sampling scheme required will depend strongly on the particular characteristics of the species studied.

Using molecular marker data to support conservation planning

Molecular marker data may be used to support conservation planning and action in a number of ways (Frankham *et al.* 2002, Lowe *et al.* 2004):

- Defining the taxonomic identity of the species of conservation concern, and whether hybridization is taking place (see, for example, Rieseberg and Swensen 1996, Robertson *et al.* 2004).
- Examining the portion of genetic variation within and among populations as a guide to sampling strategies for *ex situ* conservation, or for defining the

extent of variation within *ex situ* populations (see, for example, Allnutt *et al.* 1998, Ehtisham-Ul-Haq *et al.* 2001).
- Investigating the influence of evolutionary or biogeographic history on patterns of genetic variation (see, for example, Allnutt *et al.* 1999, Bekessy *et al.* 2002a, Premoli *et al.* 2002).
- Estimating the extent of gene flow between populations (see, for example, Bacles *et al.* 2004, 2006, White and Boshier 2000).
- Identifying whether a threatened species or population has lost genetic diversity, for example as a result of bottleneck events, population fragmentation (see, for example, Premoli *et al.* 2003), or overharvesting (see, for example, Gillies *et al.* 1999).
- Identifying genetically distinctive populations, or the distribution of rare alleles, in order to identify priorities for conservation (see, for example, Bekessy *et al.* 2002a, Premoli *et al.* 2001) or management units (Newton *et al.* 1999a).
- Identifying suitable sources of germplasm for reafforestation or restoration (see Box 6.1).
- Understanding the ecological characteristics of species, for example the breeding system (Boshier *et al.* 1995), or the pattern of seed dispersal (Bacles *et al.* 2004, 2006).

An important application of molecular markers relates to identifying whether genetic issues, such as inbreeding or genetic drift, are increasing the risk of extinction of a species or population. Similarly, it may be desirable to know whether a protected area is large enough to support a genetically viable population. To address such issues, genetic information can potentially be incorporated in a PVA (see section 5.3) to forecast the influence of both demographic and genetic factors on extinction risk. Some PVA software programs (such as GAPPS and VORTEX; see Table 5.2) are designed explicitly to achieve this. In practice, however, most PVA analyses either ignore genetic processes entirely, or consider inbreeding depression as the only genetic threat (Frankham *et al.* 2002). To incorporate inbreeding depression in PVA, information is needed on the breeding system of the species, the genetic mechanism responsible for inbreeding depression, and its relationship to fitness (Frankham *et al.* 2002). Collecting all of this information represents a substantial challenge, but the advantage of a modelling approach such as PVA is that various assumptions about genetic effects can potentially be explored, even if only partial information is available.

How may molecular marker data best be used to support conservation planning? A number of different approaches have been proposed. *Evolutionarily significant units* (ESUs) have been defined as historically isolated populations, which may require separate genetic management (Moritz 1994, 1995). The precise definition of the term has been the subject of some discussion, with respect to the genetic criteria adopted and the degree of difference that is considered to be 'significant' (Vogler and Desalle 1994). Initially the concept was developed for animals on the basis of differentiation in mitochondrial (mt) DNA, and explicitly on analysis of the spatial distributions of alleles, taking account of their phylogenetic relationships ('*molecular phylogeography*') (Moritz 1994). However, the mutation rate of mtDNA is significantly lower in plants, and therefore different criteria for defining ESUs need to be developed for tree species (Newton *et al*. 1999a). One possibility is to use cpDNA markers, which have been widely used in phylogeographic studies of trees, although the mutation rate of cpDNA is still relatively low (Newton *et al*. 1999). As an illustration, Petit *et al*. (2003) describe cpDNA variation in 22 widespread European trees and shrubs sampled in the same forests, which revealed that the genetically most diverse populations were located at intermediate latitudes, reflecting the mixture of different lineages during postglacial migration from southern refugia. Such analyses could potentially be used to define ESUs for multiple species.

The concept of ESUs has been criticized because molecular marker variation is usually selectively neutral, and therefore it ignores patterns of adaptive variation. As an alternative to ESUs, Crandall *et al*. (2000) suggested that populations be classified according to whether they show recent or historical ecological or genetic exchangeability. This concept is based on whether gene flow is currently occurring between populations, or occurred in the past, and takes into account patterns of adaptive variation. For example, evidence from differences in life history traits, morphology, habitats, and genetic loci under selection can be used to infer adaptive differentiation; where this is identified, the hypothesis of ecological exchangeability can be rejected. Management recommendations are then based on this assessment. An evaluation of this approach, and the ESU concept overall, is provided by Fraser and Bernatchez (2001).

Other terms that have been proposed include the *management unit* (MU), defined as a population with significant divergence of allele frequencies at nuclear or mitochondrial loci, regardless of the phylogenetic distinctiveness of the alleles (Moritz 1994). Potentially any molecular marker information, ideally supported by assessments of adaptive variation, could be used to define a management unit. Another term that has been widely used specifically in relation to forests is *gene resource management unit* (GRMU), which may be defined as an area of land chosen to include a representative sample of the genetic diversity of a species within a particular region, and designated for a particular genetic management objective (Ledig 1988, Millar and Libby 1991). Such GRMUs may form a central part of the genetic conservation strategy of a species, and may be managed in a particular way to maintain the genetic characteristics of the population (for example, by preventing timber harvesting or by protecting against the inflow of potentially deleterious genes from other populations, perhaps by preventing plantation forestry in the vicinity) (Millar and Libby 1991).

Caution is always required when using molecular evidence as a basis for proposing conservation action. All marker systems have limitations, and their different properties must be taken into account before the data that they provide can be interpreted properly. One of the main problems in using molecular marker data to support conservation planning is that they are generally assumed to be selectively neutral. Although differences in molecular markers may be indicative of adaptive differentiation between populations, this is usually not the case. Conservation decisions should be based on knowledge about the pattern of adaptive variation within species (Ennos 1996). Techniques by which this information can be obtained are described in the following section.

6.5.2 Quantitative variation

Quantitative genetic variation refers to variation in quantitative characters, or those characters that display continuous variation. Examples include growth rate, stem size, and form. Genetic variation in quantitative characters is due to the segregation of multiple polymorphic Mendelian loci, referred to as *quantitative trait loci* (QTL) (Frankham *et al.* 2002). The most important quantitative traits from a conservation perspective are those that determine adaptive potential, by influencing reproductive fitness, or the number of fertile offspring produced by an individual that reach reproductive age (Frankham *et al.* 2002). The main genetic concerns relating to conservation of threatened species relate to such quantitative traits, for example a reduction in reproductive fitness as a result of inbreeding (*inbreeding depression*), and loss of evolutionary potential caused by overharvesting or a reduction in population size (Frankham *et al.* 2002). As noted above, molecular marker variation is often poorly correlated with quantitative measures of variation, yet it is the latter that are closely related to evolutionary potential. Therefore conservation decisions should be based primarily on analysis of quantitative traits rather than molecular analyses (Ennos 1996). This crucial point appears to be overlooked by many researchers using molecular markers to study genetic variation (see for example Schaal *et al.* 1991, Young *et al.* 2000).

The assessment of genetic variation in tree species has a long history in forestry. This is illustrated by the review by Langlet (1971) entitled 'Two hundred years of genecology', a memorable riposte to a prior review by Heslop-Harrison (1964) that described only 40 years of genecology with herbaceous plants (omitting forestry studies altogether). Briggs and Walters (1997) provide a useful introduction to studies of quantitative genetic variation in plants. Reference texts describing techniques used in quantitative forest genetics in greater depth include Adams *et al.* (1992), White *et al.* (2002), Young *et al.* (2000), and Zobel and Talbert (1984), on which this account is based. It is regrettable that forest conservationists and ecologists have not made greater use of the information generated by such approaches, although it is primarily tree species of commercial value that have been studied to date.

The simplest method of studying variation in quantitative traits is to compare the growth and morphology of individuals of a species growing in different locations. Most tree species are genetically very variable, and some insights into this variation can be obtained by measuring morphological characteristics such as leaf size and shape, surface texture of the bark, crown form, stem height, etc., in populations sampled throughout the geographic range of a species (Schaal *et al.* 1991). Such morphological variation may be classified by taxonomists as subspecies or varieties. The problem with this approach is that any variation detected may have a purely environmental, rather than genetic, origin. For this reason, a provenance or progeny test is preferred.

Provenance test

To determine whether variation in the trait of interest has a genetic basis, some form of experiment or growth trial is required. Ecological geneticists or '*genecologists*' refer to such experiments as *common garden experiments*, which involve cultivating plant material sampled from a variety of different locations at the same site (Briggs and Walters 1997). This enables the growth and performance of the plants to be analysed, and variation that has a genetic basis to be separated from environmental variation. Foresters adopt just the same type of approach, but refer to the trials conducted as *provenance* or *progeny* tests.

The term *provenance* (or geographic race) refers to the geographic area from which the seed or other propagules were obtained (Zobel and Talbert 1984). Many provenance tests have been established by forestry agencies and private companies, in many parts of the world. The number of provenances included in such tests may be very large, reaching into the hundreds. Results of provenance tests are available through the publications of the International Union of Forest Research Organizations (IUFRO, ⟨www.iufro.org⟩) and the FAO ⟨www.fao.org⟩, as well as forestry journals such as *Silvae Genetica*, *Journal of Forestry*, *Forest Science*, *Forest Genetics*, *Forestry*, etc.

Results of provenance tests highlight the high degree of variation that occurs within many tree species, especially those with wide geographic ranges, wide altitudinal ranges, or species with a wide tolerance range of different site conditions (soil type, soil moisture availability, slope, and aspect) (Zobel and Talbert 1984). Another interesting finding from such tests is that variation is often physiological rather than morphological. Assessments based on morphology alone may fail to detect the substantial variation that commonly exists in traits relating to survival, growth, and reproduction (Zobel and Talbert 1984). This has clear implications for conservation, because it is these very same traits that determine the evolutionary or adaptive potential of a population.

Provenance tests are generally established with conventional randomized block designs. Trees of different geographic origin may be established in lines or plots (for example a 5×5 array), which are assigned a random location within a block. Where plots are used, the outer trees in each plot are usually considered as a border row and are not included in the measurements, to reduce the influence of competition with neighbouring trees on the results. The entire block, incorporating all of the provenances to be tested, is then replicated repeatedly. At least five replicates are usually included, often many more. The provenance test may be replicated across a range of different sites to provide an assessment of genotype×environment interactions. Typically, the growth and survival of the trees is regularly monitored, and measurements may also be made of other characteristics of interest, such as the incidence of insect damage or disease, the form of the tree, and wood characteristics. Data collected in this way can be subjected to analysis of variance (ANOVA), which enables the extent of genetic variation in the traits of interest to be estimated.

Progeny test

The best method to assess the extent of quantitative variation in a tree species is a progeny test. Pronounced genetic differences are often observed between the progeny of different parents. This comparison enables the heritability of different traits to be estimated (see next section). The simplest method of establishing a progeny test is to collect seed from selected parents and establish the seedlings in an appropriate experimental design (Figure 6.16). Such offspring are referred to as *open-pollinated* or '*half-sib*(ling)', where only one of the parents is known. An alternative approach is to cross-pollinate trees so that both parents are known, providing '*full-sib*' progeny. The latter approach provides greater information about patterns of inheritance and is widely used in tree breeding programmes

Fig. 6.16 A progeny test of *Cedrela odorata*, established in a nursery in Costa Rica. Open-pollinated progeny from a variety of different mother trees have been established in lines of six plants, in a randomized complete block design. The replicate blocks are orientated perpendicular to the lines featured in the photograph. Genetic differences in growth and morphology are evident after just a few months' growth. (Photo by Adrian Newton.)

(Zobel and Talbert 1984), but the former approach is easier and is more commonly adopted in studies of ecological genetics.

Trees may be established as single-tree plots, in lines or rows, or in square or rectangular plots. Statistically, the single-tree plot is the most efficient, but competitive effects between trees can influence the results obtained, and for this reason lines or plots are often preferred. When plots are used, as for provenance tests, the outer trees are considered as a border row and not included in the measurements, to reduce the effect of competition with neighbouring trees of different genetic origin. However, large numbers of families (> 20) are typically included in progeny tests, and in such cases the size of the trial can become very large if plots are used. Consequently lines or rows (typically of five or six trees) are generally preferred for this type of test.

Most commonly, progeny tests are established with a randomized complete-block design. Each family or seedlot is represented singly (as an individual tree, line, or plot) within a block, its position within the block being located randomly. The block, incorporating all of the families, is then replicated as many times as practicable; the statistical power of the design resides in the number of replicate blocks established. Blocking can usefully take account of any environmental variation in the experimental area (soil type, light availability, topography, etc.). Other experimental designs, such as latin square, lattice designs, and split-plot designs, are also sometimes used with progeny tests (Zobel and Talbert 1984), although their analysis may be more complicated. Split-plot designs may be employed when where is interest in comparing variation of families within provenances.

Once the seedlings have been established according to an appropriate design, growth variables of interest (such as growth rate, tree form, or incidence of pest attack) are measured at different time intervals. Data analysis methods are described in the following section.

Heritability

A key concept in quantitative genetics is *heritability*. This is the proportion of the total phenotypic variation observed in a population that can be attributed to genetic differences among individuals, which can be passed on to the next generation. It is therefore the heritability of a character in a population that determines its evolutionary potential (Frankham *et al.* 2002).

Two types of heritability are commonly used by forest geneticists. *Broad-sense heritability* (H^2) is the ratio of total genetic variation in a population to the phenotypic variation, or:

$$H^2 = \frac{\sigma_G^2}{\sigma_P^2} = \frac{\sigma_A^2 + \sigma_{NA}^2}{\sigma_A^2 + \sigma_{NA}^2 + \sigma_E^2}$$

where σ_G^2 is genetic variation (or the variance due to the influence of genes) and is the phenotypic variation (or all of the variance recorded in the trait of interest). Genetic variation can be further divided in to additive (σ_A^2) and non-additive (σ_{NA}^2) components. *Additive variance* refers to the proportion of the genetic variation due to the average effects of alleles (Frankham *et al.* 2002); σ_E^2 refers to variation caused by the environment.

Values of H^2 can range from 0 to 1, a value of 0 indicating that none of the variation in a population is attributable to genetics. A value of 1 would indicate that all of the variation observed is attributable to genetics.

Narrow-sense heritability (h^2) is the ratio of additive genetic variance to total variance:

$$h_2 = \frac{\sigma_A^2}{\sigma_P^2} = \frac{\sigma_A^2}{\sigma_A^2 + \sigma_{NA}^2 + \sigma_E^2}$$

A value of $h^2 = 0$ would indicate no additive variance, and a value of $h^2 = 1$ would indicate no environmental or non-additive variance. Most heritability estimates in the forest genetics literature are for h^2, because of the interest in the variation that might be inherited through sexual variation. Broad-sense variability is relevant if the species is able to reproduce clonally.

It is important to note that heritability estimates apply only to a particular population growing in a particular environment at a particular time. In other words, they are a function of the environment under which the tree species is grown. Also, estimates of heritability are not without error, and should be interpreted as a relative indication of genetic influence on a trait (Zobel and Talbert 1984). Heritability is generally estimated from the results of a progeny test (but not a provenance test). This can be done by ANOVA. Some statistical programs enable variance components to be determined individually, which is of great help in estimating heritability. For example the program SAS (⟨*http://support.sas.com*⟩; SAS 2002) has a specific procedure for this (Proc VARCOMP). Examples of heritability estimates that have been obtained for tree species for a variety of traits are presented by Cornelius (1994).

Nursery and glasshouse experiments

Conservation researchers rarely establish provenance or progeny tests, despite the useful information that can be gained from them, perhaps because of the large land areas required and the cost of maintaining the trials. However, both types of experiment can readily be carried out with tree seedlings under glasshouse or nursery conditions. For example, Bekessy *et al.* (2002b) established a progeny test under glasshouse conditions with seedlings of the threatened conifer *Araucaria araucana*, and after 21 months' growth was able to detect genetic differences in allocation of dry mass to roots and in values of carbon isotope ratio ($\delta^{13}C$), both indicators of drought tolerance. The results highlighted the importance of treating populations from either side of the Andes as separate management units in conservation planning. Neutral DNA markers (RAPD) failed to detect these adaptive traits (Bekessy *et al.* 2003). Similarly, progeny tests with seedlings of the timber tree *Cedrela odorata* established in a nursery detected genetic variation in leaf form and branching characteristics within a few months' growth (Newton *et al.* 1995) (Figure 6.16); these differences were subsequently confirmed in more extensive field trials (Newton *et al.* 1999b). In this species, neutral DNA markers (RAPD, AFLP, and cpDNA) have revealed substantial variation between Mesoamerican populations, which correlates with patterns of morphological variation, drought tolerance and pest resistance and has been proposed as the basis of defining ESUs (Cavers *et al.* 2003, 2004, Gillies *et al.* 1999). Such studies highlight the advantage of combining information from molecular marker studies with assessments of quantitative genetic variation when producing conservation recommendations.

7
Forest as habitat

7.1 Introduction

Forests are, of course, much more than just trees. Much of the challenge relating to forest conservation and management rests in understanding how different interventions, such as the harvesting of trees or the construction of a road, are likely to impact the other species living there. This requires a clear understanding of the habitat requirements of the species concerned, but unfortunately this is often lacking. Identifying and measuring those characteristics of a forest that determine its habitat value for different groups of organisms remains a key challenge for ecological researchers. In practice, managers often have to use proxies or indicators of forest condition to assess trends in habitat quality. However, the relation between such indicators and the abundance of individual species, or the species richness of different groups, often remains untested.

Many of the techniques used to assess forest habitat are described earlier in this book. Typically an assessment of habitat extent, condition or quality is based primarily on an assessment of forest extent and spatial distribution (see Chapter 2), vegetation structure and composition (see Chapter 3), and the prevailing disturbance regime (see Chapter 4). This chapter presents methods for assessing a series of habitat variables that are not covered by previous chapters, including deadwood volume, vertical stand structure, characteristics of forest fragments and edge effects, and the characteristics of habitat trees. In addition, an account is provided of modelling procedures that are increasingly being used to produce habitat maps for individual species. It should be remembered that many species that occur in forests also use or require non-forest habitats, such as grassland, shrubland, mires, lakes, or streams, which may also need to be considered in a comprehensive habitat assessment. Finally some suggestions are presented on how to undertake an assessment of forest biodiversity.

7.2 Coarse woody debris

The term *coarse woody debris* (CWD) refers to a range of different sizes and types of woody material that can be found in forests, including logs, snags, chunks of wood, large branches, and coarse roots. The definition of what constitutes coarse, as oppose to fine, woody debris varies between researchers, but a typical minimum diameter is 2.5 cm (Harmon *et al.* 1986). CWD is a very important habitat feature, particularly for fungi, invertebrates, mosses, and lichens, and may also provide sites for establishment of tree seedlings and sites for storage of carbon, water, and nutrients. Snags are used by many animal species, particularly birds, as sites for nesting, perching, and roosting. The ecology of CWD in temperate forests is reviewed by Harmon *et al.* (1986). Methods for assessing CWD are described by Harmon *et al.* (1986) and by Ståhl *et al.* (2001), on which this account is partly based.

The simplest method to assess the input of CWD from living trees is to determine rates of tree mortality by using permanent sample plots or tagged trees (see Chapter 4). This may underestimate inputs, because large branches and broken tops of trees are not included (Harmon *et al.* 1986). Input to CWD can also be measured on cleared plots, or by marking or mapping pieces present at the beginning of the period of observations. Alternatively, historical reconstructions of forest disturbance history can be used to estimate CWD inputs from catastrophic disturbance events (Henry and Swan 1974, Oliver and Stephens 1977).

When surveying, a distinction is generally made between standing dead trees (snags) and lying dead trees (or downed logs). The species, volume, biomass, decay status, number of woody fragments, surface area, and ground cover (projected area of fragments) are variables that are commonly recorded. Size class distributions and decay class distributions of wood pieces may be reported, along with descriptions of spatial patterns of distribution (such as accumulation in riparian environments) and orientation (for example, alignment in relation to slopes and direction of prevailing wind, an issue that should be borne in mind when sampling). Information on CWD is now routinely collected in many national or other large-scale forest inventories, stimulated by the common inclusion of CWD volume as an indicator of sustainable forest management (see Chapter 8). Simple, subjective assessments can be made by visually assessing CWD at locations within a forest stand. However, if there is a need to compare different areas or to monitor changes over time, then more objective methods are preferable. Note that when measuring CWD, variances tend to be high and therefore a large number of samples is typically required.

7.2.1 Assessing the volume of a single log or snag

For measurements of CWD volume, it is generally assumed that tree stems are circular in cross-section. This assumption is often not met, particularly when the logs are heavily decayed, a point that should be remembered when interpreting results.

Three methods are commonly used for volume determination (Ståhl *et al.* 2001):

- *Standard volume functions.* In many countries, standard functions relating volume to tree diameter and height are available for commercially important timber tree species. For such species, measurements of dbh and height can be used for estimating volume of the CWD fragment using these functions, if the tree is relatively intact. Note that this approach ignores the volume of stumps.

- *Sectioning.* This involves dividing a tree into a number of sections, then determining the volume of each section separately. Total volume is then estimated by adding these values together. The method is most easily used for tree trunks that are lying down, although measurements can also be made on standing trees by using instruments such as the optical relascope (see section 3.6.2). The formulae for Huber's method, Smalian's method, and Newton's method, presented in section 3.7.4, can be used to estimate volume by using this approach. Newton's formula is generally considered to be the most accurate and flexible of the three, providing unbiased estimates of volume for a cylinder, cone, paraboloid, or neiloid, the most common volumetric forms in trees and logs (Wiant *et al.* 1992). In cases where only parts of downed logs are included in the sample, Huber's method is simple to implement. In such cases it may often be sufficient to measure the length of the CWD fragment included in the sample and the diameter at its midpoint. Where entire trees are included, Smalian's formula can easily be applied. For a tree shorter than 30 m, the tree need be divided into no more than five sections. The thickest sections should be made shorter to enhance accuracy of volume estimates. For trees with a large number of branches, sectioning can be difficult, so a randomized branch sampling method may need to be used (Gove *et al.* 2002).
- *Taper functions.* Taper functions are available for a small number of tree species, describing the shape of the trunk. In cases where such functions are available, the cross-sectional area at any height can be derived by using the function and the volume obtained by integrating over the cross-sectional areas up to a certain height.

If pieces of CWD are highly irregular, calculation of volume from length and diameter measurements can be difficult; in such cases the displacement of water can be measured instead, by immersing the fragment.

For calculating the volume of standing dead trees, Harmon and Sexton (1996) recommend that the diameter at breast height be measured for intact stems, and the diameters at the base and top be measured for boles that have broken, in addition to height. For stumps, height should be measured together with either the midpoint diameter, or both the base and top diameters (Harmon and Sexton 1996).

7.2.2 Survey methods for forest stands

Six different survey methods are considered by Ståhl *et al.* (2001), as described below. These authors consider use of probability sampling approaches for each of these methods, using the Horvitz–Thompson (HT) estimation principle. The general HT formula is:

$$\hat{Y} = \sum_{i=1}^{n} \frac{y_i}{u_i}$$

where $\hat{Y}$ is an estimator of the population total (Y) of some measured quantity y_i, n is the number of sampled elements, and u_i is the inclusion probability of element i.

When using this approach, a key issue is how population elements (CWD fragments) close to the boundary of a forest stand are treated. To avoid bias, the sampling approach must ensure that elements close to the boundary receive identical inclusion probabilities as elements in the interior of a stand. One method widely used to achieve this is the *mirage method* (Gregoire 1982), in which a sampling unit (such as a plot or strip) close to the edge of the stand is 'reflected' in the boundary, so that some parts of the unit may be measured twice. The method is further discussed by Ducey *et al.* (2001). Alternatively, some sampling units can be laid slightly outside the forest stand of interest, but only those population elements lying within the stand are measured. For example, the centre of some randomly located sample plots may lie outside the forest stand, but some part of these plots may lie within the stand; within this area all the population elements (CWD fragments) are measured.

Sample plot inventory

Field plots are widely used to survey CWD (see Chapter 3 for details). Plots should ideally be randomly located, although some studies have employed systematic sampling designs. Circular plots are commonly used, although some researchers suggest that long, thin rectangular plots may be preferable. Sample plots can be divided into subplots for measuring the smaller diameter classes of logs and snags. For example, in a chronosequence of Douglas fir stands (from 40 to 900 years), Spies *et al.* (1988) used a nested plot design, with a 0.05 ha plot for logs, 0.1 ha plot for all snags, and a 0.2 ha plot for snags >50 cm dbh and >15 m tall. Line transects (see later) can be incorporated within plots to sample forest floor CWD.

A key issue is how to treat CWD fragments lying on the plot boundary. Two methods are used: (1) where a CWD fragment is entirely included in the sample if a well-defined part of it (for example the butt end of a trunk) is located within the plot, and (2) where only that part of a fragment that lies within a plot is included within the sample. In the latter case, many trees will only be partly sampled; deciding how trees should be divided can add to the time required to carry out the survey. The total volume of CWD in the plot can be calculated simply by summing the measurements of individual fragments recorded within the plot. This measure can be used to provide an estimate for the entire forest stand by multiplying with an appropriate scaling factor. There is no need to apply the HT formula when using this approach. However, in cases where trees are partly sampled by using method (2), this can lead to a severe bias in the estimates obtained. This problem can be overcome by using the appropriate HT estimator, as presented by Ståhl *et al.* (2001).

Strip surveying

This is also called belt inventory, transect inventory, strip cruising, or strip transect (see section 3.5). Typically, strips of a certain width are located randomly within the forest stand, in a random or predetermined direction. Alternatively, strips with a predetermined length and orientation can be randomly distributed within a stand. This method is equivalent to using very elongated, rectangular sample plots.

Line intercept (or intersect) sampling

In this method, all CWD fragments crossed by an inventory line are sampled. The probability that an object is included in the sample is proportional to the length or width of the object. An estimate of the total length of CWD fragments can therefore be obtained simply by counting the number of intersections. The method is only useful for downed logs, not for standing dead trees. As with strip plots, the transect lines may be laid out as segments with a particular spacing and orientation, or traverse the entire forest stand under investigation. A typical design is illustrated in Figure 7.1.

The method of estimation depends on whether the orientation of the survey lines is fixed or random. Often the orientation of the lines is fixed, and the lines are laid out systematically with spacing L, to traverse the entire stand (Figure 7.1). Following Ståhl et al. (2001), to determine the inclusion probability of a given downed log according to this sampling design, the width of its projection perpendicular to the orientation of the survey line should first be determined. This value is $l_i \sin w_i$, with l_i being the length of the log and w_i the acute angle between the log and a survey line. The estimator of the population total derived from the HT formula is:

$$\hat{Y} = L \sum_{i=1}^{n} \frac{y_i}{l_i \sin w_i}$$

The following version of the formula (which is a ratio estimator) will generally provide greater precision:

$$\hat{Y}_R = \frac{\sum_{i=1}^{n} \frac{y_i}{l_i \sin w_i}}{\sum_{i=1}^{m} s_i} A$$

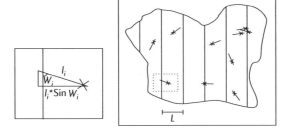

Fig. 7.1 An example of line intercept sampling with survey lines laid out systematically with spacing L. The short lines represent downed logs and the crosses on the lines are reference points of the downed logs. The length of a log is denoted l_i and the width of the projection perpendicular to the survey lines is $l_i \sin w_i$, where w_i is the acute angle between the log and a survey line. (From Ståhl et al. 2001.)

where A is the stand area, s_i is the length of the ith survey line, and m is the number of lines. Alternative estimators for different sampling designs are presented by Ståhl *et al.* (2001).

If measurement of the total volume is of interest, this can be estimated simply by recording diameter measurements at the point of intersection on each log sampled. In other words, the total volume of downed logs can be estimated simply by walking along survey lines, measuring the diameter at the point of intersection with logs, and recording the total length of the survey lines. In this case the total volume of the downed CWD can be estimated as (Ståhl *et al.* 2001):

$$\hat{Y} = \frac{\frac{\pi^2}{8} \sum_{i=1}^{n} dia_i^2}{\sum_{i=1}^{m} s_i} A$$

where dia_i is the diameter of log i at the point of intersection, A is the stand area, s_i is the length of the ith survey line, and m is the number of lines. The method works regardless of how crooked the stems are, although corrections should be made for logs not lying horizontally. However, this is valid only for the estimation of CWD volume or total length, not for total number of CWD fragments (Ståhl *et al.* 2001).

Harmon *et al.* (1986) present a relatively simple, widely used formula for estimating CWD volume (V) from data collected by using the line-intersect method, assuming that the lines are random and that the CWD fragments themselves are cylindrical, horizontal and randomly oriented:

$$V = (\pi^2/8L) \times \sum d^2$$

where d is the diameter of a CWD fragment and L is the transect length. When V is in m^3, L is in m and d is in cm, the formula becomes:

$$V = (1.234/L) \times \sum d^2$$

Where fragments of CWD are not lying horizontally, this volume estimation formula may be amended to:

$$V = (1.234/L) \times \sum (d^2 a)$$

where a is the secant (reciprocal of cosine) of the tilt angle (away from the horizontal) of each CWD piece sampled.

Although the line intersect method has been widely used, opinion varies regarding how the technique can best be implemented. A key decision, for example, is how long the transect lines should be. Harmon and Sexton (1996) considered that transect lengths in many studies have often been too short (<100 m), particularly for the larger fraction of forest floor CWD. The required transect length will depend on the density of wood pieces within the area to be surveyed.

Adaptive cluster sampling

This method is most appropriate for surveying sparse, clustered populations, for example when CWD is patchily distributed within a forest. In the first stage, a primary sample of plots, strips, lines, or other sampling unit is selected. Whenever an element is encountered for measurement (i.e. a CWD fragment), a second subsampling procedure is carried out. This may involve either sampling all plots (with predefined locations) in the vicinity, or searching within a fixed radius. This kind of approach is widely used in national forest inventories (see section 3.2).

Point and transect relascope sampling

Gove *et al.* (1999, 2001) describe a wide-angle relascope, an instrument that can be used to estimate CWD volume for woody fragments lying on the ground (Figure 7.2). The method is based on *horizontal point sampling* (HPS), a method widely used by foresters for estimating the volumes of standing trees, including those that are dead. In HPS, an angle gauge or wedge prism is used to select trees from a given point, with a probability proportional to their stem basal areas (Avery and Burkhart 2002). The angle of the gauge used determines the basal area factor (F), such that each tree sampled represents $F\,\text{m}^2$ of basal area per hectare, regardless of tree size (Gove *et al.* 2001). For a given sample point in the field, an estimate of volume or biomass per unit area is determined by using the formula:

$$\hat{e} = F \sum_{i=1}^{n} \frac{e_i}{k d_i^2}$$

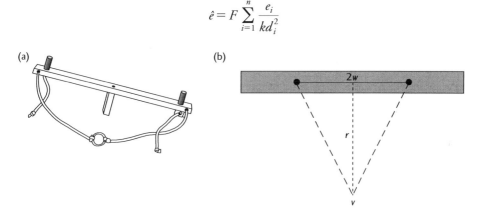

Fig. 7.2 (a) An angle gauge, or wide-angle relascope, can easily be constructed from nylon strapping, painted nails and a straight piece of well-seasoned wood. (b) The wooden portion of the gauge is shown shaded; the closed circles represent nails defining twice the width w. The observer's eye is at the apex of angle generation v, which is directly over the sample point. The reach, r, is the distance from the observer's eye to the instrument. $w = r \tan(v/2)$. All CWD fragments whose length fills the gap between the two nails are included in the sample. (From Gove *et al.* 2001. Point relascope sampling: a new way to assess downed coarse woody debris. *Journal of Foresty*, 99, 4–11.)

where n is the number of sampled trees, $k = 0.000\,078\,54$ (a correction factor), d_i is stem diameter in cm (therefore kd_i^2 is the basal area in m^2), and e_i is the quantity of interest for the ith tree. Individual point estimates are then averaged to provide an estimate for the whole stand.

Point relascope sampling uses similar methods for estimating the amount of CWD on the ground. At each sample point, the angle gauge is used to project a large fixed acute angle originating at the sample point (Figure 7.3). If the length of the individual piece of downed CWD appears longer than the width of the projected angle, it is included in the sample. Otherwise, it is not included. The logs that are included in a sample from a given point represent L square metres of log length per hectare. The factor L is equivalent to the basal area factor F in traditional HPS (Gove *et al*. 1999). Estimates per unit area can be obtained by using the formula (Gove *et al*. 2001):

$$\hat{Y} = L \sum_{i=1}^{n} \frac{y_i}{l_i^2}$$

where m is the number of logs sampled, y_i is the quantity of interest (volume, biomass, etc.), and l_i is the length in metres for the ith log in the sample. Ståhl *et al*. (2001) present further estimators based on the HT formula. Values of L for different relascope angles (ν) are presented by Gove *et al*. (1999). For example, if $\nu = 45°$, $L = 3501.162$ m^2 ha^{-1}. The method can be used to estimate the total length squared of downed logs in an area, by only counting logs. For estimates of volume or biomass, additional measurements are required.

Transect relascope sampling uses the same instrument and measurement principles while walking along survey lines, taking measurements in all directions from the line. As in line intercept sampling, the total length of logs in a stand can be estimated

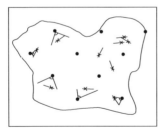 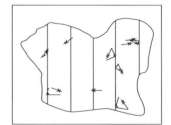

Fig. 7.3 Examples of point relascope sampling (left) and transect relascope sampling (right) for assessing CWD on the ground. All CWD elements whose length fills the gap of the instrument are included in the sample. The short lines represent downed logs and the crosses on the lines are the reference points of downed logs. Rays, with the angle of the relascope between them, emanate from sample points (in point relascope sampling) or survey lines (in transect relascope sampling) to provide an illustration of what elements would be included, or not included, in the sample. (From Ståhl *et al*. (2001). Assessment of coarse woody debris—a methodological overview. Ecological Bulletins, 49, 57–70.)

easily by counting the number of logs included in the sample and measuring the total length of the survey line (Figure 7.3). A ratio estimator appropriate for transect relascope sampling, when survey lines are orientated randomly, is as follows (Ståhl et al. 2001):

$$\hat{Y}_R = \frac{\sum_{i=1}^{n} \frac{y_i}{l_i(\frac{1}{\sin v} + \frac{2}{\pi}\cot v)}}{\sum_{i=1}^{m} s_i} \cdot A$$

Guided transect sampling typically includes two stages. First, wide strips are sampled within the forest stand. Then, these strips are subsampled using prior information for determining the route of the surveyor through the strip. A GPS device is used to follow the selected route. As in adaptive cluster sampling, a variety of different methods can be used in the second, subsampling phase (Ståhl et al. 2001).

Which of these methods should be used for assessing CWD? According to Ståhl et al. (2001), issues that should be considered when choosing a method include efficiency, robustness with regard to measurement errors, simplicity, and the completeness of the information obtained (for example, with line intercept sampling, and point and transect relascope sampling, information is obtained only for downed logs and not standing dead trees). A comparison of these different methods is presented in Table 7.1, which indicates that strip surveying and line intercept sampling may often provide the best compromise between these different issues. Strip surveying may be preferred because it works well with both standing and lying trees. In areas where the vegetation is dense, it can be difficult to locate and measure all pieces of CWD within a plot, and therefore the line intersect method may be preferred (Grove 2001). Whichever method is used, measurement errors can be substantial, and therefore training of surveyors may be required.

Table 7.1 Subjective comparison of different methods for assessing volume of CWD. After Ståhl et al. (2001).

Sampling method	Efficiency	Robustness	Simplicity
Sample plot inventory	–	+	+
Strip surveying	+	0	+
Line intercept sampling	+	+	0
Adaptive cluster sampling	+	0	–
Point relascope sampling	0	0	–
Transect relascope sampling	+	–	–
Guided transect sampling	+	0	–

+ indicates that the method performs well, 0 indicates intermediate performance, and – indicates that it does not perform well.

Ringvall and Ståhl (1999a) note that when CWD is sparsely distributed, plot-based methods tend to produce imprecise estimates. These authors performed a field test of the line intersect method with 11 surveyors in 4 coniferous forest stands in northern Sweden. Results indicated relatively little difference in the results obtained by different surveyors, but identified a negative bias in one stand among the surveyors, probably resulting from a tendency to avoid some logs when the latter occur at high density. Despite the possibility of such biases, line intersect methods have been very widely used for assessments of CWD (Harmon *et al.* 1986). Ringvall and Ståhl (1999b) further compared results obtained by different surveyors using the transect relascope sampling technique. Substantial differences were recorded between the systematic errors of different surveyors, suggesting that this method is not appropriate when there is a particular need for unbiased estimates (in long-term monitoring, for example). In a comparison of line-intersect, fixed-area, and point relascope sampling for downed CWD, Jordan *et al.* (2004) found significant differences among estimates in some stands, indicating that the methods differ in terms of bias. In terms of relative sampling efficiency, point relascope sampling displayed time efficiency comparable or superior to that of the other methods in most stands.

7.2.3 Assessing decay class and wood density

The state of CWD decay is often assessed by classifying CWD pieces into different decay classes. Such decay classes refer to the progressive change in solidity, integrity of shape, and characteristics of the log surface that occur as a result of the decay process (Pyle and Brown 1999). Typically, CWD is placed into three or five decay classes, but as many as eight or ten have been used. Such classes have been defined in a variety of different ways by different investigators (see Table 7.2). Commonly, classes have been defined on the external characteristics of the CWD, such as bark cover; the presence, colour, and abundance of attached needles, twigs, and branches; the cover of bryophytes and lichens; species and size of fungal sporocarps; the colour, crushability, moisture, and structure of the wood; the type of decay present (brown vs white rot, for example); whether the exposed wood is bleached; whether the log supports itself or has collapsed under its own weight; the age, size, and density of seedlings and saplings growing on the log; the presence and distribution of roots growing in the wood; and the presence of various decay processes such as sapwood sloughing (Harmon *et al.* 1986). Multivariate statistical analyses such as cluster analysis can be used as an objective method to identify decay classes on the basis of measurements of such variables.

Other classification systems have been developed for standing dead trees (see, for example, Spies *et al.* 1988). For example, Fridman and Walheim (2000) used a four-class decay system to categorize both forest floor CWD and snags in Swedish forests, whereas Clark *et al.* (1998a) recorded the hardness of snags, top condition (intact or broken), and percentage of bark remaining on the stem in a sub-boreal spruce forest.

Table 7.2 Characteristics of wood in classes used for determining state of decay of CWD lying on the ground (based on Pyle and Brown 1999, Sollins *et al.* 1987, Spetich *et al.* 1999, and USDA Forest Service 2001; after Woldendorp *et al.* 2002).

Decay class	Characteristics
I	Most of the bark is present Branches retain twigs Solid wood Fresh wood Original colour.
II	Some bark may be present Twigs absent Decay beginning to occur but wood still solid Invading roots are absent.
III	Bark is generally absent Log still supports own weight More extensive decay throughout but structurally sound Moss, herbs, fungal bodies, may be present Some invading roots may be present Some termite damage (in warm climates).
IV	Log cannot support its own weight, all of log on ground Kicked log will cleave into pieces or can be crushed May be partly solid or some large chunks (sometimes quite hard) still remain Bark absent Small soft blocky pieces Branch stubs rotted down, can be removed by hand Moss, herbs, fungal bodies may be present Invading roots (when present) are throughout More extensive termite damage, producing hollows (in warm climates).
V	Soft and powdery (when dry), often just a mound Log does not support own weight Does not hold original shape, flattened and spread out on ground Moss, herbs, fungal bodies may be present Invading roots (when present) are throughout Hollow log from termite damage may have collapsed or be a thin shell (in warm climates).

A *penetrometer* can be used to classify woody fragments according to decay class. This instrument comprises a short, pointed metal rod and a measurement gauge. The penetration depth of the instrument can be used to define decay classes; for example, Lambert *et al.* (1980) defined three classes as follows:

- slightly decayed, rod penetrates <0.5 cm
- moderately decayed, rod penetrates >0.5 cm to half the length
- advanced decay, more than half of the length of the rod can be pushed through the log.

Measurements of the density of CWD in the various decay classes can be obtained from samples of wood of each class. Such samples can be collected by cutting sections with a pruning saw or chain-saw, if the whole piece is too large to collect. The volume of the sample can be determined by measuring its dimensions, or by measuring the amount of water displaced on immersion (Stewart and Burrows 1994). It may be necessary to place samples in thin plastic bags before immersion, to prevent porous samples from absorbing water (Stewart and Burrows 1994, Grove 2001). After the volume has been measured, the dry mass of the samples is determined by drying in an oven at temperatures of 60–80 °C. Samples should be dried until they reach constant mass, which may take many days (Lambert *et al.* 1980, Grove 2001). In the case of large CWD fragments, dry mass can be calculated from a subsample, by first measuring the fresh mass of the entire sample, then measuring both the fresh and the dry mass of the subsample. The dry mass of the original fragment can be estimated by multiplying its wet mass by the dry: wet mass ratio of the subsample (Grove 2001).

7.2.4 Estimating decay rate

As decomposition of CWD is usually slow, long time periods are generally required in order to accurately measure the decay rate. Alternatively, the length of time that snags or logs have been dead can be determined, and changes in volume or density measured over time. The age of logs is difficult to determine, but can be achieved by ageing scars left on live trees adjacent to fallen trees, or determining (by ring counts) the age of the oldest seedling established on a fallen tree (which provides a minimum estimate of log age) (Harmon *et al.* 1986). Records of logging and thinning operations, or records of fire, insect outbreaks, or catastrophic windthrow, can also be used to estimate the age of woody debris. If a variety of fragments of different ages are available, reasonably accurate measurements of decay rate can be obtained by monitoring decomposition over relatively short periods of time. It is also possible to estimate decay rates from the ratio of CWD input to biomass, in cases where long-term data on tree mortality are available (see, for example, Sollins 1982).

Models can be used to project changes in the availability of CWD over time. The most commonly used model for this purpose is the single-exponential model:

$$Y_t = Y_0 e^{-kt}$$

where Y_0 is the initial quantity of material, Y_t is the amount left after time t, and k is the decay rate constant. Wood density is generally used as the Y variable, although volume or mass can also be used. The times to decompose 50% and 95% of the material ($t=0.5$ and $t=0.95$ respectively) are often reported. More complex models are also available that take into account the different decay rates of different fractions of the woody material, and the fact that loss of mass occurs by fragmentation of the woody pieces, as well as by respiration and leaching (Harmon et al. 1986).

7.3 Vertical stand structure

The vertical structure of a forest stand can be defined as the bottom-to-top configuration of above-ground vegetation (Brokaw and Lent 1999), and can be characterized in terms of variation in canopy density; tree size; branching patterns; the distribution of twigs, branches, and leaves; the density of the understorey; and the presence of snags and fallen trees (Hansen et al. 1991). Methods for measuring some of these variables are presented elsewhere (see, for example, sections 3.6.2 and 3.6.3). Here, further consideration is given specifically to techniques that can be used to characterize the vertical structure of forest stands in relation to its value as habitat. The importance of vertical structure for forest biodiversity is reviewed by Brokaw and Lent (1999), and its definition and measurement are reviewed by McElhinny et al. (2005).

MacArthur and MacArthur (1961) described a measure of *foliage height diversity* (FHD), referring to the arrangement of foliage within different vertical strata of the vegetation canopy. FHD is defined by the relation:

$$\text{FHD} = -\sum p_i \ln p_i$$

where p_i is the proportion of total foliage that lies in the ith of the chosen foliage layers. This measure was found to be related to a measure of bird species diversity. A number of authors have since used this measure of forest structure, although it has not always been found to be related to bird diversity (Brokaw and Lent 1999, Fuller 1995) and its relation to diversity of groups other than birds has been little tested. A further problem with the method is that different studies have defined and assessed vertical strata in different ways, raising concerns about the value of the approach, and hindering comparisons between studies (Erdelen 1984, Fuller 1995, Parker and Brown 2000).

Methods can also be used to compare foliage cover of selected tree species in different strata (Bebi et al. 2001). Ferris-Kaan et al. (1998) describe a technique in which a 10 m radius is measured out from the observer, defining an arc 20 m across. The canopy is then subdivided into four vertical height bands (see Figure 7.4). Four vegetation strata are defined as follows: S1 (field) 10 cm–1.9 m in height; S2 (shrub) 2–5 m; S3 (lower canopy) 5.1–15 m; and S4 (upper canopy)

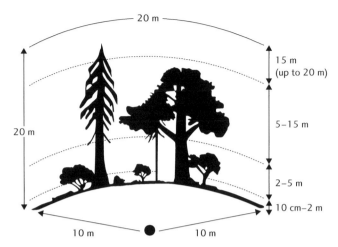

Fig. 7.4 Method for assessing vertical structure of forest stands. (After Ferris-Kaan *et al.* 1998. Assessing Structural Diversity in Managed Forests. Assessment of forest biodiversity for improved forest management, eds. Bachmann, Köhl, and Päivinnen. Kluwer Academic Publishers, with kind permission of Springer Science and Business Media.)

15.1–20 m (Figure 7.4). The percentage cover of each species within each height band is estimated visually to the nearest 5%; any plant bisecting the arc is recorded.

This method was deployed in an assessment of the biodiversity of forests in the UK (see section 7.9) by carrying out 16 vertical structure assessments within each 1 ha permanent sample plot (Humphrey *et al.* 2003a). Percentage cover was expressed as a mean of the 16 stand structure measures. To convert these cover values to a unified measure of stand structure, a *cover index* (CI) was calculated by using the formula:

$$CI = 1.9s_1 + 3s_2 + 10s_3 + 5s_4$$

where s_1–s_4 are the values for field, shrub, lower canopy, and upper canopy strata, and numbers refer to the depth of each stratum in metres. The CI therefore ranges in possible values from 0 to 1990 (assuming a maximum cover value of 100% in each layer) (Humphrey *et al.* 2003a). Whereas estimates of cover in the understorey are relatively easy, estimation of cover in the lower and upper canopy levels is much more difficult. Experience suggests that it is necessary to traverse the 20 m arc, making a visual assessment of canopy cover in the horizontal plane directly above the observer (Ferris-Kaan *et al.* 1998).

An alternative approach involves specifying the number of strata, on the basis that multi-layered stands are likely to offer a greater variety of habitats for species. This is based on the assumption that strata within a canopy can be clearly and consistently defined. In practice, this assumption is often not met. Parker and Brown (2000) provide a critical review of the concept of 'canopy stratification' and highlight the fact that the concept has been interpreted and measured variously by different authors. These authors suggest that, rather than attempting to describe the arrangement of foliage, it may be more meaningful to characterize ecological gradients created by the foliage (such as light availability; see section 4.9).

A number of attempts have been made to develop quantitative measures of canopy stratification. Ashton and Hall (1992) defined a *stratification index* as the ratio of plant biomass in the height class with the greatest biomass, to that of the height class with the least biomass. The index provides a measure that can be compared among stands, but does not identify how many strata exist, nor where they are found in the vertical profile of the stand (Baker and Wilson 2000). More recently, Latham *et al.* (1998) described a quantitative model, TSTRAT, for identifying stratification within stands. The method was developed for assessment of temperate conifer forests in the north-western USA. The TSTRAT algorithm defines strata on the basis of an assumption related to a competition cut-off point among tree crowns in a given area (Latham *et al.* 1998). The trees are sorted by height and crown ratio, and the upper 60% of the tallest tree crown is used as the basis of inclusion in the first stratum. All trees with heights equal to or greater than the lower limit are included in the first stratum. The tallest tree not included in the first stratum is then used to define the next stratum, and so on, until all trees have been classified according to the strata. The method provides a repeatable estimate of the number of strata, based on height and live crown ratio measures. However, Baker and Wilson (2000) suggest that the number of strata may often be overestimated by this method.

As an alternative, Baker and Wilson (2000) describe a simple technique for identifying stratification of individual tree crowns, based on comparing sorted tree heights to a moving average of height at the base of the live crown. The method requires only measures of the total height (HT) and height to the base of the live crown (HBLC) for each tree:

1. Sort the trees by HT and HBLC in descending order.
2. Beginning with the tallest tree (t_1), calculate the mean HBLC (for t_1, mean HBLC = HBLC (t_1), for later trees within the same stratum mean HBLC is the mean HBLC of all preceding trees).
3. Compare the height of the next tallest tree (HT(t_2)) plus the constant of overlap (k_o) to the mean HBLC. (The constant, k_o, defines a threshold distance between the mean HBLC and HT(t_2)).
4. If HT(t_2) + k_o is greater than the mean HBLC, then t_2 is in the same stratum as t_1. The mean HBLC is recalculated by using t_1 and t_2.
5. If HT(t_2) + k_o is less than the mean HBLC, then t_2 is in a stratum below t_1. The calculation of mean HBLC is re-initialized beginning with t_2, ignoring HBLC values from the preceding stratum.
6. The decision rules (steps 4 and 5) are repeated for all trees in the plot.

McElhinny *et al.* (2005) review the structural indices that have been used previously by researchers, noting that no single index is preferred over others.

These authors present the following guidelines for the development of an index of structural complexity:

- Start with a comprehensive set of structural attributes, in which there is a demonstrated association between the attributes and the elements of biodiversity that are of interest.
- Use a simple mathematical system to construct the index; this facilitates the use of multiple attributes and interpretation of the index in terms of real stand conditions.
- Score attributes relative to the range of values occurring in stands of a comparable vegetation community.
- Try different weightings of attributes in the index, adopting those weightings that most clearly distinguish between stands.

As noted by McElhinny *et al.* (2005), the choice of method to use for stand structure characterization should ideally be informed by an understanding of the habitat requirements of the species of interest. Unfortunately, this information is often lacking. Brokaw and Lent (1999) describe how the vertical organization of forest vegetation can have a variety of effects on animals and plants, influencing the availability of food for animals, as well as the arrangements of sites for nesting, resting, perching, basking, and mating. Indirect effects include impacts on microclimate and the distribution of animal prey. For epiphytic plants, stranglers, and vines, vertical structure can influence the availability of substrate for attachment. Consideration should be given to such mechanisms when choosing which variables to measure when characterizing vertical structure. It may also be difficult to separate out effects of vertical structure from other variables that can be correlated with it, such as tree composition or bark characteristics.

7.4 Forest fragmentation

The fragmentation of forest habitat is widely considered to be one of the main threats to biodiversity. Methods for assessing the pattern and extent of forest fragmentation are described in section 2.7. Here, approaches for assessing the impact of fragmentation on biodiversity are considered. This account is based on that of Fahrig (2003), who has usefully reviewed the substantial literature that exists on the topic.

First, it is important to differentiate between the effects of habitat loss and fragmentation. Many studies fail to differentiate between these two different processes. It is well established that habitat loss has large, negative effects on biodiversity. The effects of fragmentation, or the 'breaking apart' of habitat, are much less well defined, but can be either positive or negative. Research intended to examine the effects of fragmentation should therefore take care to separate out the impacts of fragmentation and habitat loss, something that has often not been achieved previously.

Many studies of habitat fragmentation compare some aspect of biodiversity at 'reference' sites within a continuous landscape to the same aspect(s) of biodiversity

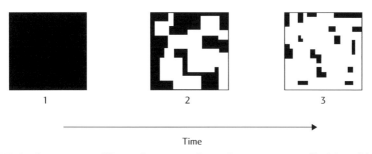

Fig. 7.5 In the process of forest fragmentation, a large expanse of habitat (black area) is converted into a smaller number of patches, isolated from each other by a non-forest matrix (white areas). Time 1 represents a landscape before fragmentation and times 2 and 3 represent a landscape following fragmentation. (From Fahrig 2003. Reprinted, with permission, from the Annual Review of Ecology, Evolution, and Systematics, vol 34 © 2003 by Annual Reviews www.annualreviews.org.)

at sites within a fragmented landscape (Figure 7.5). This approach has two main problems:

- The sample size in such studies is typically only 2 (i.e. one continuous landscape and one fragmented landscape), greatly limiting the statistical inferences that can be drawn.
- The design does not permit the relation between the degree of habitat fragmentation and the magnitude of effects on biodiversity to be studied, as each landscape can be in only one of two states, continuous or fragmented.

As noted in section 2.7, many different measures of fragmentation are available, referring to the number, size, shape, and isolation of habitat patches. Many of these measures are strongly related to the amount of habitat as well as the degree of fragmentation, so there is a risk that in using these measures the two processes of habitat loss and fragmentation can be confused. Fragmentation measurements should be made at the landscape scale, rather than at the scale of individual patches. When a study is at the patch scale, inferences at the landscape scale are not possible, because the sample size at the landscape scale is only 1. Therefore studies should seek to compare a number of different landscapes, rather than multiple patches within a single landscape (Figure 7.6).

How can forest fragmentation be measured independently of forest area? One approach is to use statistical methods to control for the amount of habitat. For example, Villard *et al.* (1999), in a study of 33 landscapes, measured the number of forest patches, total length of edge, mean nearest-neighbour distance, and percentage of forest cover. They then used regression approaches to relate each of the first three variables to the extent of forest cover. The residuals of these statistical models were then used as measures of fragmentation, taking into account their relations with forest area. Other approaches that have been used by researchers include the construction of experimental landscapes to independently control

302 | Forest as habitat

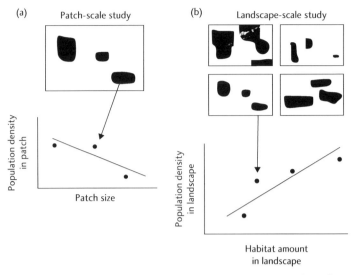

Fig. 7.6 In patch-scale studies (a), each observation represents the information from a single patch but only one landscape is studied. In landscape-scale studies (b), each observation represents the information from a single landscape. In this illustration, four landscapes have been studied. (From Fahrig 2003. Reprinted, with permission, from the Annual Review of Ecology, Evolution, and Systematics, vol 34, ©2003 by Annual Reviews www.annualreviews.org.)

habitat amount and fragmentation, and comparison of the response variable in one large patch versus several small patches (keeping the amount of habitat constant) (Fahrig 2003).

Results from those studies that have successfully managed to differentiate between the effects of habitat loss and fragmentation suggest that the effects of the former are far more pronounced than the latter. This implies that conservation research should focus on determining the amount of habitat required for conservation of the species of concern, and that conservation actions that attempt to minimize fragmentation (for a given habitat amount) may often be ineffectual (Fahrig 2003). However, there is evidence that forest fragmentation can have significant impacts on biodiversity as a result of edge effects, considered in the following section.

7.5 Edge characteristics and effects

Forested habitats are often distributed as patches within a landscape. The characteristics of the edges of such patches, where they meet other habitat or land-cover types, often differ from those of patch interiors. Characterization of habitat edges has become a major theme in ecological research and, in response to the widespread concern about the impacts of forest fragmentation, has also become important for practical forest conservation and management. Most investigations of edges focus on describing ecological patterns, such as changes in abundance of species with respect to the edge, but edge characteristics may influence a number of ecological processes such as dispersal, migration, predation, competition, and gene flow. A useful introduction to forest edges is provided by Matlack and Litvaitis (1999).

Variation in forest structure and composition can most readily be assessed by using plots or transects placed across habitat edges, and by then employing the methods used for assessing these variables described in Chapter 3. Remote sensing methods such as aerial photographs can also be used to provide descriptions of edge characteristics. Forest edges can be described by various attributes such as length, width, shape, vertical, and horizontal structure, density, or interior to edge boundary. Brändli *et al.* (1995) describe an assessment procedure for the forest edge that was operationally applied in the second Swiss national forest inventory (NFI). In the Swiss NFI, sample plots are distributed in a systematic grid. Whenever a forest margin lies within 25 m of the centre of a field plot, an assessment of the forest margin is conducted. A 50 m line forms the basis for the forest edge assessment. Along the transect, floristic diversity, habitat features (especially for birds and insects), and the aesthetic value for recreational purposes are assessed.

To provide an example of a typical research investigation, the approach used by López-Barrera and Newton (2005) for their work in the montane forests of Chiapas, Mexico, is described here in some detail. To describe edge characteristics, transects 80 m long and 10 m wide were established in each of six study sites. The transects were established perpendicular to the forest edge, running from the forest (60 m) through the edge into the forest exterior (20 m into the neighbouring grassland). The choice of transect length was influenced by the scale of heterogeneity observed within forest fragments and neighbouring forest-free areas, and to avoid the proximity of other nearby edges influencing the results. The edge was defined as the line coinciding with the base of bordering mature (> 30 cm dbh) tree stems (following Oosterhoorn and Kappelle 2000). Within each transect, all the woody plants were counted and measured, to provide an assessment of stand structure and composition.

This investigation was also designed to examine the effects of edge contrast: the difference between 'hard' forest edges, defined as those where adjacent vegetation has a simple homogeneous structure, thus creating a sharp contrast with adjacent forest cover, and 'soft' edges, which where characterized by more structurally complex successional vegetation maintained for 15–20 m from forest edges (Figure 7.7). Particular care was taken in this study to establish genuinely independent replicates, for example by establishing transects across edges of six different forest fragments. Lack of replication, or pseudoreplication, is a feature of many edge studies.

Once the edge characteristics have been described, typically an investigation also examines the distribution or behaviour of species across the edges, or some particular ecological process (Figure 7.8). For example, López-Barrera *et al.* (2005, 2006a, b) describe the effects of different forest edge types on seed dispersal, predation, and germination and seedling establishment in relation

304 | Forest as habitat

Fig. 7.7 Vegetation may differ in mean height and density across edges, leading to edges with different characteristics. For example, 'soft' (low-contrast) edges (a) may be differentiated from 'hard' (high-contrast) edges (b). The degree of edge contrast may affect the permeability of the edge to flows of energy, material and organisms (Ries *et al.* 2004). Edges may also be classified with respect to whether they are 'open' or 'closed'. A recently formed edge may be relatively permeable ('open'), but with time may become 'closed' as a result of regrowth of vegetation (Matlack and Litvaitis 1999).

to plant–animal interactions and the dynamics of forest patches within the landscape.

A thorough review of recent ecological research into habitat edges is provided by Ries *et al.* (2004), who highlight the following methodological points:

- Edges are generally defined as boundaries between distinct patch types, so the identification of edges depends on how patches are defined within a landscape. Patch definition can occur at a variety of scales, and obviously varies with the characteristics of the study area.

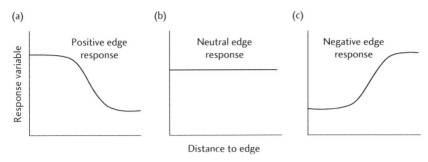

Fig. 7.8 Three classes of ecological edge responses with respect to distance from the closest habitat edge. Responses are generally categorized as (a) positive edge responses, where the variable of interest increases near the edge; (b) neutral responses, where there is no pattern with respect to the edge; and (c) negative responses, where the variable decreases near the edge. (From Ries *et al*. 2004. Reprinted, with permission, from the Annual Review of Ecology, Evolution, and Systematics, vol 35 © 2004 by Annual Reviews www.annualreviews.org.)

- Studies of edge effects in which the variables that determine organism abundance are assessed are much more powerful than purely correlational studies, but they are relatively rare in the scientific literature.
- Most edge studies have low replication, few distance categories, and little penetration into patches, and this limits their ability to detect responses.

Little progress has been made in extrapolating responses measured at local scales to larger scales, despite the value of the latter to management and conservation strategies. Two models are available for 'scaling up' edge responses to the landscape scale: the *core area* model (Laurance and Yensen 1991) and the *effective area* model (Sisk *et al*. 1997). Core area models use an estimated distance of edge influence to determine the amount of habitat in a patch that is not affected by edges (the core area); this approach is often used when considering reserve designs. In situations where most core habitat has already disappeared, the effective area model may be preferred. This method describes density (or other variables) as a function of distance from edge, allowing quantitative predictions of distributions to be made for an entire landscape.

It is important to determine the distance that edge effects extend into habitat patches. This value is often referred to as the *depth of edge influence* (DEI). In many investigations, DEI is determined by visual inspection, and results are influenced by study design (for example, the length of survey transects and the number of distance categories). Robust methods for analysis of DEI including use of appropriate statistical analyses are described by Cadenasso *et al*. (1997), Fraver (1994), Harper and MacDonald (2001), Laurance *et al*. (1998b), Mancke and Gavin (2000), and Toms and Lesperance (2003). Abiotic and plant responses are generally reported to extend up to 50 m into patches, invertebrate responses up to 100 m, and bird responses 50–200 m.

Most studies ignore the potential effect of being near more than one edge. Linear distance to the closest edge is generally used as the main explanatory variable, and therefore care should be taken to avoid placing plots near corners or other converging edge types, to limit their potential influence. It is possible that edge responses are different near multiple edges, or where edges converge, but these aspects have rarely been examined.

Ideally, the effect of edges on population dynamics should be examined by assessing the key parameters that ultimately determine distributions (birth, death, immigration and emigration rates), but few studies have achieved this to date. Many studies have measured community changes near edges, with the most common result being an increase in species diversity, but the causes and implications of such changes in diversity near edges often remain unclear.

Ries *et al.* (2004) also provide a mechanistic model to provide a framework for edge studies (see Figure 7.9).

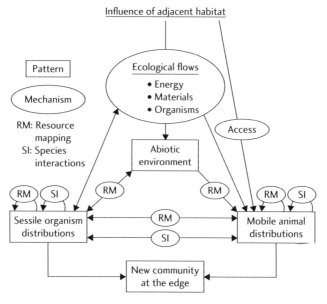

Fig. 7.9 A mechanistic model describing how the distributions of organisms are altered near habitat edges. Patterns in the abiotic environment, organism distributions, and community structure (*boxes*) are influenced by four principal mechanisms (*ovals*). Ecological flows of energy, material, and organisms across the edge influence the abiotic environment as well as organism distributions. Organisms map onto changes in the distribution of their resources. Changes in species' distributions near edges can lead to novel species interactions that can further influence abundance and distributions. All these changes in species distribution lead to altered community structure near edges. (From Ries *et al.* 2004. Reprinted, with permission, from the Annual Review of Ecology, Evolution, and Systematics, vol 35 © 2004 by Annual Reviews www.annualreviews.org.)

7.6 Habitat trees

The shape or form of a tree can have a major influence its ability to provide suitable habitat for other organisms. For example, the shape of the stem can influence the availability of suitable substrate for epiphytes, and the size, distribution, and orientation of branches affects the provision of nesting sites for birds and mammals, as well as the availability of suitable microsites for lichens, mosses, and invertebrates living on the surface of the tree.

Foresters have developed a range of methods for assessing stem form, primarily because of its importance for determining timber quality and value. A variety of form factors, form quotients, curves and formulae have been developed to describe stem form to provide accurate estimates of stem volume; these are described by Husch *et al.* (2003). For example, the cylindrical form factor f_c may be described by the equation:

$$f_c = V/gh$$

where V is the volume of tree (in cubic units), g is the cross-sectional area of a cylinder whose diameter equals tree dbh, and h is the height of cylinder whose height equals tree height. A form quotient is the ratio of diameter at some height above breast height to dbh. However, these methods have limited value for describing the characteristics of tree form relevant to provision of habitat, being designed primarily for assessment of timber value.

In the UK during the 1990s, a conservation campaign was developed focusing on 'veteran' trees, which may be defined as 'trees that are of interest biologically, culturally or aesthetically because of their age, size or condition' (Read 2000). It is important to note that this concept embraces trees that are of particular cultural or aesthetic value, as well as habitat value. An international conservation initiative has also been developed focusing on 'ancient' trees (i.e. individuals of a relatively advanced age for the species concerned) by the Ancient Tree Forum (*www.woodland-trust.org.uk/ancient-tree-forum/atfaboutus/vision.htm*). Veteran and ancient trees share many characteristics and are considered to be synonymous by some authors, although some trees of relatively young age can display 'veteran' characteristics (for example, if they have grown in an open environment).

In practice, surveys of veteran or ancient trees tend to focus on those characteristics that confer habitat value (Figure 7.10), such as (Read 2000):

- decay holes, hollows, or cavities, which can develop through limb loss and bark wounds and are expanded by microorganisms and invertebrates
- rot sites, where wood has been colonized by decay fungi
- dead wood, including dead limbs or trunk sections, often colonized by decay fungi with fallen and attached dead wood supporting a different suite of species
- hollowing in the trunk or major limbs
- naturally forming water pools
- physical damage to the trunk

308 | Forest as habitat

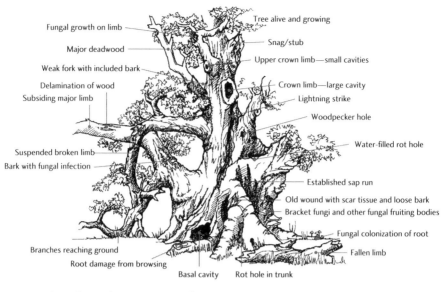

Fig. 7.10 Habitat characteristics of a 'veteran' or ancient tree. (From Read 2000.)

- bark loss/loose bark
- sap runs
- crevices in the bark, under branches, or in the root plate.

The presence of hollows or cavities is a particularly important characteristic influencing the value of a tree as habitat for wildlife. As an illustration, up to 40% of the bird species of North American forests nest in cavities (McComb and Lindenmayer 1999), and in Australia some 300 vertebrate species use tree hollows for nesting, roosting, or shelter (Gibbons *et al.* 2002). Factors such as the size and shape of the entrance, and the depth and degree of insulation of the hollow, can affect the frequency and seasonality of use. Some cavities can be difficult for human observers to detect, because many species prefer hollows with small entrances. Sites used by bats, lizards, and invertebrates may not be cavities within tree trunks, but small spaces between the bark and the wood.

Gibbons *et al.* (2002) describe a survey of the types of hollow, and types of hollow-bearing tree, used by vertebrate fauna in the temperate eucalypt forests of south-eastern Australia. Individual trees were sampled by walking along randomly oriented transects. Trees felled during silvicultural operations were examined for occupancy of hollows by animals. Hollows were defined as having a minimum entrance width of 2 cm, a minimum depth of 5 cm, and a minimum height above ground of 3 m. In this investigation, the following variables were measured to characterize the hollows (Gibbons *et al.* 2002):

- minimum entrance width (the smallest dimension of the entrance to the hollow)
- minimum width at half depth (the smallest internal dimension of the hollow at half depth)

- depth (distance from the base of the entrance to the bottom of the hollow)
- branch diameter (branch diameter measured half way between the entrance and base of the hollow)
- branch order (assigned by using an opposite ordering system, where the main stem is first-order, primary branches are second-order, etc.)
- branch health (assigned to one of three classes: living, part dead, dead)
- height (distance from hollow entrance to the base of the tree)
- occupation (whether the hollow showed signs of being occupied or not)
- species occupying hollow (determined from hair, feathers, scats, nesting material, or pellets).

Gibbons *et al.* (2002) noted that the use of hollow-bearing trees by fauna was associated primarily with the number of hollows in the tree, a result supported by other studies, suggesting that counts of the number of hollows should be made in such surveys. Tree diameter (dbh) was also found to be related to use of hollows, reflecting the larger number and size of cavities in older and larger trees. Other variables that might usefully be recorded include signs of physiological weakness (such as dead branches in the crown), the characteristics of vegetation surrounding the hollow-bearing tree, and the position of the tree in the landscape (Sedgeley and O'Donnell 1999). With respect to the last, Rhodes *et al.* (2006) applied network analysis methods to patterns of use of habitat trees by bats, an approach that clearly merits further consideration.

Working in the eucalypt forests of south-eastern Australia, Lindenmayer *et al.* (2000) examined the relations between the number, size and type of cavities in six tree species and tree diameter and height. Binoculars were used to survey the trees. Seven forms of cavity were recognized by these authors, reflecting both the location and the type of feature (Figure 7.11). Lindenmayer *et al.* (2000) recognize the problem of classifying cavities in standing trees from observations made from ground level; it is possible that some cavities may be 'blind' and not extend very far into the tree. This can only be determined by measuring the cavities directly, by either climbing the tree or felling it and dissecting it with a chainsaw, which can both be time-consuming and in the latter case undesirable because of the environmental disturbance caused.

Lindenmayer *et al.* (2000) found that both the number of cavities and cavity size were proportional to tree diameter, but inversely proportional to the square root of tree height. This proportionality differed between tree species. This simple relation offers the possibility of making rapid estimates of cavity abundance across large areas of forest by measuring simple tree attributes such as tree diameter and tree height, although whether such relations occur in other forest types remains to be tested. Crown shape (classified by visual assessment) was also found to be related to the abundance of cavities and cavity size, reflecting the fact that crown shape changes with the age and degree of senescence of the tree.

Trees continue to be of value as habitat after they have died, and therefore surveys of habitat trees should include snags or standing dead trees (see section 7.2). A number of different systems have been developed to classify standing trees

310 | Forest as habitat

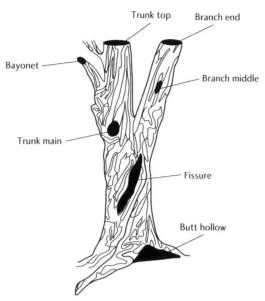

Fig. 7.11 Different types of tree cavity identified by Lindenmayer *et al.* (2000). *Fissures* are long, narrow cracks in the main trunk, *branch-end hollows* are located at the end of lateral branches, and *bayonet hollows* occur in the upper limbs of trees with an opening that is vertical or at a slight angle to vertical. (Reprinted from *Biodiversity in Britain's Planted Forests*. The Biodiversity assessment project: objectives, site selection, and survey methods, Lindenmayer *et al.* pp 11–18, (2003) with permission from Elsevier.)

according to their stage of senescence, death and collapse. For example, Maser *et al.* (1979) defined nine classes from a living tree to a decomposed stump (see Figure 7.12). In contrast Spetich *et al.* (1999) defined four decay classes for standing dead trees, namely:

I   recently dead: branches and twigs present; bark intact and tight on bole
II bark loose and/or partly absent; large branches present, much of crown broken; bole still standing and firm
III large branch stubs may be present; top may have broken; bark generally absent; bole still standing but decayed
IV branches and crown absent; bark absent; broken top; wood is heavily decayed or hollow.

Habitat features such as the presence of cavities or hollows can continue to be used after death of the tree, and therefore should be included in field surveys.

A number of methods are available that can be used to assess the decay status of trees (see Bucur 2003 for details). These are widely used by arboriculturalists to assess the risks of treefall and collapse. Simple techniques include sounding the tree by hitting it sharply with a hammer or mallet, increment boring (see section 3.6.1),

Habitat trees | 311

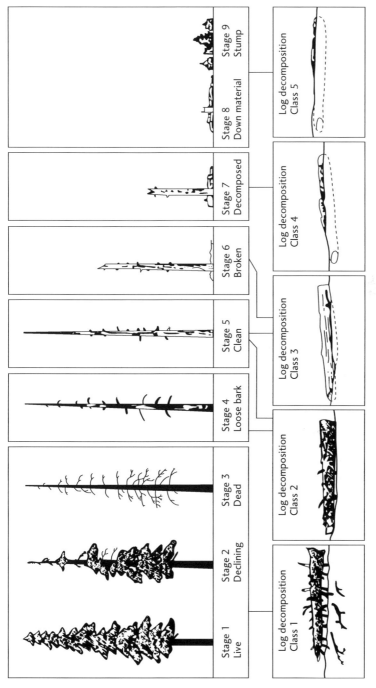

Fig. 7.12 Decay class of snags and logs, based on assessment of forests in the Pacific Northwest of the USA. (From Hunter (1999). Maintaining biodiversity in forest ecosystems, Cambridge University Press. Originally after Maser et al. 1979.)

and using a cordless drill with a long bit to determine the pattern of decay within the trunk. Other options include a *shigometer*, which delivers a pulsed electric current to the wood tissue and measures tissue resistance to the current (Butin 1995). The electrodes are inserted into a narrow hole drilled towards the centre of the stem (Shortle and Smith 1987). An alternative method is *impedance tomography*, which provides a non-destructive method of assessing stem resistivity (Weihs *et al*. 1999), although a large number of sensors is required and the approach is time-consuming (Larsson *et al*. 2004). Other non-invasive techniques are based on *X-ray tomography, microwave scanning, magnetic resonance imaging*, and *acoustical methods* (Larsson *et al*. 2004). Another instrument developed recently is the Tree Radar Unit (TRU) radar imaging system ⟨*www.treeradar.com/*⟩, which creates a high-resolution, non-invasive image of the internal structure of a tree by using a novel application of ground-penetrating radar technology. Such methods have been used relatively rarely to assess the habitat value of individual trees, but have potential for this type of application.

McComb and Lindenmayer (1999) provide guidance on how habitat should be managed for species that require cavities, snags or logs. A key decision facing such managers relates to the location, number, and characteristics of habitat trees that should be retained during management interventions. This is often difficult, because few studies have established relations between the availability of these habitat features and animal abundance. Ball *et al*. (1999) describe a deterministic computer model (HOLSIM) for predicting the long-term dynamics of hollow-bearing trees that occur in a single-species forest stand, providing a planning tool for forest and wildlife managers. In some cases trees may be deliberately killed to increase the number of snags or cavity trees, by using methods such as topping the trees with a chainsaw, girdling, herbicides, or explosives, and artificial cavities have been created by excavating holes in live trees (McComb and Lindenmayer 1999).

7.7 Understorey vegetation

The vegetation of the forest understorey can include mosses, lichens and ferns, as well as herbaceous plants and shrubs. Variables that are typically measured include:

- species composition
- relative cover by species or species groups
- density (number of stems or plants per unit area)
- frequency (the proportion of samples in which a species occurs)
- abundance (the number of stems or plants per sample)
- sizes (height and diameter, particularly for woody plants).

Frequency is based on the presence or absence of a species in sample units (plots, transects, or points) and is defined as the number of times a species is present in a given number of sample units. To obtain an estimate of frequency, therefore, only the presence of a species within a sample unit needs to be noted. *Density* refers to the number of plants in a given area. To obtain this measure, all the individuals

Table 7.3 The Domin and Braun–Blanquet scales used for visual estimation of plant cover (after Bullock 1996).

Value	Cover (%) Braun–Blanquet	Domin
+	<1	1 individual, with no measurable cover
1	1–5	<4, with few individuals
2	6–25	<4, with several individuals
3	26–50	<4, with many individuals
4	51–75	4–10
5	76–100	11–25
6		26–33
7		34–50
8		51–75
9		76–90
10		91–100

encountered within the boundary of the sample unit have to be identified and counted. Determining the number of individuals is complicated where species are multi-stemmed or clonal; in such cases it may be preferable to count the number of individual stems or shoots rather than the number of individuals.

Cover is defined as the proportion of the ground surface that is covered by the vegetation, and is usually expressed as a percentage of total ground area. *Relative cover* is the cover of a particular species expressed as a percentage of total vegetation cover. Cover is a very widely used measure of vegetation, as it enables measures of different life forms to be expressed in a comparable way. Percentage cover can be estimated by eye, either by creating percentage classes or by using the Domin or Braun–Blanquet scales (Table 7.3). As vegetation is often layered, cover may sum to more than 100%; alternatively, estimates can be produced for each layer separately (Bullock 1996). The main problem of estimating cover visually is that such subjective assessments can be inaccurate, and values produced by different observers can be very different. Such biases should be estimated, for example by repeating assessments, and care should be taken to ensure that different observers use a consistent approach to estimating cover.

Understorey vegetation is usually assessed by using small fixed-area plots or *quadrats*. A frame quadrat can be used, which divides an area into smaller subplots (Figure 7.13). Frame quadrats can be constructed from four strips of wood, metal, or rigid plastic that are tied, glued, welded, or bolted together to form a square (Bullock 1996). If relatively large quadrats are used ($>4\,m^2$), use of a frame can be difficult and instead quadrats can be laid out by using tape measures or rulers, with corners marked by using posts or pins. Quadrats can be divided into a grid of squares or subplots by using string or wire, to facilitate the survey process.

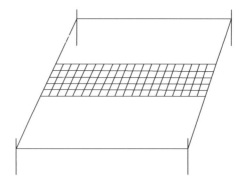

Fig. 7.13 A frame quadrat. (From Husch *et al.* 2003.)

Photographic frame quadrats can also be used, where vertical photographs are obtained at a constant height above the ground and a grid is then superimposed on the image, either by using filter on the camera lens or by overlaying a grid on the processed image. The image can be assessed in the same way as a normal frame quadrat.

Cover can be estimated visually for the whole quadrat. When gridded quadrats are used, cover can be estimated visually for each grid square, or each grid square can be classified as either covered or not covered by a particular species. The cover of the larger plot is then derived by summing the values of the grid squares. Density is measured by simply counting the total number of individuals of each species occurring within the quadrat; here, the main difficulty is determining whether those species lying on the edge of the quadrat should be included or not. Generally the criterion used is to include only those individuals that are rooted within the quadrat (Bullock 1996). Frequency is calculated as the percentage of quadrats in which a particular species is present; in this case, the number of plants in the quadrats is ignored. Alternatively, 'local frequency' can be calculated, which is the percentage of grid squares containing a species, when gridded quadrats are used (Bullock 1996).

The appropriate quadrat size to use depends on the characteristics of the vegetation being surveyed, including the size and spacing of individuals. The following plot areas have been suggested (Cain and de Oliveira Castro 1959, Bullock 1996):

- mosses, lichens, algae, and small ferns: $0.01–0.25 \, m^2$
- herbs, grasses, small seedlings, and low shrubs: $0.25–16 \, m^2$
- tall shrubs and low trees: $25–100 \, m^2$.

Quadrats of different sizes can be nested within each other to survey different components of the vegetation. Note that when measuring frequency, the result is influenced by quadrat size, and therefore care should be taken when interpreting the results. Larger quadrat sizes tend to give higher frequency estimates than smaller quadrats. Optimum quadrat sizes can be determined by examining frequency

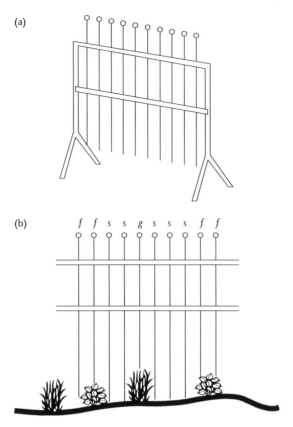

Fig. 7.14 Point quadrats: (a) Point frame with 10 pins. (b) Point frame intersections illustrating f, first hit on forb; g, first hit on grass; and s, first hit on soil. (After Husch *et al.* 2003.)

estimates obtained with different quadrat sizes; further details are provided by Goldsmith *et al.* (1986).

Alternatively, point quadrats can be constructed from a thin metal rod, sharpened at the tip (Figure 7.14). Examples of suitable materials are thick-gauge wire, knitting needles, welding rods, or bicycle spokes (Bullock 1996). Diameter should be as narrow as possible (1.5–2 mm). The point quadrat is lowered vertically through the vegetation and implanted in the soil, and the species touching the quadrat are recorded. Percentage cover is obtained by dividing the number of point quadrats for which a particular species is encountered by the total number of samples taken, and multiplying by 100. The technique is particularly useful for short vegetation, but can be laborious for dense vegetation (Bullock 1996). This method can also be applied to vertical photographs, by superimposing a grid of dots, although this method is unlikely to detect individual plants growing underneath the canopy of another plant.

316 | **Forest as habitat**

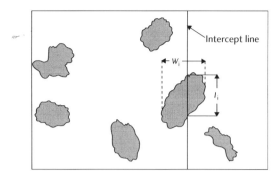

Fig. 7.15 Schematic illustration of the line intercept method for assessing the cover or density of shrubs and trees. The shaded areas represent the canopy coverage of a shrub or young tree. The measurements l_i are for cover estimates (fraction of the line covered by the canopy of this species), and the intercept distances w_i are the maximum perpendicular distance coverage and are used to estimate the numbers and density of the plants. (From Ecological Methodology, 2nd ed. by Charles J. Krelos. Copyright © 1999 by Addison-Wesley Educational Publishers, Inc. Reprinted by permission of Pearson Educational, Inc.)

The line transect or line intercept technique involves measuring the length of a transect that is intercepted by a particular species. The percentage cover of each species is estimated as the ratio of interception length to total transect length (Figure 7.15). Alternatively, the number of individual plants touching the transect line can be recorded to give a measure of density. In the point transect technique, points along a transect are established at predetermined intervals and each point is assessed to determine whether or not it is covered by the species present. Percentage cover is estimated as the number of covered points divided by the total number of points. Frequencies and densities can also be estimated by using distance methods (see section 3.5.3).

Whichever method is used to assess understorey vegetation, careful attention must be paid to the sampling design. It is important that many replicated samples are taken, which should ideally be located by using random or stratified random approaches. Many species have a patchy or clumped distribution, which should be taken into account when sampling (Bullock 1996). Sampling designs and approaches relevant to the assessment of trees are discussed in section 3.3; the same principles apply to assessment of understorey vegetation.

7.8 Habitat models

Understanding what constitutes suitable habitat for a particular species is of critical importance to conservation management and planning, yet the specific habitat requirements of many species are poorly understood. Forest managers typically require habitat maps to support management decisions, and such maps may also be required at regional or national scales to inform conservation planning. This

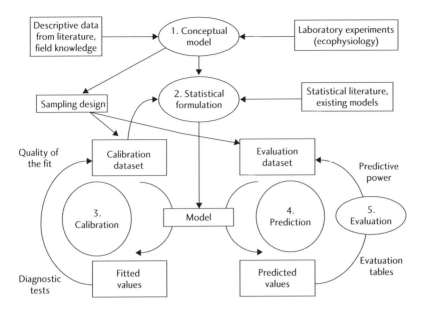

Fig. 7.16 An illustration of the process of habitat modelling, when two data sets—one for fitting and one for evaluating the model—are available. Model evaluation can either be made (a) on the calibration data set by using bootstrap, cross-validation, or jackknife techniques; or (b) on the independent data set, by comparing predicted to observed values by using approaches such as ROC curves for presence–absence models. (Reprinted from *Ecological Modelling*. Predictive habitat distribution models in ecology. 135. Guisan and Zimmerman, pp 147–186, (2000). With permission from Elsevier.)

section focuses on methods that can be used to generate habitat maps from a variety of modelling approaches. The information presented here is drawn largely from Elith and Burgman (2003) and Guisan and Zimmermann (2000). According to the former authors, there are seven main approaches to habitat modelling, which are considered separately below. The overall process of habitat modelling is illustrated in Figure 7.16.

7.8.1 Conceptual models based on expert opinion

The simplest approach to developing a habitat map is to consult relevant experts. These may be members of local communities, naturalists, taxonomists, or ecological researchers familiar with the species of interest. Forest managers or rangers may also possess anecdotal or systematically collected information about the distribution or abundance of species of interest. These may be compiled or collated to produce preliminary distribution maps, which can be refined through further discussion.

The challenges of working with expert knowledge are described in detail by Burgman (2005). Key issues are:

- *Who qualifies as an expert?* How should the experts be selected? How can a competent expert be differentiated from an incompetent one? How can the accuracy of information provided by experts be verified? The answers to these questions will depend on the particular circumstances of the study being undertaken.
- *How can information best be elicited from the selected experts?* Methods may include questionnaire surveys carried out by mail, email, or telephone, face-to-face individual interviews, structured group interviews aimed at achieving consensus, and the development and revision of conceptual models.
- *How can information from experts be aggregated?* One commonly used approach is the Delphi technique, which involves calculation of summary statistics of the information provided (such as medians and interquartile ranges) that are then distributed to participants for further comment.
- *What should be done when experts disagree?* How can uncertainty best be captured and presented? One approach is to describe such uncertainty by using subjective probability distributions, with values in a continuous range. A number of analytical procedures are available for combining and analysing such distributions, and for weighting the beliefs of different experts (Burgman 2005).

Conceptual models may be described as abstractions about how we believe the world works (Burgman 2005). Such models provide a useful tool for identifying what is known about a site, as well as for eliciting information from experts. The simplest form is a diagram that illustrates the key features of the site of interest. An example would be a sketch map or block diagram of the site, with key features and processes illustrated upon it. Alternatively, *influence diagrams* can be used to provide a visual representation of the components and dependencies of a system. Different shapes (such as ellipses and rectangles) can be used to represent variables, data, and parameters, which are connected by arrows to indicate causal relations and dependencies (Figure 7.17). Such diagrams can form the basis of analytical models, using software tools such as Analytica (Lumina Decision Systems, ⟨www.lumina.com⟩) and Hugin Expert (⟨www.hugin.com⟩). Conceptual models can readily be produced through a process of discussion with relevant experts, or based on a review of relevant literature, providing a description of the current understanding of a system. In a conservation context, such approaches can be of value in helping to identify the factors influencing the distribution and abundance of the species of interest, including threats or reasons for decline (see section 8.4).

The most commonly used type of conceptual model used in habitat mapping is the *habitat suitability index* (HSI), which was originally developed by the US Fish and Wildlife Service and is now widely used, particularly in the USA. The HSI approach depends on identifying important components of habitat for the entire life cycle of the target species (such as food availability, cover, and breeding requirements), based on

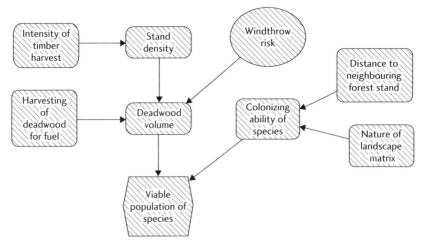

Fig. 7.17 A simple conceptual model. The example depicted is an influence diagram constructed to inform conservation of a threatened fungus species, which depends on deadwood as habitat. Different shapes (or nodes) on the diagram illustrate different kinds of variable. A rectangular node depicts a *decision variable*, which the decision-maker can control directly (in this case, harvesting of trees for timber, and deadwood for fuelwood). The rounded rectangular nodes represent *general variables*, which may represent factors or processes that can potentially be measured. The oval node depicts a *chance variable*, a variable that is uncertain and that the decision-maker cannot control directly. The hexagonal node depicts an *objective variable*, which represents the desired outcome. The arrows in the diagram represent influences among the variables; in other words, an arrow from one variable to another indicates that the value of the first variable directly affects the value of the second variable. A conceptual model can readily be constructed as an influence diagram such as this, by consulting relevant experts or scientific literature about the system of interest. (This example was constructed by using Analytica software, from Lumina Decision Systems.)

the judgement of experts. The quality of habitat is then often assessed by using a scaled index, from 0 (very poor quality) to 1 (excellent/optimum quality), for each habitat component or variable of interest. The overall habitat quality can then be expressed by combining the *suitability indices* (SI) of the individual components, again usually on a scale of 0–1. The individual SIs can be combined by using a number of different additive, multiplicative, or logical functions, again determined by expert judgement.

The HSI approach summarizes in a simple way the main environmental factors thought to influence the occurrence and abundance of wildlife species. The habitat maps that are produced represent a form of integrated expert knowledge, and synthesize subjective interpretation of biological processes (Burgman *et al.* 2001).

However, it should be noted that the maps represent potential habitat rather than occupancy, and the accuracy of expert knowledge is often untested as no data are included in the model that would allow reliability to be examined (Elith and Burgman 2003). Also, the method does not provide information on population sizes, trends or the potential responses of species to changes in habitat availability (Morrison *et al*. 1992). A review of HSI models developed in the USA is provided by Schamberger *et al*. (1982). Methods for evaluating and improving such models are described by Brooks (1997) and Roloff and Kernohan (1999).

An advantage of the HSI approach is that the variables can be easily incorporated into GIS, enabling them to be integrated with other data relating to forest inventory and assessment, and providing forest managers with a tool for evaluating the potential impact of management interventions on wildlife habitat (Donovan *et al*. 1987). At its simplest level, the technique can be used to estimate the total quantity of suitable habitat in a landscape. More sophisticated analysis can involve assessment of the spatial arrangement of the resulting habitat. This usually achieved by calculating landscape indices or metrics (Diaz 1996, O'Neill *et al*. 1988; see section 2.7).

7.8.2 Geographic envelopes and spaces

Geographic envelopes are models used to describe and analyse the geographic range of species. There are several different forms of this kind of model (Elith and Burgman 2003):

- *Convex hulls*. This refers to the smallest polygon containing all known sites of a species in which no internal angle exceeds 180°; in other words, all the outer surfaces are convex. This approach is recommended by IUCN for estimation of the *extent of occurrence* of a taxon, which is used in assessment of threat of extinction according to the IUCN Red List criteria (IUCN 2001). The constraint of convexity on the outer surface of the polygon results in a coarse level of resolution, which can produce substantial overestimates of the geographical range of a species.
- *α-hull*. This is a generalization of the convex hull derived from a Delaunay triangulation, constructed from lines joining a set of points constrained so that no lines overlap. This can provide a more accurate estimate of habitat extent, particularly when the shape of the range is irregular (Rapoport 1982).
- *Kernel density estimators*. These are non-parametric statistical methods used for estimation of home range, but can also be used to estimate the geographical range of a species or population. In this method, a kernel (for example a normal distribution) is placed over each observation point in a sample, and then density is estimated at evaluation points. The density is the sum of all kernel values at the evaluation point (Elith and Burgman 2003). Specialist statistical programs are available to do this type of analysis (see, for example, Seaman *et al*. 1998).

These methods are used to estimate geographical ranges of species, using distribution data (records of species presence) as input, but are not appropriate for detailed maps of species distributions because the envelopes include many sites that are unsuitable for the species. They are generally used to assess distribution patterns over large areas. All three approaches are sensitive to missing data and spatial errors (Elith and Burgman 2003).

7.8.3 Climatic envelopes

As with hulls, climate envelopes use records of the presence of species at different locations, but use climate rather than geographic location to produce habitat maps. An example of this approach is provided by ANUCLIM, which incorporates two subsystems for making bioclimatic predictions, BIOCLIM and BIOMAP (Busby 1991). Rainfall, temperature, and radiation records are used together with elevation data to construct a climate profile for a particular species, based on the current pattern of distribution. A habitat map can then be produced by considering each location with respect to this climate profile. This approach can be implemented for any situation as long as appropriate environmental data and distribution data are available. Generally the method is used to define a region that is climatically suitable for a species, which can be used as a target area to search for populations or as an indication of historic extent (Elith and Burgman 2003). It is generally used over large geographic areas, but tends to include any sites where the species is not actually present within the envelope that is defined. The approach obviously has particular value for exploring the potential impacts of climate change on species distributions.

7.8.4 Multivariate association methods

This method defines the habitat of a species by using multivariate statistical analyses to relate environmental variables to species presence–absence or abundance data. The statistical method that is most widely used in this context is *canonical correspondence analysis* (CCA). Details of this approach are provided by Guisan *et al.* (1999) and ter Braak (1988). CCA can be done with appropriate software such as CANOCO (ter Braak and Šmilauer 1998; see section 3.10). In this approach, the principal ordination axes are constrained to be a linear combination of environmental descriptors (ter Braak 1988). Generally CCA is applied to data collected from field plots. Usefully, the method is appropriate for data sets with many zeros (i.e. absences) (Guisan and Zimmermann 2000). However, the method is based on a number of assumptions (for example, that species have unimodal responses to environmental gradients, and species have equal ecological tolerance and equal maxima along canonical axes) that may often fail to be met in reality, which may invalidate the results (Elith and Burgman 2003).

Examples of the use of multivariate methods to produce habitat maps include DOMAIN (Carpenter *et al.* 1993), which is based on measuring environmental similarity between the site of interest and the most similar site for which the species is known to be present; again the method requires only presence data for the species. The method has proved to be suitable for situations where available distribution data are limited (Guisan and Zimmermann 2000). Another example is the

BIOMAPPER software package (Hirzel et al. 2001a). This employs *ecological niche-factor analysis* (ENFA), which differs from CCA in that it considers only one species at a time rather than a group of species. Again it has the advantage of only requiring species presence data. ENFA compares the ecogeographical predictor distribution for a presence data set with the predictor distribution of the whole area, and summarizes all predictors into a few uncorrelated factors that retain most of the information. These factors include the *marginality*, which reflects how the environmental conditions where the species is found differ from those in the entire area being considered. The relation between species distribution and these factors is used to calculate a habitat suitability index and to produce habitat maps (Hirzel et al. 2001b, 2002). The approach has been found to compare favourably with logistic regression (see later) (Hirzel et al. 2002). The main limitation with this method is that it assumes a linear relation between observed variables and underlying factors, although this can be circumvented to some extent by transforming the data (Elith and Burgman 2003).

7.8.5 Regression analysis

The most commonly used regression technique for habitat modelling involves the use of *generalized linear models* (GLMs). These include standard linear regression and ANOVA techniques, and all have a response variable (species data), predictors (explanatory variables, such as environmental data), and a function that links the two. Regression may involve either a single explanatory variable (simple regression) or a combination (multiple regression). If the response is not linearly related to a predictor, then the latter can be transformed before analysis. When presence–absence data are being analysed, as is often the case with habitat modelling, the form of GLM used is logistic regression, as the data have a binomial distribution. In this case the link function is a logit. However, a range of other response variable distributions and link functions can be used, such as Poisson and quasi-likelihood models applied to simulated abundance data, and Gaussian models applied to plant abundance data (Elith and Burgman 2003).

Generalized additive models (GAMs) offer an alternative approach, in which one or more of the functions in a GLM is replaced by a smoothed data-dependent function. A variety of different methods of smoothing can be used, the most common of which are splines (cubic b-splines, for example) (Elith and Burgman 2003). The main advantage of this technique is that the response is not limited to a parametric function, potentially enabling the fitted response surface to be a closer approximation to reality.

Regression offers a powerful and widely used tool for habitat modelling, enabling the probability of presence or the abundance of the species to be estimated and predicted for new sites. However, the method requires more than solely presence data (such as information on absences as well), which can limit its usefulness (when only herbarium records are available for analysis, for example) (Elith and Burgman 2003). Absence data are often difficult to obtain, requiring a comprehensive and systematic survey of the areas concerned. To overcome this problem, some attempts have been made to generate pseudo-absence data from non-present sites (see, for example, Zaniewski *et al.* 2002). The resulting predictions are expressed as a relative likelihood of occurrence (Elith and Burgman 2003).

7.8.6 Tree-based methods

The 'tree' referred to here is a decision-tree, not the kind of tree you find in a forest! This approach is based on the concept of keys, such as those widely used for taxonomic identification. The user is guided through a series of steps to arrive at an outcome. In the case of habitat modelling, a set of decision rules ('splits') are created that classify presence–absence species data in relation to environmental variables (Elith and Burgman 2003). Automated algorithms are typically used to perform the analysis, which can be used to provide probabilistic predictions for site occupancy. The method is relatively easy to perform and interpret; however, more than presence-only data are required, and small changes in the data can produce very different outcomes. 'Pruning' techniques can be used to reduce the complexity of the tree, by limiting the number of terminal nodes (Guisan and Zimmermann 2000).

7.8.7 Machine learning methods

These methods employ computers to classify data. This can be achieved by using a variety of different analytical approaches, including genetic algorithms and neural networks. A *genetic algorithm* is an optimization technique based on a set of logical learning rules (Elith and Burgman 2003). An example is GARP, a program that predicts species distributions from environmental variables by using GIS (Stockwell 1999, Stockwell and Peters 1999). In GARP, the genetic algorithm is used to generate, test, and modify rules for predicting distributions, which are presented in the form of relative likelihoods of the presence of the species. GARP can be used with both presence–absence data or presence-only data, although in the latter case the program creates pseudo-absences. A desktop version of the program is available from ⟨www.lifemapper.org/desktopgarp/⟩. An example of the application of GARP is provided by Peterson *et al.* (2002), who used the method to examine the potential impacts of climate change on Mexican fauna. However, Stockman *et al.* (2006) found that this approach was not suitable for predicting the spatial distribution of a non-vagile invertebrate. Anderson *et al.* (2003) provide a detailed account of the technique.

Neural networks identify patterns by learning from a set of input or 'training' data. Predictions are produced by using non-linear, optimally weighted combinations of the original variables (Elith and Burgman 2003). The method has been widely used in analysis of remote sensing data (see, for example, Carpenter *et al.* 1999), but has not been extensively used for habitat mapping to date, at least at the species level (see, for example, Manel *et al.* 1999, Olden 2003). The networks themselves are not easily interpreted; the analytical process is complex and not widely understood, even by computer scientists (Guisan and Zimmermann 2000).

It is also easy to over-fit models (Moisen and Frescino 2002). Examples of the application of neural networks to habitat modelling are provided by Ejrnæs *et al.* (2002) and Monteil *et al.* (2005).

In addition to these seven main methods, a number of other techniques have been used to model habitat. For example, *Bayesian methods* (see section 1.2) can be used to predict the probability of a species occuring at a given site with known environmental attributes. This is achieved by combining *a priori* probabilities of observing species or communities with their probabilities of occurrence based on the value of environmental predictors. *A priori* probabilities can be based on previous survey results or the literature. Alternatively, GIS can be used to develop simple models by using overlays of environmental variables and rules for combining single probabilities of occurrence (Guisan and Zimmermann 2000). An alternative approach is to use *cellular automata* (see section 5.5.3), which have been used to predict the impact of climate change on plant distribution (Carey 1996) and to simulate the migration of plants along corridors in fragmented landscapes (van Dorp *et al.* 1997).

7.8.8 Choosing and using a modelling method

The relative strengths and weaknesses of these different modelling approaches are described in detail by Elith and Burgman (2003) and by Guisan and Zimmermann (2000), who make the following points:

- In general, regression techniques are generally preferred for producing predictions that differentiate between suitable and unsuitable habitat, although some of the emerging methods (such as neural networks) clearly have potential. The main limitation of regression approaches is the need for enough presence–absence records to provide a sufficiently robust model. Predictions based on presence-only records are often limited compared with those developed by using presence–absence data, although the use of pseudo-absence records can help overcome this limitation.
- All of the methods require some computing and statistical expertise. Although regression methods and CCA can be complex to implement, they are supported by an extensive literature and documentation, and powerful software is available to perform them. The same is not necessarily true for other methods.
- All of the methods produce predictions that can be tested against new data, something that should always be incorporated in modelling investigations. Regression methods and CCA produce statistical estimates of error and deviance, which provide a useful measure of the extent to which a model fits the data.

Generally, models are constructed by using data derived from or displayed in GIS. Typically, species data are georeferenced point locations, and predictions are developed by applying the models to raster (grid) representations of the environmental data. The resolution and accuracy of these data is of great importance in producing reliable habitat maps. Ideally, species distribution data should be collected systematically throughout the area of interest, covering the full range of environmental variation within which the species occurs. Some form of stratified sampling approach should be adopted. Alternatively, the gradsect approach (Austin and Heyligers 1989, 1991, Gillison and Brewer 1985) can be used, which involves sampling species along transects oriented along environmental gradients.

Similarly, environmental data should ideally be available at fine resolution, and data are required for all of the variables that are likely to influence distribution of the species of interest. Climate and soil data are often available only at low resolution, but *digital elevation models* (DEM) are often available at relatively high resolution and can be used to derive variables such as slope and aspect.

Particular care is needed when the species is expanding or declining in abundance or range size; in such circumstances it may be necessary to include factors such as disturbance, dispersal barriers, or successional dynamics to produce useful habitat models. Such factors may mean that suitable habitat is not occupied. It is important to remember that habitat suitability is not the same as occupancy.

The reliability of habitat models can be evaluated by using statistical approaches such as the *Mann–Whitney statistic, receiver operating characteristic (ROC) curves*, and *confusion matrices*. ROC curves are used to judge the effectiveness of predictions for repeated binary decisions (such as present/absent) and are built around confusion matrices that summarize the frequencies of false and true positives and negative predictions. Further details of these approaches are given by Burgman (2005), Fieldings and Bell (1997), and Manel *et al.* (1999).

Ideally, the model should be evaluated by using a second set of data, independent of the data set used to build or calibrate the model (see Figure 7.16). If possible, the two data sets should originate from two different sampling strategies, such as stratified random sampling and observational surveys. Where insufficient data are available for the approach, a single data set can be used to both calibrate and evaluate the model; the latter can be achieved by using cross-validation or jackknife procedures.

According to Fahrig (2003), the most important question in biodiversity conservation is probably 'How much habitat is enough?' In other words, conservation of species in a given region requires identifying which species are most vulnerable to habitat loss, and estimating the minimum habitat required for persistence of each of these relatively vulnerable species (Fahrig 2003). How can this be determined? One approach is to link the habitat models described in this section with models of PVA (see section 5.3), an approach that is receiving increasing interest from researchers (Elith and Burgman 2003). Usually, this is achieved by converting a habitat map to a representation of patches and dispersal pathways. The dynamics of populations within patches, together with the rate of movement between patches, can then be modelled. A more abstract approach includes the classical metapopulation models (see section 5.5.3), in which only presence or absence in a patch is modelled, rather than dynamics within the patch. A key decision relating to modelling habitat as patches relates to the spatial scale at which individual pixels in a habitat map should be coalesced into patches (Elith and Burgman 2003). A lower limit to this is set by the resolution of the maps.

7.9 Assessing forest biodiversity

Effective conservation of biodiversity is dependent on knowing which species are present in a particular forest area. However, a comprehensive assessment of biodiversity, including an inventory of all species present, is difficult. Hunter (1999) suggests that no one has completed a complete inventory for even a single forest ecosystem, and they are unlikely to do so any time soon, primarily because of the difficulties of sampling and identifying microorganisms. Surveys of other megadiverse groups, such as invertebrates and fungi, also present a significant challenge.

Methods for assessing plant diversity are described in previous chapters. A detailed description of methods for surveying other groups of organisms is beyond the scope of this book. Instead, the reader is referred to other publications in this series; for example, Sutherland *et al.* (2004) provide a detailed account of methods for surveying birds. Other practical handbooks for surveying different components of biodiversity are provided by Hill *et al.* (2005), Jermy *et al.* (1995), and Sutherland (1996). The publication by Jermy *et al.* (1995) is accompanied by a set of field guides. In addition, Feinsinger (2001) provides a useful introduction to field techniques for biodiversity conservation. A brief overview of some recent approaches to assessing biodiversity is presented in Box 7.1. Examples of different approaches to assessing forest biodiversity are provided by Bachmann *et al.* (1998) and Angelstam *et al.* (2004b), although these publications focus primarily on temperate and boreal forests.

Box 7.1 Biodiversity assessment approaches

An abundance of information exists relating to biodiversity within individual countries. Sources include expedition reports, natural history society journals, field study reports, impact assessment documents, taxonomic reviews of particular groups or organisms or areas, museum and herbarium specimen labels and catalogues, forest inventories, etc. A number of countries have established national centres for biodiversity assessment and information management, such as INBIO in Costa Rica and CONABIO in Mexico, and these institutions are now important information sources themselves. Although an enormous body of pertinent information exists, considerable effort is required to create harmonized sets of data that can be readily analysed, and used as a basis for presentation of information to a non-technical audience. Many data, often collected with difficulty and at great expense, remain entirely in specialized and technical literature and have never been applied to practical biodiversity conservation and forest management. The collation, integration, and analysis of patchy, inadequate data is one of the most significant challenges to biodiversity assessment, at any scale. This reflects the fact that relatively few systematic surveys of biodiversity are currently being undertaken; most information is collected on an *ad hoc* or opportunistic basis. Consultation with national and international experts is often an explicit and integral part of the data compilation process. Once data have been compiled, priorities for conservation or further data collection can be identified. The collection of new data is generally achieved through some form of field survey, which may be supported by the use of remote sensing techniques (see, for example, Nagendra 2001, Nagendra and Gadgil 1999).

In recent years, a number of different approaches to assessing biodiversity have been developed, some of which are briefly outlined below. These examples include assessments that have been applied at both national and subnational levels, often to identify priority areas for conservation (i.e. those areas of high species diversity or possessing large numbers of restricted-range or threatened species). The examples given here differ in terms of scope and objectives, as well as depth of coverage.

Gap analysis ⟨http://gapanalysis.nbii.gov/⟩

Originally developed by US Fish and Wildlife Service and now widely implemented in the USA, gap analysis is used to identify gaps in the representation of biodiversity within reserves (i.e. areas managed solely or primarily for the purpose of biodiversity conservation). Once identified, such gaps can potentially be addressed through the creation of new reserves, changes in the designation of existing reserves, or changes in management practices. Gaps in the protection of biodiversity can be identified by superimposing relevant data layers in GIS, and analyses at local, regional, or national scales. Gap analysis therefore aims to give land managers and policy-makers the information they need to make better-informed decisions when identifying priority areas for conservation. Further information and a handbook are available from the website given above.

BioRAP: for rapid assessment of biological diversity
⟨*www.amonline.net.au/systematics/faith5.htm#introduction*⟩

The BioRap Toolbox consists of a set of analytical tools that can be used to identify, with high spatial resolution, priority areas for the conservation and sustainable management of biodiversity. These tools were developed by the Australian Museum, CSIRO and other partners, initially for application in Papua New Guinea. The principal components of the BioRap Toolbox are spatial modelling tools and classification and biodiversity-priority setting tools. These tools support high-spatial-resolution biodiversity assessments that are readily integrated with existing spatially distributed planning information, as was available for Papua New Guinea in the form of PNGRIS, the Papua New Guinea Resource Information System. Further, the BioRap approach introduces socioeconomic factors along with biodiversity at the earliest stage of analysis. Further information is provided by Faith *et al.* (2001) and Nix *et al.* (2000).

World Bank toolkit ⟨*www.worldbank.org/biodiversity*⟩

This set of documents summarizes best practice in treatment of biodiversity within an environmental assessment, with a particular focus on determining the potential impacts of development projects.

All taxa biodiversity inventory (ATBI) ⟨*www.dlia.org/atbi/*⟩
The aim of an ABTI, originally developed by staff at the University of Pennsylvania in conjunction with INBIO (Costa Rica), is to make a thorough inventory or description of all the species present in a particular area, using highly trained taxonomic specialists recruited internationally and nationally. The goals of ATBI are to recognize and describe species and assign stable scientific binomial names (facilitating information exchange between researchers in different parts of the world); determine where at least some of the members of each taxon or species live and can be found; and, through accumulation of ecological and behavioural information, determine their role in the ecosystem. Undertaking an ATBI can be a costly and long-term endeavour. Efforts are currently under way at a number of locations in the USA, including the Great Smoky Mountains.

Rapid biodiversity assessment (RBA)
Developed by researchers at MacQuarie University (Australia) and partners, RBA is based on the premise that certain aspects of biological diversity can be quantified without knowing the scientific names of the species involved. Data are gathered on selected groups of organisms. Several groups, chosen as good 'predictor sets' or 'biodiversity surrogates' of biodiversity, are needed at each location inventoried. The main characteristic of RBA is reduction of the formal taxonomic content in the classification and identification of organisms, to enable assessments to be done relatively quickly. The units of variety recorded by such a scheme may be referred to as morphospecies, operational taxonomic units (OTUs) or recognizable taxonomic units (RTUs). Biodiversity technicians (such as local people living within the study area) can be trained by taxonomists and employed to separate specimens into RTUs. Further details are provided by Oliver (1996) and Oliver and Beattie (1996).

Rapid assessment programme (RAP)
⟨*www.biodiversityscience.org/xp/CABS/research/rap/methods/*⟩
Conservation International created the RAP in 1989 to fill the gaps in regional knowledge of the world's biodiversity 'hotspots'. The RAP process assembles teams of experts to conduct preliminary assessments of the biological value of poorly known areas. RAP teams usually consist of experts in taxonomically well-known groups such as higher vertebrates (birds and mammals) and vascular plants, so that ready identification of organisms to the species level is achieved. The biological value of an area can be characterized by species richness, degree of species endemism (i.e. percentage of species that are found nowhere else), special habitat types, threatened species, degree of habitat degradation, and the presence of introduced species. RAP teams use standardized methods to survey the diversity of plants, mammals, birds, reptiles, amphibians and selected insect groups. The RAP methodology is not a substitute for more in-depth inventories or monitoring, but it is designed to provide critical scientific information quickly.

Often, rather than survey all groups of organisms, forest biodiversity is assessed by using indicators. The design and use of such indicators is considered in section 8.6. There has been particular interest among both researchers and practitioners in combining biodiversity assessments with standard approaches to forest inventory, an issue that is also considered in the following chapter. An example of a plot-based method for simultaneously assessing a range of different components of biodiversity is described in Box 7.2.

Box 7.2 Example of a method for assessing biodiversity within forest stands: the Biodiversity Assessment Project of the Forestry Commission, UK

As part of a national research programme to characterize the biodiversity of forest stands in the UK, sample plots were established at a range of sites distributed throughout the country and surveyed over a 2–4 year period. At each site, measurements were made of structural aspects of biodiversity (for example vertical foliage cover and deadwood), taxa important in ecosystem functioning (for example fungi), and a range of different groups that make up the 'compositional'

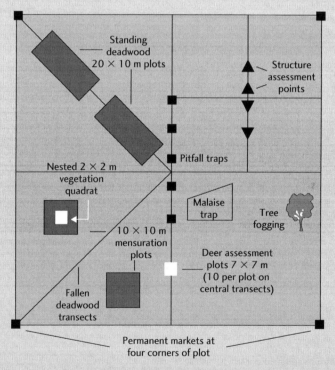

Fig. 7.18 Sample plot design used to produce an inventory of biodiversity in forest stands as implemented in the Biodiversity Assessment Project of the UK Forestry Commission. (After Humphrey et al. 2003a © Crown Copyright. Reproduced with permission of the Forestry Commission.)

aspect of biodiversity (for example higher and lower plants, invertebrates). Plots were established stands at different stages of development, and in both semi-natural and plantation forests.

Sample plots were established by using the design illustrated in Figure 7.18. Plots were situated in areas selected to minimize internal heterogeneity in terms of stand structure, species composition, topography, and hydrology. The plots were permanently marked with concrete posts. A standardized system of assessment was developed to maximize potential comparisons between measured attributes and to minimize disturbance during sampling (Table 7.4).

Table 7.4 Summary of variables measured and methods used for assessment of the biodiversity of forest stands, as implemented in the Biodiversity Assessment Project of the UK Forestry Commission (after Humphrey *et al.* 2003a). These methods were implemented by using the plot design illustrated in Figure 7.18.

Variable of interest	Assessment method
Deer	Densities estimated from 10 7 × 7 m faecal pellet group clearance plots per 1 ha.
Small mammals	Live capture/release with paired Longworth traps.
Songbirds	Point counts within and adjacent to each 1 ha plot, plus territory mapping.
Invertebrates	Sampling stratified by ground, subcanopy, canopy strata, and deadwood. Five pitfall traps per 1 ha, one Malaise trap per 1 ha, one tree fogged, deadwood emergence traps.
Bryophytes and lichens growing on pieces of deadwood	Species frequency and abundance estimates on individuals growing on deadwood.
Macrofungi	Frequency and abundance of sporocarps recorded 3 times yearly over 4 years in each mensuration plot (note that microfungi were not recorded).
Deadwood: fallen (logs), standing (snags) and stumps	Volume and length of logs recorded on two diagonal transects by using the line intercept method, volume of snags and stumps recorded in 8 20 × 10 m plots.
Soil microbial communities	32 soil samples taken from each 1 ha plot, 4 in each mensuration plot.
Ground vegetation (bryophytes, lichens and vascular plants)	Percentage cover and frequency in 2 × 2 m quadrats nested within the 8 10 × 10 m mensuration plots.
Soil seed banks	One sample per plot bulked from collections in each mensuration plot.

Natural regeneration of seedlings (<1.3 m in height)	Height of all seedlings recorded in 10 randomly located 40 × 40 cm plots within each mensuration plot.
Mensuration	Dbh, height to live crown, height of all trees within the 8 10 × 10 m plots.
LAI	Estimated from light measurements along transects with each 1 ha plot (by using a handheld Decagon sunfleck ceptometer; see section 3.7.3).
Vertical structure	Percentage cover of foliage estimated in four vertical strata—ground, shrub, lower and upper canopy layers—at 16 sampling points (see section 7.3).
Soil chemistry and litter	One soil pit dug per 1 ha plot and described, chemical analysis of two strata—32 bulked samples per strata per 1 ha sample plot. Mean litter depth ha^{-1} estimated from 32 random samples (4 in each mensuration plot).
Climate variables	Obtained from national climate model.

This example is provided as an illustration of how one group of researchers addressed the challenge of assessing forest biodiversity at the stand scale; it is not intended to provide a blueprint for other studies. There were some particular difficulties with the chosen design: for example, using faecal pellet counts to estimate deer densities is thought not to provide an accurate assessment at the scale of 1 ha adopted here. The problem of scale is also evident from the songbird data, as bird territories range over a spatial extent much greater than 1 ha. Other problems were encountered with the small-mammal sampling, which was abandoned after only 1 year owing to excessive costs and logistical difficulties. The assessment of diverse groups such as fungi and insects was reliant on specialist taxonomic expertise; the samples collected took far longer to identify than to simply collect, adding significantly to the cost of the project (which was substantial). However, the project is notable for providing a rare example of an attempt to assess multiple components of biodiversity simultaneously, and illustrates how this can be achieved in practice. Results of the project are summarized by Humphrey *et al.* (2003b).

8

Towards effective forest conservation

8.1 Introduction

Effective conservation is not achieved simply by following a well-defined scientific method or set of protocols. What works well in one area may not work well in another. As noted in Chapter 1, conservation management is a complex social, economic, and political process, requiring recognition of the values held by different people and an ability to identify the trade-offs and compromises that need to be addressed in reaching a practical solution. As it is largely about dealing with people, conservation management is more of an art than a science. So what contribution can scientific methods make to practical conservation efforts?

Many ecological researchers are very concerned about the current divide between conservation science and practice. There is a widespread belief that the effectiveness of much conservation action could be improved by strengthening the scientific foundation on which it is based, for example by implementing scientifically rigorous approaches to environmental survey and monitoring, and by selecting management interventions on the basis of the best scientific evidence available. Unfortunately, such evidence is often lacking, partly because the scientific community has traditionally failed to address research questions of direct relevance to management practice, and partly because the scientific information that is available is often not readily accessible by conservation managers. Although this is now gradually changing, it is certain that conservation managers are often faced with making decisions without access to any reliable scientific information about what the potential outcomes of alternative management actions might be.

Are there lessons to be learned from forestry? I believe so. Certainly, how forestry is generally practised provides some interesting contrasts with forest management that has purely conservation objectives. There is arguably a much longer tradition of research informing forestry practice than is the case for conservation management (even if today the distinction between these two disciplines is increasingly becoming blurred). Foresters have long understood the value of a quantitative approach to estimating growth and yield, and the importance of using experimental evidence from silviculture to inform management practice. The economic value of timber has prompted substantial investment in the forest sector in many countries, which has supported research, inventory, and monitoring. The technical, financial, and information resources available to many practising foresters are enough to make any conservation manager deeply envious. It is no coincidence that when it comes to reporting on the state of the global environment (for example the Millennium Ecosystem Assessment ⟨*www.maweb.org/*⟩) the most reliable and comprehensive statistics are always those relating to estimates of forest cover. Significantly, forestry research has an enviable tradition of being related directly to the practical problems facing managers, even if the research has traditionally neglected some areas (such as biodiversity conservation) that should have played a much greater role in informing management practice in the past.

This chapter provides a brief overview of some techniques that can contribute to more effective forest conservation. The aim here is to identify those approaches that can help bridge the gap between conservation research and management, and help communicate the results of research to those making decisions about how forests are managed. Methods are also presented that can help ensure that management decisions are based on scientifically rigorous information. Inevitably, this chapter is more a reflection of my own personal beliefs and prejudices than the others. This reflects the fact there is no single method by which effective conservation can be achieved; choice of approach will depend strongly on local circumstances. As everywhere else in this book, the reader is encouraged to critically evaluate the methods presented. However, it is in this area of linking scientific methods with conservation practice that there lies greatest scope for improvement, and for identifying new approaches that can make a real difference to forest conservation.

8.2 Approaches to forest conservation

How can a forest be conserved? There are three main approaches:

- *protection*, through designation and management of some form of protected area
- *sustainable forest management*, involving sustainable harvesting of forest products to provide a source of financial income
- *restoration* or rehabilitation.

The choice of which approach to adopt will be governed by local socioeconomic, political, and ecological circumstances. These approaches are not mutually exclusive: a forest management plan might potentially incorporate elements of all three.

If the principal goal is biodiversity conservation, designation of a protected area will always be the preferred approach. Harvesting of forest products may be preferred if there is a need to generate financial income, beyond the amount that can be provided by a protected area (which can be substantial, for example through tourism revenues). Restoration is appropriate on sites that have been degraded or perhaps deforested entirely. Each of the three approaches is briefly considered here, by referring to methods and information resources that can help ensure effectiveness. In following sections, methods and tools are described that are relevant to the implementation of all three approaches.

8.2.1 Protected areas

The development of protected area networks is widely recognized to be the most important approach for forest conservation. Protected areas range from strict reserves to areas where some degree of harvesting is permitted. IUCN (1994) has developed a classification system for protected areas that is widely used:

I *strict nature reserve/wilderness area*: protected area managed mainly for science of wilderness protection
II *national park*: protected area managed mainly for ecosystem protection and recreation
III *natural monument*: protected area managed mainly for conservation of specific natural features
IV *habitat/species management area*: protected area managed mainly for conservation through management intervention
V *protected landscape/seascape*: protected area managed mainly for landscape/ seascape protection and recreation.
VI *managed resource protected area*: protected area managed mainly for the sustainable use of natural ecosystems.

This classification system can be difficult to apply in some situations. For example, how should sacred groves or watershed protection forests be classified? Guidance on application of the IUCN classification system explicitly to forests, and a consideration of the many different kinds of protected forests that exist, is provided by Dudley and Phillips (2006).

Although the global forest area designated as protected continues to grow, protected areas are not always successful at preventing losses of biodiversity. The problems faced by protected areas are considered in detail by Brandon *et al.* (1998), through consideration of a series of case studies. Widespread threats include infrastructural development close to reserve boundaries, colonization, and mineral extraction, as well as policy-related issues such as weak government institutions, conflicting policies and resource tenure (Brandon *et al.* 1998). The extent to which protected areas are effective in conserving biodiversity is receiving increasing attention from researchers (see, for example, Bruner *et al.* 2001, DeFries *et al.* 2005, Román-Cuesta and Martínez-Vilalta 2006). For example, in Nepal, Bajracharya *et al.* (2005) were able to evaluate the effectiveness of community-based approaches to management of the Annapurna Conservation Area, by comparing forest structure and composition inside and outside the reserve. This investigation considered effectiveness in terms of impacts on biodiversity, but other authors have considered effectiveness in terms of the management process and in terms of the coverage of the protected area network (Ervin 2003b).

Protected area management can be envisaged as a cycle of planning, implementation and evaluation, consisting of a series of stages (Hockings *et al.* 2000, Ervin 2003a) (Figure 8.1). Each of these stages should be assessed as part of an overall process of evaluation. Tools for this kind of evaluation are provided by

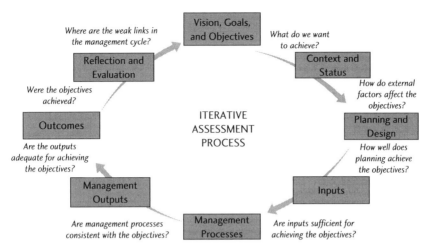

Fig. 8.1 Proposed assessment and management cycle for protected areas This is used as a basis of the RAPPAM methodology (see Box 8.1). The stages in the cycle can include: (a) vision, including goals and objectives, describing what the programme is trying to achieve; (b) assessment of how context—existing status, threats, and external factors—affects the ability to achieve the objectives; (c) assessment of the suitability of planning and design for achieving the objectives; (d) assessment of the adequacy of resources and inputs for achieving the objectives; (e) assessment of management processes, and their consistency with the objectives; (f) assessment of the management outputs, and their adequacy for achieving objectives; (g) assessment of the actual outcomes, and whether or not objectives were met; (h) reflection on the system as a whole, including an assessment of the weakest links and the most important areas for improvement. (After Hockings *et al.* 2000 and Ervin 2003a.)

Hockings *et al.* (2000) and by the WWF/World Bank Alliance (the RAPPAM methodology; Box 8.1). Such evaluations are usually achieved by holding interactive workshops in which protected area managers, policy-makers, and other stakeholders participate in evaluating the protected areas, analysing the results, and identifying subsequent next steps (Ervin 2003a). RAPPAM provides a series of questionnaires and simple scoring tools to assist in this process.

Other approaches for evaluating the effectiveness of protected areas are described by Lü *et al.* (2003) and Pressey *et al.* (2002). The latter authors applied six measures to assess the effectiveness to conservation areas in north-eastern New South Wales, Australia, namely: (1) number of conservation areas; (2) total extent of conservation areas; (3) representativeness (the proportion of natural features such as forest types represented in conservation areas), (4) efficiency or representation bias (the extent to which some features are protected above target levels at the expense of others that remain poorly protected); (5) relative protection of vulnerable areas in public land (i.e. those suitable for agriculture); and (6) relative protection of vulnerable areas across all land tenures.

The coverage of protected area networks, in terms of the species and habitats that they conserve, is another issue that has attracted increasing attention from researchers (Rodrigues *et al.* 2004). This reflects the recent development of systematic approaches to conservation planning (Margules and Pressey 2000) such as gap analysis (Jennings 2000) (see section 7.8). Such systematic approaches seek to address the biases inherent in many protected area networks with respect to the

---

Box 8.1 Sources of information on effective management of protected areas.

A useful overview is provided by Lockwood *et al.* (2006).

IUCN Programme on Protected Areas (PPA)
⟨*www.iucn.org/themes/wcpa/ppa/programme.htm*⟩
Supports the work of the IUCN World Commission on Protected Areas (WCPA), and shares the same mission and vision as WCPA. A useful online source of information, including the Guidelines for Protected Area Management Categories.

UNEP World Conservation Monitoring Centre
⟨*www.unep-wcmc.org/protected_areas/protected_areas.htm*⟩
Manages the World Database of Protected Areas (WDPA), among other data sets, some of which can be accessed on-line. Also produces a range of publications relevant to protected area management and biodiversity conservation (see, for example, Chape *et al.* 2005).

World Commission on Protected Areas (WCPA)
⟨*www.iucn.org/themes/wcpa/* ⟩
Describes itself as 'the world's premier network of protected area expertise', with over 1000 members, spanning 140 countries. It is administered by IUCN's Programme on Protected Areas. The WCPA produces a range of publications that can be freely downloaded, including Best Practice Guidelines for the planning and management of protected areas (see, for example, Hockings *et al.* 2000).

WWF and World Bank Alliance
⟨*www.panda.org/about_wwf/what_we_do/forests/our_solutions/protection/ rappam/tracking_tool/index.cfm*⟩
These organizations have developed the Rapid Assessment and Prioritization of Protected Areas Management (RAPPAM) methodology, which provides tools for assessing the effectiveness of protected area management, and a Tracking Tool to monitor progress towards this goal (Ervin 2003a). These are based on the WCPA framework and can be freely downloaded.

species or habitats incorporated within them. Margules and Pressey (2000) identified six main stages in systematic conservation planning:

- Compile data on the biodiversity of the planning region, by reviewing existing data and carrying out new field surveys if necessary.
- Identify conservation goals for the planning region, by setting quantitative conservation targets for species, habitat types, the minimum size and connectivity of habitat patches, etc.
- Review existing conservation areas, to assess the extent to which conservation goals are currently being met. This should include an assessment of threats to biodiversity within the planning region.
- Select additional conservation areas, for example by using reserve selection algorithms or decision-support tools.
- Implement conservation actions in the areas that have been selected.
- Maintain the required values of conservation areas, by setting conservation goals for each individual conservation area and monitoring progress towards these goals by using appropriate indicators.

A number of software tools are now available to assist with systematic conservation planning (Box 8.2). It is important that such tools are not used uncritically. There has been some debate about the practical applicability of such approaches, and the reliability of the results obtained. For example, the outcome of the analyses can be influenced by the characteristics of the data that are entered. Critical assessments of different analytical approaches have been undertaken by a variety of different authors, such as Csuti *et al.* (1997), Fischer and Church (2005), Moore *et al.* (2003), Pressey *et al.* (1997, 1999), and Williams *et al.* (2004). Meir *et al.* (2004) question the value of optimal approaches to designing protected area networks, suggesting that simple decision rules, such as protecting the available site with the highest species richness, may be more effective when implementation occurs over many years.

Box 8.2 Selected software tools and analytical approaches available to support systematic conservation planning.

C-Plan ⟨*www.uq.edu.au/~uqmwatts/cplan.html*⟩

C-Plan (Conservation Planning System) was developed by the National Parks and Wildlife Service in Australia to identify conservation areas in heavily influenced landscapes. Although it links with ESRI GIS software (ArcView), it requires the Borland Database Engine and the user is required to build a database using GIS layers.

MARXAN ⟨*www.ecology.uq.edu.au/marxan.htm*⟩

When provided with data on species, habitats, and/or other biodiversity features and information on planning units, MARXAN minimizes the cost (calculated as a weighted sum of area and boundary length) while meeting user-defined biodiversity targets. The software employs a simulated annealing optimization algorithm for this task. See Ball and Possingham (2000), Possingham *et al.* (2000). Available as a free download.

Reserves.xla ⟨*www.lifesciences.napier.ac.uk/staff/rob/software/reserves/*⟩

This is a freely downloadable add-in for Microsoft Excel that implements a variety of reserve selection algorithms (Briers 2002).

ResNet ⟨*http://uts.cc.utexas.edu/~consbio/Cons/ResNet.htm*⟩
⟨*www.consnet.org/manuals/ResNet.mnl-1.2.htm*⟩

The algorithms implemented in ResNet consider three aspects: rarity, complementarity between sites, and species richness. The aim of the algorithms is to achieve the set targets by selecting as few places as possible that together reach the stipulated conservation goal (Pressey and Nicholls 1989). The software is available as a free download (Garson *et al.* 2002).

Sites ⟨*www.biogeog.ucsb.edu/projects/tnc/overview.html*⟩

Developed by the Nature Conservancy, this program employs two different algorithms for reserve selection (Andelman *et al.* 1999). The 'greedy heuristic' algorithm adds planning units in a stepwise fashion to reduce overall cost. The 'simulated annealing' algorithm compares the costs of whole sets of sites to each other. Has some similarities to MARXAN.

WORLDMAP ⟨*www.nhm.ac.uk/research-curation/projects/worldmap/*⟩

A GIS-based software program developed to examine geographic patterns of biological diversity and identify areas of conservation priority (Williams *et al.* 1996).

ZONATION ⟨*www.helsinki.fi/science/metapop/*⟩

The Metapopulation Research Group at the University of Helsinki have in progress a number of research projects investigating different approaches to reserve selection, one of which is ZONATION. This method is based on iterative removal of least valuable areas with the value of remaining habitat being updated simultaneously. The method has recently been applied to determine conservation priority areas for threatened butterflies in the UK (Moilanen *et al.* 2005).

8.2.2 Sustainable forest management

Sustainable forest management (SFM) has been the central issue in international forest policy since the statement of Forest Principles and Chapter 11 of Agenda 21, which were formulated at the 1992 United Nations Conference on Environment and Development (UNCED) (see Chapter 1.6). The Forest Principles aim to 'contribute to the management, conservation and sustainable development of forests' and note the need for setting relevant standards for forest use. Sustainability concepts have in fact been recognized by foresters for at least 200 years (Wiersum 1995). Traditionally, the concept was equated with the principle of sustained yield of timber; in recent years the concept has broadened to include environmental and socio-economic aspects (Wiersum 1995). SFM has been defined in a variety of different ways but most authors agree that it comprises three main components: environmental, social, and economic (Nussbaum and Simula 2005). Environmental sustainability requires that an ecosystem be able to support healthy organisms, while maintaining its productivity, adaptability and capability for renewal; social sustainability requires that an activity not stretch a community beyond its tolerance for change; and economic sustainability requires that some form of equivalent capital (such as a natural resource) be handed down from one generation to the next. Here the focus is on the first of these three elements, and on biodiversity in particular.

International policy dialogue has led to the development of a wide variety of different criteria and indicators (C&I), designed to assess progress towards SFM. Criteria may be defined as the essential elements or major components that define SFM, whereas indicators are qualitative or quantitative parameters of a criterion, which provide a basis for assessing the status of, and trends in, forests and forest management. The C&I have been developed under a series of international processes, including ITTO, the Pan-European (or 'Helsinki') Process, the Montreal Process, and the Tarapoto, Lepaterique, Near East, Dry Zone Asia, and Dry Zone Africa processes, each of which have generated sets of C&I (FAO 2001a). Although the processes share similar objectives and overall approach, the C&I they have developed are different.

These processes provide a valuable source of information on the indicators that are considered important for forests in different regions. However, it is important to note that most processes have focused on developing C&I for application at the regional or national level. Only four of the nine processes—ATO, ITTO, Lepaterique (as a follow up to the 1997 process), and Tarapoto—have produced sets of C&I for application at the forest management unit (FMU) level, which is the scale most likely to be of value in supporting practical forest management. The development of indicators at the FMU level has primarily been driven by the growth of interest in forest certification. Certification is a tool for promoting responsible forestry practices, and involves certification of forest management operations by an independent third party against a set of standards. Typically, forest products (generally timber but also non-timber forest products) from certified forests are labelled so that consumers can identify them as having been derived from well-managed sources. There are now many different organizations certifying forests against a variety of different standards (Box 8.3), including the Sustainable Forestry Initiative (SFI) Program, American Tree Farm System (ATFS), Canadian Standards Association (CSA) Sustainable Forest Management Program and the Forest Stewardship Council (FSC). Holvoet and Muys (2004) provide a detailed comparison of forest standards developed to date, drawing on those generated by both certification bodies and C&I processes.

Box 8.3 Information sources on SFM.

Higman *et al.* (2005) provide a useful guide to how SFM can be achieved in practice, and Nussbaum and Simula (2005) provide an authoritative account of forest certification, including a description of relevant standards. Internet resources are listed below:

Food and Agriculture Organization of the United Nations (FAO) ⟨*www.fao.org/forestry/*⟩

FAO plays a major role in supporting SFM initiatives among member states, and also provides a substantial on-line information resource. The FAO is also responsible for producing the Global Forest Resources Assessment (FAO 2001b), which is regularly updated, and is increasingly being designed to support efforts at SFM.

International Tropical Timber Organization (ITTO) ⟨*www.itto.or.jp/*⟩

ITTO assists tropical member countries plan for SFM, including the development and implementation of criteria and indicators, and aspects such as reduced impact logging, community forestry, fire management and biodiversity and transboundary conservation, forest law enforcement and the sustainable use and conservation of mangrove ecosystems.

Centre for International Forestry Research (CIFOR) ⟨*www.cifor.cgiar.org/*⟩

An international research organization that has played a major role in undertaking research into the development and implementation of criteria and indicators for SFM. Extensive information resources including publications and decision-support tools available online.

Forest certification: Programme for the Endorsement of Forest Certification (PEFC) schemes ⟨*www.pefc.org/*⟩

A non-profit, non-governmental organization that promotes sustainably managed forests through independent third-party certification. PEFC is a global umbrella organization for the assessment of national forest certification schemes developed in a multi-stakeholder process. PEFC has in its membership 32 independent national forest certification systems, making it the world's largest certification scheme.

Forest Stewardship Council (FSC) ⟨*www.fsc.org/en/*⟩

The FSC is an international network that promotes responsible management of the world's forests. It sets international standards for responsible forest management and accredits independent third-party organizations who can certify forest managers and forest producers to FSC standards. Over the past 10 years, over 73 million ha in more than 72 countries have been certified according to FSC standards and several thousand products carry the FSC trademark.

How can SFM be assessed in practice? Essentially, the process involves checking whether the standards are being met, through some form of field survey or evaluation. The process of verification differs between organizations and from place to place (Rametsteiner and Simula 2003). Further details are given by Nussbaum and Simula (2005). Has the process of certification actually improved forest management? Based on a review of the available evidence, Rametsteiner and Simula (2003) conclude that there is little direct evidence that forest certification is an effective instrument for biodiversity conservation. The main impact of certification appears to have been some improvement in internal auditing and monitoring of forest organizations, and increased sensitivity of forest managers to issues such as natural regeneration/afforestation, reduced impact harvesting, road construction, and the use of fertilizers and pesticides. However, impacts on biodiversity conservation appear to have been slight.

One of the main problems is that most, if not all, of the proposed indicators relating to forest biodiversity for use at the local level are in some sense deficient (Prabhu *et al.* 1996, Stork *et al.* 1997). Many are difficult to measure in practice. There is scope for research here, to develop indicators that can readily be applied in the field, and that genuinely capture or represent important aspects of ecological processes relevant to biodiversity conservation (see section below).

Newton and Kapos (2002) examined the biodiversity indicators that have been proposed by existing C&I processes, with the aim of identifying how data required for such indicators could be derived through the use of standard forest inventory approaches. Although a large number of different biodiversity indicators have been proposed previously, they can be divided into eight generalized groups:

- forest area by type, and successional stage relative to land area
- protected forest area by type, successional stage and protection category relative to total forest area
- degree of fragmentation of forest types
- rate of conversion of forest cover (by type) to other uses
- area and percentage of forests affected by anthropogenic and natural disturbance
- complexity and heterogeneity of forest structure
- numbers of forest-dependent species
- conservation status of forest-dependent species.

Brief suggestions for how these can be measured in practice are provided in Table 8.1. This list does not consider indicators of genetic variation, which generally require sophisticated laboratory-based analyses (see section 6.5 and Namkoong *et al.* 1996; but see Jennings *et al.* 2001). Appropriate analysis and presentation of the data collected are of critical importance. As many of the biodiversity indicators considered here (Table 8.1) relate directly to forest area, GIS is of particular value for both data analysis and communicating results (see section 2.6). For example, spatial data relating to species distributions or protected areas can be overlaid on to maps of forest cover, to examine the linkages between them, and to generate statistics relevant for use as indicators.

Table 8.1 Methods for collecting information required for biodiversity indicators proposed for assessment of sustainable forest management (adapted from Newton and Kapos 2002).

Indicator	Methods	Considerations
Forest area by type, and successional stage relative to land area	Remote sensing or aerial survey combined with carefully designed sample of ground inventory plots. Remote survey (see Chapter 2) provides estimate of forest extent. Plots provide ground truth and refinement of estimated forest extent derived from remote survey, as well as data on composition and structure (see Chapter 3) that in turn can be used to identify forest types and successional stages.	Remote survey data need to be of appropriate scale and resolution. Sampling design of inventory needs to be of adequate intensity and representativeness. Remote survey can be used for stratification. Ground inventory must incorporate measures that elucidate forest type and successional stage, such as diameter class distribution, species composition, and occurrence of distinctive structural elements such as vines and epiphytes. Forest types need to be defined in a national context, but with reference to international systems such as UNESCO or IGBP to facilitate regional and global assessments.
Protected forest area by type, successional stage and protection category relative to total forest area	Remote sensing or aerial survey combined with sample of ground inventory plots and mapped data on protected areas (and/or PA inventory data). (See section 8.2.1 for information on protected areas.) Remote survey provides estimate of forest extent. Plots provide ground truth and refinement of estimated forest extent derived from remote survey, as well as data on composition and structure that in turn can be used to identify forest types and successional stages. Overlay of protected areas boundaries to determine proportion protected	Protected area boundary maps need to be available in (or converted to) electronic form. IUCN management categories are the most widely accepted classification of protection (see section 8.2.1). Mapped boundaries need to be attached to category. Care needs to be taken to avoid double counting of forest in overlapping protected areas.
Degree of fragmentation of forest types	GIS analysis of forest cover data derived from above approaches to provide summary statistics on forest area belonging to different classes or categories of fragmentation (see section 2.7 for methods).	Care needs to be taken in the selection and interpretation of fragmentation metrics, in relation to the scale and resolution of the data used for analysis (see sections 2.7 and 7.4).

Rate of conversion of forest cover (by type) to other uses	Reiteration over time of above approaches and comparison of results. Initial estimate will require use of historical data, which may require calibration for comparison. Land use data are needed if specific changes in use are to be reported. Change in cover is more easily verified.	The timescale of re-assessment needs to be decided. Reassessment methods need to be consistent over time, including with respect to scale and resolution (or cross-calibration required). The possibility of re-establishment of forest cover needs to be included.
Area and percentage of forests affected by anthropogenic and natural disturbance	Recording in ground inventory of frequency or intensity of characteristic evidence of principal forms of disturbance, e.g. paths, cut stumps, fire scars, evidence of grazing animals (see section 4.2.3). Extrapolation via remote survey and spatial analysis (see section 8.4).	Factors recorded and disturbance classes will need to be determined according to local conditions and needs. Spatial analysis of exposure or accessibility to human activity can serve as useful indicator of anthropogenic pressure related to disturbance.
Complexity and heterogeneity of forest structure	Ground-based forest inventory that includes measures of stand structure and canopy openness (see Chapter 3).	The importance of various structural characteristics varies with management priorities. Therefore specific measures may need to be decided in the national (or local) context and aggregation based on classification of forest area.
Numbers of forest-dependent species	Ground-based forest inventory can provide tree species richness, and could be used to express forest area in terms of tree species richness classes. Other species groups require purpose-designed sampling of their own and skilled survey teams—likely to be outside the scope of standard forest inventory (see section 7.9). Estimates can be derived from review of national fauna lists combined with distribution data and/or habitat requirement information.	Dbh thresholds will determine the richness recorded; data are more complete if broader dbh ranges are adopted. For non-tree species, defining and confirming forest dependence is problematic. Measurement of species numbers in relation to survey area and/or sampling effort is essential for monitoring or cross-comparison of the data.
Conservation status of forest-dependent species	Species lists (see above) cross-referenced to national and global assessments of conservation status (e.g. Red Lists, CITES) and/or specific assessments.	Endemic species should be among national priorities for inventory and assessment. Global and national conservation status may be very different.

344 | Towards effective forest conservation

For forest management to be genuinely sustainable, appropriate indicators should be used as a basis for monitoring, to support a process of adaptive management (see section 8.3). Further consideration of how appropriate biodiversity indicators might be developed and implemented in this context is given later in this chapter (section 8.6).

8.2.3 Sustainable use of tree species

Whereas SFM considers the impact of tree harvesting on the entire forest ecosystem, it is also useful to consider the impact of harvesting on the population dynamics of the individual tree species concerned. It should not be assumed that all species sourced from a forest certified as 'sustainably managed' are necessarily themselves sustainably harvested. For example, management of peat swamp forests in Sarawak achieved a sustainable yield in terms of total timber volume, but the most valuable timber species in this forest type (ramin, *Gonystylus bancana*) was severely overcut (ITTO 1990). Similarly, although forest management in areas of Quintana Roo, Mexico, has been certified as sustainable, regeneration of the main species of economic value (mahogany, *Swietenia macrophylla*) is inadequate to maintain current population size (Snook 1996) (Figure 8.2).

Sustained yield may be defined as maintaining a regular and continuing supply of forest products without impairing the capacity of the land to support production

Fig. 8.2 Community-based forest management of mahogany (*Swietenia macrophylla*) in southern Mexico (Quintana Roo). This was the first forest in Latin America to be certified, but despite careful management, concerns remain about whether timber harvesting is genuinely sustainable, because of the difficulty of ensuring adequate regeneration of mahogany. (Photo by Adrian Newton.)

(after Matthews 1989). The key objective is the achievement of an appropriate ('normal') distribution of size classes of trees within the area under management (the inverse-J shape; section 3.7.1). Two main silvicultural approaches are used to regulate which trees should be cut annually, to achieve the required age or size distribution. These are (D.M. Smith 1986):

- *the area method*, which consists of dividing the forest area into as many equally productive units as there are years in the planned rotation, and harvesting one unit each year
- *the volume method*, in which the allowable cut is defined in terms of the volume of wood, based on assessments of current and future growth rate, and the existing and desired volume of growing stock.

The volume method usually depends on regulating diameter distributions, assessed by measuring dbh and calculating the size class distribution. The *allowable annual cut* (AAC) is then allocated among the various diameter classes following comparison of the actual size class distribution with that desired for sustainable yield. The AAC is defined by whatever volume the remaining growing stock will yield in annual growth (Smith 1986). In practice this requires data on growth rates assessed in sample plots, and often computer simulation techniques.

Although the area method is the most dependable technique of achieving sustained yield, it does not provide sufficient flexibility for dealing with the non-uniform stands that are often encountered in natural forests, or in forests managed for environmental objectives. For this reason, many management schemes combine elements of both approaches (Smith 1986). Whatever method is used, the volume of timber that may be cut in one year in any given area (the AAC) should be set at a level that ensures that no deterioration occurs in the prospects for future harvests.

The following points highlight some of aspects relating to management for sustained yield (based in part on Smith 1986):

- An accurate inventory of the forest stand is essential, both to determine the current size class distribution and to estimate future growth (see Chapters 2 and 3).
- The achievement of an inverse-J-shaped diameter distribution does not necessarily guarantee sustained yield. The approach is based on the assumption that trees of each size class will continue to grow, which may not occur if (for example) the smaller trees are suppressed. The continuing recruitment of seedlings and saplings is essential for sustained yield, but can be easily neglected.
- It is often assumed that the allowable annual harvest is equal to the annual increment of the stand under management, but in fact this only holds if the stand has a 'normal' age or size class distribution.
- Usually the whole forest rather than an individual stand of trees is managed for sustained yield. By integrating many stands, the deficiencies in size classes in

some stands may be compensated for in others. Therefore, the fact that an individual stand does not demonstrate the required size class distribution does not indicate that sustained yield will not be obtained from the forest as a whole.
- When only the largest (or most valuable) trees are cut ('high-grading'), sustained yield is possible but difficult to achieve. Careful scrutiny is required to ensure that the rate of removal of large trees is kept in balance with the ability of the remaining stand to maintain a supply of large trees of adequate quality in the future.
- The minimum diameter for cut has often been based on industry requirements rather than precise calculations of what is required for sustained yield, because long-term data on growth rates are often lacking, particularly in tropical forests. Growth, yield, and regeneration data from permanent sample plots are required to determine the pattern of harvesting, the detailed marking of trees for felling and for retention, the silvicultural system to be applied, the length of the cutting cycle, and the nature of the future crop. Such data may therefore be seen as an essential prerequisite for sustainable harvesting (Palmer and Synnott 1992).

Although management for sustained yield is a traditional objective in forestry, it may conflict with other components of sustainability. For example, large ancient trees of primary (or 'old growth') forests may be felled to produce a size structure appropriate for sustained yield, despite their exceptionally high ecological value. Sustained yield concepts have also been criticized for failing to take account of natural ecological processes in forest ecosystems, leading to ecologically inappropriate management (Mladenoff and Pastor 1993). Although some authors have suggested that sustained yield (resulting in a continuous flow of forest products) is an essential component of sustainability (ITTO 1992), it is arguable whether this is in fact the case.

In order for the use of a particular tree species to be sustainable, the species must retain sufficient genetic variation to be able to adapt to changing environmental conditions, and the processes enabling this adaptation to occur must be maintained (see Chapter 6). In addition, sufficient numbers of individuals must also be maintained to avoid extinction. This requires that key regenerative processes, such as pollination, seed development and dispersal, seedling establishment and growth, should be maintained (see Chapters 4 and 5). Methods of modelling population dynamics of tree species, such as transition matrix models (Chapter 5) can be used to assess the potential impact of harvesting on population viability, and thereby to define harvesting regimes that are sustainable.

Peters (1994) makes the important observation that tree species differ in their potential for sustainable use, as a result of their contrasting biological characteristics. This variation in potential provides an alternative basis for assessment of sustainability. The biological characteristics that determine the ability of a species to withstand use are those that enable the species to tolerate or to recover after harvesting. The key characteristics, therefore, are those that determine the regeneration capacity of the species, including reproductive biology and regeneration characteristics. Differences between species in such characteristics can be used to classify them according to their potential for sustainable use (see Table 8.2).

Species with relatively regular and copious production of flowers and fruits, and with relatively abundant pollinators and seed dispersers, are more likely to be able to maintain populations through natural regeneration. Such species may be considered to have a relatively high potential for sustainable use (Table 8.2). In contrast, natural regeneration of species with highly specific pollinators, seed dispersers, or requirements for seedling establishment, may be irregular or highly sporadic, leading to populations with few young individuals or a high proportion of individuals within a narrow range of age classes. Such species are much less able to maintain population size if individuals are removed by harvesting; these species may therefore be considered to have a relatively low potential for sustainable use (Table 8.2). The same applies to species that occur at low density, particularly with uneven or 'clumped' distributions (Peters 1994).

Peters (1994) also describes a useful a strategy for ensuring that tree species are harvested sustainably (Figure 8.3). Although it was developed for non-timber forest products, the strategy applies equally well to timber trees. The strategy represents a form of adaptive management: critically, the impact of harvesting the resources is regularly monitored, then harvesting levels are adjusted to minimize impacts on the resource.

8.2.4 Forest restoration

Forest restoration refers to the process of assisting the recovery of a forest ecosystem that has been degraded, damaged, or destroyed (Mansourian 2005). This may involve the re-establishment of the characteristics of a forest ecosystem, such as composition, structure, and function, which were prevalent before its degradation (Higgs 1997, Hobbs and Norton 1996, Jordan *et al.* 1987). Ecological restoration has been defined in a variety of ways in the past decade; earlier definitions indicated that the purpose of restoration is the comprehensive re-creation of a specified historical ecosystem, including structural, compositional, and functional aspects (Jordan 1994). Such definitions emphasize the importance of historical fidelity as an endpoint of restoration. By contrast, more recent definitions allow a more flexible set of objectives, noting that cultural values may be important and that a range of ecological variables can be acceptable as endpoints (Higgs 1997).

A number of related terms are widely used in the literature, but are interpreted variously by different authors. Care should therefore be taken when using them.

- *Rehabilitation* emphasizes ecosystem recovery, without including the re-establishment of some pre-existing state as a management goal.
- *Reclamation* generally refers to the environmental improvement of mined lands, and may incorporate soil stabilization and aesthetic improvement. In this case there may be less emphasis on restoring the original biodiversity present at a degraded site, and greater emphasis on restoring productivity.

Table 8.2 The potential for sustainable use of different tree species, based on their biological characteristics (adapted from Peters 1994, http://www.panda.org/).

		Low	Moderate	High
Reproductive characteristics	Flower number, size	Few, large	Intermediate	Many, small
	Fruit number, size	Few, large	Intermediate	Many, small
	Reproductive phenology	Irregular, supra-annual	Regular, supra-annual	Regular, annual
	Pollination system	Biotic, with specialized	Biotic, with generalist	Abiotic
	Pollinator abundance	Low (bats, hummingbirds)	Moderate (beetles, moths)	High (small insects)
	Sprouting ability	None	Low	High
Regeneration processes	Seed dispersal	Biotic, with specialized vector	Biotic, with generalist vector	Abiotic
	Disperser abundance	Low (large birds, primates)	Moderate (small mammals)	High (small birds)
	Seed germination	Low viability; recalcitrant	Intermediate	High viability; orthodox
	Shade tolerance	Pioneer	Intermediate	Shade tolerant
	Regeneration niche	Narrow; specialized	Intermediate	Broad; generalist
Population structure*	Size-class distribution	Type III curve (low representation in more than one size class)	Type II curve (low representation of reproductive adults)	Type I curve (inverse-J; exponential decay)
	Tree density	Low (0–5 adults ha^{-1})	Moderate (5–10 adults ha^{-1})	High (>10 adults ha^{-1})
	Spatial distribution	Scattered	Clumped	Evenly distributed

* See section 4.74.

Approaches to forest conservation | 349

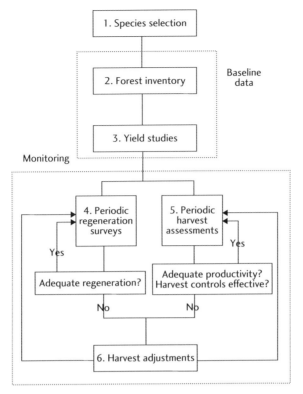

Fig. 8.3 Flow chart of the basic strategy for exploiting tree species on a sustainable basis. The process is composed of six steps: (1) species selection; (2) forest inventory, to assess the extent of the resource; (3) yield studies, to determine productivity; (4) regeneration surveys, to ascertain whether the tree species is regenerating adequately from seed; (5) harvest assessments, to assess the impact of harvesting on the individual trees; and (6) serial harvest adjustments, by which the results of monitoring are used to adjust harvesting levels. (From Peters 1994.)

- *Forest landscape restoration (FLR)* is a term that has been developed by WWF and IUCN (see Box 8.4), referring explicitly to restoration at the landscape scale. The concept of FLR embraces improvement of social, cultural, and economic benefits to people, as well as ecological recovery.
- *Afforestation* and *reforestation* refer to the establishment of trees on a site, in the former case where no trees existed before, and in the latter case following deforestation (Mansourian 2005).

One of the main concerns among forest restoration ecologists has been the need to differentiate between interventions aiming at ecological recovery and conventional plantation forestry undertaken with purely commercial aims. For example, it is widely assumed that establishment of native tree species will confer greater

Box 8.4 Information sources for forest restoration.

Mansourian *et al.* (2005) provide a useful overview of forest restoration, including a number of practical tools and techniques. Other useful information resources are provided by Lamb and Gilmour (2003) and Perrow and Davy (2002). Internet resources are listed below.

Forest Restoration Information Service (FRIS)
⟨*www.unep-wcmc.org/forest/restoration/*⟩

FRIS is a web-based information service developed by the UNEP World Conservation Monitoring Centre, in collaboration with a range of partners. The site includes an online database of forest restoration projects throughout the world, and a number of tools to facilitate the prioritization, design and execution of forest restoration efforts.

Global Partnership on Forest Landscape Restoration
(*www.unep-wcmc.org/forest/restoration/globalpartnership/index.htm*)

The Global Partnership is a network of governments, organizations, communities and individuals active in forest restoration at the landscape scale (FLR). The partnership is designed to support international efforts at forest restoration by fostering information exchange (for example through FRIS), and by linking policy and practice.

IUCN Forest Conservation Programme
(*www.iucn.org/themes/fcp/experience_lessons/flr.htm*)

One of the key elements of the work of the IUCN Forest Programme focuses on FLR, which aims to 'bring people together to identify, negotiate and put in place practices that optimize the environmental, social and economic benefits of forests and trees within a broader pattern of land uses'. This is achieved through a number of field-based projects.

WWF's Forest for Life Programme
⟨*www.panda.org/forests/restoration/*⟩

WWF has established a global network of over 300 forest conservation projects in nearly 90 countries, including a portfolio of forest landscape restoration programmes undertaken in collaboration with IUCN. WWF has adopted a target to restore forests in 20 landscapes of outstanding importance within priority ecoregions by 2020.

Society for Ecological Restoration International
⟨*www.ser.org/*⟩

A non-profit organization with more than 2000 members worldwide, widely recognized as a source for expertise on restoration science, practice, and policy. Although it does not engage in restoration projects directly itself, the Society supports dialogue and information exchange through its website and through publication of academic journals such as *Restoration Ecology*.

environmental benefits than establishing the non-native tree species often employed in commercial plantation forestry. However, plantation forests of exotic tree species can be of significant habitat value for wildlife (Humphrey *et al.* 2003b), and in some cases it may be necessary to first establish non-native tree species to act as a 'nurse' for establishing native species. In such circumstances restoration goals might focus most usefully on improving habitat quality, rather than striving to achieve ecological fidelity (*sensu* Higgs 1997). This may be achieved by targeting the specific attributes of ecosystems, with an emphasis on manipulating processes that have changed, rather than on a comprehensive re-creation of historical ecosystem composition, something that can often be difficult to define (Newton *et al.* 2001).

Although the number of practical restoration projects being implemented has increased rapidly in recent years, there is a perception that these activities have been undertaken with relatively little input from the scientific community. This is despite the fact that many of the decisions which need to be made in restoration projects are essentially based on scientific concepts. This raises questions about what constitutes 'good' or appropriate restoration practice, and the role of ecological theory in informing the process of restoration (Higgs 1997, Moore *et al.* 1999, Stephenson 1999). It has been suggested that practical restoration projects need to be based much more firmly on an understanding of ecological processes, and that such an understanding could help resolve many key questions facing restoration practitioners, such as selecting an appropriate baseline for restoration, determining the evolutionary viability of restored populations, predicting whether species will be able to colonize newly available habitat, and measuring the ecological success of restoration efforts (Clewell and Rieger 1997).

Different approaches to forest restoration vary in terms of their relative cost, their benefits to biodiversity, and their potential impact on provision of other ecological services, such as water regulation and nutrient cycling (Table 8.3). In general, the preferred method is to allow the forest to recover naturally through a process of succession ('passive restoration', Table 8.3). For such successional recovery to occur, the following conditions must be met (Lamb and Gilmour 2003):

- The disturbing agent or agents must be removed. If disturbances such as fire, timber harvesting, or grazing continue, succession is interrupted and recovery is unlikely.
- Plants and animals must remain at the site or in the region as a source of new colonists, and must be able to move across the landscape and recolonize the degraded area. The more distant these source populations are, the slower the recolonization process. Potentially, connecting habitat fragments or 'stepping stones' can increase the rate of the recovery process. This is an argument for planning forest restoration at the landscape scale (Humphrey *et al.* 2003c).
- Soils at the site must remain reasonably intact. If severe erosion has taken place or if fertility has been depleted the soils may no longer be suitable for the

original species, and other species (perhaps not native to the area) may come to dominate.
- Weed species, invasive exotic species, or animal pests must be excluded or controlled if the original community is to be re-established successfully.

In order to estimate the rate of forest recovery, the modelling techniques described in Chapter 5 can potentially be used. Similarly, the techniques described in Chapter 4 are of value in analysing the factors influencing successional processes, which are of crucial importance to forest recovery.

In some situations the forest may be so degraded that natural recovery will be very limited (such as the case of Carrifran, Box 6.1). In such situations there may be a need to establish trees artificially, through methods such as direct seeding or planting of tree seedlings (Table 8.3). Key decisions include (Newton and Ashmole 1998):

- *Which tree species should be established?* This can be determined by reference to results from pollen analysis or historical records.
- *What should their spatial distribution and relative abundance be?* Field surveys of soil characteristics and topographic variation across the site to be restored, together with autoecological information about the tree species selected, can be used to ensure that species are correctly matched to the microsites on which

Table 8.3 Relative costs and benefits of various methods of overcoming forest degradation (from Lamb and Gilmour 2003 http://www.panda.org.).

Method	Relative direct cost	Relative rate of biodiversity gain	Potential ecological services benefit
(a) *Prime focus of biodiversity restoration*			
Passive restoration	Low	Slow	High
Enrichment planting	Low–medium	Slow–medium	High
Direct seeding	Low-medium	Medium	High
Scattered plantings	Low	Slow	Medium
Close plantings of a few species	Medium	Medium	High
Intensive plantings after mining	High	Fast	High
(b) *Prime focus on productivity and biodiversity*			
Managing secondary forests	Low–Medium	Medium	High
Enrichment plantings	Low–medium	Medium	Medium–high
Agroforestry	Medium–high	Medium	Medium–high
Monoculture plantations with buffers	High	Slow	Medium
Mosaics of monocultures	High	Slow	Low–medium
Mixed species plantations	High	Slow	Medium
Enhanced understorey development	Low	Slow	Medium–high

they are to be established. If the aim is to mimic natural forest structures, then spacing of trees should be irregular, and individuals of each species should be grouped (Rodwell and Patterson 1994).
- *How should planting stock be sourced?* For native woodland establishment to be successful, planting material should be well adapted to the site conditions, and therefore local sources should be used. Genetic effects such as genetic drift and inbreeding may reduce the viability of planting stock obtained from isolated forest fragments; genetic analyses may be required to assess such risks (see Chapter 6).
- *How much of the area should be forested?* Open spaces within forest areas are important as wildlife habitat, and therefore as much as 30–40% of an area may be left unplanted initially, both to provide long-term open space and to provide areas for future tree establishment by natural regeneration or planting.

Mapping tools such as GIS (see Chapter 2.6) are likely to be of value when planning restoration projects. Increasingly, forest restoration projects are being planned and implemented at large spatial scales (as in the case of FLR projects, Box 8.4), and in such cases GIS becomes particularly valuable (Humphrey *et al.* 2003c). Another key issue is the need to monitor restoration progress, ideally as part of an adaptive management cycle (see section 8.3). Many restoration projects neglect this important aspect. Once restoration targets or management goals have been identified, indicators will have to be developed (see section 8.6) to help track progress towards these goals (Figure 8.4). This can be achieved by examining the threats (pressures) affecting the forest, and identifying the specific management

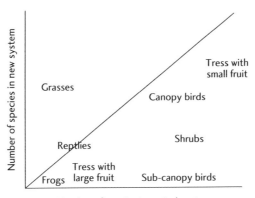

Fig. 8.4 One approach to monitoring forest restoration. Forest recovery can be monitored by comparing the number of species in different life forms in the recovering forest with those present originally. In this illustration, the numbers of reptiles, birds found in the forest canopy, and trees with small fruit have recovered to near their original condition, whereas there are fewer species of most other life forms than in the original forest. On the other hand, the site now has more grass species. (After Lamb and Gilmour 2003.)

interventions needed to address these threats (O'Connor *et al.* 2005). Further information on monitoring techniques is provided below.

8.3 Adaptive management

Whichever approach to forest management and conservation is adopted, *adaptive management* techniques should be employed. Particularly useful resources on this topic have been developed by Foundations of Success, a not-for-profit organization committed to improving the practice of conservation through the process of adaptive management. A number of information resources are accessible from their website ⟨*http://fosonline.org/*⟩; see Margoluis and Salafsky (1998) and Salafsky *et al.* (2001, 2002). This brief account is based on these sources, which should be consulted for more details.

Adaptive management can be defined as the integration of design, management, and monitoring to systematically test assumptions in order to adapt and learn. It also offers a method by which research can be incorporated into conservation action. The approach includes the following elements:

- *Testing assumptions*. This involves systematically trying different management actions to achieve a desired outcome. This depends on first thinking about the situation at the specific project site, developing a specific set of assumptions about what is occurring, and considering what actions could be taken to affect these events. These actions are then implemented and the results are monitored to assess how they compare to the ones predicted at the outset, on the basis of the assumptions.
- *Adaptation*. If the expected results were not obtained, then the assumptions were wrong, the actions were poorly executed, the conditions at the project site have changed or the monitoring was faulty. Adaptation involves changing assumptions and interventions in response to the information obtained as a result of monitoring. This is the defining feature of adaptive management.
- *Learning*. This refers to the process of systematically documenting the management process, and the results achieved. This helps avoid repeating the same mistakes in the future.

The adaptive management process involves six steps (Figure 8.5):

Start *Define the objectives clearly.* This is a crucially important first step in any conservation programme, yet is surprisingly often neglected, and as a result management objectives are often vague or poorly defined. This makes it very difficult to ascertain whether the programme is being successful or not.

Step A *Design an explicit model of the system*. A model is a simplified version of reality. Such models can help to integrate and organize information, provide a framework for identifying and comparing management interventions, and support discussion about how the management should be undertaken. The models used need not be quantitative; methods of developing conceptual models are described in section

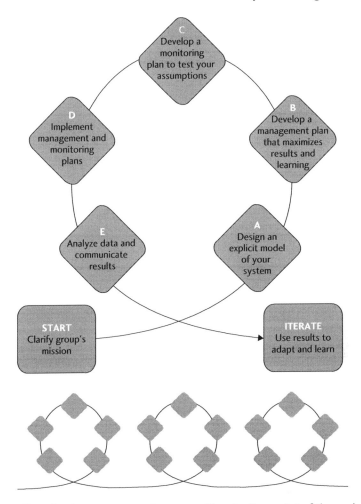

Fig. 8.5 The adaptive management process. The starting point of the cycle involves identifying the overall mission. Step A involves assessing the conditions and determining the major threats to biodiversity at the project site. Using a conceptual model, the project team defines relations between key threats, other factors and elements of biodiversity at the project site. Step B involves using this model to develop a project management plan that outlines the results that the project team would like to accomplish and the specific actions that will be undertaken to reach them. Step C involves developing a monitoring plan for assessing progress. Step D involves implementing the management actions and monitoring plan. Step E involves analysing the data collected during monitoring and communicating this information to the appropriate audiences. Results of this analysis are used to change the project and learn how to improve it in the future. Based on feedback information, there may be a need to modify the conceptual model, management plan, or monitoring plan. (From Salafsky *et al*. 2001, http://www.panda.org.)

7.8.1. The model should incorporate the assumed linkages between the various direct and indirect threats or factors that affect the conservation targets of interest.

Step B *Develop a management plan*. This is a crucial element of any conservation programme. The plan should outline the factors or threats that need to be addressed and the specific actions that will be undertaken to change them. Selection of actions should be based on the results that are desired. Experiments can be undertaken to help define which actions are successful. The plan can usefully be informed by the conceptual model developed under Step A. A useful first step is to rank the various threats identified in the model and decide which cause the biggest problems and therefore need addressing first. Then a specific *objective* should be developed for each threat or factor, and proposed *activities* defined to achieve each objective. Further details of how to develop a conservation management plan are presented by Sutherland (2000).

Step C *Develop a monitoring plan*. Monitoring is essential to the adaptive management process, to assess whether the management actions are being effective. Adaptive management requires testing explicit *assumptions* about the area of interest and collecting only the data that are needed to test these assumptions. Development of the conceptual model under Step A can greatly help decide what should be monitored. *Indicators* will need to be chosen for measurement, representing both the threats or factors influencing the conservation targets, and the targets themselves. A *monitoring plan* must be produced describing which methods will be used to collect data on the indicators. Many of the techniques described in this book could be of value here. The challenge is to avoid collecting too much data, and instead focus on the critical factors that are most relevant to the project. Further information on indicators and monitoring is given in the following sections.

Step D Implement the management and monitoring plans. The most important step!

Step E *Analyse data and communicate results*. Many practitioners feel that they are too busy with day-to-day work and problems to analyse the data that they collect. Yet it is very important to include analysis and communication efforts in the work plan. If the project is planned properly, then the conceptual model and management plan should contain the objectives set, the assumptions made, and the interventions being undertaken. The monitoring plan should outline the data being collected. The main additional need is to analyse and interpret what these results mean and then communicate them in a way that addresses the needs of key audiences. Analysis is most effectively done in the context of specific questions being asked or assumptions being tested.

Iterate *Use results to adapt and learn*. As illustrated in Figure 8.4, the adaptive management process is iterative. It is important to return to the original conceptual model and to the assumptions that were identified at the outset and then tested experimentally, through the management activities. If the management interventions or experiments turn out exactly as predicted, then the assumptions will have been confirmed. If not, the results need to be used to change the actions being undertaken. The model itself will also need to be changed. This will capture the learning accomplished and incorporate it into the project's institutional knowledge. A new round of assumptions can then be identified for future testing. Over time, the cycle may be passed through many times, learning more each time, ultimately leading to better conservation.

There is no doubt that adaptive management approaches can be very successful if implemented properly. One of the main challenges is to undertake management interventions in the form of experiments; ideally this requires appropriate controls and replication, something often very difficult to achieve in practice. Often the results of interventions or experiments may be difficult to interpret or ambiguous; in such cases it can be difficult to decide how best to amend the management action. There can also be resistance among practitioners in visualizing their systems as conceptual models, and their interventions as experiments (Sutherland 2000). Given the potential benefits of the approach, there is enormous scope for researchers to work closely alongside practitioners to provide support in its implementation, and thereby to improve the effectiveness of conservation.

8.4 Assessing threats and vulnerability

Effective conservation depends strongly on a full appreciation of the causes of biodiversity loss. Yet identifying such threats or 'threatening processes' has received remarkably little attention from researchers. Surprisingly, the issue is ignored by many conservation biology textbooks. There is enormous scope for improving methods for assessing threats and diagnosing their impacts, and improving the quality of information available to conservation managers. However, identifying the precise causes of decline in the abundance of a particular species can often be surprisingly difficult.

Different types of threat may be identified. Direct threats are those that are directly responsible for loss or degradation of forests, or their associated biodiversity. Indirect threats are the underlying causes of such direct threats. For example, an underlying threat such as a government policy may be responsible for the direct threat of forest conversion to agriculture. Other terms used to describe threats include 'drivers' or 'pressures'; these terms may be preferred because they imply that effects on biodiversity can be either negative or positive, whereas the term 'threat' implies only negative impacts. However, it should be remembered that different authors interpret these terms in different ways. Salafsky *et al.* (2002) list many of the most widespread threats to biodiversity.

Robinson (2005) identifies the following main methods of assessing threats:

- *Conceptual modelling*, used to illustrate the relation between threats and their impacts, and for providing a strategic framework for identifying appropriate management interventions (see section 7.8.1).
- *Threat matrices.* Matrices can vary from simple tables to complex logical frameworks linking different threats and interventions to conservation

targets. Matrices are relatively simple to implement and can readily be updated, but their dependence on subjective information is a weakness. An example of a threat matrix is illustrated in Table 8.4.
- *Participatory threat mapping*, which can involve use of pictorial maps or three-dimensional models to elicit information about changes in forest habitat quality or quantity, when working with community groups.
- *GIS-based mapping*, incorporating quantitative spatial data. Direct threats, such as habitat fragmentation (see section 2.7), can be assessed and displayed by using GIS. Spatial models of forest dynamics (see section 5.5) can be used to explore and illustrate the potential impacts of different threats on forest extent, structure and composition.

Wilson *et al.* (2005b) provide a detailed review of the concept of vulnerability in conservation planning, noting that information on threatening processes and the relative vulnerability of areas and features to these threats pervades the process of conservation planning. Pressey *et al.* (1996) defined vulnerability as 'the likelihood or imminence of biodiversity loss to current or impending threatening processes'. Wilson *et al.* (2005b) extend this definition by distinguishing three dimensions of vulnerability, *exposure*, *intensity*, and *impact*, and provide the following information regarding their measurement:

- *Exposure* can be measured either as the probability of a threatening process affecting an area over a specified time, or the expected time until an area is affected. Exposure is commonly measured categorically, for example as 'high', 'medium', or 'low' suitability for agriculture, but has also been measured on continuous scales by some authors. Maps can be produced illustrating the relative exposure of different areas to a particular threat.
- *Intensity* measures might include magnitude, frequency, and duration of the threat. Examples include livestock density, volume of timber extracted per hectare of a forest type, or the density of an invasive plant species. Intensity can also be estimated categorically, and can be mapped across whole planning areas.
- *Impact* refers to the effects of a threatening process on particular features such as the distribution, abundance, or likelihood of persistence of a species of interest. For example, logging may have much greater impact on animal species dependent on old growth forests than on species that inhabit a variety of post-logging stages. Impact might also depend on the spatial pattern of the threatening process, for example on the degree of connectivity between old growth patches retained after logging operations.

As conservation planning is generally spatial, a key issue is whether vulnerability can be mapped, and therefore integrated with other spatial information such as forest cover, boundaries of management units etc. According to Wilson *et al.* (2005b), this requires mapping of spatial predictions of the future distribution of threatening processes. Maps of exposure can be based on the current distributions of threats and knowledge of variables that could predispose areas or features to those

Table 8.4 Example of a simple method of scoring the different pressures or threats affecting a forest area (from Ervin 2003a, http://www.panda.org/). Here, different human activities have been scored on a simple scale with respect to three variables: extent, impact, and permanence. Extent is the area across which the impact of the activity occurs. Impact is the degree, either directly or indirectly, to which the threat affects overall forest resources. Permanence is the length of time needed for the affected area to recover with or without human intervention. A combined score ('degree') has been produced by multiplying the individual scores together.

Activity	Extent	Impact	Permanence	Degree	Description and rationale
NTFP collection	Localized (1)	Mild (1)	Short term (1)	1	NTFP collection consists primarily of mushroom harvesting for consumption by local residents. Harvesting occurs near an adjacent village, and harvesters generally leave large areas undisturbed.
Road	Scattered (2)	Moderate (2)	Medium term (2)	8	A road is planned through a portion of a protected area. The actual impact of construction will be minimized by using environmental best practices. It is a gravel access road, and will only be used seasonally by park staff and visitors with permits.
Tourism	Localized (3)	High (3)	Short term (1)	9	Tourists have recently begun to drive motorized off-road vehicles through sensitive wetlands. Springtime vehicle use has already disrupted the mating and denning habits of large numbers of bears, considered a key species in this protected area.
Poaching	Widespread (3)	High (3)	Medium term (2)	18	The main species poached is tiger, which is extensively poached in the protected area. A large percentage of the tiger population is killed annually.
Alien species	Widespread (3)	High (3)	Long term (3)	27	An invasive plant species (*Chromolaena*) covers a quarter of the park. It has rendered large areas of rhino and elephant habitat unsuitable, and is extremely difficult to control or eradicate.
Dam building	Throughout (4)	Severe (4)	Permanent (4)	64	A large-scale hydro-electric dam is planned that would flood at least half of the protected area.

threats. For example, the likelihood of forest conversion to agriculture is often related to the suitability of the soil for agricultural crops, topography, and proximity to infrastructure or population centres (see, for example, Mertens and Lambin 1997). Spatial predictions of intensity are less common in the literature than predictions of exposure; for some threats, such as forest clearing, intensity is considered as binary (either cleared or not). Impact is the most difficult dimension of vulnerability to map, as this may require feature-specific information on the effects of different levels of intensity, spatial information on features relative to variations in intensity, and ways of integrating this information across assemblages of species, sets of vegetation types, or other groups of features (Wilson *et al.* 2005b).

Wilson *et al.* (2005b) also reviewed the different methods that have been used by previous researchers to assess the three dimensions of vulnerability. The methods were grouped into four groups based on the types of data used (Table 8.5). Little information is available in the literature to indicate which approach would be preferable in a given situation; it is likely that a combination of approaches will give the best overall result (Wilson *et al.* 2005b). As noted by these authors, a comprehensive assessment of vulnerability would consider all of the threats affecting an area and also include the dynamic responses of threats to conservation actions. Combining vulnerability scores for multiple threats is analytically tractable and can be achieved by differentially weighting threats to reflect their relative importance, ideally informed by their respective impacts. An example of the latter approach is provided by Miles *et al.* (2006), in their global assessment of the conservation status of tropical dry forests. Here, GIS techniques were used to analyse and combine maps of different threats derived from remote sensing data to identify those forest areas vulnerable to multiple threats. Such methods, coupled with spatial analysis and statistical modelling techniques, offer the potential for powerful and sophisticated analyses of threats at a range of different scales.

It should be remembered that risk analysis and hazard assessment are widely practiced by foresters, and a substantial literature exists on this topic (see also Burgman 2005). This literature has been little used by the conservation community to date. Methods for assessing different types of forest disturbance are described in Chapter 4. Such measurements can be used as a basis for assessing risk, as described briefly below for some of the most important types of disturbance.

- *Fire.* Forest fire risk is generally assessed by identifying the potentially contributing variables and integrating them into a mathematical expression or index. The index is used to indicate the level of risk, and can be mapped. A wide variety of different approaches are used to produce such indices, which vary particularly in terms of the timescales involved. Estimates of the probability of fire occurrence are typically based on variables such as the amount and type of fuel available for burning, topographic variables, vegetation characteristics, and meteorological variables. Further details of the most common methods can be found in Chuvieco (1999) and San-Miguel-Ayanz *et al.* (2003).
- *Wind.* A number of different approaches to analysing and modelling windthrow risk have been developed. Statistical models use empirical information

Table 8.5 Methods used to assess vulnerability (adapted from Wilson *et al.* 2005b). The table is based on a thorough review of the literature; for full references, see Wilson *et al.* (2005b).

Concept	Method of measurement	Dimension of vulnerability	Example reference
Group 1: Methods based on tenure and land use	Vulnerability is estimated from coverage in existing conservation areas	Exposure	Castley and Kerley (1996)
	Vulnerability is estimated from permitted or projected land uses	Exposure and intensity	Abbitt *et al.* (2000)
Group 2: Methods based on environmental or spatial variables	The past impacts of threatening processes are used to indicate the vulnerability of features. These values are then given to presently unaffected areas that contain the same features	Exposure	Myers *et al.* (2000)
	Characteristics of areas or features exposed to threats in the past are used in qualitative or informal quantitative analyses to predict vulnerability	Exposure and intensity	Sisk *et al.* (1994)
	Characteristics of areas or features exposed to threats in the past are used in spatially explicit, quantitative models to predict vulnerability	Exposure and intensity	Wilson *et al.* (2005a)
Group 3: Threatened species are used to indicate vulnerability	The number of threatened species and their relative threat ratings are combined to indicate vulnerability	Exposure, intensity, and impact	Brooks *et al.* (2001)
Group 4: Experts decide on relative vulnerability	Opinions are sought from experts on the relative vulnerability of areas or features	Exposure, intensity, and impact	Ricketts *et al.* (1999)

on damage collated over a number of years in selected areas, whereas deterministic models consider tree or stand characteristics and the windiness of a site or the critical wind speed (Lanquaye-Opoku and Mitchell 2005, Quine 1995). An example of a widely used mechanistic model is GALES, developed by the Forestry Commission in the UK (Gardiner *et al.* 2004), which estimates the threshold wind speeds required for overturning and breaking the mean tree of a stand. The resulting *decision support system* (DSS), ForestGALES, can be used to predict the wind damage risk for 19 different conifer species, growing in homogenous, even-aged stands in Europe. Spatially explicit models of forest dynamics such as LANDIS (see section 5.5) also offer tools for exploring the potential impacts of different wind regimes.

- *Herbivory*. The determination of browsing damage on forest regeneration at a given time can be used to forecast impacts in the future, for example when the timber is harvested or when the function of the forest is seen to have been compromised. The effects of browsing and the resulting damage can be decades apart. In order to estimate the long-term impacts of browsing, indicators may need to be specified for young forest stands at the time when top-twig browsing is no longer possible (Reimoser *et al.* 1999). Ultimately the risk of damage by herbivores is a function of animal behaviour, something that is difficult to predict but is also an active area of research.

- *Deforestation*. Research has indicated that deforestation can be related to a range of factors, such as population density, population growth, agricultural expansion, income levels, and amount of timber harvesting (Allen and Barnes 1985, Uusivuori *et al.* 2002). Typically, deforestation is assessed by comparing the forest cover at different times by means of remote sensing images (see Chapter 2). To calculate deforestation rate, the following formula can be used (Puyravaud, 2003) (see FAO (1995) for an alternative):

$$P = \frac{100}{t_2 - t_1} \ln \frac{A_2}{A_1}$$

where A_1 and A_2 are the forest cover at times t_1 and t_2, respectively and P is percentage per year. Deforestation can then be analysed by using multivariate statistics and regression techniques, to identify the influence of different pressures or drivers. For example, Wilson *et al.* (2005a) analysed deforestation in southern Chile, by first identifying where native forests had been converted to plantations, then using a multivariate, spatially explicit statistical model to identify the variables responsible for this conversion. Predictions were then made of where native forest conversion is likely to occur in the future, as a function of patterns of climate, topography, soils and proximity to roads and towns, providing an assessment of relative vulnerability to deforestation. Similarly, in Cameroon, Mertens and Lambin (1997) found a negative relation between deforestation and proximity to roads, which could potentially be used to identify areas vulnerable to future forest loss.

Modelling of land-use changes such as deforestation requires combining spatially explicit ecological data with information on socio-economic as well as biophysical variables. A range of models of land-use change have been developed, based on a variety of analytical approaches (Mladenoff and Baker 1999, Veldkamp and Lambin 2001). For example, GEOMOD is a GIS-based model that simulates the location of deforested cells by using geophysical attributes as well as spatial data of forest cover at different time intervals (Pontius *et al.*, 2001), and is now commercially available as part of the IDRISI software program (⟨*www.clarklabs.org*⟩). In a review of the current understanding of land-cover change in tropical regions, Lambin *et al.* (2003) emphasize the complexity of the issue and highlight the role of market forces, policy interventions, and changes in social organization and attitudes. Relatively simple statistical models are unlikely to capture this complexity, and instead these authors highlight the role of agent-based systems and narrative perspectives to provide an explanation and prediction of land-use changes.

Although important advances have been made in modelling the spatial pattern of deforestation, ecosystem heterogeneity and lack of data (particularly for relevant socioeconomic drivers) continue to constrain analyses and the development of deforestation scenarios (Grainger 1999). As deforestation cuts across a range of disciplines (sociology, economics, ecology, and geography), some researchers have encouraged a focus on the interactions between these disciplinary approaches rather than on the details of each subcomponent (Dale and Pearson 1997).

8.5 Monitoring

Monitoring is the process of periodically collecting and using data to inform management decisions (O'Connor *et al.* 2005). Monitoring is a critically important aspect of any conservation project, to help ensure that management interventions are effective. As noted above, monitoring is an essential component of adaptive management approaches. The monitoring should not be initiated only at the end of the project, but should be integrated into the overall project cycle. Detailed guidance on monitoring is provided by Hurford and Schneider (2006), Spellerberg (1991), and Sutherland (2000).

Many of the techniques described in this book can be used as part of a monitoring programme. Field-based forest monitoring is usually achieved by establishing and repeatedly surveying permanent sample plots (section 4.7), incorporating assessment of forest stand structure and composition (Chapter 3) and habitat characteristics (Chapter 7). Such field surveys may be complemented by use of remote sensing analyses (Chapter 2). Here, some general issues and principles are described that should be borne in mind when designing and implementing a monitoring programme.

Yoccoz *et al.* (2001) highlight the fact that many biodiversity monitoring programmes are inadequate. Specifically, inadequate attention is given to three basic questions:

- Why monitor?
- What should be monitored?
- How should monitoring be carried out?

The following recommendations are made by these authors, in a critical review of current practice:

- *It is essential to clearly define the objectives of monitoring.* Many programmes are based on the assumption that collecting any information about a system will be useful, but such an approach is inefficient and can result in large amounts of irrelevant information being collected. Two main types of objective can be differentiated: *scientific objectives*, focusing on understanding of the behaviour and dynamics of the system, and *management objectives*, designed to provide information for informing management decisions.
- *Monitoring, management and research should be closely integrated*, as explicitly required in adaptive management approaches (see above). Research can play a critical role in identifying appropriate management actions, and it can be argued that monitoring is trivial unless it is linked to experimental work to understand the mechanisms underpinning system changes. In practice, monitoring is often limited to the assessment of management policies.
- *Decisions regarding what to monitor* should be determined by the *a priori* hypotheses to be addressed by the programme and by the relative values of different components of diversity as specified in management objectives.
- *The sampling design* will be dependent on the choice of biological diversity measures. For some objectives, it might be adequate to focus on species richness of some groups. For others it will be necessary to estimate the abundances of each species in the community of interest, which requires greater effort. It is also important that monitoring programmes provide data to estimate not only the state variables of interest, but also the rate parameters that determine system dynamics.
- *It is important to estimate detection error*, which arises because few survey methods permit the detection of all individuals, or even all species, in surveyed areas. Distance sampling methods (see section 3.5.3) can be used to estimate detection probabilities associated with count statistics. Monitoring programmes should estimate the detection probabilities associated with the selected survey methods; without these, it is not possible to draw strong inferences about the monitored system.
- *Spatial variation* is a second source of error because of the inability to survey large areas entirely. Most surveys are not based on an appropriate spatial sampling scheme, and therefore do not provide unbiased estimates of biodiversity at larger spatial scales. Sample locations for monitoring programmes should not be selected arbitrarily, but be selected to permit inference to the larger area of interest. Sampling designs should be chosen with respect to their efficiency, in terms of the precision of resulting estimates.

Legg and Nagy (2006) similarly point out that much current monitoring of biodiversity is a waste of time, because of the lack of detailed goal and hypothesis formulation, inadequate survey design and data quality, and lack of assessment of statistical power at the outset. As a result, most programmes are unlikely to be capable of rejecting a false null hypothesis with reasonable power. Along with a number of other recommendations (Box 8.5), these authors suggest that in any monitoring programme a power analysis shoule be done at the outset, to ensure that the proposed design is sufficiently robust. Statistical power is 1.0 minus the probability of a type II error, i.e. the probability of rejecting the null hypothesis when it is false. The power of a statistical test depends on effect size, error variance, sample size, and the type I error rate (Legg and Nagy 2006). For example,

---

Box 8.5 Recommendations for the design and implementation of biodiversity monitoring programmes.

Recommendations for good management of a monitoring programme
- Secure long-term funding and commitment.
- Develop flexible goals.
- Refine objectives.
- Pay adequate attention to information management.
- Train personnel and ensure commitment to careful data collection.
- Locations, objectives, methods, and recording protocols should be detailed in the establishment report.
- Obtain peer review and statistical review of research proposals and publications.
- Obtain periodic research programme evaluation and adjust sampling frequency and methodology accordingly.
- Develop an extensive outreach programme.

Recommendations for good design and field methods in monitoring
- Take an experimental approach to sampling design.
- Select methods appropriate to the objectives and habitat type.
- Minimize physical impact to the site.
- Avoid bias in selection of long-term plot locations.
- Field markings must be adequate to guard against loss of plots.
- Ensure adequate spatial replication.
- Ensure adequate temporal replication.
- Blend theoretical and empirical models with the means (including experiments) to validate both.
- Synthesize retrospective, experimental and related studies
- Integrate and synthesize with larger- and smaller-scale research, inventory, and monitoring programmes.

(after Legg and Nagy 2006).

following these authors, the power of a *t*-test is derived from the *t*-distribution and the value of *t* given by:

$$t_{\beta(1),\nu} = \frac{\delta}{\sqrt{s^2/n}} - t_{\alpha,\nu}$$

The sample size required to detect a difference between means of δ with power (1–β) is:

$$n = \frac{s^2}{\delta^2}(t_{\alpha,\nu} + t_{\beta(1),\nu})^2$$

where *n* is sample size, s^2 is an estimate of variance, δ is the minimum detectable difference, $t_{a,\nu}$ is the critical value of *t* for a probability of α (one-tailed or two-tailed as appropriate), $t_{b,\nu}$ is the critical value of *t* for a one-tailed probability level β, β is the probability of type II error and ν is the degrees of freedom. Note that *t* is a function of *n* and so the solution must be obtained by iteration (Legg and Nagy 2006).

The simplest way to increase power is to increase sample size, but this costs time and money. There is therefore a trade-off between sample size and the quality of information that can be obtained from each observation (Legg and Nagy 2006). For example, estimates of plant cover made by averaging the visual estimates of cover in subunits within gridded quadrats show much less between-observer and within-observer error than visual estimates from ungridded quadrats. If the between-quadrat variance is high, then large numbers of low-precision ungridded quadrats give greater power than the same amount of time spent on a few high-quality gridded quadrats (Legg and Nagy 2006). Prior knowledge or a pilot study will therefore often be required to find the optimal method.

A simple approach to defining required sample sizes is provided by Manley (1992). First, the maximum size of sample that can be collected is estimated, given the resources available. From this, the power of the test that one wishes to apply can be estimated. If the estimated power is inadequate, then a decision needs to be made regarding whether to proceed or to abandon the study altogether. There is little point in a monitoring programme that cannot reject a null hypothesis that is false (Legg and Nagy 2006). Ideally, monitoring programmes should be designed around a simple and powerful statistical model such as analysis of variance (ANOVA) that can make use of all of the information available to reduce residual errors. Parametric statistical tests are usually more powerful than non-parametric tests (Legg and Nagy 2006).

Sheil (2001b) has provided a particularly valuable critique of biodiversity monitoring in tropical countries. He points out that monitoring activities can actually hinder, rather than improve, conservation action, as limited resources are diverted away from practical management activities. Sheil makes the following recommendations, which should be considered whenever assessments are being planned:

- Monitoring and assessment activities must be allocated with sensitivity to local priorities and limitations, especially when local resources are involved.

- Researchers should ensure that they are familiar with local management issues before they become general advisers on local conservation needs.
- Care must be exercised whenever monitoring activities are promoted at the possible expense of important conservation actions.
- Managers should only be required to collect data that are useful to them in ways that they understand.
- High-level monitoring is vital: information is needed on threats to biodiversity, and conservation priorities should be continually refined in the light of such information. However, the costs and responsibilities for generating such information must be allocated with care.
- Interventions should bolster, and not undermine, the attainment of conservation goals; case-by-case assessment is needed.

8.6 Indicators

Indicators are surrogates for properties or responses of a system that is too difficult or costly to measure in its entirety (Hyman and Leibowitz 2001). They provide insight into the state and dynamics of the environment, and may help make detectable a trend or phenomenon of interest (Niemeijer 2002). They can help isolate key aspects from complex situations to help decision-makers see what is happening, and help them determine what action is appropriate (Niemeijer 2002). Indicators can be derived from measurements of ecological features, either by field survey or by using remote sensing data. Therefore in their broadest sense, ecological indicators include anything that can be measured by using any of the techniques described in this book. These measurements can be used directly or combined into summary values (as some form of index, for example).

Indicators are generally used to monitor trends. At the local scale they are an important tool for monitoring and adaptive management, as described in previous sections, enabling progress towards management objectives to be assessed. At national and international scales they are used to monitor progress towards achieving policy goals and as the basis of environmental reporting. The types of indicator required for these different purposes may be quite different. However, information gathered at the local scale can be aggregated for reporting at the national scale. Given that the choice of potential indicators is so broad, a key issue is how to make an appropriate choice and to make sure that the right things are measured.

The use of ecological indicators has recently been reviewed by Niemi and McDonald (2004), who point out that most attempts to use ecological indicators have rightly been heavily criticized. As noted earlier, many attempts at environmental monitoring suffer from a lack of clear objectives and a failure to consider different sources of error (Yoccoz *et al.* 2001). Other criticisms levied at indicators have focused on the lack of: (1) identification of the appropriate context (spatial and temporal) for the indicator, (2) a conceptual framework for what the indicator is supposed to indicate, and (3) validation of the indicator (Niemi and McDonald

2004). Development and testing of biodiversity indicators is area of active research, designed to address such problems.

This section first addresses the issue of conceptual frameworks, then provides some guidance on selection and implementation of appropriate indicators. Further information on the topic is provided by McKenzie *et al.* (1992) and Noss (1990, 1999), and a range of examples drawn from European forests is presented by Angelstam *et al.* (2004a).

8.6.1 Indicator frameworks

It is widely recognized that some form of framework is required in order for meaningful indicators to be developed. The most widely used framework is *pressure–state–response* (PSR), which was developed by the Organisation for Economic Co-operation and Development (OECD 1993) on the basis of the *stress–response model* developed by Friend and Rapport (1979). The PSR framework states that human activities exert pressures on the environment (such as clearance of forest for agriculture), which can induce changes in the state of the environment (for example, the extent of forest cover). Society may then respond to changes in pressures or state with policies and programmes intended to prevent, reduce or mitigate pressures and thereby reduce environmental damage. The PSR framework has been widely applied to indicator development, particularly for monitoring progress towards policy goals; for example, it is explicitly recognized by the CBD.

This approach was further expanded by the European Environment Agency to include drivers (D) and impacts (I), forming the DPSIR framework (EEA 1998). Both the PSR and the extended DPSIR frameworks are based on the fact that different societal activities (drivers) cause a pressure on the environment, causing quantitative and qualitative changes of it (changing state and impact). Society has to respond to these changes in order to achieve sustainable development. According to the DPSIR framework, different indicators of sustainability may be developed, relating to drivers, pressure, state, impact, and response.

A number of other indicator frameworks have been proposed by researchers. For example, Noss (1990) presents a hierarchical framework for development of biodiversity indicators, recognizing that three attributes of biodiversity, composition, structure, and function, can be considered at a number of different levels of organization. This framework is relevant only to biodiversity 'state' indicators. Stork *et al.* (1997) provide a framework based on a conceptual model of the relationship between anthropogenic activities affecting forests, and the processes that influence biodiversity. Indicators may therefore be developed for particular human interventions or mediators (pressure indicators), as well as processes maintaining biodiversity, and biodiversity itself (state indicators).

Frameworks can be of great value in helping to organize information. Environmental measurements can readily be categorized as representing either 'pressure', 'state' or 'response' variables, according to the PSR framework. Information on threats (see section 8.4), for example, can be considered as pressures, whereas forest structure and composition can be considered as 'state' indicators. Yet often the use of frameworks is neglected in practice, leading to confusion. For monitoring the effectiveness of management at the local scale, it is 'state' indicators that are of paramount importance, yet often the indicators used in practice are those representing management responses (mostly because these are easier to measure). The DPSIR framework is somewhat more difficult to implement in practice than PSR, mostly because of the difficulty of separating out 'state' from 'impact' variables. However, it also draws attention to the 'drivers' or underlying causes of biodiversity loss, which ultimately need to be addressed if loss of biodiversity is to be prevented.

Development of a conceptual model (see section 7.8.1) can also be of great value in identifying the relevance of different indicators, by identifying the relationships between proposed indicators the features of interest, such as biodiversity (Hyman and Leibowitz 2001). Note that there is some confusion in the scientific literature; the terms 'framework' and 'conceptual model' are often used interchangeably. In fact a conceptual model and a framework are *both* required for indicator development, the former to define the relationship between the indicator and the feature of interest (the 'endpoint'), and the latter to categorize the variables and help organize the collection and reporting of information. Of course a conceptual model can adopt an indicator framework (such as PSR) as its basic structure, and in fact this is a logical and useful way of going about developing such a model.

Could quantitative models of forest dynamics (see Chapter 5) also be of value in identifying suitable indicators? At the very least, such models could enable the sensitivity of indicators to environmental pressures to be explored, and their relevance to be tested. Remarkably little research seems to have been done in this area to date.

8.6.2 Selection and implementation of indicators

The choice of indicator will depend primarily on the objectives of the investigation. People working on forests have an enormous range of potential indicators to choose from, because of all the work undertaken on development of criteria and indicators for sustainable forest management (see section 8.2.2). A first step might therefore usefully be to examine the lists of indicators that have been developed by the various SFM C&I processes, and those developed for the purposes of forest certification (section 8.2.2). The aim should be to identify those indicators that relate most closely to the management and monitoring objectives of the area being investigated.

It may often be necessary to develop new indicators that are specific to some particular local circumstance. According to Noss (1990), indicators should be:

- sufficiently sensitive to provide an early warning of change
- widely applicable
- capable of providing a continuous assessment over a wide range of conditions
- relatively independent of sample size
- easy and cost-effective to measure, collect, assay and/or calculate
- able to differentiate between natural cycles or trends and those induced by human activities
- relevant to ecologically significant phenomena

Noss (1990) also notes that, because no single indicator possesses all of these properties, a set of complementary indicators will be required. The most important issue to remember is that the indicators must be readily measurable in a repeatable manner over time. The problem with many indicators proposed by the C&I processes is that they are simply not practicable or are stated so vaguely that it is unclear how to measure them.

Forest biodiversity indicators have also been proposed by the research community. Examples of sets of biodiversity indicators proposed for European and North American forests are presented by Angelstam and Dönz-Breuss (2004) (Table 8.6) and Keddy and Drummond (1996) (Table 8.7), respectively. There are some similarities between the two lists. The work of Keddy and Drummond (1996) is particularly notable, as they sought to identify potential targets or indicator values

Table 8.6 Basic variables collected by field survey for measuring compositional, structural and functional elements of biodiversity. From Angelstam and Dönz-Breuss (2004), based on research in northern Europe.

Elements	Description of basic variables
Composition	Basal area of all living tree species of different diameter classes
	Basal area of standing and lying dead wood of different decay stages and diameter classes
	Specialized pendant lichens (>20 cm) and conspicuous lichen species (e.g. *Lobaria* spp.).
	Insect specialist signs (exit holes) in standing/lying wood without bark
	Direct and indirect signs of specialized vertebrates (e.g. grouse, woodpeckers).
Structure	Canopy height
	Site type as determined by ground vegetation and its cover
	Stand structure
	Vertical layering
	Tree age structure
	Tree regeneration
	Shrub species
	Special trees with important microhabitats (e.g. 'veteran' or 'habitat' trees); trees with cavities.
Function	Land management
	Land abandonment (e.g. indicated by harvested stumps, archaeological features)
	Abiotic processes (fire, flooding, wind)
	Biotic processes (wood-living bracket fungi, bark beetle outbreaks)
	Damage by large mammal herbivores (browsing, bark-peeling, etc.)
	Predation (carnivore scats, corvid observations, etc.)
	Human disturbance.

Table 8.7 Ecological attributes for the evaluation, management and restoration of temperate deciduous forest ecosystems, based on forests in eastern USA (adapted from Keddy and Drummond 1996).

Property	Potential values
Tree size	Old growth forests tend to be characterized by relatively high numbers of large trees. A mean basal area of $29 \pm 4 \text{ m}^2 \text{ ha}^{-1}$ was recorded on 10 pristine sites.
Canopy composition	Mature forests tend to be dominated by only a few relatively shade-tolerant species. Successional forests tend to incorporate a larger number of tree species, including shade-intolerant species.
Coarse woody debris	Includes fallen logs, snags, and large branches. An important habitat component for many organisms including birds, mammals, invertebrates and fungi. Highest volumes tend to be recorded in old growth stands (a mean of 27 Mg ha^{-1} recorded on 10 pristine sites).
Herbaceous layer	Many temperate deciduous forests are characterized by a diverse herbaceous flora, which may be sensitive to logging and especially grazing.
Epiphytic bryophytes and lichens	Diverse communities of cryptogams (mosses and lichens) may typically be present on the trunks and branches of trees, particularly in undisturbed forests unaffected by aerial pollution, in humid environments.
Wildlife trees	Many birds, mammals, and invertebrates require trees with particular characteristics for habitat (e.g. as sites for nesting, perching, roosting or foraging). Large-diameter snags (standing dead trees) and cavity trees (live trees with central decay) are of particular importance. Old growth forests tend to be characterized by ≥ 4 wildlife trees per 10 ha.
Fungi	Temperate forests are often characterized by diverse communities of larger fungi, which play a critical role in decomposition and nutrient cycling. Many temperate trees form associations with ectomycorrhizal fungi, which assist in nutrient uptake and form an important food resource for many other organisms. The composition of fungal communities remains poorly documented, but diversity in old growth forests may exceed 100 species ha^{-1}.
Birds	The composition of bird communities appears to be particularly sensitive to the area of forest patches, some species being dependent on large areas of intact forest.

Table 8.7 (*Cont.*)

Property	Potential values
Large carnivores	As large carnivores tend to be at the top of food chains, their presence indicates an intact food web. They may play an important role in keeping herbivore numbers in check, preventing overgrazing and browsing. Large carnivores have explicitly been exterminated in many temperate forests and may therefore need to be considered as an explicit objective of restoration action.
Forest area	In many areas, once continuous tracts of forest have been highly fragmented as a result of human activity. Fragmentation reduces species diversity and changes species composition in remaining forests. Mammals and birds are most affected because of their large territorial requirements. For a forest to contain the full complement of species, it must be large enough to accommodate those species with largest area requirements (i.e. >100 000 ha).

associated with relatively undisturbed or pristine sites. This is much more difficult to achieve in heavily deforested areas such as Europe, but the approach is a valuable one and should be replicated elsewhere. These lists are presented for illustrative purposes only: appropriate indicators for forests in other regions, such as the lowland tropics and Mediterranean regions, might be quite different. It should also be remembered that these examples represent biodiversity 'state' indicators, not pressure or response indicators. Most of these indicators can be measured by using techniques described in previous chapters.

Another possibility is to monitor the populations of selected species, termed *indicator species*. Noss (1999) identifies the types of species that might be suitable for forest monitoring as follows:

- *Area-limited species*, which require large patch sizes to maintain viable populations. These species typically have large home ranges and/or low population densities, such as many mammalian carnivores.
- *Dispersal-limited species*, which are limited in their ability to move between habitat patches, or face a high mortality risk in trying to do so. Examples include flightless insects limited to forest interiors, lungless salamanders, small forest mammals, and large mammals subject to roadkill or hunting.
- *Resource-limited species*, which require specific resources that are often in critically short supply. Resources include large snags, nectar sources, fruits, etc. Examples are hummingbirds, frugivorous birds, and cavity-nesting birds and mammals.
- *Process-limited species*, which are sensitive to the level, rate, spatial characteristics or timing of some ecological process, such as flooding, fire, grazing or predation. Examples include plant species that require fire for germination.

- *Keystone species*, which are ecologically pivotal species whose impact on a community or ecosystem is disproportionately large for their abundance. Examples include dominant tree species, cavity-excavating birds, and herbivorous insects subject to outbreaks.
- *Narrow endemic species*, which are species restricted to a small geographic range (say <50 000 km^2).
- *Flagship species*, which are those that can be used to promote public support for conservation efforts, such as the giant panda.

Once target or indicator species have been chosen, then demographic or genetic attributes will need to be selected for measurement. Important issues include whether a decline in abundance is occurring, and how much habitat needed for persistence (see Chapter 7.8), but these can be challenging to determine. The populations most in need of tracking are often the most difficult to monitor (Noss 1999).

How may indicators be validated? This is an area neglected by much previous research (Noss 1990), yet it is critically important. For example, measures of forest fragmentation may be obtained fairly readily (see sections 2.7 and 7.4). But how do these measures relate to the potential impact of fragmentation on organisms? Angelstam *et al.* (2004b) provide some suggestions regarding how indicators might be validated:

- Examine how the value of a given indicator correlates with other elements of biodiversity; for example, the presence of a habitat might correlate with the presence of a particular species, and some habitat features might correlate with species richness.
- Compare their level in gradients from intensively managed forests to natural or near-natural forests. If an indicator shows a consistent trend when measured across that gradient, then it can be considered a reliable indicator of natural forest conditions.

Ultimately the validation of indicators depends on a comprehensive understanding of the links between biodiversity patterns and the processes responsible for producing them. This is a job for the research community, and there remains enormous scope for further research in this area. The job is an important one. Consider the Natura 2000 network of protected areas, one of the main coordinated actions being undertaken at the European level towards biodiversity conservation. The main aim of the Natura 2000 network is to maintain a *favourable conservation status* (FCS) to help ensure the maintenance of biodiversity of natural habitats and of wild flora and fauna in the European territory. A laudable goal, but how will anyone be sure that this target is being met? The only way of providing such information is to monitor FCS, enabling the performance of Natura 2000 network to be evaluated. This requires development of appropriate indicators. Astonishingly, few indicators have been proposed to date, and their validation is in its infancy (Cantarello 2006).

It is also important to remember that indicators are essentially tools for communicating results. They can provide a useful way of making a link between research and management practice, and between research and policy. Once data have been collected, they have to be presented in a form that can be readily interpreted by the decision-makers for whom they are intended. Newton and Kapos (2002) suggest that this can most readily be achieved by summarizing data in categorical form, and presenting them in relation to forest area. For example, species richness could be presented as the area of forest possessing more than a certain number of tree species, or forest fragmentation could be presented as forest areas with particular values of fragmentation indices. This approach enables results to be mapped, facilitating communication and integration with other data. The approach can also assist in the aggregation of data across different scales for monitoring and reporting purposes.

8.7 Scenarios

Conservation actions are generally based upon some expectation about what might happen in the future. For example, if a species is declining in abundance, then a conservation intervention might be planned based on an assumption that this decline is likely to continue unless action is taken. However, the future is highly unpredictable. Even if models of ecological dynamics are available (Chapter 5), they will be based on a range of assumptions and uncertainties. Scenario planning offers a framework for developing conservation approaches under such uncertain conditions. A scenario can be defined in this context as an account of a plausible future (Peterson *et al*. 2003). The development of scenarios is a recognized tool in business planning and economic forecasting, but has only recently been applied to conservation. Peterson *et al*. (2003) provide a valuable introduction to the use of scenarios in a conservation context. An attempt to develop global biodiversity scenarios is described by Sala *et al*. (2000), elaborated further by Chapin *et al*. (2001) (see also Carpenter *et al*. 2005).

Scenario planning provides a tool to explore the uncertainty surrounding the future consequences of a decision, by developing a small number of contrasting scenarios. Generally scenarios are developed by a diverse group of people in a systemic process of collecting, discussing, and analysing information, through a series of workshops. The group might involve research scientists, forest managers, conservation practitioners or activists, policy-makers, local community members, and other stakeholders. The scenarios may draw upon a variety of quantitative and qualitative information, such as the results of ecological surveys and outputs from modelling exercises. Peterson *et al*. (2003) suggest that the major benefits of using scenario planning for conservation are (1) increased understanding of key uncertainties, (2) incorporation of alternative perspectives into conservation planning, and (3) greater resilience of decisions to surprise.

Peterson *et al.* (2003) describe scenario planning as consisting of six interacting stages:

- *Identification of a focal issue or problem.* This may usefully be phrased as a question, such as how to maintain a functional oak forest in a protected area for the next 20 years, or how to make an existing reserve network more robust.
- *Assessment.* This should include an assessment of the people, institutions, ecosystems, and linkages among them that define a system. External changes should also be identified, including both ecological and social factors that influence system dynamics. This might usefully include an assessment of threats, pressures, or driving forces (see previous section). Uncertainties that may have a large impact on the focal issue should be identified. For example, climatic change may be a key uncertainty in the planning of a park system.
- *Identification of alternatives.* The aim of this stage is to identify alternative ways that the system could evolve. The alternatives should be plausible and relevant to the original question. A commonly used way of defining a set of alternatives is to choose two or three uncertain or uncontrollable driving forces, such as population growth and settlement pattern, which can be used to define alternatives such as increased migration to cities, increased migration to rural areas, and general population decline. The uncertainties chosen to define the alternatives should have differences that are directly related to the defining question or issue. The set of alternatives provides a framework around which scenarios can be constructed.
- *Building scenarios.* Scenarios are developed around some of the alternatives defined in the previous step, considering the key uncertainties. The scenarios should usefully expand and challenge current thinking about the system. The number of scenarios is generally three or four. The scenarios are built by converting the key alternatives into narratives by describing a credible series of external forces and actors' responses, linking historical and present events with hypothetical future events. Each scenario should be clearly anchored in the past, with the future emerging from the past and present in a seamless way, and should track key indicator variables (such as percentage of intact old-growth forest). Usually each scenario is given a name that evokes its main features (see Box 8.6)—this is often the part that is most fun! Successful scenarios are vivid and different, can be told easily, and plausibly capture possible future change.
- *Testing scenarios.* The dynamics of scenarios must be plausible. The scenarios should be tested for consistency; this can be achieved through expert opinion or by comparison with other scenarios. Simulation models can be used to test the dynamics of a scenario, but models are not the primary tool for scenario development. The key issue to check is the behaviour of actors; expected changes within each scenario should be examined from the perspective of each key actor within it.
- *Policy screening.* Once developed, scenarios can be used to test, analyse, and create policies. This is achieved by assessing how existing policies would fare under different scenarios, enabling the properties of relatively weak or strong policies

to be identified. For example, scenarios could be used to help identify land-management strategies that produce protected areas that are resilient in response to change. It is important to identify traps and opportunities and aspects of the current situation that could influence these scenario features. The process may suggest novel policies, areas for research, and issues to monitor.

Although scenarios have not been used widely in conservation planning to date, they offer a tractable and stimulating method of engaging in a debate about what might happen in the future and how to deal with it. The method therefore has great potential. Scenarios are increasingly being used in environmental assessment at global and regional scales (Box 8.6), but the approach can potentially be implemented at the local scale, such as an individual forest management unit or protected area.

> **Box 8.6** Information resources on scenarios.
>
> **Millennium Ecosystem Assessment** ⟨*/www.maweb.org/*⟩
>
> This international environmental assessment developed a set of four scenarios through a major collaborative effort. The report produced can be freely downloaded from this website. The scenarios were defined as:
>
> - *Global Orchestration*, defined as socially conscious globalization, in which equity, economic growth, and public goods are emphasized, reacting to ecosystem problems when they reach critical stages.
> - *Order from Strength*, representing a regionalized approach, in which the emphasis is on security and economic growth, again reacting to ecosystem problems only as they arise.
> - *Adapting Mosaic*, defined also as a regionalized approach, but one that emphasizes proactive management of ecosystems, local adaptation, and flexible governance.
> - *TechnoGarden*, which is a globalized approach with an emphasis on green technology and a proactive approach to managing ecosystems.
>
> **Global Environment Outlook** ⟨*www.unep.org/geo/*⟩
>
> Another international environmental assessment process, coordinated by UNEP, which has used scenarios extensively in the past. Materials freely downloadable from the website.
>
> **Shell Global Scenarios** ⟨*www.shell.com/scenarios/*⟩
>
> Shell has been involved in scenario development for some 30 years. The website offers introductory information about scenario planning as well as several downloadable scenarios.
>
> **Scenarios for Sustainability** ⟨*www.scenariosforsustainability.org*⟩
>
> A very useful online resource offering practical tools for scenario development.

8.8 Evidence-based conservation

Concern has been growing in recent years that much conservation practice is based on tradition or the experience of practitioners, rather than on the results of scientific research. To increase the effectiveness of conservation, some researchers have recently proposed that conservation action should become more 'evidence-based', by drawing more heavily on the results of scientific research (Pullin and Knight 2001, 2003, Pullin *et al.* 2004, Sutherland *et al.* 2004b). This has been inspired by the 'effectiveness revolution' that has occurred in medicine over the past two decades, aimed at linking medical research with medical practice (Sutherland 2000).

A key element of the evidence-based approach is *systematic review*, which involves reviewing the scientific literature and unpublished sources, critically assessing the methods used and synthesizing the evidence in relation to a research question or management issue. The approach differs from conventional literature reviews in following a strict methodological protocol, often incorporating some form of meta-analysis. The aim is to provide a comprehensive, unbiased and objective assessment of available evidence.

Results of the first systematic reviews of conservation evidence are now becoming available (Stewart *et al.* 2005), although to date, none of the reviews has focused on forest ecosystems. Websites have been created aiming to support the dissemination of research results to conservation practitioners (examples are ⟨www.cebc.bham.ac.uk⟩ and ⟨www.conservationevidence.com⟩). Although such initiatives are clearly to be welcomed, one of the problems facing the conservation community is just how little well-designed research has been undertaken on many conservation problems. The most reliable form of evidence is a fully replicated experiment, but such experiments may be difficult or even impossible under field conditions (Diamond 1986), and for almost any conservation issue such experiments are few in number. Conservation management is likely to continue to depend on a blend of experience, tradition and scientific evidence, but the challenge for researchers is to help ensure that their research results are communicated effectively to practitioners in a form that can directly support decision-making. Indicators, models, GIS, and other decision-support tools (see ⟨http://ncseonline.org/NCSSF/DSS⟩ for a useful compendium) can all help in this process, but there is ultimately no substitute for researchers and practitioners collaborating more closely, perhaps implementing an adaptive management approach together.

8.9 Postscript: making a difference

So, at the end of the day, how can we make a difference to conservation? In other words, how can we ensure that the work we do really has an impact on the ground, and helps save the forests that we care about? Make no mistake, working on forest conservation is often a depressing business. Devastated forests are all too easy to find. The problems can often seem so overwhelming that the contribution we can make as individuals may seem trivial or insignificant. It can seem as if our work is just a drop in the ocean. 'But what is an ocean but a multitude of drops?' (Mitchell 2004).

There is no doubt that individuals can make a difference. Think of Wangari Maathai, recently awarded the Nobel Peace prize for having planted 30 million trees. In my own experience, I have been inspired by individuals such as Alan Watson of Trees for Life ⟨www.treesforlife.org.uk⟩ and Philip Ashmole of the Carrifran Wildwood Project ⟨www.carrifran.org⟩, who through their own personal vision and dedication have successfully developed and implemented large-scale forest restoration projects in the UK, despite having to overcome many obstacles in the process. It is possible to make a difference through environmental activism, campaigning, or advocacy; by being a politician or government official; by running a large business; or simply by being rich and spending your money on creating a protected area.

But what can we achieve as researchers? Can science help conservation? Or, as some have suggested (Whitten *et al.* 2001), is conservation biology merely a 'displacement activity for academics'? There are many dedicated conservation practitioners who make a real difference on a daily basis, with no scientific training at all. On the other hand, some of them make mistakes. If asked to provide evidence of the impacts of management interventions they have undertaken, most practitioners are unable to provide anything beyond anecdotal observations. Scientifically robust forest monitoring is actually very rare. Consequently, adaptive management is very rare. Truly sustainable forest management is also very rare (perhaps it doesn't exist at all). When we start to consider all the organisms other than trees—the insects, fungi, lichens, mammals, reptiles, amphibians, birds, and many other creatures that inhabit forests—we start to realize how very little we know about most forest communities. In fact, we cannot even estimate with reasonable accuracy how many species are being lost as a result of current deforestation. This is something that should shame us all.

The need for scientific research into forest ecology and conservation is immense. The challenge is often to convince non-scientists—whether they be local communities, politicians, business leaders, funding agencies, or the public at large—that this is the case. Furthermore, it is essential to ensure that the scientific work that is being carried out is relevant to the problems that need addressing. The responsibility for ensuring this lies with just one person. The person holding this book.

References

Abbitt, R. J. F., Scott, J. M., and Wilcove, D. S. (2000). The geography of vulnerability: incorporating species geography and human development patterns into conservation planning. *Biological Conservation*, 96, 169–175.

Acevedo, M. F., Urban, D. L., and Ablan, M. (1995). Transition and gap models of forest dynamics. *Ecological Applications*, 5(4), 1040–1055.

Acevedo, M. F., Urban, D. L., and Ablan, M. (1996a). Landscape scale forest dynamics: GIS, gap and transition models. In *GIS and environmental modeling: Progress and research issues*, eds. M. F. Goodchild, L. T. Steyaert, and B. O. Parks, pp. 181–185. GIS World Books, Fort Collins, CO.

Acevedo, M. F., Urban, D. L., and Shugart, H. H. (1996b). Models of forest dynamics based on roles of tree species. *Ecological Modelling*, 87, 267–284.

Adams, J. B., Sabol, D. E., Kapos, V., Filho, R. A., Roberts, D., Smith, M. O., and Gillespie, A. R. (1995). Classification of multispectral images based on fractions of endmembers: applications to land-cover change in the Brazilian Amazon. *Remote Sensing of Environment*, 52, 137–152.

Adams, W. T., Strauss, S. H., Copes, D. L., and Griffin, A. R. (eds.) (1992). *Population genetics of forest trees*. Springer-Verlag, Berlin.

Agee, J. K. (1993). *Fire ecology in Pacific Northwest forests*. Island Press, Washington DC.

Agyeman, V. K., Swaine, M. D., and Thompson, J. (1999). Responses of tropical forest tree seedlings to irradiance and the derivation of a light response index. *Journal of Ecology*, 87, 815–827.

Akçakaya, H. R. and Root, W. (2002). *RAMAS Metapop: Viability analysis for stage-structured metapopulations* (ver. 4.0). Applied Biomathematics, Setauket, NY.

Akçakaya, H. R. and Sjögren-Gulve, P. (2000). Population viability analysis in conservation planning: an overview. *Ecological Bulletins*, 48, 9–21.

Akçakaya, H. R., Radeloff, V. C., Mladenoff, D. J., and He, H. S. (2004). Integrating landscape and metapopulation modelling approaches: viability of the sharp-tailed grouse in a dynamic landscape. *Conservation Biology*, 18, 526–537.

Alder, D. and Synnott, T. J. (1992). *Permanent sample plot techniques for mixed tropical forest*. Tropical Forestry Papers 25, Oxford Forestry Institute, Oxford.

Aldred, A. H. and Alemdag, I. S. (1988). *Guidelines for forest biomass inventory*. Petawawa National Forestry Institute Report PI-X-77. Canadian Forest Service, Natural Resources Canada, Ottawa.

Aldrich, P. R., Parker, G. R., Ward, J. S., and Michler, C. H. (2003). Spatial dispersion of trees in an old-growth temperate hardwood forest over 60 years of succession. *Forest Ecology and Management*, 180(1–3), 475–491.

Allen, J. and Barnes, D. (1985). The causes of deforestation in developing countries. *Annals of the Association of American Geographers*, 75, 163–184.

Allnutt, T. R., Thomas, P., Newton, A. C., and Gardner, M. F. (1998). Genetic variation in *Fitzroya cupressoides* cultivated in the British Isles, assessed using RAPDs. *Botanical Journal of Edinburgh*, 55(3), 329–341.

Allnutt, T. R., Newton, A. C., Lara, A., Premoli, A., Armesto, J. J., Vergara, R., and Gardner, M. (1999). Genetic variation in *Fitzroya cupressoides* (alerce), a threatened South American conifer. *Molecular Ecology*, 8, 975–987.

Alonso, D. and Sole, R. V. (2000). The DivGame Simulator: a stochastic cellular automata model of rainforest dynamics. *Ecological Modelling*, 133(1–2), 131–141.

Alvarez-Aquino, C., Williams-Linera, G., and Newton, A. C. (2005). Composition of the seed bank in disturbed fragments of Mexican cloud forest. *Biotropica*, 37(3), 337–342.

Alvarez-Buylla, E. R. and Slatkin, M. (1991). Finding confidence limits on population growth rates. *Trends in Ecology and Evolution*, 6, 221–224.

Alves, D. S., Pereira, J. L. G., De Sousa, C. L., Soares, J. V., and Yamaguchi, F. (1999). Characterising landscape changes in central Rondonia using Landsat TM imagery. *International Journal of Remote Sensing*, 20, 2877–2882.

Amarnath, G., Murthy, M. S. R., Britto, S. J., Rajashekar, G., and Dutt, C. B. S. (2003). Diagnostic analysis of conservation zones using remote sensing and GIS techniques in wet evergreen forests of the Western Ghats—an ecological hotspot, Tamil Nadu, India. *Biodiversity and Conservation*, 12(12), 2331–2359.

Amaro, A., Reed, D., and Saores, P. (eds.) (2003). *Modelling forest systems*. CABI Publishing, Wallingford.

Andelman, S., Ball, I., Davis, F., and Stoms, D. (1999). *Sites 1.0: An analytical toolbox for designing ecoregional conservation portfolios*. Manual prepared for The Nature Conservancy, Arlington, VA.

Andersen, M. (1991). Mechanistic models for the seed shadows of wind-dispersed plants. *American Naturalist*, 137, 476–497.

Anderson, M. C. (1964). Studies of the woodland light climate. I. The photographic computation of light conditions. *Journal of Ecology*, 52, 27–41.

Anderson, R. P., Lew, D., and Peterson, A. T. (2003). Evaluating predictive models of species' distributions: criteria for selecting optimal models. *Ecological Modelling*, 162, 211–232.

Angelstam, P. and Dönz-Breuss, M. (2004). Measuring forest biodiversity at the stand scale—an evaluation of indicators in European forest history gradients. *Ecological Bulletins*, 51, 305–332.

Angelstam, P., Dönz-Breuss, M., and Roberge, J.-M. (2004a). *Targets and tools for the maintenance of forest biodiversity*. (Ecological Bulletins No. 51) Blackwell Science, Oxford.

Angelstam, P., Roberge, J.-M., Dönz-Breuss, M., Burfield, I. J., and Ståhl, G. (2004b). Monitoring forest biodiversity—from the policy level to the management unit. *Ecological Bulletins*, 51, 295–304.

Armenteras, D., Gast, F., and Villareal, H. (2003). Andean forest fragmentation and the representativeness of protected natural areas in the eastern Andes, Colombia. *Biological Conservation*, 113, 245–256.

Arno, S. F. and Sneck, K. M. (1977). *A method for determining fire history in coniferous forests of the Mountain West*. US Department of Agriculture, Forest Service, General Technical Report INT-42.

Ashley, C. and Carney, D. (1999). *Sustainable livelihoods: Lessons from early experience*. DFID, London.

Ashton, P. M. S. (1995). Seedling growth of co-occurring *Shorea* species in the simulated light environments of a rain forest. *Forest Ecology and Management*, 72, 1–12.

Ashton, P. M. S., Gunatilleke, C. V. S., and Gunatilleke, I. A. U. N. (1995). Seedling survival and growth of four Shorea species in Sri Lankan rainforest. *Journal of Tropical Ecology*, 11, 263–279.

Ashton, P. S. and Hall, P. (1992). Comparisons of structure among mixed dipterocarp forests of north-western Borneo. *Journal of Ecology*, 80, 459–481.

Asner, G. P., Hicke, J. A. and Lobell, D. B. (2003). Per-pixel analysis of forest structure. In *Remote sensing of forest environments*, eds. M. A. Wulder and S. E. Franklin, pp. 209–254. Kluwer, Dordrecht.

Attiwill, P. M. (1994). The disturbance of forest ecosystems: the ecological basis for conservative management. *Forest Ecology and Management*, 63, 247–300.

Augspurger, C. K. (1983). Seed dispersal of the tropical tree, *Platypodium elegans*, and the escape of its seedlings from fungal pathogens. *Journal of Ecology*, 71, 759–771.

Augspurger, C. K. (1984). Light requirements of neotropical tree seedlings: a comparative study of growth and survival. *Journal of Ecology*, 72, 777–795.

Augspurger, C. K. and Hogan, K. P. (1983). Wind dispersal of fruits with variable seed number in a tropical tree (*Lonchocarpus pentaphyllus*, Leguminosae). *American Journal of Botany*, 70, 1031–1037.

Augspurger, C. K. and Kelly, C. K. (1984). Pathogen mortality of tropical tree seedlings: experimental studies of the effects of dispersal distance, seedling density, and light conditions. *Oecologia*, 61, 211–217.

Austin, M. P. and Heyligers, P. C. (1989). Vegetation survey design for conservation: gradsect sampling of forests in Northeast New South Wales. *Biological Conservation*, 50, 13–32.

Austin, M. P. and Heyligers, P. C. (1991). New approach to vegetation survey design: gradsect sampling. In *Nature conservation: Cost effective biological surveys and data analysis*, eds. C. R. Margules and M. P. Austin, pp. 31–36. CSIRO, Canberra, Australia.

Avery, T. E. (1968). *Interpretation of aerial photographs*, 2nd edition. Burgess, Minneapolis, MN.

Avery, T. E. (1978). *Forester's guide to aerial photo interpretation*. USDA Forest Service, Agriculture Handbook no. 308. US Department of Agriculture, Washington, DC.

Avery, T. E. and Berlin, G. L. (1992). *Fundamentals of remote sensing and airphoto interpretation*, 5th edition. Macmillan, NY.

Avery, T. E. and Burkhart, H. E. (2002). *Forest measurements*, 5th edition. McGraw-Hill, NY.

Bachmann, P., Köhl, M., and Päivinen, R. (eds.). (1998). *Assessment of biodiversity for improved forest planning*. Kluwer, Dordrecht.

Bacles, C. F. E., Lowe, A. J., and Ennos, R. A. (2004). Genetic effects of chronic habitat fragmentation on tree species: the case of *Sorbus aucuparia* in a deforested Scottish landscape. *Molecular Ecology*, 13, 573–584.

Bacles, C. F. E., Lowe, A. J., and Ennos, R. A. (2006). Effective seed dispersal across a fragmented landscape. *Science*, 311, 628.

Baguette, M. (2004). The classical metapopulation theory and the real, natural world: a critical appraisal. *Basic and Applied Ecology*, 5, 213–224.

Bajracharya, S. B., Furley, P. A., and Newton, A. C. (2005). Effectiveness of community involvement in delivering conservation benefits to the Annapurna Conservation Area, Nepal. *Environmental Conservation*, 32(3), 239–247.

Baker, A. (ed.) (2000). *Molecular methods in ecology*. Blackwell Science, Oxford.

Baker, P. J. and Wilson, J. S. (2000). A quantitative technique for the identification of canopy stratification in tropical and temperate forests. *Forest Ecology and Management*, 127, 77–86.

Baker, T. R., Swaine, M. D., and Burslem, D. F. R. P. (2003). Variation in tropical forest growth rates: combined effects of functional group composition and resource availability. *Perspectives in Plant Ecology, Evolution and Systematics*, 6/1(2), 21–36.

Baldocchi, D. and Collineau, S. (1994). The physical nature of solar radiation in heterogeneous canopies: spatial and temporal attributes. In *Exploitation of environmental heterogeneity by plants*, eds. M. M. Caldwell and R. W. Pearcy, pp. 21–71. Academic Press, San Diego, CA.

Ball, I. R. and Possingham, H. P. (2000). *MARXAN (V1.8.2): Marine reserve design using spatially explicit annealing, a manual*. University of Queensland, Australia.

Ball, I. R., Lindenmayer, D. B., and Possingham, H. P. (1999). A tree hollow dynamics simulation model. *Forest Ecology and Management*, 123(2–3), 179–194.

Balmford, A., Bennun, L., ten Brink, B. *et al*. (2005). The Convention on Biological Diversity's 2010 target. *Science*, 307(5707), 212–213.

Balzter, H., Braun, P. W., and Kohler, W. (1998). Cellular automata models for vegetation dynamics. *Ecological Modelling*, 107(2–3), 113–125.

Barabesi, L. and Fattorini, L. (1995). Kernel plant density estimation by a ranked set sampling of point-to-plant distances. In *The Monte Verita conference on forest survey designs*, eds.

M. Köhl, P. Bachmann, P. Brassel, and G. Preto, pp. 71–80. WSL/ETHZ, Birmensdorf, Switzerland

Bardon, R. E., Countryman, D. W., and Hall, R. B. (1995). A reassessment of using light-sensitive diazo paper for measuring integrated light in the field. *Ecology*, 76, 1013–1016.

Barnes, B. V., Zak, D. R., Denton, S. R., and Spurr, S. H. (1998). *Forest ecology*, 4th edition. Wiley, New York.

Barot, S., Gignoux, J., and Menaut, J.-C. (1999). Demography of a savanna palm tree: predictions from comprehensive spatial pattern analyses. *Ecology*, 80, 1987–2005.

Baskent, E. (1999). Controlling spatial structure of forested landscapes: a case study towards landscape management. *Landscape Ecology*, 14, 83–97.

Baskent, E. Z. and Jordan, G. A. (1995). Characterizing spatial structure of forest landscapes. *Canadian Journal of Forest Research*, 25, 1830–1849.

Baskin, C. C. and Baskin, J. M. (1998). *Seeds: Ecology, biogeography and evolution of dormancy and germination*. Academic Press, San Diego, CA.

Bass, B., Hansell, R., and Choi, J. (1998). Towards a simple indicator of biodiversity. *Environmental Monitoring and Assessment*, 49, 337–347.

Batista, W. B., Platt, W. J., and Macchiavelli, R. E. (1998). Demography of a shade-tolerant tree (*Fagus grandifolia*) in a hurricane-disturbed forest. *Ecology*, 79, 38–53.

Bawa, K. S. and Hadley, M. (eds.) (1990). *Reproductive ecology of tropical forest plants*. Man and the Biosphere Series vol. 7. UNESCO, Paris.

Bebber, D., Brown, N., Speight, M., Moura-Costa, P., and Sau Wai, Y. (2002). Spatial structure of light and dipterocarp seedling growth in a tropical secondary forest, *Forest Ecology and Management*, 157(1–3), 65–75.

Bebi, P., Kienast, F., and Schonenberger, W. (2001). Assessing structures in mountain forests as a basis for investigating the forest's dynamics and protective function. *Forest Ecology and Management*, 145, 3–14.

Becker, M. (1971). Une technique nouvelle d'utilization des photographies hémisphériques pour la mesure du climat lumineux en forêt. *Annales des Sciences Forestières*, 28, 425–442.

Beers, T. W. (1969). Slope correction in horizontal point sampling. *Journal of Forestry*, 67, 188–192.

Begon, M. and Mortimer, M. (1981). *Population ecology: a unified study of animals and plants*. Blackwell Scientific, Oxford.

Begon, M., Harper, J. L., and Rownsend, C. R. (1996). *Ecology: Individuals, populations and communities*. Blackwell Science, Oxford.

Behera, M. D., Kushwaha, S. P. S., and Roy, P. S. (2005). Rapid assessment of biological richness in a part of Eastern Himalaya: an integrated three-tier approach. *Forest Ecology and Management*, 207(3), 363–384.

Bekessy, S. A., Allnutt, T. R., Premoli, A. C. *et al.* (2002a). Genetic variation in the vulnerable and endemic Monkey Puzzle tree, detected using RAPDs. *Heredity*, 88, 243–249.

Bekessy, S. A., Sleep, D., Stott, A. *et al.* (2002b). Adaptation of Monkey Puzzle to arid environments reflected by regional differences in stable carbon isotope ratio and allocation to root biomass. *Forest Genetics*, 9(1), 63–70.

Bekessy, S. A., Ennos, R. A., Ades, P. K., Burgman, M. A., and Newton, A. C. (2003). Neutral DNA markers fail to detect divergence in an ecologically important trait. *Biological Conservation*, 110, 267–275.

Bekessy, S. A., Newton, A. C., Fox, J. C. *et al.* (2004). The monkey puzzle tree in southern Chile. In *Species conservation and management: Case studies using RAMAS GIS*, eds. H. R. Açkakaya, M. A. Burgman, O. Kindval, P. Sjögren-Gulve, J. Hatfield and M. McCarthy, pp. 48–63. Oxford University Press, Oxford.

Bellingham, P. J., Tanner, E. V. J., and Healey, J. R. (1994). Sprouting of trees in Jamaican montane forests after a hurricane. *Journal of Ecology*, 82, 747–758.

Bellingham, P. J., Tanner, E. V. J., and Healey, J. R. (1995). Damage and responsiveness of Jamaican montane tree species after disturbance by a hurricane. *Ecology*, 76, 2562–2580.

Bertault, J. G. and Sist, P. (1997). An experimental comparison of different harvesting intensities with reduced-impact and conventional logging in East Kalimantan, Indonesia. *Forest Ecology and Management*, 94(1–3), 209–218.

Bhadresa, R. (1986). Faecal analysis and exclosure studies. In *Methods in plant ecology*, 2nd edition, eds. P. D. Moore and S. B. Chapman, pp. 61–71. Blackwell Scientific, Oxford.

Bierzychudek, P. (1999). Looking backwards: assessing the projections of a transition matrix model. *Ecological Applications*, 9, 1278–1287.

Blasco, F., Whitmore, T. C., and Gers, C. (2000). A framework for the worldwide comparison of tropical woody vegetation types. *Biological Conservation*, 95, 175–189.

Bond, W. J. (1997). Functional types for predicting changes in biodiversity: a case study in Cape fynbos. In *Plant functional types*, eds. T. M. Smith, H. H. Shugart, and F. I. Woodward, pp. 174–194. Cambridge University Press, Cambridge.

Bond, W. J. (1998). Ecological and evolutionary importance of disturbance and catastrophes in plant conservation. In *Conservation in a changing world*, eds. G. M. Mace, A. Balmford, and J. R. Ginsberg, pp. 87–106. Cambridge University Press, Cambridge.

Boose, E. R., Chamberlin, K. E., and Foster, D. R. (2001). Landscape and regional impacts of hurricanes in New England. *Ecological Monographs*, 71, 27–48.

Boot, R. G. A. and Gullison, R. E. (1995). Approaches to developing sustainable extraction systems for tropical forest products. *Ecological Applications*, 5(4), 896–903.

Borrough, P. A. and McDonnell, R. (1998). *Principles of geographical information systems (spatial information systems)*. Clarendon Press, Oxford.

Boshier, D. H., Chase, M. R., and Bawa, K. S. (1995). Population genetics of *Cordia alliodora* (Boraginaceae), a neotropical tree. 2. Mating system. *American Journal of Botany*, 82, 476–483.

Bossard, C. C. (1990). Tracing of ant-dispersed seeds: a new technique. *Ecology*, 71, 2370–2371.

Botkin, D. B. (1993). *Forest dynamics: An ecological model*. Oxford University Press. Oxford.

Botkin, D. B., Janak, J. R., and Wallis, J. R. (1972a). Rationale, limitations and assumptions of a northeast forest growth simulator. *IBM Journal of Research and Development*, 16, 101–116.

Botkin, D. B., Janak J. F., and Wallis, J. R. (1972b). Some ecological consequences of a computer model of forest growth. *Journal of Ecology*, 60, 849–872.

Boucher, D. H. and Mallona, M. A. (1997). Recovery of the rain forest tree *Vochysia ferruginea* over 5 years following Hurricane Joan in Nicaragua: a preliminary population projection matrix. *Forest Ecology and Management*, 91, 195–204.

Bower, D. R. and Blocker, W. W. (1966). Accuracy of bands and tape for measuring diameter increments. *Journal of Forestry*, 64, 21–22.

Bradbeer, J. W. (1988). *Seed dormancy and germination*. Blackie, Glasgow.

Bradstock, R. A. and Kenny, B. J. (2003). Application of plant functional traits to fire management in a conservation reserve in southeastern Australia. *Journal of Vegetation Science*, 14, 345–354.

Brändli, U. B., Kaufmann, E., and Stierlin, H. R. (1995). Survey of biodiversity at the forest margin in the second Swiss NFI. In *The Monte Verita conference on forest survey designs*, eds. M. Köhl, P. Bachmann, P. Brassel, and G. Preto, pp. 141–150. WSL/ETHZ, Birmensdorf, Switzerland.

Brandon, K., Redford, K. H., and Sanderson, S. E. (eds.) (1998). *Parks in peril. People, politics and protected areas*. Island Press, Washington, DC.

Bréda, N. J. J. (2003). Ground-based measurements of leaf area index: a review of methods, instruments and current controversies. *Journal of Experimental Botany*, 54(392), 2403–2417.

Briers, R. A. (2002). Incorporating connectivity in reserve selection procedures. *Biological Conservation*, 103, 77–83.

Briggs, D. and Walters, S. M. (1997). *Plant variation and evolution*, 3rd edition. Cambridge University Press, Cambridge.

Brigham, C. A. and Schwartz, M. A. (eds.) (2003). *Population viability in plants. Conservation, management and modeling of rare plants*. Ecological Studies 165, Springer-Verlag, Berlin.

Brigham, C. A. and Thomson, D. M. (2003). Approaches to modeling population viability in plants: an overview. In *Population viability in plants. Conservation, management and modeling of rare plants*, eds. C. A. Brigham and M. A. Schwartz, pp. 145–171. Ecological Studies 165, Springer-Verlag, Berlin.

Brockhaus, J. A., Khorram, S., Bruck, R., Campbell, M. V., and Stallings, C. (1992). A comparison of Landsat TM and SPOT HRV data for use in the development of forest defoliation models. *International Journal of Remote Sensing*, 13, 3235–3240.

Brokaw, N. V. (1985). Treefalls, regrowth, and community structure in tropical forests. In *The ecology of natural disturbance and patch dynamics*, eds. S. T. A. Pickett and P. S. White, pp. 53–69. Academic Press, Orlando, FL.

Brokaw, N. V. L. and Lent, R. A. (1999). Vertical structure. In *Maintaining biodiversity in forest ecosystems*, ed. M. L. Hunter, pp. 373–399. Cambridge University Press, Cambridge.

Brommit, A., Charbonneau, N., Contreras, T. A., and Fahrig, L. (2004). Crown loss and subsequent branch sprouting of forest trees in response to a major ice storm. *Journal of the Torrey Botanical Society*, 131, 169–176.

Brook, B. W., O'Grady, J. J., Chapman, A. P., Burgman, M. A., Akçakaya, H. R., and Frankham, R. (2000). Predictive accuracy of population viability analysis in conservation biology. *Nature*, 404, 385–387.

Brook, B. W., Burgman, M. A., Akçakaya, H. R., O'Grady, J. J., and Frankham, R. (2002). Critiques of PVA as the wrong questions: throwing the heuristic baby out with the numerical bath water. *Conservation Biology*, 16(1), 262–263.

Brooks, R. P. (1997). Improving habitat suitability index models. *Wildlife Society Bulletin*, 25, 163–167.

Brooks, T., Balmford, A., Burgess, N., Fjeldså, J., Hansen, L. A., Moore, J., Rahbek, C., and Williams, P. (2001). Toward a blueprint for conservation in Africa. *Bioscience*, 51, 613–624.

Brown, A. H. D. (1990). Genetic characterization of plant mating systems. In *Plant Population genetics, breeding and genetic resources*, eds. A. H. D. Brown, M. T. Clegg, A. L. Kahler, and B. S. Weir, pp. 145–162. Sinauer Associates, Sunderland, MA.

Brown, N. D. and Jennings, S. (1998). Gap-size niche differentiation by tropical rainforest trees: a testable hypothesis or a broken-down bandwagon? In *Dynamics of tropical communities*, eds. D. M. Newbery, H. H. T. Prins, and N. Brown, pp. 79–94. Blackwell Science, Oxford.

Brown, N., Jennings, S., Wheeler, P., and Nabe-Nielson, J. (2000). An improved method for the rapid assessment of forest understorey light environments. *Journal of Applied Ecology*, 37, 1044–1053.

Brown, S. L., Schroeder, P., and Kern, J. S. (1999). Spatial distribution of biomass in forests of the eastern USA. *Forest Ecology and Management*, 123, 81–90.

Brown, V. K. and Gange, A. C. (1989). Differential effects of above- and below-ground insect herbivory during early plant succession. *Oikos*, 54, 67–76.

Brown, V. K., Gange, A. C., Evans, I. M., and Storr, A. L. (1987). The effect of insect herbivory on the growth and reproduction of two annual *Vicia* species at different stages in plant succession. *Journal of Ecology*, 75, 1173–1189.

Bruner, A. G., Gullison, R. E., Rice, R. E., and da Fonseca, G. A. B. (2001). Effectiveness of parks in protecting tropical biodiversity. *Science*, 291(5501), 125–128.

Bryant, D., Nielsen, D., and Tangley, L. (1997). *The last frontier forests. Ecosystems and economies on the edge*. World Resources Institute, Washington, DC.

Buckland, S. T., Anderson, D. R., Burnham, K. P., Laake, J. L., Borchers, D. L., and Thomas, L. (2001). *Introduction to distance sampling: Estimating abundance of biological populations*. Oxford University Press, Oxford.

Bucur, V. (2003). *Nondestructive characterization and imaging of wood*. Springer-Verlag, New York.
Bugmann, H. (2001). A review of forest gap models. *Climatic Change*, 51(3–4), 259–305.
Bugmann, H. (2003). Predicting the ecosystem effects of climate change. In *Models in ecosystem science*, eds. C. D. Canham, J. J. Cole, and W. K. Lauenroth, pp. 385–409. Princeton University Press, Princeton, NJ.
Bullock, J. (1996). Plants. In *Ecological census techniques. A handbook*, ed. W. J. Sutherland, pp. 111–138. Cambridge University Press, Cambridge.
Bullock, J. M. and Clarke, R. T. (2000). Long distance seed dispersal by wind: measuring and modelling the tail of the curve. *Oecologia*, 124, 506–521.
Bunnell, F. L. and Vales, D. J. (1990). Comparison of methods for estimating forest overstory cover: differences among techniques. *Canadian Journal of Forest Research*, 20, 101–107.
Bunyavejchewin, S., LaFrankie, J. V., Baker, P. J., Kanzaki, M., Ashton, P. S., and Yamakura, T. (2003). Spatial distribution patterns of the dominant canopy dipterocarp species in a seasonal dry evergreen forest in western Thailand. *Forest Ecology and Management*, 175(1–3), 87–101.
Burgess, N., D'Amico Hales, J., Underwood, E. *et al.* (2005). *The terrestrial ecoregions of Africa and Madagascar*. Island Press, Washington, DC.
Burgman, M. (2005). *Risks and decisions for conservation and environmental management*. Cambridge University Press, Cambridge.
Burgman, M. A., Breininger, D. R., Duncan, B. W., and Ferson, S. (2001). Setting reliability bounds on habitat suitability indices. *Ecological Applications*, 11, 70–78.
Burk, T. E. and Newberry, J. D. (1984). A simple algorithm for moment-based recovery of Weibull distribution parameters. *Forest Science*, 30, 329–332.
Burkhart, H. E. (2003). Suggestions for choosing an appropriatge level for modelling forest stands. In *Modelling forest systems*, eds. A. Amaro, D. Reed, and P. Saores, pp. 3–10. CABI Publishing, Wallingford.
Burslem, D. F. R. P., Turner, I. M., and Grubb, P. J. (1994). Mineral nutrient status of coastal hill dipterocarp forest and adinandra belukar in Singapore: bioassays of nutrient limitation. *Journal of Tropical Ecology*, 10, 579–599.
Burslem, D. F. R. P., Whitmore, T. C., and Brown, G. C. (2000). Short-term effects of cyclone impact and long-term recovery of tropical rain forest on Kolombangara, Solomon Islands. *Journal of Ecology*, 88(6), 1063–1078.
Busby, J. R. (1991). BIOCLIM—a bioclimate analysis and prediction system. In *Nature conservation: Cost-effective biological surveys and data analysis*, eds. C. R. Margules, and M. P. Austin, pp. 64–68. CSIRO, Canberra, Australia.
Busing, R. T. and Mailly, D. (2004). Advances in spatial, individual-based modelling of forest dynamics. *Journal of Vegetation Science*, 15(6), 831–842.
Butin, H. (1995). *Tree diseases and disorders: Causes, biology, and control in forest and amenity trees*. Oxford University Press, Oxford.
Byram, G. M. (1959). Combustion of forest fuels. In *Forest fire: Control and use*, ed. K. P. Davis, pp. 61–89. McGraw-Hill, New York.
Cadenasso, M. L., Traynor, M. M., and Pickett, S. T. A. (1997). Functional location of forest edges: gradients of multiple physical factors. *Canadian Journal of Forest Research*, 27, 774–782.
Cahill, J. F., Castelli, J. P., and Casper, B. B. (2001). The herbivore uncertainty principle: visiting plants can alter herbivory. *Ecology*, 82, 307–312.
Cailliez, F. (1980). *Forest volume estimation and yield prediction*. Vol. 1. *Volume estimation*. FAO Forestry Paper 22/1. FAO, Rome.
Cain, M. L., Milligan, B. G., and Strand, A. E. (2000). Long-distance seed dispersal in plant populations. *American Journal of Botany*, 87, 1217–1227.
Cain, S. A. and de Oliveira Castro, G. M. (1959). *Manual of vegetation analysis*. Harper & Row, New York.

Calcote, R. R. (1995). Pollen source area and pollen productivity: evidence from forest hollows. *Journal of Ecology*, 83, 591–602.

Caldecott, J. and Miles, L. (eds.) (2005). *World atlas of great apes and their conservation*. University of California Press, Berkeley, CA.

Campbell, D. R. and Waser, N. M. (1989). Variation in pollen flow within and among populations of *Ipomopsis aggregata*. *Evolution*, 43, 1444–1455.

Campbell, J. E. and Gibson, D. J. (2001). The effect of seeds of exotic species transported via horse dung on vegetation along trail corridors. *Vegetatio*, 157(1), 23–35.

Campbell, J. B. (1996). *Introduction to remote sensing*, 2nd edition. Taylor and Francis, London.

Campbell, P., Comiskey, J., Alonso, A. et al. (2002). Modified Whittaker plots as an assessment and monitoring tool for vegetation in a lowland tropical rainforest. *Environmental Monitoring and Assessment*, 76(1), 19–41.

Canham, C. D., Finzi, A. C., Pacala, S. W., and Burbank, D. H. (1994a). Causes and consequences of resource heterogeneity in forests: Interspecific variation in light transmission by canopy trees. *Canadian Journal of Forest Research*, 24, 337–349.

Canham, C. D., McAninch, J. B., and Wood, D. W. (1994b). Effects of the frequency, timing, and intensity of simulated browsing on growth and mortality of tree seedlings. *Canadian Journal of Forest Research*, 24(4), 817–825.

Canham, C. D., Cole, J. J., and Lauenroth, W. K. (eds.) (2003). *Models in ecosystem science*. Princeton University Press, Princeton, NJ.

Cannas, S. A., Paes, S. A., and Marco, D. E. (1999). Modeling plant spread in forest ecology using cellular automata. *Computer Physics Communications*, 122, 131–135.

Cannell, M. G. R. and Grace, J. (1993). Competition for light: detection, measurement, and quantification. *Canadian Journal of Forest Research*, 23, 1969–1979.

Cant, E. T., Smith, A. D., Reynolds, D. R., and Osborne, J. L. (2005). Tracking butterfly flight paths across the landscape with harmonic radar. *Proceedings of the Royal Society of London*, B272, 785–790.

Cantarello, E. (2006). *Towards cost-effective indicators to maintain Natura 2000 sites in a favourable conservation status*. PhD thesis, University of Parma, Italy.

Carey, P. D. (1996). DISPERSE: a cellular automaton for predicting the distribution of species in a changed climate. *Global Ecology and Biogeography Letters*, 5, 217–226.

Carle, J. and Holmgren, P. (2003). *Definitions related to planted forests*. Forest Resources Assessment Programme Working Paper 79. FAO, Rome.

Carney, D. (2002). *Sustainable livelihoods approaches: Progress and possibilities for change*. DFID, London.

Carpenter, G., Gillison, A. N., and Winter, J. (1993). DOMAIN: a flexible modelling procedure for mapping potential distributions of plants and animals. *Biodiversity and Conservation*, 2, 667–680.

Carpenter, G. A., Gopal, S., Macomber, S., Martens, S., Woodcock, C. E., and Franklin, J. (1999). A neural network method for efficient vegetation mapping. *Remote Sensing of Environment*, 70, 326–338.

Carpenter, S. R., Pingali, P. L., Bennett, E. M., and Zurek, M. B. (eds.) (2006). *Millennium ecosystem assessment. Ecosystems and human well-being: scenarios. Findings of the scenarios working group*. Millennium Ecosystem Assessment Series, Island Press, Washington, DC. Available at ⟨www.maweb.org/⟩.

Carter, R. E., MacKenzie, M. D., and Gjerstad, D. H. (1999). Ecological land classification in the Southern Loam Hills of south Alabama. *Forest Ecology and Management*, 114, 395–404.

Castañeda, F. (2001). Collaborative action and technology transfer as means of strengthening the implementation of national-level criteria and indicators. In *Criteria and indicators for sustainable forest management*, ed. R. J. Raison, A. G. Brown, and D. W. Flinn, pp. 145–163. IUFRO Research Series No. 7. CABI Publishing, Wallingford.

Castley, J. G. and Kerley, G. I. H. (1996). The paradox of forest conservation in South Africa. *Forest Ecology and Management*, 85, 35–46.

Caswell, H. (1989). *Matrix population models: Construction, analysis and interpretation*. Sinauer Associates, Sunderland, MA.

Caswell, H. (2000). Prospective and retrospective analyses: their roles in conservation biology. *Ecology*, 81, 619–627.

Caswell, H. (2001). *Matrix population models: Construction, analysis and interpretation*, 2nd edition. Sinauer Associates, Sunderland, MA.

Cavers, S., Navarro, C., and Lowe, A. J. (2003). Chloroplast DNA phylogeography reveals colonization history of a Neotropical tree, *Cedrela odorata* L., in Mesoamerica. *Molecular Ecology*, 12, 1451–1460.

Cavers, S., Navarro, C., and Lowe, A. J. (2004). A combination of molecular markers (cpDNA, PCR-RFLP, AFLP) identifies evolutionarily significant units in *Cedrela odorata* L. (Meliaceae) in Costa Rica. *Conservation Genetics*, 4, 571–580.

Cavers, S., Degen, B., Caron, H., Hardy, O., Lemes, M., and Gribel, R. (2005). Optimal sampling strategy for estimation of spatial genetic structure in tree populations. *Heredity*, 95, 281–289.

Cayuela, L., Golicher, D. J., Salas Rey, J., and Rey Benayas, J. M. (2006). Classification of a complex landscape using Dempster-Shafer theory of evidence. *International Journal of Remote Sensing*, 27(10), 1951–1971.

Chambers, J. Q. and Trumbore, S. E. (1999). An age old problem. *Trends in Plant Sciences*, 4, 385–386.

Chambers, R. (1992). *Rural appraisal: Rapid, relaxed, and participatory*. Discussion Paper 311, Institute of Development Studies, Brighton.

Chambers, R. (2002). *Participatory workshops: A sourcebook of 21 sets of ideas and activities*. Earthscan, London.

Chambers, R. and Conway, G. (1992). *Sustainable rural livelihoods: practical concepts for the 21st century*. Discussion Paper 296, Institute of Development Studies, Brighton.

Chao, A., Hwang, W.-H., Chen, Y.-C., and Kuo, C. Y. (2000). Estimating the number of shared species in two communities. *Statistica Sinica*, 10, 227–246.

Chape, S., Harrison, J., Spalding, M., and Lysenko, I. (2005). Measuring the extent and effectiveness of protected areas as an indicator for meeting global biodiversity targets. *Philosophical Transactions of the Royal Society*, B360, 443–455.

Chapin, F. S., Sala, O. E., and Huber-Sannwald, E. (eds.) (2001). *Global biodiversity in a changing environment. Scenarios for the 21st century*. Ecological Studies 152, Springer-Verlag, New York.

Chapman, C. A., Chapman, L. J., Wangham, R., Hunt, R., Gebo, D., and Gardner, L. (1992). Estimators of fruit abundance of tropical trees. *Biotropica*, 24(4), 527–531.

Chapman, C. A., Wangham, R., Chapman, L. J., Kennard, D. K., and Zanne, A. E. (1999). Fruit and flower phenology at two sites in Kibale National Park, Uganda. *Journal of Tropical Ecology*, 15, 189–211.

Charlesworth, D. (1988). A method for estimating outcrossing rates in natural populations of plants. *Heredity*, 61, 469–471.

Chase, M. W. and Hills, H. H. (1991). Silica gel: an ideal material for field preservation of leaf samples for DNA studies. *Taxon*, 40, 215–220.

Chason, J. W., Baldocchi, D. D., and Huston, M. A. (1991). A comparison of direct and indirect methods for estimating forest canopy leaf area. *Agricultural and Forest Meteorology*, 57, 107–128.

Chazdon, R. L., Colwell, R. K., Denslow, J. S., and Guariguata, M. R. (1998). Statistical methods for estimating species richness of woody regeneration in primary and secondary rain forests of NE Costa Rica. In *Forest biodiversity research, monitoring and modeling: Conceptual background and Old World case studies*, eds. F. Dallmeier and J. A. Comiskey, pp. 285–309. Parthenon, Paris.

Chazdon, R. L. and Field, C. B. (1987). Photographic estimation of photosynthetically active radiation: evaluation of a computerized technique. *Oecologia*, 73, 525–532.

Chen, J. M. and Cihlar, J. (1995). Quantifying the effect of canopy architecture on optical measurements of leaf area index using two gap size analysis methods. *IEEE Transactions on Geoscience and Remote Sensing*, 33, 777–787.

Chiariello, N. R., Mooney, H. A., and Williams, K. (1989). Growth, carbon allocation and cost of plant tissues. In *Plant physiological ecology*, eds. R. W. Pearcy, J. R. Ehleringer, H. A. Mooney, and P. W. Rundel, pp. 327–365. Chapman & Hall, London.

Chirici, G., Corona, P., Marchetti, M., and Travaglini, D. (2003). Testing Ikonos and Landsat 7 ETM+ potential for stand-level forest type mapping by soft supervised approaches. In *Advances in forest inventory for sustainable forest management and biodiversity monitoring*, eds. P. Corona, M. Köhl, and M. Marchetti, pp. 71–86. Kluwer, Dordrecht.

Chuvieco, E. (ed.) (1999). *Remote sensing of large wildfires in the European Mediterranean basin*. Springer-Verlag, Berlin.

Clark, D. A. and Clark, D. B. (1992). Life history diversity of canopy and emergent trees in a neotropical rainforest. *Ecological Monographs*, 62, 314–344.

Clark, D. F., Kneeshaw, D. D., Burton, P. J., and Antos, J. A. (1998a). Coarse woody debris in sub-boreal spruce forests of west-central British Columbia. *Canadian Journal of Forest Research*, 28, 284–290.

Clark, J. S. and Royall, P. S. (1995). Transformation of a northern hardwood forest by aboriginal (Iroquois) fire: charcoal evidence from Crawford Lake, Ontario, Canada. *The Holocene*, 5, 1–9.

Clark, J. S., Macklin, E., and Wood, L. (1998b). Stages and spatial scales of recruitment limitation in southern Appalachian forests. *Ecological Monographs*, 68, 213–235.

Clark, J. S., Beckage, B., Camill, P. et al. (1999a). Interpreting recruitment limitation in forests. *American Journal of Botany*, 86, 1–16.

Clark, J. S., Silman, M., Kern, R., Macklin, E., and Hillerislambers, J. (1999b). Seed dispersal near and far: patterns across temperate and tropical forests. *Ecology*, 80, 1475–1494.

Clark, J. S., Ladeau, S., and Ibanez, I. (2004). Fecundity of trees and the colonization–competition hypothesis. *Ecological Monographs*, 74(3), 415–442.

Clark, N. A., Wayne, R. H., and Schmidt, D. L. (2000). A review of past research on dendrometers. *Forest Science*, 46, 570–576.

Clark, P. J. and Evans, F. C. (1954). Distance to nearest neighbour as a measure of spatial relationships in populations. *Ecology*, 35, 445–453.

Clark, T. L., Radke, L., Coen, J., and Middleton, D. (1999c). Analysis of small-scale convective dynamics in a crown fire using infrared video camera imagery. *Journal of Applied Meteorology*, 38, 1401–1420.

Clarke, K. R. and Warwick, R. M. (2001). *Change in marine communities: An approach to statistical analysis and interpretation*, 2nd edition. Primer-E, Plymouth Marine Laboratory, Plymouth.

Clewell, A. and Rieger, J. (1997). What practitioners need from restoration ecologists. *Restoration Ecology*, 5(4), 350–354.

Cochran, W. G. (1977). *Sampling techniques*, 3rd edition. Wiley, New York.

Cohen, W. B., Fiorella, M., Gray, J., Helmer, E., and Anderson, K. (1998). An efficient and accurate method for mapping forest clearcuts in the Pacific Northwest using Landsat imagery. *Photogrammetric Engineering and Remote Sensing*, 64, 293–300.

Cohen, W. B. and Goward, S. N. (2004). Landsat's role in ecological applications of remote sensing. *BioScience*, 54(6), 535–545.

Cohen, W. B. and Spies, T. A. (1992). Estimating structural attributes of Douglas-fir/western hemlock forest stands from Landsat and SPOT imagery. *Remote Sensing of Environment*, 41, 1–17.

Cohen, W. B., Spies, T. A., and Fiorella, M. (1995). Estimating the age and structure of forests in a multiownership landscape of western Oregon, USA. *International Journal of Remote Sensing*, 16, 721–746.

Colpaert, N., Cavers, S., Bandou, E., Caron, H., Gheysen, G., and Lowe, A. J. (2005). Sampling tissue for DNA analysis of trees: trunk cambium as an alternative to canopy leaves. *Silvae Genetica*, 54(6), 265–269.

Colwell, R. K. (2004a). *EstimateS 7 User's Guide (online)*. University of Connecticut, Storrs, ⟨http://viceroy.eeb.uconn.edu/EstimateS7Pages/EstS7UsersGuide/EstimateS7UsersGuide.htm⟩.

Colwell, R. K. (2004b). *EstimateS 7.00: Statistical estimation of species richness and shared species from samples*. ⟨http://viceroy.eeb.uconn.edu/EstimateS⟩.

Colwell, R. K. and Coddington, J. A. (1994). Estimating terrestrial biodiversity through extrapolation. *Philosophical Transactions of the Royal Society of London* B345, 101–118.

Colwell, R. K., Mao, C. X., and Chang, J. (2004). Interpolating, extrapolating, and comparing incidence-based species accumulation curves. *Ecology*, 85(10), 2717–2727.

Comeau, P. G., Gendron, F., and Letchford, T. (1998). A comparison of several methods for estimating light under a paper birch mixed-wood stand. *Canadian Journal of Forest Research*, 28, 1843–1850.

Condit, R. (1998). *Tropical forest census plots*. Springer-Verlag, Berlin.

Condit, R., Hubbell, S. P., Lafrankie, J. V. *et al.* (1996a). Species-area and species-individual relationships for tropical trees: A comparison of three 50-ha plots. *Journal of Ecology*, 84(4), 549–562.

Condit, R., Hubbell, S. P., and Foster, R. B. (1996b). Assessing the response of plant functional types to climatic change in tropical forests. *Journal of Vegetation Science*, 7(3), 405–416.

Condit, R., Foster, R., Hubbell, S. P. *et al.* (1998). Assessing forest diversity on small plots: calibration using species-individual curves from 50-ha plots. In *Forest biodiversity research, monitoring and modeling: Conceptual background and Old World case studies*, eds. F. Dallmeier and J. A. Comiskey, pp. 247–268. UNESCO/Parthenon, Paris.

Condit, R., Ashton, P. S., Baker, P. *et al.* (2000). Spatial patterns in the distribution of tropical tree species. *Science*, 288, 1414–1418.

Connell, J. H. (1971). On the role of natural enemies in preventing competitive exclusion in some marine animals and in rain forest trees. In *Dynamics of populations*, eds. P. J. dem Boer and G. R. Gradwell, pp. 298–312. Centre for Agricultural Publishing and Documentation, Wageningen.

Cook, E. R. and Kairiukstis, L. A. (1990). *Methods of dendrochronology: Applications in the environmental sciences*. Kluwer, New York.

Cook, J. G., Stutzman, T. W., Bowers, C. W., Brenner, K. A., and Irwin, L. L. (1995). Spherical densiometers produce biased estimates of forest canopy cover. *Wildlife Society Bulletin*, 23, 711–717.

Coombs, J., Hall, D. O., Long, S. P., and Scurlock, J. M. O. (eds.) (1985). *Techniques in bioproductivity and photosynthesis*, 2nd edition. Pergamon, Oxford.

Coomes, D. A. and Grubb, P. J. (1998). Responses of juvenile trees to above- and belowground competition in nutrient-starved Amazonian rain forest. *Ecology*, 79, 768–782.

Cooney, R. (2004). *The precautionary principle in biodiversity conservation and natural resource management. An issue paper for policy-makers, researchers and practitioners*. Policy and Global Change Series, IUCN, Gland.

Corona, P., Köhl, M., and Marchetti, M. (eds.) (2003). *Advances in forest inventory for sustainable forest management and biodiversity monitoring*. Kluwer, Dordrecht.

Comeau, P. G., Gendron, F., and Letchford, T. (1998). A comparison of several methods for estimating light under a paper birch mixedwood stand. *Canadian Journal of Forest Research*, 28, 1843–1850.

Cornelissen, J. H. C., Lavorel, S., Garnier, E. *et al.* (2003). A handbook of protocols for standardized and easy measurement of plant functional traits worldwide. *Australian Journal of Botany*, 51, 335–380.

Cornelius, J. (1994). Heritabilities and additive genetic coefficients of variation in forest trees. *Canadian Journal of Forest Research*, 24, 372–379.

Cottrell, T. R. (2004). Seed rain traps for forest lands: Considerations for trap construction and study design. *BC Journal of Ecosystems and Management* 5(1). ⟨*www.forrex.org/jem/2004/vol5/no1/art1.pdf*⟩.

Coulson, T., Mace, G. M., Hudson, E., and Possingham, H. (2001). The use and abuse of population viability analysis. *Trends in Ecology and Evolution*, 16, 219–221.

Coutts, M. P. and Grace, J. (eds.) (1995). *Wind and trees*. Cambridge University Press, Cambridge.

Crandall, K. A., Bininda-Edmonds, O. R. P., Mace, G. M., and Wayne, R. K. (2000). Considering evolutionaty processes in conservation biology: an alternative to 'evolutionary significant units'. *Trends in Ecology and Evolution*, 15, 290–295.

Crawley, M. J. (1993). *GLIM for ecologists*. Blackwell Scientific, Oxford.

Crawley, M. J. (1997). Plant-herbivore dynamics. In *Plant ecology*, ed. M. J. Crawley, 2nd edition, pp.401–474. Blackwell Science, Oxford.

Croker, B. H. (1959). A method for estimating the botanical composition of the diet of sheep. *New Zealand Journal of Agricultural Research*, 2, 72–85.

Csuti, B., Polaski, S., Williams, P. H. *et al.* (1997). A comparison of reserve selection algorithms using data on terrestrial vertebrates in Oregon. *Biological Conservation*, 80(1), 83–97.

Cumming, S. and Vernier, P. (2002). Statistical models of landscape pattern metrics, with applications to regional scale dynamic forest simulations. *Landscape Ecology*, 17, 433–444.

Cumming, S. G., Schmiegelow, F. K. A., and Burton, P. J. (2000). Gap dynamics in boreal aspen stands: is the forest older than we think? *Ecological Applications*, 10(3), 744–759.

Curran, P. J. and Williamson, H. D. (1985). The accuracy of ground data used in remote sensing investigations. *International Journal of Remote Sensing*, 6, 1637–1651.

Curran, P. J., Dungan, J., and Gholz, H. L. (1992). Seasonal LAI measurements in slash pine using Landsat TM. *Remote Sensing of Environment*, 39, 3–13.

Cushman, S. A. and Wallin, D. O. (2000). Rates and patterns of landscape change in the Central Sikhote-alin Mountains, Russian Far East. *Landscape Ecology*, 15(7), 643–659.

Czaplewski, R. L. (2003). Accuracy assessment of maps of forest condition. In *Remote sensing of forest environments*, eds. M. A. Wulder and S. E. Franklin, pp. 115–140. Kluwer, Dordrecht.

Dafni, A. (1992). *Pollination ecology. A practical approach*. IRL Press/Oxford University Press, Oxford.

Dale, M. R. T. (1999). *Spatial pattern analysis in plant ecology*. Cambridge University Press, Cambridge.

Dale, V. H. and Pearson, S. M. (1997). Quantifying habitat fragmentation due to land use change in Amazonia. In *Tropical forest remnants*, eds. W. Laurance and R. Bierregaard, pp. 400–414. University of Chicago Press, Chicago.

Dallmeier, F. and Comiskey, J. (eds.) (1998a). *Forest biodiversity research, monitoring and modeling: Conceptual background and Old World case studies*. Man and the Biosphere Series, Vol. 20. UNESCO/Parthenon, Paris.

Dallmeier, F. and Comiskey, J. (eds.) (1998b). *Forest biodiversity in North, Central, and South America, and the Caribbean: Research and monitoring*. Man and the Biosphere Series, Vol. 21. UNESCO/Parthenon, Paris.

Danson, F. M. and Curran, P. J. (1993). Factors affecting the remotely sensed response of coniferous forest plantations. *Remote Sensing of Environment*, 43, 55–65.

D'eca-Neves, F. F. and Morellato, L. P. C. (2004). Methods applied for sampling and estimate tropical forest phenology. *Acta Botanica Brasiliana*, 18(1), 99–108.

DeFries, R., Hansen, A., Newton, A. C., and Hansen, M. C. (2005). Increasing isolation of protected areas in tropical forests over the past twenty years. *Ecological Applications*, 15(1), 19–26.

Degen, B., Gregorius, H.-R., and Scholtz, F. (1996). ECO-GENE, a model for simulation studies on the spatial and temporal dynamics of genetic structure of tree populations. *Silvae Genetica*, 45, 323–329.

Degen, B., Petit, R. J., and Kremer, A. (2001). SGS—Spatial Genetic Software: a computer program for analysis of spatial genetic and phenotypic structures of individuals and populations. *Journal of Heredity*, 92, 447–448.

De Kroon, H., Plaiser, A., van Groenendael, J. M., and Caswell, H. (1986). Elasticity as a measure of the relative contribution of demographic parameters to population growth rate. *Ecology*, 67, 1427–1431.

De Kroon, H., van Groenendael, J. M., and Ehrlen, J. (2000). Elasticities: a review of methods and model limitations. *Ecology*, 81, 607–618.

Del Tredici, P. (2001). Sprouting in temperate trees: A morphological and ecological review. *Botanical Reviews*, 67, 21–140.

DeMers, M. N. (2005). *Fundamentals of geographic information systems*. Wiley, New York.

Dennis, B. (1996). Discussion: should ecologists become Bayesians? *Ecological Applications*, 6(4), 1095–1103.

Denslow, J. S. (1980). Gap partitioning among tropical rainforest trees. *Biotropica*, 12(Suppl.), 47–55.

Dettmers, R. and Bart, J. (1999). A GIS modeling method applied to predicting forest songbird habitat. *Ecological Applications*, 9, 152–163.

de Vries, P. G. (1986). *Sampling theory for forest inventory: A teach yourself course*. Springer-Verlag, New York.

Diamond, J. (1986). Overview: laboratory experiments, field experiments, and natural experiments. In *Community Ecology*, eds. J. Diamond and T. K. Case, pp. 3–22. Harper & Row, New York.

Diaz, N. M. (1996). Landscape metrics: a new tool for forest ecologists. *Journal of Forestry*, 94, 12–16.

Digby, P. G. N. and Kempton, R. A. (1987). *Multivariate analysis of ecological communities*. Chapman & Hall, London.

Diggle, P. J. (1979). On parameter estimation and goodness-of- fit testing for spatial point patterns. *Biometrics*, 35, 87–101.

Diggle, P. J. (1983). *Statistical analysis of spatial point patterns*. Academic Press, New York.

Dinerstein, E., Olson, D. J., Graham, D. M. *et al*. (1995). *A conservation assessment of the terrestrial ecoregions of Latin America and the Caribbean*. World Bank, Washington, DC.

Donovan, M. L., Rabe, D. L., and Olson, C. E. (1987). Use of Geographic Information Systems to develop habitat suitability models. *Wildlife Society Bulletin*, 15, 574–579.

Dow, B. D. and Ashley, M. V. (1996). Microsatellite analysis of seed dispersal and parentage of saplings in bur oak, *Quercus macrocarpa*. *Molecular Ecology*, 5, 615–627.

Drechsler, M., Lamont, B. B., Burgman, M. A., Akçakaya, H. R., Witkowski, E. T. F., and Supriyadi (1999). Modelling the persistence of an apparently immortal *Banksia* species after fire and land clearing. *Biological Conservation*, 88, 249–259.

Droege, S., Cyr, A., and Larivée, J. (1998). Checklists: an under-used tool for the inventory and monitoring of plants and animals. *Conservation Biology*, 12, 1134–1138.

Dubayah, R. O. and Drake, J. B. (2000). Lidar remote sensing for forestry. *Journal of Forestry*, 98, 44–46.

Ducey, M. J., Gove, J. H., Ståhl, G., and Ringvall, A. (2001). Clarification of the mirage method for boundary correction, with possible bias in plot and point sampling. *Forest Science*, 47, 242–245.

Dudley, N. and Phillips, A. (2006). *Forests and protected areas. Guidance on the use of the IUCN protected area management categories*. World Commission on Protected Areas (WCPA). Best Practice Protected Area Guidelines Series No. 12. IUCN, Gland.

Durrett, R. and Levin, S. A. (1994). Stochastic spatial models: a user's guide to ecological applications. *Philosophical Transactions of the Royal Society of London*, B343, 329–350.

Dyck, W. J., Cole, D. W., and Comerford, N. B. (1994). *Impacts of forest harvesting on long-term site productivity*. Chapman & Hall, London.

Dytham, C. (2003). *Choosing and using statistics. A biologist's guide*, 2nd edition. Blackwell, Oxford.

Easter, M. J. and Spies, T. A. (1994). Using hemispherical photography for estimating photosynthetic photon flux density under canopies and in gaps in Douglas-fir forests of the Pacific Northwest. *Canadian Journal of Forest Research*, 24, 2050–2058.

Eberhardt, L. L. (1978). Transect methods for population studies. *Journal of Wildlife Management*, 42, 1–31.

Ebert, T. A. (1999). *Plant and animal populations: Methods in demography*. Academic Press, New York.

Echeverría, C. M. (2003). *Deforestation and forest fragmentation of temperate forests in Chile*. MPhil thesis, University of Cambridge.

Echeverría, C. M. (2005). *Fragmentation of temperate rain forests in Chile: patterns, causes, and impacts*. PhD thesis, University of Cambridge.

EEA (1998). *Europe's environment—The 2nd assessment*. European Environment Agency, Office for Publications of the European Communities, Brussels.

Ehtisham-Ul-Haq, M., Allnutt, T. R., Armesto, J. J., Smith, C., Gardner, M., and Newton, A. C. (2001). Patterns of genetic variation in the threatened Chilean vine *Berberidopsis corallina* Hook. f. sampled *in* and *ex situ*, detected using RAPD markers. *Annals of Botany*, 87(6), 813–821.

Ejrnæs, R., Aude, E., Nygaard, B., and Münier, B. (2002). Prediction of habitat quality using ordination and neural networks. *Ecological Applications*, 12(4), 1180–1187.

Ek, A. R., Shifley, S. R., and Burk, T. E. (eds.) (1988). *Forest growth modelling and prediction*. General Technical Report NC 120. USDA Forest Service, Washington, DC.

Elderd, B. D., Shahani, P., and Doak, D. F. (2003). The problems and potential of count-based population viability analyses. In *Population viability in plants. Conservation, management and modeling of rare plants*, eds. C. A. Brigham and M. A. Schwartz, pp. 173–202. Ecological Studies 165, Springer-Verlag, Berlin.

Elith, J. and Burgman, M. A. (2003). Habitat models for population viability analysis. In *Population viability in plants. Conservation, management and modeling of rare plants*, eds. C. A. Brigham and M. A. Schwartz, pp. 203–235. Ecological Studies 165, Springer-Verlag, Berlin.

Ellison, A. M. (1996). An introduction to Bayesian inference for ecological research and environmental decision-making. *Ecological Applications*, 6(4), 1036–1046.

Ellner, S. P., Fieberg, J., Ludwig, D., and Wilcox, C. (2002). Precision of population viability analysis. *Conservation Biology*, 16(1), 258–261.

Engeman, R. M. and Sugihara, R. T. (1998). Optimization of variable area transect sampling using Monte Carlo simulation. *Ecology*, 79(4), 1425–1434.

Engeman, R. M., Sugihara, R. T., Pank, L. F., and Dusenberry, W. E. (1994). A comparison of plotless density estimators using Monte Carlo simulation. *Ecology*, 75(6), 1769–1779.

Englund, S. R., O'Brien, J. J., and Clark, D. B. (2000). Evaluation of digital and film hemispherical photography and spherical densiometry for measuring forest light environments. *Canadian Journal of Forest Research*, 30, 1999–2005.

Ennos, A. R. (1997). Wind as an ecological factor. *Trends in Ecology and Evolution*, 12(3), 108–111.

Ennos, R. A. (1994). Estimating the relative rates of pollen and seed migration among plant populations. *Heredity*, 72, 250–259.

Ennos, R. A. (1996). Utilising genetic information in plant conservation programmes. In *Aspects of the genesis and maintenance of biological diversity*, eds. M. E. Hochberg, J. Clobert, and R. Barbault, pp. 278–291. Oxford University Press, Oxford.

Ennos, R. (1998). Genetic constraints on native woodland restoration. In *Native Woodland restoration in southern Scotland: Principles and practice*, eds. A. C. Newton and P. Ashmole, pp. 27–34. Borders Forest Trust, Ancrum, Jedburgh.

Ennos, R. A., Worrell, R., and Malcolm, D. C. (1998). The genetic management of native species in Scotland. *Forestry*, 71, 1–23.

Enright, N. J. and Ogden, J. (1979). Application of transition matrix models in forest dynamics: *Araucaria* in Papua New Guinea and *Nothofagus* in New Zealand. *Australian Journal of Ecology*, 4, 3–23.

Erdelen, M. (1984). Bird communities and vegetation structure: I. Correlations and comparisons of simple diversity indices. *Oecologia*, 61, 277–284.

Erickson, A. A., Saltis, M., Bell, S. S., and Dawes, C. J. (2003). Herbivore feeding preferences as measured by leaf damage and stomatal ingestion: a mangrove crab example. *Journal of Experimental Marine Biology and Ecology*, 289(1), 123–138.

Ervin, J. (2003a). *WWF: Rapid assessment and prioritization of protected area management (RAPPAM) methodology*. WWF, Gland.

Ervin, J. (2003b). Protected area assessments in perspective. *BioScience*, 53(9), 819–822.

Estabrook, G. F., Winsor, J. A., Stephenson, A. G., and Howe, H. F. (1982). When are two phenological patterns different? *Botanical Gazette*, 143(3), 374–378.

Eva, H. and Lambin, E. F. (2002). Fires and land-cover change in the tropics: a remote sensing analysis at the landscape scale. *Journal of Biogeography*, 27, 765–776.

Evans, G. C. (1972). *The quantitative analysis of plant growth*. Studies in Ecology, vol. 1. University of California Press, Berkeley, CA.

Evans, G. C. and Coombe, D. E. (1959). Hemispherical and woodland canopy photography and the light climate. *Journal of Ecology*, 47, 103–113.

Fahrig, L. (2003). Effects of habitat fragmentation on biodiversity. *Annual Review of Ecology, Evolution and Systematics*, 34, 487–515.

Fairbanks, D. H. K. and McGwire, K. C. (2004). Patterns of floristic richness in vegetation communities of California: regional scale analysis with multi-temporal NDVI. *Global Ecology and Biogeography*, 13(3), 221–235.

Faith, D. P., Nix, H. A., Margules, C. R. *et al.* (2001). The BioRap biodiversity assessment and planning study for Papua New Guinea. *Pacific Conservation Biology*, 6, 279–288.

FAO (1995). *Forest Resources Assessment 1990. Global Synthesis*. FAO, Rome.

FAO (2001a). *Criteria and indicators for sustainable forest management: a compendium*. Forest Management Working Papers no. 5, FAO, Rome.

FAO (2001b). *Global forest resources assessment. Main report*. FAO, Rome.

FAO (2003). *Workshop on the FAO approach to National Forest Resources Assessment and ongoing projects*. Forest Resources Assessment WP 70, FAO, Rome.

FAO (2005). *Proceedings of the third expert meeting on harmonizing forest-related definitions for use by various stakeholders*, 17–19 January 2005. FAO, Rome.

Fassnacht, K. S., Gower, S. T., MacKenzie, M. D., Nordheim, E. V., and Lillesand, T. M. (1997). Estimating the leaf area index of north central Wisconsin forests using Landsat Thematic Mapper. *Remote Sensing of Environment*, 61, 229–245.

Fazakas, Z. and Nilsson, M. (1996). Volume and forest cover estimation over southern Sweden using AVHRR data calibrated with TM data. *International Journal of Remote Sensing*, 17, 1701–1709.

Fazakas, Z., Nilsson, M., and Olsson, H. (1999). Regional forest biomass estimation by use of satellite data and ancilliary data. *Agriculture and Forest Metereology*, 98/99, 417–425.

Feinsinger, P. and Busby, W. H. (1987). Pollen carryover: experimental comparisons between morphs of *Palicourea lasiorrachis* (Rubiaceae), a distylous, bird-pollinated, tropical treelet. *Oecologia*, 73(3), 231–235.

Feldpausch, T. R., Jirka, S., Passos, C. A. M., Jasper, F., and Riha, S. J. (2005). When big trees fall: damage and carbon export by reduced impact logging in southern Amazonia. *Forest Ecology and Management*, 219(2–3), 199–215.

Fenner, M. (Ed.) (2000). *Seeds: The ecology of regeneration in plant communities*. CABI, Wallingford.

Ferment, A., Picard, N., Gourlet-Fleury, S., and Baraloto, C. (2001). A comparison of five indirect methods for characterizing the light environment in a tropical forest. *Annals of Forest Science*, 58, 877–891.

Ferris-Kaan, R., Peace, A. J., and Humphrey, J. W. (1998). Assessing structural diversity in managed forests. In *Assessment of forest biodiversity for improved forest management*, eds. P. Bachmann, M. Köhl and R. Päivinnen, pp. 331–342. Kluwer, Dordrecht.

Fieberg, J. and Ellner, S. P. (2000). When is it meaningful to estimate an extinction probability? *Ecology*, 81(7), 2040–2047.

Fieberg, J. and Ellner, S. P. (2001). Stochastic matrix models for conservation and management: a comparative review of methods. *Ecology Letters*, 4, 244–266.

Fieldings, A. H. and Bell, J. F. (1997). A review of methods for the assessment of prediction errors in conservation presence: absence models. *Environmental Conservation*, 24(1), 38–49.

Feinsinger, P. (2001). *Designing field studies for biodiversity conservation*. Island Press, Washington, DC.

Fiorella, M. and Ripple, W. J. (1993). Determining the successional stage of temperate coniferous forest with Landsat satellite data. *Photogrammetric Engineering and Remote Sensing*, 59, 239–246.

Fischer, D. T. and Church, R. L. (2005). The SITES reserve selection system: A critical review. *Environmental Modeling and Assessment*, 10(3), 215–228.

Fisher, B. L., Howe, H. F., and Wright, S. J. (1991). Survival and growth of *Virola surinamensis* yearlings: water augmentation in gap and understorcy. *Oecologia*, 86, 292–297.

Flinn, K. M. and Vellend, M. (2005). Recovery of forest plant communities in post-agricultural landscapes. *Frontiers in Ecology and Evolution*, 3, 243–250.

Foody, G. M. (1996). Fuzzy modeling of vegetation from remotely sensed imagery. *Ecological Modelling*, 85, 3–12.

Foody, G. M. (2003). Remote sensing of tropical forest environments: towards the monitoring of environmental resources for sustainable development. *International Journal of Remote Sensing*, 24(20), 4035–4046.

Foody, G. M. and Cutler, M. E. J. (2003). Tree biodiversity in protected and logged Bornean tropical rain forests and its measurement by satellite remote sensing. *Journal of Biogeography*, 30(7), 1053–1066.

Foody, G. M., Boyd, D. S., and Cutler, M. E. J. (2003). Predictive relations of tropical forest biomass from Landsat TM data and their transferability between regions. *Remote Sensing of Environment*, 85(4), 463–474.

Foody, G., Palubinskas, G., Lucas, R. M., Curran, P. J., and Honzak, M. (1996). Identifying terrestrial carbon sinks: classification of successional stages in regenerating tropical forests from Landsat Thematic Mapper data. *Remote Sensing of Environment*, 55, 205–216.

References | 395

Ford, E. D. (2000). *Scientific method for ecological research.* Cambridge University Press, Cambridge.

Forget, P.-M. and Wenny, D. (2005). How to elucidate seed fate? A review of methods used to study seed removal and secondary seed dispersal. In *Seed fate. Predation, dispersal and seedling establishment*, eds. P.-M. Forget, J. E. Lambert, P. E. Hulme, and S. B. Vander Wall, pp. 379–393. CABI Publishing, Wallingford, Oxford.

Forman, R. T. T. and Godron, M. (1986). *Landscape ecology.* Wiley, New York.

Fortin, M. J. and Dale, M. R. T. (2005). *Spatial analysis: a guide for ecologists.* Cambridge University Press, Cambridge.

Fournier, R. A., Mailly, D., Walter, J.-M. N., and Soudani, K. (2003). Indirect measurement of forest canopy structure from in situ optical sensors. In *Remote sensing of forest environments*, eds. M. A. Wulder and S. E. Franklin, pp. 77–113. Kluwer, Dordrecht.

Frank, K., Lorek, H., Köster, F., Sonnenschein, M., Wissel, C., and Grimm, V. (2003). *META-X—Software for metapopulation viability analysis.* Springer-Verlag, Berlin.

Frankham, R., Ballou, J. D., and Briscoe, D. A. (2002). *Introduction to conservation genetics.* Cambridge University Press, Cambridge.

Franklin, J., Phinn, S. R., Woodcock, C. E., and Rogan, J. (2003). Rationale and conceptual framework for classification approaches to assess forest resources and properties. In *Remote sensing of forest environments*, eds. M. A. Wulder and S. E. Franklin, pp. 279–300. Kluwer, Dordrecht.

Franklin, S. E. (2001). *Remote sensing for sustainable forest management.* Lewis Publishers/CRC Press, Boca Raton, FL.

Franklin, S. E. (2006). *Forest disturbance and spatial pattern: Remote sensing and GIS approaches.* CRC Press, Boca Raton, FL.

Franklin, S. E. and Raske, A. (1994). Satellite remote sensing of spruce budworm defoliation in western Newfoundland. *Canadian Journal of Remote Sensing*, 21, 299–308.

Fransson, J. E. S., Walter, F., and Olsson, H. (1999). Identification of clear-felled areas using SPOT P and Almaz-1 SAR data. *International Journal of Remote Sensing*, 20, 3583–3593.

Fraser, D. J. and Bernatchez, L. (2001). Adaptive evolutionary conservation: towards a unified concept for defining conservation units. *Molecular Ecology*, 10, 2741–2752.

Fraver, S. (1994). Vegetation responses along edge-to-interior gradients in the mixed hardwood forests of the Roanoke River basin, North Carolina. *Conservation Biology*, 8, 822–832.

Frazer, G. W., Canham, C. D., and Lertzman, K. P. (1999). *Gap Light Analyzer (GLA), Version 2.0: Imaging software to extract canopy structure and gap light transmission indices from true-colour fisheye photographs, users manual and program documentation.* Simon Fraser University, Burnaby, BC and the Institute of Ecosystem Studies, Millbrook, NY.

Frazer, G. W., Fournier, R. A., Trofymow, J. A., and Hall, R. J. (2001). A comparison of digital and film fisheye photography for analysis of forest canopy structure and gap light transmission. *Agricultural and Forest Meteorology*, 109, 249–263.

Freckleton, R. P., Silva Matos, D. M., Bovi, M. L. A., and Watkinson, A. R. (2003). Predicting the impacts of harvesting using structured population models: the importance of density-dependence and timing of harvest for a tropical palm tree. *Journal of Applied Ecology*, 40, 846–858.

Frelich, L. E. (2002). *Forest dynamics and disturbance regimes. Studies from temperate evergreen-deciduous forests.* Cambridge University Press, Cambridge.

Frelich, L. E. and Lorimer, C. G. (1991). A simulation of landscape-level stand dynamics in the northern hardwood region. *Journal of Ecology*, 79, 223–233.

Frelich, L. E., Sugita, S., Reich, P. B., Davis, M. B., and Friedman, S. K. (1998). Neighbourhood effect in forests: implication for within patch structure. *Journal of Ecology*, 86, 149–161.

Fridman, J. and Walheim, M. (2000). Amount, structure, and dynamics of dead wood on managed forestland in Sweden. *Forest Ecology and Management*, 131, 23–36.

Friend, A. and Rapaport, D. (1979). *Towards a comprehensive framework for environment statistics: A stress-response approach*. Statistics Canada, Ottawa, Canada.

Friend, A. D., Shugart, H. H., and Running, S. W. (1993). A physiology-based model of forest dynamics. *Ecology*, 74, 792–797.

Friend, A. D., Stevens, A. K., Knox, R. G., and Cannell, M. G. R. (1997). A process-based, terrestrial biosphere model of ecosystem dynamics (Hybrid v3.0). *Ecological Modelling*, 95, 249–287.

Friend, D. T. C. (1961). A simple method of measuring integrated light values in the field. *Ecology*, 42, 577–580.

Fritts, H. C. (1976). *Tree rings and climate*. Academic Press, New York.

Fritts, N. C. and Fritts, E. C. (1955). A new dendrograph for recording radial changes of a tree. *Forest Science*, 1, 271–276.

Fujita, T., Itaya, A., Miura, M., Manabe, T., and Yamamoto, S. (2003a). Canopy structure in a temperate old-growth evergreen forest analyzed by using aerial photographs. *Plant Ecology*, 168(1), 23–29.

Fujita, T., Itaya, A., Miura, M., Manabe, T., and Yamamoto, S. (2003b). Long-term canopy dynamics analysed by aerial photographs in a temperate old-growth evergreen broad-leaved forest. *Journal of Ecology*, 91(4), 686–693.

Fuller, R. J. (1995). *Bird life of woodland and forest*. Cambridge University Press, Cambridge.

Furusawa, T., Pahari, K., Umezaki, M., and Ohtsuka, R. (2004). Impacts of selective logging on New Georgia Island, Solomon Islands evaluated using very-high-resolution satellite (IKONOS) data. *Environmental Conservation*, 31(4), 349–355.

Galiano, E. F. (1982). Pattern detection in plant populations through the analysis of plant-to-all-plants distances. *Vegetatio*, 49, 39–43.

Gardiner, B., Peltola, H., and Kellomaki, S. (2000). Comparison of two models for predicting the critical wind speeds required to damage coniferous trees. *Ecological Modelling*, 129, 1–23.

Gardiner, B., Suárez, J., Achim, A., Hale, S., and Nicoll, B. (2004). *ForestGALES. A PC-based wind risk model for British forests*. User's Guide version 2.0. Forestry Commission, Cambridge.

Gardner, R. H. and Urban, D. L. (2003). Model validation and testing: past lessons, present concerns, future prospects. In *Models in ecosystem science*, eds. C. D. Canham, J. J. Cole, and W. K. Lauenroth, pp. 185–203. Princeton University Press, Princeton, NJ.

Garrison, G. A. (1949). Uses and modifications for the 'Moosehorn' crown closure estimator. *Journal of Forestry*, 47, 733–735.

Garson, J., Aggarwal, A., and Sarkar, S. (2002). *ResNet Manual Ver. 1.2*. University of Texas at Austin, TX.

Gendron, F., Messier, C., and Comeau, P. G. (1998). Comparison of various methods for estimating the mean growing season percent photosynthetic photon flux density in forests. *Agricultural and Forest Meteorology*, 92, 55–70.

Gentry, A. H. (1982). Patterns of neotropical plant species diversity. *Evolutionary Biology*, 15, 1–84.

Gentry, A. H. (1988). Changes in plant community diversity and floristic composition on environmental and geographical gradients. *Annals of the Missouri Botanical Garden*, 75, 1–34.

Ghazoul, J. and McAllister, M. (2003). Communicating complexity and uncertainty in decision making contexts: Bayesian approaches to forest research. *International Forestry Review*, 5(1), 9–19.

Gibbons, P., Lindenmayer, D. B., Barry, S. C., and Tanton, M. T. (2002). Hollow selection by vertebrate fauna in forests of southeastern Australia and implications for forest management. *Biological Conservation*, 103, 1–12.

Gibson, D. J. (2002). *Methods in comparative plant population ecology*. Oxford University Press, Oxford.
Gill, F. B., Mack, A. L., and Ray, R. T. (1982). Competition beween hermit hummingbirds Phaethorninae and insects for nectar in a Costa Rican rain forest. *Ibis*, 124, 44–49.
Gill, R. M. A. (1992). A review of damage by mammals in north temperate forests 3. Impact on trees and forests. *Forestry*, 65(4), 363–388.
Gillet, E. and Gregorius, H. R. (2000). Qualified testing of single-locus codominant inheritance using single tree progenies. *Biometrics*, 56, 801–807.
Gillies, A. C. M., Navarro, C., Lowe, A. J. *et al.* (1999). Genetic diversity in Mesoamerican populations of mahogany (*Swietenia macrophylla*), assessed using RAPDs. *Heredity*, 83, 722–732.
Gillison, A. (2002). A generic, computer-assisted method for rapid vegetation classification and survey: tropical and temperate case studies. *Conservation Ecology*, 6(2), 3 ⟨www.consecol.org/vol6/iss2/art3⟩.
Gillison, A. N. and Brewer, K. R. W. (1985). The use of gradient directed transects or gradsects in natural resource surveys. *Journal of Environmental Management*, 20, 103–127.
Ginzburg, L. R., Ferson, S., and Akçakaya, H. R. (1990). Reconstructability of density dependence and the conservative assessment of extinction risks. *Conservation Biology*, 4, 63–70.
Gitay, H. and Noble, I. R. (1997). What are functional types and how should we seek them? In *Plant functional types*, eds. T. M. Smith, H. H. Shugart, and F. I. Woodward, pp. 3–19. Cambridge University Press, Cambridge.
Glaubitz, J. C. and Moran, G. F. (2000). Genetic tools: the use of biochemical and molecular markers. In *Forest conservation genetics. Principles and practice*, eds. A. Young, D. Boshier, and T. Boyle, pp. 39–59. CSIRO, Collingwood, Australia and CABI, Wallingford.
Goldingay, R. L., Carthew, S. M., and Whelan, R. J. (1991). The importance of non-flying mammals in pollination. *Oikos*, 61(1), 79–87.
Goldsmith, F. B., Harrison, C. M., and Morton, A. J. (1986). Description and analysis of vegetation. In *Methods in plant ecology*, eds. P. D. Moore and S. B. Chapman, pp. 437–524. Blackwell Scientific, Oxford.
Golicher, D. J., O'Hara, R. B., Ruíz-Montoya, L., and Cayuela, L. (2006). Lifting a veil on diversity: a Bayesian approach to fitting relative-abundance models. *Ecological Applications*, 16(1), 202–212.
Gong, P. and Xu, B. (2003). Remote sensing of forests over time. In *Remote sensing of forest environments*, eds. M. A. Wulder and S. E. Franklin, pp. 301–333. Kluwer, Dordrecht.
Goovaerts, P. (1997). *Geostatistics for natural resources evaluation*. Oxford University Press, Oxford.
Gordon, J. E. (2005). *Biodiversity conservation and non-governmental organisations in Oaxaca, Mexico*. PhD thesis, University of Durham.
Gordon, J. E., Hawthorne, W. D., Reyes-García, A., Sandoval, G., and Barrance, A. J. (2004). Assessing landscapes: a case study of tree and shrub diversity in the seasonally dry tropical forests of Oaxaca, Mexico and southern Honduras. *Biological Conservation*, 117, 429–442.
Gotelli, N. J. and Colwell, R. K. (2001). Quantifying biodiversity: procedures and pitfalls in the measurement and comparison of species richness. *Ecology Letters*, 4, 379–391.
Gould, W. (2000). Remote sensing of vegetation, plant species richness, and regional biodiversity hotspots. *Ecological Applications*, 10(6), 1861–1870.
Gove, J. H., Ringvall, A., Ståhl, G., and Ducey, M. J. (1999). Point relascope sampling of coarse woody debris. *Canadian Journal of Forest Research*, 29(11), 1718–1726.
Gove, J. H., Ducey, M. J., Ståhl, G., and Ringvall, A. (2001). Point relascope sampling: a new way to assess downed coarse woody debris. *Journal of Forestry*, 99, 4–11.
Gove, J. H., Ducey, M. J., and Valentine, H. T. (2002). Multistage point relascope and randomized branch sampling for downed coarsewoody debris estimation. *Forest Ecology and Management*, 155, 153–162.

Grabe, D. F. (ed.) (1970). *Tetrazolium testing handbook for agricultural seeds*. Association of Official Seed Analysts, Stillwater, OK.

Grace, J. (1977). *Plant responses to wind*. Academic Press, London.

Grace, S. L. and Platt, W. J. (1995). Effects of adult tree density and fire on the demography of pregrass stage juvenile longleaf pine (*Pinus palustris* Mill.). *Journal of Ecology*, 83, 75–86.

Grainger, A. (1999). Constraints on modelling the deforestation and degradation of tropical open woodlands. *Global Ecology and Biogeography*, 8, 179–190.

Gray, A. (2003). Monitoring stand structure in mature coastal Douglas-fir forests: effect of plot size. *Forest Ecology and Management*, 175(1–3), 1–16.

Greene, D. F. and Calogeropoulos, C. (2002). Measuring and modelling seed dispersal of terrestrial plants. In *Dispersal ecology*, eds. J. M. Bullock, R. E. Kenward, and R. S. Hails, pp. 3–23. Blackwell, Oxford.

Greene, D. F. and Johnson, E. A. (1989). A model of wind dispersal of winged or plumed seeds. *Ecology*, 70, 339–347.

Greene, D. F. and Johnson, E. A. (1994). Estimating the mean annual seed production of trees. *Ecology*, 75, (3), 642–647.

Greene, D. F. and Johnson, E. A. (1997). Secondary dispersal of tree seeds on snow. *Journal of Ecology*, 85, 329–340.

Greenpeace International (2002). *The last of the world's ancient forests*. Greenpeace International, Amsterdam.

Greenwood, J. J. D. (1996). Basic techniques. In: *Ecological census techniques. A handbook*, ed. W. J. Sutherland, pp. 11–110. Cambridge University Press, Cambridge.

Gregoire, T. C. (1982). The unbiasedness of the mirage correction procedure for boundary overlap. *Forest Science*, 28, 504–508.

Grieg-Smith, P. (1957). *Quantitative plant ecology*. Academic Press, New York.

Grimm, V., Lorek, H., Finke, J. *et al*. (2004). META-X: generic software for metapopulation viability analysis. *Biodiversity and Conservation*, 13, 165–188.

Grove, S. J. (2001). Extent and composition of dead wood in Australian lowland tropical rainforest with different management histories. *Forest Ecology and Management*, 154, 35–53.

Guisan, A. and Zimmermann, N. E. (2000). Predictive habitat distribution models in ecology. *Ecological Modelling*, 135, 147–186.

Guisan, A., Weiss, S. B., and Weiss, A. D. (1999). GLM versus CCA spatial modeling of plant species distribution. *Plant Ecology*, 143, 107–122.

Gutzwiller, K. J. (ed.) (2002). *Applying landscape ecology in biological conservation*. Springer-Verlag, Berlin.

Haig, S. M. (1998). Molecular contributions to conservation. *Ecology*, 79, 413–425.

Hale, S. E. and Brown, N. (2005). Use of the canopy-scope for assessing canopy openness in plantation forests. *Forestry*, 78(4), 365–371.

Hall, D. O., Scurlock J. M. O., Bolhar-Nordenkampf, H. R., Leegood, R., and Long, S. P. (eds.) (1992). *Photosynthesis and production in a changing environment: A field and laboratory manual*. Kluwer, Dordrecht.

Hall, J. B. (1991). Multiple-nearest-tree sampling in an ecological survey of Afromontane catchment forest. *Forest Ecology and Management*, 42, 245–266.

Hall, P. and Bawa, K. (1993). Methods to assess the impact of extraction of non-timber tropical forest products on plant-populations. *Economic Botany*, 47(3), 234–247.

Hall, P., Ashton, P. S., Condit, R., Manokaran, N., and Hubbell, S. P. (1998). Signal and noise in sampling tropical forest structure and dynamics. In *Forest Biodiversity research, monitoring and modeling: Conceptual background and Old World case studies*, eds. F. Dallmeier and J. A. Comiskey, pp. 63–77. UNESCO/Parthenon Paris.

Hall, R. J. (2003). The roles of aerial photographs in forestry remote sensing image analysis. In *Remote sensing of forest environments*, eds. M. A. Wulder and S. E. Franklin, pp. 47–76. Kluwer, Dordrecht.

Hall, R. J., Kruger, A. R., Scheffer, J., Titus, S. J., and Moore, W. C. (1989a). A statistical evaluation of Landsat TM and MSS data for mapping forest cutovers. *Forestry Chronicle*, 65, 441–449.

Hall, R. J., Morton, R., and Nesby, R. (1989b). A comparison of existing models for dbh estimation from large-scale photos. *Forestry Chronicle*, 65, 114–120.

Hall, R. J., Dams, R. V., and Lyseng, L. N. (1991). Forest cutover mapping from SPOT satellite data. *International Journal of Remote Sensing*, 12, 2193–2204.

Hamrick, J. L. and Godt, M. J. W. (1990). Allozyme diversity in plant species. In *Plant population genetics, breeding and genetic resources*, eds. A. H. D. Brown, M. T. Clegg, A. L. Kahler, and B. S. Weir, pp. 43–63. Sinauer Associates, Sunderland, MA.

Hamrick, J. L. and Nason, J. D. (2000). Gene flow in forest trees. In *Forest conservation genetics. Principles and practice*, eds. A. Young, D. Boshier, and T. Boyle, pp. 81–90. CSIRO, Collingwood, Australia and CABI, Wallingford.

Hamrick, J. L., Godt, M. J. W., Murawski, D. A., and Loveless, M. D. (1991). Correlations between species traits and allozyme diversity: implications for conservation biology. In *Genetics and conservation of rare plants*, eds. D. A. Falk and K. E. Holsinger, pp. 75–86. Oxford University Press, Oxford.

Hamrick, J. L., Godt, M. J. W., and Sherman-Broyles, S. L. (1992). Factors influencing levels of genetic diversity in woody plant species. *New Forests*, 6, 95–124.

Hanan, N. P. and Bégué, A. (1995). A method to estimate instantaneous and daily intercepted photosynthetically active radiation using a hemispherical sensor. *Agriculture Forest Meteorology*, 74, 155–168.

Hansen, A. J., Spies, T. A., Swanson, F. J., and Ohmann, J. L. (1991). Conserving biodiversity in managed forests. *BioScience*, 41, 382–392.

Hansen, M. J., Franklin, S. E., Woudsma, C. G., and Peterson, M. (2001). Caribou habitat mapping and fragmentation analysis using Landsat MSS, TM, and GIS data in the North Columbia Mountains, British Columbia, Canada. *Remote Sensing of Environment*, 77, 50–65.

Hanski, I. (1994). A practical model of metapopulation dynamics. *Journal of Animal Ecology*, 63, 151–162.

Hanski, I. (1999). *Metapopulation ecology*. Oxford University Press, Oxford.

Hanski, I. (2004). Metapopulation theory, its use and misuse. *Basic and Applied Ecology*, 5, 225–229.

Hargis, C. D., Bissonette, J. A., and David, J. L. (1998). The behavior of landscape metrics commonly used in the study of habitat fragmentation. *Landscape Ecology*, 13, 167–186.

Haridasan, M. and de Araújo, G. M. (1988). Aluminium-accumulating species in two forest communities in the cerrado region of central Brazil. *Forest Ecology and Management*, 24, 15–26.

Harmon, M. E. and Sexton, J. (1996). *Guidelines for measurements of woody detritus in forest ecosystems*. Publication No. 20, US Long-term Ecological Research Network Office, University of Washington 〈http://old.lternet.edu/research/pubs/woodydetritus/〉.

Harmon, M. E., Franklin, J. F., Swanson, F. J. *et al.* (1986). Ecology of coarse woody debris in temperate ecosystems. *Advances in Ecological Research*, 15, 133–302.

Harper, K. A. and MacDonald, S. E. (2001). Structure and composition of riparian boreal forest: new methods for analyzing edge influence. *Ecology*, 82, 649–659.

Harrison, S. and Ray, C. (2002). Plant population viability and metapopulation-level processes. In *Population viability analysis*, eds. S. Beissinger and D. McCullogh, pp. 109–122. University of Chicago Press, Chicago.

Hawthorne, W. D. (1995). *Ecological profiles of Ghanaian forest trees*. Tropical Forestry Paper 29, Oxford Forestry Institute, Oxford.

Hawthorne, W. D. (1996). Holes in sums of parts in Ghanaian forest: regeneration scale and sustainable use. *Proceedings of the Royal Society of Edinburgh*, 104(B), 75–176.

Hawthorne, W. D. and Abu-Juam, M. (1995). *Forest protection in Ghana*. IUCN, Gland.

Hayes, D. J. and Sader, S. A. (2001). Comparison of change-detection techniques for monitoring tropical forest clearing and vegetation regrowth in a time series. *Photogrammetric Engineering and Remote Sensing*, 67(9), 1067–1075.

He, F. L. and Duncan, R. P. (2000). Density-dependent effects on tree survival in an old-growth Douglas fir forest./*Journal of Ecology*, 88, 676–688.

He, H. S. and Mladenoff, D. J. (1999). Spatially explicit and stochastic simulation of forest landscape fire disturbance and succession. *Ecology*, 80, 81–99.

He, H. S., Mladenoff, D. J., and Boeder, J. (1999). Object-oriented design of LANDIS, a spatially explicit and stochastic landscape model. *Ecological Modelling*, 119, 1–19.

Hemingway, C. A. and Overdorff, D. J. (1999). Sampling effects on food availability estimates phenological method, sample size, and species comparition. *Biotropica*, 31(2), 354–364.

Henderson, P. A. (2003). *Practical methods in ecology*. Blackwell, Oxford.

Hendricks, W. A. (1956). *The mathematical theory of sampling*. Scarecrow Press, New Brunswick, NJ.

Henle, K., Sarre, S., and Wiegand, K. (2004). The role of density regulation in extinction processes and population viability analysis. *Biodiversity and Conservation*, 13, 9–52.

Henry, J. D. and Swan, J. M. A. (1974). Reconstruction of forest history from live and dead plant material: an approach to the study of forest succession in southwest New Hampshire. *Ecology*, 55, 772–783.

Heslop-Harrison, J. (1964). Forty years of genecology. *Advances of Ecological Research*, 2, 159–247.

Higgs, E. S. (1997). What is good ecological restoration? *Conservation Biology*, 11(2), 338–348.

Higman, S., Mayers, J., Bass, S., Judd, N., and Nussbaum, R. (2005). *Sustainable forestry handbook*, 2nd edition. Earthscan, London.

Hilborn, R. and Mangel, M. (1997). *The ecological detective: Confronting models with data*. Princeton University Press, Princeton, NJ.

Hill, D., Fasham, M., Tucker, G., Shewry, M., and Shaw, P. (eds.) (2005). *Handbook of biodiversity methods. Survey, evaluation and monitoring*. Cambridge University Press, Cambridge.

Hill, M. O. (1979). *TWINSPAN—A FORTRAN programme for arranging multivariate data in an ordered two-way table by classification of individuals and attributes*. Cornell University, Ithaca, NY.

Hill, M. O., Bunce, R. G. H., and Shaw, M. W. (1975). Indicator species analysis, a divisive polythetic method of classification and its application to a survey of native pinewoods in Scotland. *Journal of Ecology*, 63, 597–613.

Hill, R. A. (1999). Image segmentation for humid tropical forest classification in Landsat TM data. *International Journal of Remote Sensing*, 20, 1039–1044.

Hill, R. A. and Foody, G. M. (1994). Separability of rain-forest types in Tambopata-Candamo reserved zone, Peru. *International Journal of Remote Sensing*, 15, 2687–2693.

Hinckley, T. M., Sprugel, D. G., Batista, J. L. F. *et al.* (1996). Use of the JABOWA family of individual-tree based models for exploration of forest responses to global climate change. *NCASI Technical Bulletin No. 717*, II1–II54.

Hinsley, S. A., Hill, R. A., Gaveau, D. L. A., and Bellamy, P. E. (2002). Quantifying woodland structure and habitat quality for birds using airborne laser scanning. *Functional Ecology*, 16, 851–857.

Hirata, Y., Akimaya, Y., Saito, H., Miyamoto, A., Fukuda, M., and Nishizono, T. (2003). Estimating forest canopy structure using helicopter-borne Lidar measurement. In *Advances *in forest inventory for sustainable forest management and biodiversity monitoring*, eds. P. Corona, M. Köhl, and M. Marchetti, pp. 125–134. Kluwer, Dordrecht.

Hirzel, A. H., Hausser, J., and Perrin, N. (2001a). *BIOMAPPER 1.0—A new software to compute habitat-suitability maps*. Laboratory for Conservation Biology, University of Lausanne, Switzerland ⟨www.unil.ch/biomapper⟩.

Hirzel, A. H., Helfer, V., and Metral, F. (2001b). Assessing habitat suitability models with a virtual species. *Ecological Modelling*, 145, 111–121.

Hirzel, A., Hausser, J., and Perrin, N. (2002). Ecological-niche factor analysis: How to compute habitat-suitability maps without absence data? *Ecology*, 83(7), 2027–2036.

Hobbs, J. E., Currall, J. E., and Gimingham, C. H. (1984). The use of 'thermocolor' pyrometers in the study of heath fire behaviour. *Journal of Ecology*, 72, 241–250.

Hobbs, R. J. and Norton D. A. (1996). Towards a conceptual framework for restoration ecology. *Restoration Ecology*, 4(2), 93–110.

Hochberg, M. E., Menaut, J. C., and Gignoux, J. (1994). The influences of tree biology and fire in the spatial structure of the West African savannah. *Journal of Ecology*, 82, 217–226.

Hockings, M., Stolton, S., and Dudley, N. (2000). *Evaluating effectiveness. A framework for assessing the management of protected areas*. World Commission on Protected Areas. Best Practice Protected Area Guidelines no. 6. IUCN, Gland.

Hodgson, J. G., Wilson, P. J., Hunt, R., Grime, J. P., and Thompson, K. (1999). Allocating C-S-R plant functional types: a soft approach to a hard problem. *Oikos*, 85, 282–294.

Hohn, M. E., Liebhold, A. M., and Gribko, L. S. (1993). A geostatistical model for forecasting the spatial dynamics of defoliation caused by the gypsy moth, *Lymantria dispar* (Lepidoptera: Lymantriidae). *Environmental Entomology*, 22, 1066–1077.

Holland, J. and Campbell, J. (eds.) (2005). *Methods in development research. Combining qualitative and quantitative approaches*. ITDG, Rugby.

Holmgren, J., Joyce, S., Nilsson, M., and Olsson, H. (1999). Estimating stem volume and basal area in forest compartments by combining satellite image data with field data. *Scandinavian Journal of Forest Research*, 15, 103–111.

Holmgren, P. and Thuresson, T. (1998). Satellite remote sensing for forestry planning—a review. *Scandinavian Journal of Forest Research*, 13S, 90–110.

Holvoet, B. and Muys, B. (2004). Sustainable forest management worldwide: a comparative assessment of standards. *International Forestry Review*, 6(2), 99–122.

Hope, J. C. E. (2003). *Modelling forest landscape dynamics in Glen Affric, northern Scotland*. PhD thesis, University of Stirling.

Horler, C. N. H. and Ahern, F. J. (1986). Forestry information content of Thematic Mapper data. *International Journal of Remote Sensing*, 7, 405–428.

Horn, D. J. (1988). *Ecological approach to pest management*. Elsevier, London.

Horn, H. S. (1975). Markovian properties of forest succession. In *Ecology and evolution of communities*, eds. M. L. Cody and J. M. Diamond, pp. 196–211. Harvard University Press, Cambridge, MA.

Horn, H. S. (1976). Succession. In *Theoretical ecology*, ed. R. M. May, pp. 187–204. Blackwell Scientific, Oxford.

Horn, H. S., Nathan, R., and Kaplan, S. R. (2001). Long-distance dispersal of tree seeds by wind. *Ecological Research*, 16, 877–885.

Horning, N. (2004). *Remote sensing guides*. Center for Biodiversity and Conservation. American Museum of Natural History ⟨http://cbc.rs-gis.amnh.org/guides/⟩.

Houle, A. C., Chapman, C., and Vickery, W. L. (2004). Tree climbing strategies for primate ecological studies. *International Journal of Primatology*, 25, 237–260.

House, S. M. (1989). Pollen movement to flowering canopies of pistillate individuals of three rain forest tree species in tropical Australia. *Australian Journal of Ecology*, 14(1), 77–93.

Hubbell, S. P. (2001). *The unified neutral theory of biodiversity and biogeography.* Princeton University Press, Princeton, NJ.

Hubbell, S. P. and Foster, R. B. (1986). Biology, chance and history, and the structure of tropical tree communities. In *Community ecology*, ed. J. M. Diamond and T. J. Case, pp. 314–324. Harper & Row, New York.

Hubbell, S.P., Foster, R.B., O'Brien, S.T. et al. (1999). Light-gap disturbances, recruitment limitation, and tree diversity in a neotropical forest. *Science*, 283(5401), 554–557.

Hudson, W. D. (1991). Photo interpretation of montane forests in the Dominican Republic. *Photogrammetric Engineering and Remote Sensing*, 57, 79–84.

Hughes, J. W., Fahey, T. J., and Brown, B. (1987). A better seed and litter trap. *Canadian Journal of Forest Research*, 17, 1623–1624.

Humphrey, J., Ferris, R., and Peace, A. (2003a). The biodiversity assessment project: objectives, site selection and survey methods. In *Biodiversity in Britain's planted forests. Results from the Forestry Commission's Biodiversity Assessment Project*, eds. J. Humphrey, R. Ferris, and C. Quine, pp. 11–18. Forestry Commission, Edinburgh.

Humphrey, J., Ferris, R., and Quine, C. (eds.) (2003b). *Biodiversity in Britain's planted forests. Results from the Forestry Commission's Biodiversity Assessment Project*, Forestry Commission, Edinburgh.

Humphrey, J., Newton, A., Latham, J. et al. (eds.) (2003c). *The restoration of wooded landscapes*. Forestry Commission, Edinburgh.

Hurlbert, S. H. (1984). Pseudoreplication and the design of ecological field experiments. *Ecological Monographs*, 54, 187–211.

Hunt, R. (1978). *Plant growth analysis*. Edward Arnold, London.

Hunt, R. and Parsons, I. T. (1974). A computer program for deriving growth functions in plant growth-analysis. *Journal of Applied Ecology*, 11, 297–307.

Hunter, M. L. (ed.) (1999). *Maintaining biodiversity in forest ecosystems*. Cambridge University Press, Cambridge.

Hurford, C. and Schneider, M. (2006). *Monitoring nature conservation in cultural habitats: a practical guide and case studies*. Springer-Verlag, Berlin.

Husch, B., Beers, T. W., and Kershaw, J. A. (2003). *Forest mensuration*. Wiley, New York.

Hutchings, M. J. (1986). Plant population biology. In *Methods in plant ecology*, eds. P. D. Moore and S. B. Chapman, pp. 377–435. Blackwell Scientific, Oxford.

Hutchings, M. J., Booth, K. D., and Waite, S. (1991). Comparison of survivorship by the logrank test: criticisms and alternatives. *Ecology*, 72, 2290–2293.

Huth, A. and Ditzer, T. (2000). Simulation of the growth of a lowland Dipterocarp rain forest with FORMIX3. *Ecological Modelling*, 134(1), 1–25.

Huth, A., Ditzer, T., and Bossel, H. (1998). *The rain forest growth model FORMIX3—Model description and analysis of forest growth and logging scenarios for the Deramakot forest reserve (Malaysia)*, Erich Goltze, Göttingen.

Hyman, J. B. and Leibowitz, S. G. (2001). JSEM: a framework for identifying and evaluating indicators. *Environmental Monitoring and Assessment*, 66, 207–232.

Hyyppä, J., Hyyppä, H., Inkinen, M., Engdahl, M., Linko, S., and Zhu, Y. H. (2000). Accuracy comparison of various remote sensing data sources in the retrieval of forest stand attributes. *Forest Ecology and Management*, 128, 109–120.

Imbernon, J. and Branthomme, A. (2001). Characterization of landscape patterns of deforestation in tropical rain forests. *International Journal of Remote Sensing*, 22, 1753–1765.

Ingram, J. C., Dawson, T. P., and Whittaker, R. J. (2005). Mapping tropical forest structure in southeastern Madagascar using remote sensing and artificial neural networks. *Remote Sensing of Environment*, 94(4), 491–507.

Innes, J. L. (1993). *Forest health: Its assessment and status*. CABI, Wallingford.

Isaaks, E. H. and Srivastava, R. M. (1989). *Applied geostatistics*. Oxford University Press, Oxford.
ITTO (1990). *Guidelines for the sustainable management of natural tropical forests*. Policy Development Series No.1, ITTO, Yokohama.
ITTO (1992). *Criteria for the measurement of sustainable tropical forest management*. Policy Development Series No.3, ITTO, Yokohama.
IUCN (1994). *Guidelines for protected area management categories*. IUCN and the World Conservation Monitoring Centre, Gland.
IUCN (2001). *Red List categories and criteria. Version 3.1*. IUCN Species Survival Commission, Gland.
IUCN, PROFOR, and World Bank (2004). *Ecosystem approaches and sustainable forest management*. A discussion paper for the UNFF Secretariat, 28 February 2004. IUCN, Gland.
IUCN/WWF International. (1999). *Forest Quality. An Introductory Booklet*. IUCN/WWF International, Gland.
Iverson, L. R., Yaussy, D. A., Rebbeck, J., Hutchinson, T. F., Long, R. P., and Prasad, A. M. (2004). A comparison of thermocouples and temperature paints to monitor spatial and temporal characteristics of landscape-scale prescribed fires. *International Journal of Wildland Fire*, 13(3), 1–12.
Izhaki, I., Walton, P. B., and Safriel, U. N. (1991). Seed shadows generated by frugivorous birds in an eastern Mediterranean scrub. *Journal of Ecology*, 79, 575–590.
Jack, W. H. and Saville, P. S. (1973). The causes of tattering of flags under natural conditions. *International Journal of Biometeorology*, 17, 185–192.
Jackson, L. J., Trebitz, A. S., and Cottingham, K. L. (2000). A simulation primer for ecological modelling. *BioScience*, 50, 694–706.
Jacoby, P. W., Ainsley, R. J., and Trevino, B. A. (1992). An improved method for measuring temperatures during range fires. *Journal of Range Management*, 45, 216–220.
Janzen, D. H. (1970). Herbivores and the number of tree species in tropical forests. *American Naturalist*, 104, 501–528.
Jennings, M. D. (2000). Gap analysis: concepts, methods, and recent results. *Landscape Ecology*, 15, 5–20.
Jennings, S. B., Brown, A. G., and Sheil, D. (1999). Assessing forest canopies and understorey illumination: canopy closure, canopy cover and other measures. *Forestry*, 72(1), 59–73.
Jennings, S. B., Brown, N. D., Boshier, D. H., Whitmore, T. C., and Lopes, J. do. C. A. (2001). Ecology provides a pragmatic solution to the maintenance of genetic diversity in sustainably managed tropical rain forests. *Forest Ecology and Management*, 154, 1–10.
Jennings, S., Nussbaum, R., Judd, N., and Evans, T. (2003). *The high conservation value forest toolkit*. Edition 1. Proforest, Oxford.
Jermy, A. C., Long, D., Sands, M. J. S., Stork, N. E., and Winser, S. (eds.) (1995). *Biodiversity assessment: A guide to good practice*. Department of the Environment/HMSO, London.
Johansson, T. (1985). Estimating canopy density by the vertical tube method. *Forest Ecology and Management*, 11, 139–144.
Johns, A. D. (1992). Species conservation in managed tropical forests. In *Tropical Deforestation and Species Extinction*, eds. T. C. Whitmore and J. A. Sayer, pp. 15–53. IUCN Forest Conservation Programme, Chapman & Hall, London.
Johnsen, K. L., Johnsen, K., Samuelson, L., Teskey, R., McNulty, S., and Fox, T. (2001). Process models as tools in forestry research and management. *Forest Science*, 47(1), 2–8.
Johnson, E. A. (1992). *Fire and vegetation dynamics: Studies from the North American boreal forest*. Cambridge University Press, Cambridge.
Johnson, E. W. (2001). *Forest sampling desk reference*. CRC Press, Boca Raton, FL.
Johnson, N. C., Milk, A. J., Sari R. C., and Sexton, W. T. (1999). *Ecological stewardship: a common reference for ecosystem management*. Elsevier Science, Oxford.

Johnston, C. A. (1998). *Geographic information systems in ecology*. Blackwell Science, Oxford.
Jongman, R. H. G., ter Braak, C. J. F., and van Tongeren, O. F. R. (eds.) (1995). *Data analysis in community and landscape ecology*. Cambridge University Press, Cambridge.
Jones, F. A., Chen, J., Weng, J., and Hubbell, S. P. (2005). A genetic evaluation of seed dispersal in the neotropical tree *Jacaranda copaia* (Bignoniaceae). *American Naturalist*, 166(5), 543–555.
Jordan, G. J., Ducey, M. J., and Gove, J. H. (2004). Comparing line-intersect, fixed-area, and point relascope sampling for dead and downed coarse woody material in a managed northern hardwood forest. *Canadian Journal of Forest Research*, 34(8), 1766–1775.
Jordan, W. R. (1994). 'Sunflower Forest': ecological restoration as the basis for a new environmental paradigm. In *Beyond preservation: Restoring and inventing landscapes*, eds. A. D. Baldwin, J. de Luce, and C. Pletsch, pp. 17–34. University of Minnesota Press, Minneapolis, MN.
Jordan, W., Gilpin, M., and Aber, J. (eds.) (1987). *Restoration ecology: A synthetic approach to ecological research*. Cambridge University Press, Cambridge.
Jorge, L. A. B. and Garcia, G. J. (1997). A study of habitat fragmentation in Southern Brazil using remote sensing and geographic information systems (GIS). *Forest Ecology and Management*, 98, 35–47.
Jorgensen, A. F. and Nohr, H. (1996). The use of satellite images for mapping landscape and biological diversity. *International Journal of Remote Sensing*, 17, 99–109.
Kammesheidt, L., Köhler, P., and Huth, A. (2001). Sustainable timber harvesting in Venezuela: a modelling approach. *Journal of Applied Ecology*, 38, 756–770.
Katul, G. G., Porporato, A., Nathan, R. *et al.* (2005). Mechanistic analytical models for long-distance seed dispersal by wind. *American Naturalist*, 166(3), 368–381.
Kearns, C. A. and Inouye, D. W. (1993). *Techniques for pollination biologists*. University Press of Colorado, Niwot, CO.
Keddy, P. A. and Drummond, C. G. (1996). Ecological properties for the evaluation, management, and restoration of temperate deciduous forest ecosystems. *Ecological Applications*, 6(3), 748–762.
Keeley, J. E. and Fotheringham, C. J. (2005). Plot shape effects on plant species diversity measurements. *Journal of Vegetation Science*, 16(2), 249–256.
Kellomäki, S. and Väisänen, H. (1991). Application of a gap model for the simulation of forest ground vegetation in boreal conditions. *Forest Ecology and Management*, 42, 35–47.
Kenkel, N. C., Juhász-Nagy, P., and Podani, J. (1989). On sampling procedures in population and community ecology. *Vegetatio*, 83, 195–207.
Kent, M. and Coker, P. (1992). *Vegetation description and analysis*. Wiley, Chichester.
Kerr, J. T. and Ostrovsky, M. (2003). From space to species: ecological applications for remote sensing. *Trends in Ecology and Evolution*, 18(6), 299–305.
Kershaw, K. A. and Looney, J. H. H. (1985). *Quantitative and dynamic plant ecology*, 3rd edition. Edward Arnold, London.
Kesteven, J. L., Brack, C. L., and Furby, S. L. (2003). Using remote sensing and a spatial plant productivity model to assess biomass change. In *Advances in forest inventory for sustainable forest management and biodiversity monitoring*, eds. P. Corona, M. Köhl, and M. Marchetti, pp. 33–56. Kluwer, Dordrecht.
Kilkki, P. and Päivinen, R. (1987). Reference sample plots to combine field measurements and satellite data in forest inventories. *University of Helsinki, Department of Forest Mensuration and Management, Research Notes*, 19, 209–215.
Kimmins, J. P. (1997). *Forest ecology. A foundation for sustainable management*. 2nd edition. Prentice Hall, Englewood Cliffs, NJ.
Kindt, R. and Coe, R. (2005). *Tree diversity analysis. A manual and software for common statistical methods for ecological and biodiversity studies*. World Agroforestry Centre (ICRAF), Nairobi.

References

Kint, V., Robert, D. W., and Noel, L. (2004). Evaluation of sampling methods for the estimation of structural indices in forest stands. *Ecological Modelling*, 180(4), 461–476.

Kleinn, C. (2003). New technologies and methodologies for national forest inventories. *Unasylva*, 53, 10–15.

Kleinn, C. and Traub, B. (2003). Describing landscape pattern by sampling methods. In *Advances in forest inventory for sustainable forest management and biodiversity monitoring*, eds. P. Corona, M. Köhl, and M. Marchetti, pp. 175–189. Kluwer, Dordrecht.

Knight, F. B. and Heikkenen, H. J. (1980). *Principles of forest entomology*, 5th edition. McGraw-Hill, New York.

Kobe, R. K. (1999). Light gradient partitioning among tropical tree species through differential seedling mortality and growth. *Ecology*, 80, 187–201.

Kobe, R. K., Pacala, S. W., Silander, J. A., and Canham, C. D. (1995). Juvenile tree survivorship as a component of shade tolerance. *Ecological Applications*, 5, 517–532.

Koenig, W. D., Knops, J. M. H., Carmen, W. J., Stanback, M. T., and Mumme, R. L. (1994a). Estimating acorn crops using visual surveys. *Canadian Journal of Forest Research*, 24, 2105–2112.

Koenig, W. D., Mumme, R. L., Carmen, W. J., and Stanback, M. T. (1994b). Acorn production by oaks in central coastal California: variation within and among years. *Ecology*, 75, 99–109.

Köhl, M. and Gertner, G. Z. (1997). Geostatistics in evaluating forest damage surveys: considerations on methods for describing spatial distributions. *Forest Ecology and Management*, 95, 131–140.

Köhl, M. and Lautner, M. (2001). Erfassung von Waldökosystemen durch Hyperspektraldaten, Photogrammetrie-Fernerkundung. *Geoinformation*, 2, 107–117.

Köhler, P. and Huth, A. (1998). The effect of tree species grouping in tropical rain forest modelling—simulation with the individual based model Formind. *Ecological Modelling*, 109, 301–321.

Kohyama, T. (1993). Size-structured tree populations in gap-dynamic forest: the forest architecture hypothesis for the stable coexistence of species. *Journal of Ecology*, 81, 131–143.

Kohyama, T. (2005). Scaling up from shifting gap mosaic to geographic distribution in the modeling of forest dynamics. *Ecological Research*, 20, 302–312.

Kohyama, T. and Shigesada, N. (1995). A size-distribution-based model of forest dynamics along a latitudinal environmental gradient. *Vegetatio*, 121, 117–126.

Kollmann, J. and Goetze, D. (1998). Notes on seed traps in terrestrial plant communities. *Flora*, 193, 31–40.

Koukal, T. and Schneider, W. (2003). Mapping and monitoring of tree resources outside the forest in Central America. In *Advances in forest inventory for sustainable forest management and biodiversity monitoring*, eds. P. Corona, M. Köhl, and M. Marchetti, pp. 313–323. Kluwer, Dordrecht.

Koutsias, N. and Karteris, M. (2000). Burned area mapping using logistic regression modeling of a single post-fire Landsat-5 Thematic Mapper image. *International Journal of Remote Sensing*, 21, 673–687.

Krebs, C. J. (1999). *Ecological methodology*, 2nd edition. Addison-Wesley, Menlo Park, CA.

Kremer, A., Caron, H., Cavers, S. *et al.* (2005). Monitoring genetic diversity in tropical trees with multilocus dominant markers. *Heredity*, 95, 274–280.

Kushla, J. D. and Ripple, W. J. (1998). Assessing wildfire effects with Landsat Thematic Mapper data. *International Journal of Remote Sensing*, 19, 2493–2507.

Kwit, C., Horvitz, C. C., and Platt, W. J. (2004). Conserving slow-growing, long-lived tree species: input from the demography of a rare understory conifer, *Taxus floridana*. *Conservation Biology*, 18(2), 432–443.

Lachowski, H., Maus, P., and Roller, N. (2000). From pixels to decisions: digital remote sensing technologies for public land managers. *Journal of Forestry*, 98, 13–15.

Lacy, R. C. (1993). VORTEX: a computer simulation model for population viability analysis. *Wildlife Research*, 20(1), 45–65.

LaDeau, S. and Clark, J. S. (2001). Rising CO_2 and the fecundity of forest trees. *Science*, 292, 95–98.

Laffan, M., Jordan, G., and Duhig, N. (2001). Impacts on soils from cable-logging steep slopes in northeastern Tasmania, Australia. *Forest Ecology and Management*, 144(1–3), 91–99.

Lamb, D. and Gilmour, D. (2003). *Rehabilitation and restoration of degraded forests*. IUCN/WWF, Gland.

Lambers, H. and Poorter, H. (1992). Inherent variation in growth rate between higher plants: a search for physiological causes and ecological consequences. *Advances in Ecological Research*, 23, 187–261.

Lambert, R. C., Lang, G. E., and Reiners, W. A. (1980). Loss of mass and chemical change in decaying boles of a subalpine balsam fir forest. *Ecology*, 61, 1460–1473.

Lambin, E. F. (1999). Monitoring forest degradation in tropical regions by remote sensing: some methodological issues. *Global Ecology and Biogeography*, 8(3–4), 191–198.

Lambin, E., Turner, B., Geist, H. *et al.* (2001). The causes of land-use and land-cover change: moving beyond the myths. *Global Environmental Change*, 11, 261–269.

Lambin, E. F., Gesit, H. J., and Lepers, E. (2003). Dynamics of land-use and land-cover change in tropical regions. *Annual Review of Environment and Resources*, 28, 205–241.

Landsberg, J. (2003). Modelling forest ecosystems: state of the art, challenges, and future directions. *Canadian Journal of Forest Research*, 33(3), 385–397.

Landsberg, J. J. and Gower, S. T. (1997). *Applications of physiological ecology to forest management*. Academic Press, San Diego, CA.

Langlet, O. (1971). Two hundred years of genecology. *Taxon*, 20, 653–722.

Lanquaye-Opoku, N. and Mitchell, S. J. (2005). Portability of stand-level empirical windthrow risk models. *Forest Ecology and Management*, 216, 134–148.

Larsen, D. R. and Kershaw, J. A. (1990). The measurement of leaf area. In *Techniques in forest tree ecophysiology*, eds. J. Lassoie and T. Hinkley, pp. 465–475. CRC Press, Boca Raton, FL.

Larson, A. L. (1971). *Two-way thermogradient plates for seed germination research: construction plans and procedures*. USDA Agricultural Research Service Report 51–41. US Government Printing Office, Washington DC.

Larsson, B., Bengtsson, B., and Gustafsson, M. (2004). Nondestructive detection of decay in living trees. *Tree Physiology*, 24, 853–858.

Latham, P. A., Zuuring, H. R., and Coble, D. W. (1998). A method for quantifying vertical forest structure. *Forest Ecology and Management*, 104, 157–170.

Laurance, W. F. and Yensen, E. (1991). Predicting the impacts of edge effects in fragmented habitats. *Biological Conservation*, 55, 77–92.

Laurance, W. F., Ferreira, L. V., Rankin-de Merona, J. M., and Hutchings, R. W. (1998a). Influence of plot shape on estimates of tree diversity and community composition in central Amazonia. *Biotropica*, 30(4), 662–665.

Laurance, W. F., Ferreira, L. V., Rankin-De Merona, J. M., and Laurance, S. G. (1998b). Rain forest fragmentation and the dynamics of Amazonian tree communities. *Ecology*, 79, 2032–2040.

Leck, M. A., Parker, V. T., and Simpson, R. L. (eds.) (1989). *Ecology of soil seed banks*. Academic Press, San Diego, CA.

Leckie, D. G., Yuan, X., Ostaff, D. P., Piene, H., and MacLean, D. A. (1992). Analysis of high resolution multispectral MEIS imagery for spruce budworm damage assessment on a single tree basis. *Remote Sensing of Environment*, 40, 125–136.

Ledig, F. T. (1988). The conservation of diversity in forest trees: why and how should genes be conserved? *BioScience*, 38, 471–479.

Lee, E. T. (1992). *Statistical methods for survival data analysis*. Wiley, New York.
Lefsky, M. A., Cohen, W. B., Acker, S. A., Parker, G. G., Spies, T. A., and Harding, D. (1999a). Lidar remote sensing of canopy structure and biophysical properties of Douglas-fir western hemlock forests. *Remote Sensing of Environment*, 70, 339–361.
Lefsky, M. A., Harding, D., Cohen, W. B., Parker, G., and Shugart, H. H. (1999b). Surface lidar remote sensing of basal area and biomass in deciduous forests of eastern Maryland, USA. *Remote Sensing of Environment*, 67, 83–98.
Lefsky, M. A., Cohen, W. B., Parker, G. G., and Harding, D. J. (2002). Lidar remote sensing for ecosystem studies. *BioScience*, 52, 19–30.
Legendre, P. and Legendre, L. (1998) *Numerical ecology*, 2nd edition. Developments in Environmental Modelling. Elsevier, Amsterdam.
Legg, C. J. and Nagy, L. (2006). Why most conservation monitoring is, but need not be, a waste of time. *Journal of Environmental Management*, 78, 194–199.
Leishman, M. R. and Westoby, M. (1992). Classifying plants into groups on the basis of associations of individual traits—evidence from Australian semi-arid woodlands. *Journal of Ecology*, 80, 417–424.
Lemmon, P. E. (1956). A spherical densiometer for estimating overstorey density. *Forest Science*, 2, 314–320.
Leslie, A. D. (2004). The impacts and mechanics of certification. *International Forestry Review*, 6(1), 30–39.
Levin, S. A., Muller-Landau, H. C., Nathan, R., and Chave, J. (2003). The ecology and evolution of seed dispersal: a theoretical perspective. *Annual Review of Ecology, Evolution, and Systematics*, 34, 575–604.
Lewis, S. L., Phillips, O. L., Sheil, D. *et al.* (2004). Tropical forest tree mortality, recruitment and turnover rates: calculation, interpretation and comparison when census intervals vary. *Journal of Ecology*, 92(6), 929–944.
Lewontin, R. C. (1972). The apportionment of human diversity. *Evolutionary Biology*, 6, 381–398.
Li, H. and Reynolds, J. F. (1993). A new contagion index to quantify spatial patterns of landscapes. *Landscape Ecology*, 8, 155–162.
LI-COR Inc. (1992). *LAI-2000 plant canopy analyzer operating manual*. LI-COR Inc., Lincoln, NE.
Lieberman, D., Lieberman, M., Peralta, R., and Hartshorn, G. S. (1985). Mortality patterns and stand turnover rates in a wet tropical forest in Costa Rica. *Journal of Ecology*, 73, 915–924.
Lieberman, M., Lieberman, D., and Peralta, R. (1989). Forests are not just Swiss cheese: canopy stereogeometry of non-gaps in tropical forests. *Ecology*, 70(3), 550–552.
Lillesand, T. M. and Kiefer, R. W. (1994). *Remote sensing and image interpretation*, 3rd edition. Wiley, New York.
Lindenmayer, D. B. and Franklin, J. F. (2002). *Conserving forest biodiversity. A comprehensive multiscaled approach*. Island Press, Washington, DC.
Lindenmayer, D. B., Burgman, M. A., Akçakaya, H. R., Lacy, R. C., and Possingham, H. P. (1995). A review of the generic computer programmes ALEX, RAMAS/space and VORTEX for modeling the viability of wildlife populations. *Ecological Modelling*, 82, 161–174.
Lindenmayer, D. B., Cunningham, R. B., Pope, M. L., Gibbons, P., and Donnelly, C. F. (2000). Cavity sizes and types in Australian eucalypts from wet and dry forest types—a simple rule of thumb for estimating size and number of cavities. *Forest Ecology and Management*, 137, 139–150.
Lindner, M., Sievänen, R., and Pretzsch, H. (1997). Improving the simulation of stand structure in a forest gap model. *Forest Ecology and Management*, 95, 183–195.
Lines, R. and Howell, R. S. (1963). *The use of flags to estimate the relative exposure of trial plantations*. Forestry Commission Forest Record 51. HMSO, London.

Liu, J. G. and Ashton, P. S. (1995). Individual-based simulation-models for forest succession and management. *Forest Ecology and Management*, 73(1–3), 157–175.

Liu, J. and Ashton, P. S. (1998). FORMOSAIC: an individual-based spatially explicit model for simulating forest dynamics in landscape mosaics. *Ecological Modelling*, 106, 177–200.

Liu, Q.-H. and Hytteborn, H. (1991). Gap structure, disturbance and regeneration in a primeval *Picea abies* forest. *Journal of Vegetation Science*, 2(3), 391–402.

Lobo, A. (1997). Image segmentation and discriminant analysis for the identification of land cover units in ecology. *IEEE Transactions on Geoscience and Remote Sensing*, 35, 1136–1145.

Lockwood, M., Warboys, G., and Kothari, A. (2006). *Managing protected areas. A global guide*. IUCN/Earthscan, London.

Löfstrand, R., Folving, S., Kennedy, P. *et al.* (2003). Habitat characterization and mapping for umbrella species—an integrated approach using satellite and field data. In *Advances in forest inventory for sustainable forest management and biodiversity monitoring*, eds. P. Corona, M. Köhl, and M. Marchetti, pp. 191–204. Kluwer, Dordrecht.

Longino, J. T., Coddington, J., and Colwell, R. K. (2002). The ant fauna of a tropical rain forest: estimating species richness three different ways. *Ecology*, 83, 689–702.

Longley, P. A., Goodchild, M. F., Maguire, D. J., and Rhind, D. W. (2005). *Geographical information systems and science*, 2nd edition. Wiley, New York.

López-Barrera, F. and Newton, A. C. (2005). Edge type effect on germination of oak tree species in the Highlands of Chiapas, Mexico. *Forest Ecology and Management*, 217(1), 67–79.

López-Barrera, F., Newton, A. C., and Manson, R. (2005). Edge effects in a tropical montane forest mosaic: experimental tests of post-dispersal acorn removal. *Ecological Research*, 20(1), 31–40.

López-Barrera, F., Manson, R. H., Gonzalez-Espinosa, M., and Newton, A. C. (2006a). Effects of varying forest edge permeability on seed dispersal in a neotropical montane forest. *Landscape Ecology*. In press.

López-Barrera, F., Manson, R. H., Gonzalez-Espinosa, M., and Newton, A. C. (2006b). Effects of the type of montane forests edge on oak seedling establishment along forest–edge–exterior gradients. *Forest Ecology and Management*, 225(1–3), 234–244.

Lorimer, C. G. (1985). Methodological considerations in the analysis of forest disturbance history. *Canadian Journal of Forest Research*, 15, 200–213.

Loveless, M. D. (1992). Isozyme variation in tropical trees: patterns of genetic organisation. *New Forests*, 6, 67–94.

Lowe, A., Harris, S., and Ashton, P. (2004). *Ecological genetics. Design, analysis and application*. Blackwell, Oxford.

Lowe, A. J., Boshier, D., Ward, M., Bacles, C. F. E., and Navarro, C. (2005). Genetic resource impacts of habitat loss and degradation; reconciling empirical evidence and predicted theory for neotropical trees. *Heredity*, 95, 255–273.

Lowell, K. E., Edwards, G., and Kucera, G. (1996). Modelling the heterogeneity and change of natural forests. *Geomatica*, 50, 425–440.

Lowman, M. D. and Wittman, P. K. (1996). Forest canopies: methods, hypotheses, and future directions. *Annual Review of Ecology and Systematics*, 27, 55–81.

Lü, Y., Chen, L., Fu, B., and Liu, S. (2003). A framework for evaluating the effectiveness of protected areas: the case of Wolong Biosphere Reserve. *Landscape and Urban Planning*, 63(4), 213–223.

Ludwig, D. (1999). Is it meaningful to estimate a probability of extinction? *Ecology*, 80(1), 298–310.

Lynch, M. and Milligan, B. G. (1994). Analysis of population genetic structure with RAPD markers. *Molecular Ecology*, 3, 91–99.

MacArthur, R. H. and MacArthur, J. W. (1961). On bird species diversity. *Ecology*, 42(3), 594–598.

Mace, G. M., Balmford, A., Boitani, L. *et al.* (2000). It's time to work together and stop duplicating conservation efforts. *Nature*, 405, 393.

Machado, J.-L. and Reich, P. B. (1999). Evaluation of several measures of canopy openness as predictors of photosynthetic photon flux density in deeply shaded conifer-dominated forest understory. *Canadian Journal of Forest Research*, 29, 1438–1444.

Mack, A. L. (1995). Distance and non-randomness of seed dispersal by the dwarf cassowary *Casuarius bennetti*. *Ecography*, 18, 286–295.

Madany, M. H., Swetnam, T. W., and West, N. E. (1982). Comparison of two approaches for determining fire dates from tree scars. *Forest Science*, 28, 856–861.

Magurran, A. E. (2004). *Measuring biological diversity*. Blackwell, Oxford.

Mäkelä, A., Landsberg, J., Ek, A. R. *et al.* (2000). Process-based models for forest ecosystem management: current state of the art and challenges for practical implementation. *Tree Physiology*, 20(5–6), 289–298.

Mancke, R. G. and Gavin, T. A. (2000). Breeding bird density in woodlots: effects of depth and buildings at the edges. *Ecological Applications*, 10, 598–611.

Manel, S., Dias, J.-M., and Ormerod, S. J. (1999). Comparing discriminant analysis, neural networks and logistic regression for predicting species distributions: a case study with a Himalayan river bird. *Ecological Modelling*, 120, 337–347.

Manley, B. F. J. (1992). *The design and analysis of research studies*. Cambridge University Press, Cambridge.

Mansourian, S. (2005). Overview of forest restoration strategies and terms. In *Forest restoration in landscapes. Beyond planting trees*, eds. S. Mansourian, D. Vallauri, and N. Dudley, pp. 8–13. Springer, New York.

Mansourian, S., Vallauri, D., and Dudley, N. (eds.) (2005). *Forest restoration in landscapes. Beyond planting trees*. Springer, New York.

Margoluis, R. and Salafsky, N. (1998). *Measures of success: Designing, managing, and monitoring conservation and development projects*. Island Press, Washington, DC.

Margules, C. R. and Pressey, R. L. (2000). Systematic conservation planning. *Nature*, 405, 243–253.

Marren, P. (2002). *Nature conservation*. HarperCollins, London

Marshall, E., Schreckenberg, K., and Newton, A. C. (eds.) (2006). *Commercialization of non-timber forest products: Factors influencing success. Lessons learned from Mexico and Bolivia and policy implications for decision-makers*. UNEP World Conservation Monitoring Centre, Cambridge.

Martínez-Ramos, M. and Alvarez-Buylla, E. R. (1999). How old are tropical forest trees? *Trends in Plant Sciences*, 3, 400–405.

Maser, C., Anderson, R. G., Cromack, K. Jr., Williams, J. T., and Martin, R. E. (1979). Dead and down woody material. In *Wildlife habitats in managed forests: The Blue Mountains of Oregon and Washington*, ed. J. W. Thomas, pp. 79–85. USDA Forest Service and Agriculture Handbook 553, Washington, DC.

Matlack, G. R. (1989). Secondary dispersal of seed across snow in *Betula lenta*, a gap-colonising tree species. *Journal of Ecology*, 77, 858–869.

Matlack, G. R. and Litvaitis, J. A. (1999). Forest edges. In *Maintaining biodiversity in forest ecosystems*, ed. M. L. Hunter, pp. 210–233. Cambridge University Press, Cambridge.

Matthews, E. (2001). *Understanding the FRA 2000*. Forest Briefing No. 1. World Resources Institute, Washington, DC.

Matthews, J. D. (1989). *Silvicultural systems*. Clarendon Press, Oxford.

Mayaux, P., Holmgren, P., Achard, F., Eva, H., and Stibig, H. (2005). Tropical forest cover change in the 1990s and options for future monitoring. *Philosophical Transactions of the Royal Society of London*, B360(1454), 373–384.

Mayers, J. and Bass, S. (2004). *Policy that works for forests and people: Real prospects for governance and livelihoods*. Earthscan, Sterling, VA.

Mayhew, J. and Newton, A. C. (1998). *Silviculture of mahogany*. CABI Bioscience, Wallingford.

McCallum, H. (2000). *Population parameters. Estimation for ecological models*. Blackwell Science, Oxford.

McCarthy, M. A., Possingham, H. P., Day, J. R., and Tyre, A. J. (2001). Testing the accuracy of population viability analysis. *Conservation Biology*, 15, 1030–1038.

McComb, W. and Lindenmayer, D. (1999). Dying, dead, and down trees. In *Maintaining biodiversity in forest ecosystems*, ed. M. L. Hunter, pp. 335–372. Cambridge University Press, Cambridge.

McCracken, J. A., Pretty, J. N., and Conway, G. R. (1988). *An introduction to rapid rural appraisal for agricultural development*. International Institute for Environment and Development, London.

McDonald, R. I., Peet, R. K., and Urban, D. L. (2003). Spatial pattern of *Quercus* regeneration limitation and *Acer rubrum* invasion in a Piedmont forest. *Journal of Vegetation Science*, 14, 441–450.

McElhinny, C., Gibbons, P., Brack, C., and Bauhus, J. (2005). Forest and woodland stand structural complexity: its definition and measurement. *Forest Ecology and Management*, 218, 1–24.

McEuen, A. B. and Curran, L. M. (2005). Plant recruitment bottlenecks in temperate forest fragments: seed limitation and insect herbivory. *Plant Ecology*, 184(2), 297–309.

McGarigal, K. (2002). Landscape pattern metrics. In *Encyclopedia of environmetrics*, Volume 2A, eds. H. El-Shaarawi and W. W. Piegorsch, pp. 1135–1142. Wiley, Chichester.

McGarigal, K. and Marks, B. (1995). *Fragstats: spatial pattern analysis program for quantifying landscape structure (Fragstats NT Version 2.0)*. USDA Forest Service General Technical report PNW-GTR-351, Portland, OR.

McGarigal, K., Cushman, S. A., Neel, M. C., and Ene, E. (2002). *Fragstats: Spatial pattern analysis program for categorical maps*. University of Massachusetts, Landscape Ecology Program ⟨www.umass.edu/landeco/research/fragstats/fragstats.html⟩

McKenzie, D. H., Hyatt, D. E., and McDonald, V. J. (eds.) (1992). *Ecological indicators*. Vols. 1, 2. Elsevier Applied Science, London.

McNab, W. H., Browning, S. A., Simon, S. A., and Fouts, P. E. (1999). An unconventional approach to ecosystem unit classification in western North Carolina, USA. *Forest Ecology and Management*, 114, 405–420.

McPeek, M. A. and Kalisz, S. (1993). Population sampling and bootstrapping in complex designs: demographic analysis. In *Design and analysis of ecological experiments*, 2nd edition, eds. S. M. Scheiner and J. Gurevitch, pp. 232–252. Oxford University Press, New York.

McRoberts, N., Finch, R. P., Sinclair, W. *et al.* (1999). Assessing the ecological significance of molecular diversity data in natural plant populations. *Journal of Experimental Botany*, 50, 1635–1655.

Medhurst, J. L. and Beadle, C. L. (2002). Sapwood hydraulic conductivity and leaf area-sapwood area relationships following thinning of a *Eucalyptus nitens* plantation. *Plant, Cell and Environment*, 25, 1011–1019.

Meir, E., Andelman, S., and Possingham, H. P. (2004). Does conservation planning matter in a dynamic and uncertain world? *Ecology Letters*, 7, 615–622.

Menges, E. S. (1990). Population viability analysis for an endangered plant. *Conservation Biology*, 4(1), 52–62.

Menges, E. S. (2000). Population viability analyses in plants: challenges and opportunities. *Trends in Ecology and Evolution*, 15, 51–56.

Mergen, F. (1954). Mechanical aspects of windbreak and windfirmness. *Journal of Forestry*, 52, 119–125.

Mertens, B. and Lambin, E. F. (1997). Spatial modelling of deforestation in southern Cameroon: spatial disaggregation of diverse deforestation processes. *Applied Geography*, 17, 143–162.
Messier, C. and Puttonen, P. (1995). Spatial and temporal variation in the light environment of developing Scots pine stands: the basis for a quick and efficient method of characterizing light. *Canadian Journal of Forest Research*, 25, 343–354
Michener, W. K. and Houhoulis, P. F. (1997). Detection of vegetation changes associated with extensive flooding in a forested ecosystem. *Photogrammetric Engineering and Remote Sensing*, 63, 1363–1374.
Middleton, B. A. (1995). Sampling devices for the measurement of seed rain and hydrochory in rivers. *Bulletin of the Torrey Botanical Club*, 122, 152–155.
Miles, L., Newton, A. C., DeFries, R., Kapos, V., Blyth, S., and Gordon, J. (2006). A global overview of the conservation status of tropical dry forests. *Journal of Biogeography*, 33, 491–505.
Millar, C. I. and Libby, W. J. (1991). Strategies for conserving clinal, ecotypic and disjunct population diversity in widespread species. In *Genetics and conservation of rare plants*, eds. D. A. Falk and K. E. Holsinger, pp. 149–170. Oxford University Press, New York.
Millington, A. C., Velez-Liendo, X. M., and Bradley, A. V. (2003). Scale dependence in multi-temporal mapping of forest fragmentation in Bolivia: implications for explaining temporal trends in landscape ecology and applications to biodiversity conservation. *Photogrammetric Engineering and Remote Sensing*, 57, 289–299.
Mills, L. S., Doak, D. F., and Wisdom, M. J. (1999). Reliability of conservation actions based on elasticity analysis of matrix models. *Conservation Biology*, 13, 815–829
Mitchell, A. W. (1982). *Reaching the rain forest roof. A handbook on techniques of access and study in the canopy*. Leeds Literary and Philosophical Society, Leeds.
Mitchell, A. W. (1986). *The enchanted canopy*. Collins, Glasgow.
Mitchell, D. (2004). *Cloud atlas*. Hodder & Stoughton, London.
Mittermeier, R. A., Gil, P. R., Hoffman, M. *et al.* (2004). *Hotspots revisited: Earth's biologically richest and most threatened terrestrial ecoregions*. Conservation International, Washington, DC.
Mladenoff, D. J. (2004). LANDIS and forest landscape models. *Ecological Modelling*, 180, 7–19.
Mladenoff, D. J. and Baker, W. L. (1999). Development of forest modeling approaches. In *Spatial modelling of forest landscape change: Approaches and applications*, eds. D. J. Mladenoff and W. L. Baker, pp.1–13. Cambridge University Press, Cambridge.
Mladenoff, D. J. and He, H. S. (1999). Design and behavior of LANDIS, an object-oriented model of forest landscape disturbance and succession. In *Spatial modelling of forest landscape change: Approaches and applications*, eds. D. J. Mladenoff and W. L. Baker, pp. 125–162. Cambridge University Press, Cambridge.
Mladenoff, D. J. and Pastor, J. (1993). Sustainable forest ecosystems in the northern hardwood and conifer forest region: concepts and management. In *Defining sustainable forestry*, eds. G. H. Aplet, N. Johnson, J. T. Olson, and V. A. Sample, pp. 145–180. Island Press, Washington, DC.
Moffett, M. W. (1993). *The High Frontier*. Harvard University Press, Cambridge, MA.
Moilanen, A., Franco, A. M. A., Eary, R. I., Fox, R., Wintle, B., and Thomas, C. D. (2005). Prioritizing multiple-use landscapes for conservation: methods for large multi-species planning problems. *Proceedings of the Royal Society of London*, B272(1575), 1885–1891.
Moisen, G. G. and Frescino, T. S. (2002). Comparing five modeling techniques for predicting forest characteristics. *Ecological Modelling*, 157, 209–225.
Molofsky, J. and Fisher, B. L. (1993). Habitat and predation effects on seedling survival and growth in shade-tolerant tropical trees. *Ecology*, 74, 261–265.
Moloney, K. (1986). A generalised algorithm for determining category size. *Oecologia*, 69, 176–180.

Monserud, R. A. (2003). Evaluating forest models in a sustainable forest management context. *Forest Biometry, Modelling and Information Sciences*, 1, 35–47.

Monteil, C., Deconchat, M., and Balent, G. (2005). Simple neural network reveals unexpected patterns of bird species richness in forest fragments. *Landscape Ecology*, 20(5), 513–527.

Moore, A. D. and Noble, I. R. (1990). An individualistic model of vegetation stand dynamics. *Environmental Management*, 31, 61–81.

Moore, J. L., Folkmann, M., Balmford, A. *et al.* (2003). Heuristic and optimal solutions for set-covering problems in conservation biology. *Ecography*, 26(5), 595–601.

Moore, M. M., Covington, W. W., and Fulé, P. Z. (1999). Reference conditions and ecological restoration: a southwestern ponderosa pine perspective. *Ecological Applications*, 9(4), 1266–1277.

Moore, P. D. and Chapman, S. B. (eds.) (1986). *Methods in plant ecology*, 2nd edition. Blackwell Scientific, Oxford.

Morisita, M. (1954). Estimation of population density by spacing method. *Memoirs of the Faculty of Science, Kyushu University*, Series E 1, 187–197.

Morisita, M. (1957). A new method for the estimation of density by the spacing method applicable to nonrandomly distributed populations. *Physiology and Ecology*, 7, 134–144.

Moritz, C. (1994). Applications of mitochondrial DNA analysis in conservation: a critical review. *Molecular Ecology*, 3, 401–411.

Moritz, C. (1995). Uses of molecular phylogenies for conservation. *Philosophical Transactions of the Royal Society of London*, B349, 113–118.

Morrison, M. L., Marcot, B. G., and Mannan, R. W. (1992). *Wildlife-habitat relationships: Concepts and applications*. University of Wisconsin Press, Madison, WI.

Mueller-Dombois, D. and Ellenberg, H. (1974). *Aims and methods of vegetation ecology*. Wiley, New York.

Mukai, Y. and Hasegawa, I. (2000). Extraction of damaged areas of windfall trees by typhoons using Landsat TM data. *International Journal of Remote Sensing*, 20, 2703–2721.

Munro, D. D. (1974). Forest growth models: A prognosis. In *Growth models for tree and stand simulation*, ed. J. Fries, pp. 7–21. Research Notes 30, Department of Forest Yield Research, Royal College of Forestry, Stockholm.

Myers, N. (1988). Threatened biotas: 'Hotspots' in tropical forests. *The Environmentalist*, 8, 1–20.

Myers, N. (1990). The biodiversity challenge: Expanded hot-spots analysis. *The Environmentalist*, 10, 243–256.

Myers, N. (2003). Biodiversity hotspots revisited. *BioScience*, 53, 916–917.

Myers, N., Mittermeier, R. A., Mittermeier, C. G., da Fonseca, G. A. B., and Kent, J. (2000). Biodiversity hotspots for conservation priorities. *Nature*, 403, 853–858.

Naef-Daenzer, B., Früh, D., Stalder, M., Wetli, P., and Weise, E. (2005). Miniaturization (0.2g) and evaluation of attachment techniques of telemetry transmitters. *Journal of Experimental Biology*, 208, 4063–4068.

Naesset, E. (1997). Estimating timber volume of forest stands using airborne laser scanner data. *Remote Sensing of Environment*, 61, 246–253.

Naesset, E. and Bjerknes, K.-O. (2001). Estimating tree heights and number of stems in young stands using airborne laser scanner data. *Remote Sensing of Environment*, 78, 328–340.

Nagendra, H. (2001). Using remote sensing to assess biodiversity. *International Journal of Remote Sensing*, 22(12), 2377–2400.

Nagendra, H. and Gadgil, M. (1999). Biodiversity assessment at multiple scales: Linking remotely sensed data with field information. *Proceedings of the National Academy of Sciences of the USA*, 96(16), 9154–9158.

Namkoong, G., Boyle, T., Gregorius, H.-R. *et al.* (1996). *Testing criteria and indicators for assessing the sustainability of forest management: genetic criteria and indicators*. Working Paper No. 10, CIFOR, Bogor, Indonesia.

Nanos, N., Gonzalez-Martinez, S. C., and Bravo, F. (2004). Studying within-stand structure and dynamics with geostatistical and molecular marker tools. *Forest Ecology and Management*, 189(1–3), 223–240.

Nathan, R., Perry, G., Cronin, J. T., Strand, A. E., and Cain, M. L. (2003). Methods for estimating long-distance dispersal. *Oikos*, 103, 261–273.

National Research Council (1991). *Managing global genetic resources. Forest trees.* National Academy Press, Washington, DC.

Nei, M. (1973). Analysis of gene diversity in subdivided populations. *Proceedings of the National Academy of Sciences of the USA*, 70, 3321–3323.

Nelson, B. W., Ferreira, C., Da Silva, M., and Kawasaki, M. (1990). Endemism centers, refugia and botanical collection density in Brazilian Amazonia. *Nature*, 345, 714–716.

Newbery, D. M., Alexander, I. J., Thomas, D. W., and Gartlan, J. S. (1988). Ectomycorrhizal rainforest legumes and soil phosphorus in Korup National Park, Cameroon. *New Phytologist*, 109, 433–450.

Newstrom, L. E., Frankie, G. W., and Baker, H. G. (1994). A new classification for plant phenology based on flowering patterns in lowland tropical rain forest trees at La Selva, Costa Rica. *Biotropica*, 26(2), 141–159.

Newton, A. C. (1998). Restoration of native woodland in southern Scotland: the Carrifran Wildwood project. *Restoration and Management Notes*, 16(2), 212–213.

Newton, A. C. and Ashmole, P. (1998). How may native woodland be restored to southern Scotland? *Scottish Forestry*, 52, 168–171.

Newton, A. C. and Kapos, V. (2002). Biodiversity indicators in national forest inventories. *Unasylva*, 53, 56–64.

Newton, A. C. and Oldfield, S. (2005). Forest policy, the precautionary principle and sustainable forest management. In: *Biodiversity and the precautionary principle. Risk and uncertainty in conservation and sustainable use*, eds. R. Cooney and B. Dickson, pp. 21–38. Earthscan, London.

Newton, A. C. and Pigott, C. D. (1991). Mineral nutrition and mycorrhizal infection of seedling oak and birch. II. The effect of application of fertilizers on growth, mineral uptake and mycorrhizal infection. *New Phytologist*, 117, 45–52.

Newton, A. C., Cornelius, J. P., Mesén, J. F., and Leakey, R. R. B. (1995). Genetic variation in apical dominance of *Cedrela odorata* seedlings in response to decapitation. *Silvae Genetica*, 44(2–3), 146–150.

Newton, A. C., Allnutt, T., Gillies, A. C. M., Lowe, A., and Ennos, R. A. (1999a). Molecular phylogeography, intraspecific variation and the conservation of tree species. *Trends in Ecology and Evolution*, 14(4), 140–145.

Newton, A. C., Watt, A. D., Lopez, F., Cornelius, J. P., Mesén, J. F., and Corea, E. (1999). Genetic variation in host susceptibility to attack by the mahogany shoot borer, *Hypsipyla grandella* (Zeller). *Agriculture and Forest Entomology*, 1, 11–18.

Newton, A. C., Stirling, M., and Crowell, M. (2001). Current approaches to native woodland restoration in Scotland. *Botanical Journal of Scotland*, 53(2), 169–196.

Niemeijer, D. (2002). Developing indicators for environmental policy: data-drive and theory-drive approaches examined by example. *Environmental Science and Policy*, 5, 91–103.

Niemi, G. J. and McDonald, M. E. (2004). Application of ecological indicators. *Annual Review of Ecology, Evolution and Systematics*, 35, 89–111.

Nilsson, M., Folving, S., Kennedy, P. *et al.* (2003). Combining remote sensing and field data for deriving unbiased estimates of forest parameters over large regions. In *Advances in forest inventory for sustainable forest management and biodiversity monitoring*, eds. P. Corona, M. Köhl, and M. Marchetti, pp. 19–31. Kluwer, Dordrecht.

Nix, H. A., Faith, D. P., Hutchinson, M. F. *et al.* (2000). *The BioRap toolbox: A national study of biodiversity assessment and planning for Papua New Guinea.* Consultancy Report to the World

Bank. Centre for Resource and Environmental Studies, Australian National University, Canberra.

Noble, I. R. and Slatyer, R. O. (1980). The use of vital attributes to predict successional changes in plant communities subject to recurrent disturbance. *Vegetatio*, 43, 5–21.

Norman, J. M. and Campbell, G. S. (1989). Canopy structure. In *Plant Physiological Ecology*, eds. R. W. Pearcy, J. R. Ehleringer, H. A. Mooney, and P. W. Rundel, pp. 301–325. Chapman & Hall, London.

Normand, F., Habib, R., and Chaduf, J. (2002). A stochastic flowering model describing an asynchronically flowering set of trees. *Annals of Botany*, 90, 405–415.

Noss, R. F. (1990). Indicators for monitoring biodiversity: a hierarchical approach. *Conservation Biology*, 4(4), 355–364.

Noss, R. F. (1999). Assessing and monitoring forest biodiversity: a suggested framework and indicators. *Forest Ecology and Management*, 115, 135–146.

Noy-Meir, I. (1975). Stability of grazing systems: an application of predator graphs. *Journal of Ecology*, 63, 459–481.

Nussbaum, R. and Simula, M. (2005). *The forest certification handbook*. Earthscan, London.

Nybom, H. and Bartish, I. V. (2000). Effects of life history traits and sampling strategies on genetic diversity estimates obtained with RAPD markers in plants. *Perspectives in Plant Ecology, Evolution and Systematics*, 3(2), 93–114.

O'Connor, S., Salafsky, N., and Salzer, D. W. (2005). Monitoring forest restoration projects in the context of an adaptive management cycle. In *Forest restoration in landscapes. Beyond planting trees*, eds. S. Mansourian, D. Vallauri, and N. Dudley, pp. 145–149. Springer, New York.

OECD (1993). *OECD Core Set of Indicators for Environmental Performance Reviews*. Environmental Monograph No 83, OECD, Paris.

Okubo, A. and Levin, S. A. (1989). A theoretical framework for data analysis of wind dispersal of seeds and pollen. *Ecology*, 70, 329–338.

Olden, J. D. (2003). A species-specific approach to modeling biological communities and its potential for conservation. *Conservation Biology*, 17(3), 854–863.

Oldfield, S., Lusty, C., and MacKinven, A. (1998). *The world list of threatened trees*. World Conservation Press, WCMC, Cambridge.

Oliver, C. D. and Larson, B. C. (1996). *Forest stand dynamics*. Wiley, New York.

Oliver, C. D. and Stephens, E. P. (1977). Reconstruction of a mixed forest in central New England. *Ecology*, 58, 562–572.

Oliver, I. (1996). Rapid biodiversity assessment and its application to invertebrate conservation in production forests. *Australian Journal of Ecology*, 21, 349–350.

Oliver, I. and Beattie, A. J. (1996). Designing a cost-effective invertebrate survey: a test of some methods for the rapid assessment of invertebrate biodiversity. *Ecological Applications*, 6(2), 594–607.

Olmsted, I. and Alvarez-Buylla, E. (1995). Sustainable harvesting of tropical forest trees: demography and matrix models of two palm species in Mexico. *Ecological Applications*, 5, 484–500.

Olson, D. M. and Dinerstein, E. (1998). The global 200: a representation approach to conserving the Earth's most biologically valuable ecoregions. *Conservation Biology*, 12(3), 502–515.

Olson, D. M., Dinerstein, E., Abell, R. *et al.* (2000). The global 200: A representation approach to conserving the earth's distinctive ecoregions. Conservation Science Program, World Wildlife Fund-US, Washington, DC.

Olson, D. M., Dinerstein, E., Wikramanayake, E. D. *et al.* (2001). Terrestrial ecoregions of the world: a new map of life on Earth. *BioScience*, 51, 933–938.

Olthof, I. and King, D. J. (2000). Development of a forest health index using multispectral airborne digital camera imagery. *Canadian Journal of Remote Sensing*, 26, 166–176.

O'Neill, R. V., Krummel, J. R., Gardner, R. H. *et al.* (1988). Indices of landscape pattern. *Landscape Ecology*, 1, 153–162.

O'Neill, R. V., Hunsaker, C. T., Timmins, S. P., Jackson, B. L., Jones, K. B., and Riitters, K. H. (1996). Scale problems in reporting landscape pattern at the regional scale. *Landscape Ecology*, 11, 169–180.

Oosterhoorn, M. and Kappelle, M. (2000). Vegetation structure and composition along an interior-edge-exterior gradient in a Costa Rican montane cloud forest. *Forest Ecology and Management*, 126, 291–307.

Osawa, A., Shoemaker, C. A., and Stedinger, J. R. (1983). A stochastic model of balsam fir bud phenology utilizing maximum likelihood parameter estimation. *Forest Science*, 29, 478–490.

Osunkoya, O. O., Ash, J. E., Hopkins, M. A., and Graham, A. W. (1994). Influence of seed size and seedling ecological attributes on shade-tolerance of rain-forest tree species in northern Queensland. *Journal of Ecology*, 82, 149–163.

Ouborg, N. J., Piquot, Y., and Van Groenendael, J. M. (1999). Population genetics, molecular markers and the study of dispersal in plants. *Journal of Ecology*, 87, 551–568.

Pacala, S. W., Canham, C. D., and Silander Jr., J. A. (1993). Forest models defined by field measurements: I. The design of a northeastern forest simulator. *Canadian Journal of Forest Research*, 23, 1980–1988.

Pacala, S. W., Canham, C. D., Saponara, J., Silander, J. A., Kobe, R. K., and Ribbens, E. (1996). Forest models defined by field measurement: estimation, error analysis and dynamics. *Ecological Monographs*, 66, 1–43.

Paciorek, C. J., Condit, R., Hubbell, S. P., and Foster, R. B. (2000). The demographics of resprouting in tree and shrub species of a moist tropical forest. *Journal of Ecology*, 88(5), 765–777.

Page, B. G. and Thompson, W. T. (1997). *The insecticide, herbicide, fungicide quick guide*. Thompson, Fesno, CA.

Paine, R. T. (1980). Food webs: linkage interaction strength and community infrastructure. *Journal of Animal Ecology*, 49, 667–685.

Palmer, J. and Synnott, T. J. (1992). The management of natural forests. In *Managing the world's forests*, ed. N. P. Sharma, pp. 337–374. Kendall/Hunt, Dubuque, IA.

Parent, S. and Messier, C. (1996). A simple and efficient method to estimate microsite light availability under a forest canopy. *Canadian Journal of Forest Research*, 26, 151–154.

Parker, G. G. and Brown, M. J. (2000). Forest canopy stratification—is it useful? *The American Naturalist*, 155(4), 473–484.

Parker, K. R. (1979). Density estimation by variable area transect. *Journal of Wildlife Management*, 43(2), 484–492.

Parker, P. G., Snow, A. A., Schug, M. D., Booton, G. C., and Fuerst, P. A. (1998). What molecules can tell us about populations: choosing and using a molecular marker. *Ecology*, 79, 361–382.

Parmenter, A. W., Hansen, A., Kennedy, R. E. *et al.* (2003). Land use and land cover change in the Greater Yellowstone Ecosystem: 1975–1995. *Ecological Applications*, 13(3), 687–703.

Parrado-Rosselli, A., Machado, J.-L., and Tatiana, P.-L. (2006).Comparison between two methods for measuring fruit production in a tropical forest. *Biotropica*, 38(2), 267–271.

Patenaude, G., Milne, R., and Dawson, T. P. (2005). Synthesis of remote sensing approaches for forest carbon estimation: reporting to the Kyoto Protocol. *Environmental Science and Policy*, 8(2), 161–178.

Patil, G. P., Burnham, K. P., and Kovner, J. L. (1979). Nonparametric estimation of plant density by the distance method. *Biometrics*, 35, 597–604.

Payandeh, B. and Ek, A. R. (1986). Distance methods and density estimators. *Canadian Journal of Forest Research*, 16, 918–924.

Pearcy, R. W., Ehleringer, J. R., Mooney, H., and Rundel, P. W. (eds.) (1990). *Plant physiological ecology: Field methods and instrumentation*. Springer-Verlag, New York.

Peck, S. L. (2000). A tutorial for understanding ecological modeling papers for the nonmodeler. *American Entomologist*, 46, 40–49.

Pekkarinen, A. and Tuominen, S. (2003). Stratification of a forest area for multisource forest inventory by means of aerial photographs and image segmentation. In *Advances in forest inventory for sustainable forest management and biodiversity monitoring*, eds. P. Corona, M. Köhl, and M. Marchetti, pp. 111–123. Kluwer, Dordrecht.

Peng, C. (2000). Growth and yield models for uneven-aged stands: past, present and future. *Forest Ecology and Management*, 132, 259–279.

Pennanen, J. and Kuuluvainen, T. (2002). A spatial simulation approach to the natural forest landscape dynamics in boreal Fennoscandia. *Forest Ecology and Management*, 164, 157–175.

Pereira, J. M. C., Tomé, M., Carreiras, J. M. B. *et al.* (1997). Leaf area estimation from tree allometrics in *Eucalyptus globulus* plantations. *Canadian Journal of Forest Research*, 27, 166–173.

Peres, C. A., Baider, C., Zuidema, P. A. *et al.* (2003). Demographic threats to the sustainability of Brazil nut exploitation. *Science*, 302, 2112–2114.

Perrow, M. and Davy, A. (eds.) (2002). *Handbook of ecological restoration*. Cambridge University Press, Cambridge.

Petchey, O. L. and Gaston, K. J. (2002a). Extinction and the loss of functional diversity. *Proceedings of the Royal Society of London*, B269, 1721–1727.

Petchey, O. L. and Gaston, K. J. (2002b). Functional diversity (FD), species richness and community composition. *Ecology Letters*, 5, 402–411.

Peterken, G. F. (1996). *Natural woodland. Ecology and conservation in northern temperate regions*. Cambridge University Press, Cambridge.

Peters, C. M. (1994) *Sustainable harvest of non-timber plant resources in tropical moist forest: an ecological primer*. Biodiversity Support Programme, World Wildlife Fund, Washington DC.

Peters, C. M. (1996). *The ecology and management of non timber forest resources*. Technical Paper 322. World Bank, Washington, DC.

Peters, R. H. (1991). *A critique for ecology*. Cambridge University Press, Cambridge.

Peterson, A. T., Ortega-Huerta, M. A., Bartley, J. *et al.* (2002). Future projections for Mexican faunas under global climate change scenarios. *Nature*, 416, 626–629.

Peterson, C. J. (2000). Damage and recovery of tree species after two different tornadoes in the same old growth forest: a comparison of infrequent wind disturbances. *Forest Ecology and Management*, 135, 237–252.

Peterson, C. J. and Rebertus, A. J. (1997). Tornado damage and initial recovery in three adjacent, low-land temperate forests in Missouri. *Journal of Vegetation Science*, 8, 559–564.

Peterson, G. D., Cumming, G. S., and Carpenter, S. R. (2003). Scenario planning: a tool for conservation in an uncertain world. *Conservation Biology*, 17(2), 358–366.

Petit, R. J., Aguinagalde, I., de Beaulieu, J.-L. *et al.* (2003). Glacial refugia: hotspots but not melting pots of genetic diversity. *Science*, 300, 1563–1565.

Philip, M. S. (1994). *Measuring trees and forests*, 2nd edition. CABI, Wallingford.

Phillips, O. L. and Gentry, A. H. (1994). Increasing turnover through time in tropical forests. *Science*, 263, 954–958.

Phillips, O. L., Hall, P., Gentry, A. H., Sawyer, S. A., and Vásquez, M. R. (1994). Dynamics and species richness of tropical rainforests. *Proceedings of the National Academy of Sciences of the USA*, 91, 2805–2809.

Phillips, O. L., Malhi, Y., Higuchi, N. *et al.* (1998). Changes in the carbon balance of tropical forest: evidence from long-term plots. *Science*, 282, 439–442.

Phillips, O. L., Malhi, Y., Vinceti, B *et al.* (2002a). Changes in growth of tropical forests: evaluating potential biases. *Ecological Applications*, 12, 576–587.

Phillips, O. L., Vásquez, M. R., Arroyo, L. *et al.* (2002b). Increasing dominance of large lianas in Amazonian forests. *Nature*, 418, 770–774.

Phillips, O. L., Vásquez Martínez, R., Núñez Vargas, P. *et al.* (2003). Efficient plot-based floristic assessment of tropical forests. *Journal of Tropical Ecology* 19, 629–645.

Pickett, S. T. A. and White, P. S. (eds.) (1985). *The ecology of natural disturbance and patch dynamics.* Academic Press, New York.

Pickett, S. T. A., Kolasa, J., and Jones, C. G. (1994). *Ecological understanding.* Academic Press, San Diego, CA.

Pigott, C. D., Newton, A. C., and Zammit, S. (1991). Predation of acorns and oak seedlings by grey squirrel. *Quarterly Journal of Forestry*, 85, 173–178.

Pillar, V. D. and Sosinski, Jr., E. E. (2003). An improved method for searching plant functional types by numerical analysis. *Journal of Vegetation Science*, 14, 323–332.

Pinard, M. (1993). Impacts of stem harvesting on populations of *Iriartea deltoidea* (Palmae) in an extractive reserve in Acre, Brazil. *Biotropica*, 25(1), 2–14.

Pontius, R. G., Cornell, J., and Hall, C. (2001). Modelling the spatial pattern of land-use change with GEOMOD2: application and validation for Costa Rica. *Agriculture, Ecosystems and Environment*, 85, 191–203.

Poorter, L. (1999). Growth responses of 15 rainforest tree species to a light gradient: the relative importance of morphological and physiological traits. *Functional Ecology*, 13, 396–410.

Poorter, L. (2001). Light-dependent changes in biomass allocation and their importance for growth of rain forest tree species. *Functional Ecology*, 15, 113–123.

Poorter, L., Bongers, F., Kouame, F. N., and Hawthorne, W. D. (eds.) (2004). *Biodiversity of West African forests. An ecological atlas of woody plant species.* CABI, Wallingford.

Porte, A. and Bartelink, H. H. (2002). Modelling mixed forest growth: a review of models for forest management. *Ecological Modelling*, 150(1–2), 141–188.

Poso, S., Wang, G., and Tuominen, S. (1999). Weighting alternative estimates when using multi-source auxiliary data for forest inventory. *Silva Fennica*, 33, 41–50.

Possingham, H. P. and Davies, I. (1995). ALEX: A population viability analysis model for spatially structured populations. *Biological Conservation*, 73, 143–150.

Possingham, H. P., Ball, I. R., and Andelman, S. (2000). Mathematical methods for identifying representative reserve networks. In *Quantitative methods for conservation biology*, eds. S. Ferson and M. Burgman, pp. 291–305. Springer, New York.

Prabhu, R., Colfer, C. J. P., Venkateswarlu, P., Tan, L. C., Soekmadi, R., and Wollenberg, E. (1996). *Testing criteria and indicators for the sustainable management of forests: phase I. Final report.* CIFOR, Jakarta, Indonesia.

Premoli, A. C., Souto, C. P., Allnut, T. R., and Newton, A. C. (2001). Effects of population disjunction on isozyme variation in the widespread *Pilgerodendron uviferum. Heredity*, 87(3), 337–343.

Premoli, A. C., Souto, C. P., Rovere, A. E., Allnut, T. R., and Newton, A. C. (2002). Patterns of isozyme variation as indicators of biogeographic history in *Pilgerodendron uviferum* (D. Don) Florín. *Diversity and Distributions*, 8, 57–66.

Premoli, A. C., Vergara, R., Souta, C. P., Lara, A., and Newton, A. C. (2003). Lowland valleys shelter the ancient conifer *Fitzroya cupressoides* in the Central Depression of southern Chile. *Journal of the Royal Society of New Zealand*, 33(3), 623–631.

Pressey, R. L. and Nicholls, A. O. (1989). Efficiency in conservation evaluation: scoring versus iterative approaches. *Biological Conservation*, 50, 199–218.

Pressey, R. L., Ferrier, S., Hager, T. C., Woods, C. A., Tully, S. L., and Weinman, K. M. (1996). How well protected are the forests of north-eastern New South Wales? Analyses of forest environments in relation to formal protection measures, land tenure and vulnerability to clearing. *Forest Ecology and Management*, 85, 311–333.

Pressey, R. L., Possingham, H. P., and Day, J. R. (1997). Effectiveness of alternative heuristic algorithms for identifying indicative minimum requirements for conservation reserves. *Biological Conservation*, 80, 207–219.

Pressey, R. L., Possingham, H. P., Logan, V. S., Day, J. R., and Williams, P. H. (1999). Effects of data characteristics on the results of reserve selection algorithms. *Journal of Biogeography*, 26, 179–191.

Pressey, R. L., Whish, G. L., Barrett, T. W., and Watts, M. E. (2002). Effectiveness of protected areas in north-eastern New South Wales: recent trends in six measures. *Biological Conservation*, 106(1), 57–69.

Pretty, J. N., Guijt, I., Thompson, J., and Scoones, I. (1995). *Participatory learning and action. A trainer's guide*. IIED, London ⟨www.iied.org/sarl/pubs/particmethod.html⟩.

Price, M. (1986). The analysis of vegetation change by remote sensing. *Progress in Physical Geography*, 10, 473–491.

Price, M. V. and Reichman, O. J. (1987). Distribution of seeds in Sonoran desert soils: implications for heteromyid rodent foraging. *Ecology*, 68, 1797–1811.

Price, M. V. and Waser, N. M. (1979). Pollen dispersal and optimal outcrossing in *Delphinium nelsonii*. *Nature*, 277, 294–296.

Prins, E. and Kikula, I. S. (1996). Deforestation and regrowth phenology in miombo woodland assessed by Landsat Multispectral Scanner System data. *Forest Ecology and Management*, 84, 263–266.

Pullin, A. S. and Knight, T. M. (2001). Effectiveness in conservation practice: pointers from medicine and public health. *Conservation Biology*, 15, 50–54.

Pullin, A. S. and Knight, T. M. (2003). Support for decision making in conservation practice: an evidence-based approach. *Journal for Nature Conservation*, 11, 83–90.

Pullin, A. S., Knight, T. M., Stone, D. A., and Charman, K. (2004). Do conservation managers use scientific evidence to support their decision-making? *Biological Conservation*, 119, 245–252.

Putman, R. J. (1986). *Grazing in temperate ecosystems. Large herbivores and the ecology of the New Forest*. Croom Helm, London.

Putman, R. J. (1996). Ungulates in temperate forest ecosystems: perspectives and recommendations for future research. *Forest Ecology and Management*, 88, 205–214.

Puyravaud, J. P. (2003). Standardizing the calculation of the annual rate of deforestation. *Forest Ecology and Management*, 177, 593–596.

Pyle, C. and Brown, M. M. (1999). Heterogeneity of wood decay classes within hardwood logs. *Forest Ecology and Management*, 114, 253–259.

Quine, C. P. (1995). Assessing the risk of wind damage; practice and pitfalls. In: *Wind and trees*, eds. M. P. Coutts and J. Grace, pp. 379–403. Cambridge University Press, Cambridge.

Quine, C. P. and White, I. M. S. (1994). Using the relationship between rate of tatter and topographic variables to predict site windiness in upland Britain. *Forestry*, 67, 245–256.

Rackham, O. (1986). *History of the countryside*. Dent, London.

Rackham, O. (2003). *Ancient woodland: Its history, vegetation and uses in England*. Castlepoint Press, Dalbeattie, Kirkcudbrightshire.

Rametsteiner, E. and Simula, M. (2003). Forest certification—an instrument to promote sustainable forest management? *Journal of Environmental Management*, 67(1), 87–98.

Ramirez-Garcia, P., Lopez-Blanco, J., and Ocana, D. (1998). Mangrove vegetation assessment in the Santiago River Mouth, Mexico, by means of supervised classification using Landsat TM imagery. *Forest Ecology and Management*, 105, 217–229.

Ramsay, E. W. III., Chappell, D. K., Jacobs, D. M., Sapkota, S. K., and Baldwin, D. G. (1998). Resource management of forest wetlands: hurricane impact and recovery mapped by combining Landsat TM and NOAA AVHRR data. *Photogrammetric Engineering and Remote Sensing*, 64, 733–738.

Ranneby, B., Cruse, T., Hägglund, B., Jonasson, H., and Swård, J. (1987). *Designing a new national forest survey for Sweden*. Studia Forestalia Suecica, 177. Swedish University of Agricultural Sciences, Faculty of Forestry, Uppsala.

Ranta, P., Blom, T., Niemela, J., Joensuu, E., and Siitonen, M. (1998). The fragmented Atlantic rain forest of Brazil: size, shape and distribution of forest fragments. *Biodiversity and Conservation*, 7, 385–403.

Rapoport, E. H. (1982). *Aerography*. Pergamon, Oxford.

Read, H. (2000). *Veteran trees: A guide to good management*. English Nature, Peterborough.

Redford, K. H., Coppolillo, P., Sanderson, E. W. *et al.* (2003). Mapping the conservation landscape. *Conservation Biology*, 17(1), 116–131.

Reed, D. D. and Mroz, G. D. (1997). *Resource assessment in forested landscapes*. Wiley, New York.

Reimoser, F., Armstrong, H., and Suchant, R. (1999). Measuring forest damage of ungulates: what should be considered. *Forest Ecology and Management*, 120, 47–58.

Rhodes, M., Wardell-Johnson, G. W., Rhodes, M. P., and Raymond, B. (2006). Applying network analysis to the conservation of habitat trees in urban environments: a case study from Brisbane, Australia. *Conservation Biology*. In press.

Ribbens, E. J., Silander, J. A., and Pacala, S. W. (1994). Seedling recruitment in forests: calibrating models to predict patterns of tree seedling dispersion. *Ecology*, 75, 1794–1806.

Rich, P. M. (1989). *A manual for analysis of hemispherical canopy photography*. Report LA-11733-M, Los Alamos National Laboratory, Los Alamos, NM.

Rich, P. M. (1990). Characterizing plant canopies with hemispherical photographs. *Remote Sensing Reviews*, 5, 13–29.

Rich, P. M., Clark, D. B., Clark, D. A., and Oberbauer, S. F. (1993). Long-term study of solar-radiation regimes in a tropical wet forest using quantum sensors and hemispherical photography. *Agricultural and Forest Meteorology*, 65, 107–127.

Richards, P. W. (1996). *The tropical rain forest*, 2nd edition. Cambridge University Press, Cambridge.

Ricketts, T. H., Dinerstein, E., Olson, D. M. *et al.* (1999). *Terrestrial ecoregions of North America: A conservation assessment*. Island Press. Washington, DC.

Ries, L., Fletcher, R. J., Battin, J., and Sisk, T. D. (2004). Ecological responses to habitat edges: mechanisms, models, and variability explained. *Annual Review of Ecology, Evolution and Systematics*, 35, 491–522.

Rieseberg, L. H. and Swensen, S. M. (1996). Conservation genetics of endangered island plants. In *Conservation genetics: Case histories from nature*, eds. J. C. Avise and J. L. Hamrick, pp. 305–334. Chapman & Hall, New York.

Riitters, K. H., O'Neill, R. V., Hunsaker, C. T. *et al.* (1995). A factor analysis of landscape pattern and structure metrics. *Landscape Ecology*, 10, 23–39.

Ringvall, A. and Ståhl, G. (1999a). Field aspects of line intersect sampling for assessing coarse woody debris. *Forest Ecology and Management*, 119, 163–170.

Ringvall, A. and Ståhl, G. (1999b). On the field performance of transect relascope sampling for assessing downed coarse woody debris. *Scandinavian Journal of Forest Research*, 14, 552–557.

Ripley, B. D. (1981). *Spatial statistics*. Wiley, New York.

Ripley, B. D. (1988). *Statistical inference for spatial processes*. Cambridge University Press, Cambridge.

Ritland, K. (1986). Joint maximum likelihood estimation of genetic and mating structure using open-pollinated progenies. *Biometrics*, 42, 25–43.

Ritland, K. (1990). A series of FORTRAN computer programs for estimating plant mating systems. *Journal of Heredity*, 81, 235–237.

Ritland, K. (2002). Extension of models for the estimation of mating systems using *n* independent loci. *Heredity*, 88, 221–228.

Roberts, D. W. (1996). Landscape vegetation modelling with vital attributes and fuzzy systems theory. *Ecological Modelling*, 90, 175–184.

Roberts, H. A. (1981). Seed banks in soils. *Advances in Applied Biology*, 6, 1–55.

Robertson, A., Newton, A. C., and Ennos, R. A. (2004). Breeding systems and continuing evolution in the endemic *Sorbus* taxa on Arran. *Heredity*, 93(5), 487–495.

Robertson, G. P. (1987). Geostatistics in ecology: interpolating with known variance. *Ecology*, 68, 744–748.

Robinson, D. (2005). Assessing and addressing threats in restoration programmes. In *Forest restoration in landscapes. Beyond planting trees*, eds. S. Mansourian, D. Vallauri, and N. Dudley, pp. 73–77. Springer, New York.

Rodrigues, A. S. L., Andelman, S. J., Bakarr, M. I. *et al.* (2004) Effectiveness of the global protected area network in representing species diversity. *Nature*, 428, 640–643.

Rodwell, J. S. and Patterson, G. S. (1994). *Creating new native woodlands*. Bulletin 112. HMSO, London.

Rorison, I. H. and Robinson, D. (1986). Mineral nutrition. In *Methods in plant ecology*, 2nd edition, eds. P. D. Moore and S. B. Chapman, pp. 145–213. Blackwell Scientific, Oxford.

Roloff, G. J. and Kernohan, B. J. (1999). Evaluating reliability of habitat suitability index models. *Wildlife Society Bulletin*, 27, 973–985.

Román-Cuesta, R. M. and Martínez-Vilalta, J. (2006). Effectiveness of protected areas in mitigating fire within their boundaries: Case study of Chiapas, Mexico. *Conservation Biology*, 20, 1074–1086.

Rosenzweig, M. L. (1995). *Species diversity in space and time*. Cambridge University Press, Cambridge.

Roxburgh, J. R. and Kelly, D. (1995). Uses and limitations of hemispherical photography for estimating forest light environments. *New Zealand Journal of Ecology*, 19(2), 213–217.

Roy, P. S. and Tomar, S. (2000). Biodiversity characterization at landscape level using geospatial modelling technique. *Biological Conservation*, 95(1), 95–109.

Roy, P. S., Padalia, H., Chauhan, N. *et al.* (2005). Validation of geospatial model for biodiversity characterization at landscape level—a study in Andaman & Nicobar Islands, India. *Ecological Modelling*, 185(2–4), 349–369.

Rubino, D. L. and McCarthy, B. C. (2004). Comparative analysis of dendroecological methods used to assess disturbance events. *Dendrochronologia*, 21(3), 97 115.

Rundel, P. W. and Jarrell, W. M. (1989). Water in the environment. In *Plant physiological ecology*, eds. R. W. Pearcy, J. R. Ehleringer, H. A. Mooney, and P. W. Rundel, pp. 29–56. Chapman & Hall, London.

Runkle, J. R. (1981). Gap regeneration in some old-growth forests of the eastern United States. *Ecology*, 62, 1041–1051.

Runkle, J. R. (1982). Patterns of disturbance in some old-growth mesic forests of eastern North America. *Ecology*, 63, 1533–1546.

Runkle, J. R. (1992). *Guidelines and sample protocol for sampling forest gaps*. Gen.Tech.Rep. PNW-GTR-283. US Department of Agriculture, Forest Service, Pacific Northwest Research Station Portland, OR.

Running, S. W., Nemani, R. R., Peterson, D. L. *et al.* (1989). Mapping regional forest evapotranspiration and photosynthesis by coupling satellite data with ecosystem simulation. *Ecology*, 70, 1090–1101.

Rutter, N. (1966). Tattering of flags under controlled conditions. *Agricultural and Forest Meteorology*, 3, 153–165.

Rutter, N. (1968a). Tattering of flags at different sites in relation to wind and weather. *Agricultural and Forest Meteorology*, 5, 163–181.

Rutter, N. (1968b). Geomorphic and tree shelter in relation to surface wind conditions, time of day and season. *Agricultural and Forest Meteorology*, 5, 319–334.

Ryan, T. P. (1997). *Modern regression methods*. Wiley, New York.

Sabersky, R. H., Acosta, A. J., and Hauptman, E. G. (1989). *Fluid flow*, 3rd edition, Macmillan, New York.

Sader, S. A., Waide, R. B., Lawrence, W. T., and Joyce, A. T. (1989). Tropical forest biomass and successional age class relationships to a vegetation index derived from Landsat data. *Remote Sensing of Environment*, 28, 143–156.

Saito, K. (2001). Flames. In *Forest fires. Behaviour and ecological effects*, eds. E. A. Johnson and K. Miyanishi, pp. 11–54. Academic Press, San Diego, CA.

Sala, O. E., Chapin III, F. S., Armesto, J. J. *et al.* (2000). Global biodiversity scenarios for the year 2100. *Science*, 287, 1770–1774.

Salafsky, N., Margoluis, R., and Redford, K. (2001). *Adaptive management: A tool for conservation practitioners*. Biodiversity Support Program, Washington, DC.

Salafsky, N., Margoluis, R., Redford, K., and Robinson, J. (2002). Improving the practice of conservation: a conceptual framework and agenda for conservation science. *Conservation Biology*, 16, 1469–1479.

Salvador, R., Valeriano, J., Pons, X., and Diaz-Delgado, R. (2000). A semi-automatic methodology to detect fire scars in shrubs and evergreen forests with Landsat MSS time series. *International Journal of Remote Sensing*, 21, 655–671.

Sample, V. A. and Cheng, A. S. (2004). *Forest conservation policy: A reference handbook*. ABC Clio, Santa Barbara, CA.

San-Miguel-Ayanz, J., Carlson, J. D., Alexander, M. *et al.* (2003). Current methods to assess fire danger potential. In *Wildland fire danger estimation and mapping—The role of remote sensing data*, ed. E. Chuvieco, pp. 21–61. Series in Remote Sensing Volume 4. World Scientific, Singapore.

Sarathchandra, S. U., Dimenna, M. E., Burch, G. *et al.* (1995). Effects of plant parasitic nematodes and rhizosphere microorganisms on the growth of white clover (*Trifolium repens* L.) and perennial rye grass (*Lolium perenne* L.). *Soil Biology and Biochemistry*, 27, 9–16.

SAS (2002). *SAS online documentation, version 9*. SAS Institute, Cary, NC ⟨www.support.sas.com/⟩.

Schaal, B. A., Leverich, W. J., and Rogstad, S. H. (1991). A comparison of methods for assessing genetic variation in plant conservation biology. In *Genetics and conservation of rare plants*, eds. D. A. Falk and K. E. Holsinger, pp. 123–134. Oxford University Press, Oxford.

Schamberger, M., Farmer, A. H., and Terrell, J. (1982). *Habitat suitability index models: Introduction*. FW/OBS-82/10 USDI Fish and Wildlife Service, Washington DC.

Schirone, B., Leone, A., Mazzoleni, S., and Spada, F. (1990). A new method of survey and data analysis in phenology. *Journal of Vegetation Science*, 2, 27–34.

Schnitzer, S. A., Reich, P. B., Bergner, B., and Carson, W. P. (2002). Herbivore and pathogen damage on grassland and woodland plants: a test of the herbivore uncertainty principle. *Ecology Letters*, 5, 531–539.

Schreuder, H. T., Gregoire, T. G., and Wood, G. B. (1993). *Sampling methods for multi-resource forest inventory*. Wiley, New York.

Schulenberg, T. S., Short, C. A., and Stephenson, P. J., (eds.) (1999). *A biological assessment of Parc National de la Marahoué, Côte d'Ivoire*. RAP Working Papers 13. Conservation International, Washington, DC.

Schume, H., Jost, G., and Katzensteiner, K. (2003). Spatio-temporal analysis of the soil water content in a mixed Norway spruce (*Picea abies* (L.) Karst.)–European beech (*Fagus sylvatica* L.) stand. *Geoderma*, 112(3–4), 273–287.

Schwartz, M. W. (2003). Assessing population viability in long-lived plants. In *Population viability in plants. Conservation, management and modeling of rare plants*, eds. C. A. Brigham and M. A. Schwartz, pp. 239–266. Ecological Studies 165. Springer-Verlag, Berlin.

Schweingruber, F. H. (1988). *Tree rings: Basics and applications of dendrochronology*. Kluwer, Dordrecht.

Seaman, D. E., Griffith, B., and Powerll, R. A. (1998). KERNELHR: a program for estimating animal home ranges. *Wildlife Society Bulletin*, 26, 95–100.

Sedgeley, J. A. and O'Donnell, C. F. J. (1999). Roost selection by the longtailed bat, *Chalinolobus tuberculatus*, in temperate New Zealand rainforest and its implications for the conservation of bats in managed forests. *Biological Conservation*, 88, 261–276.

Shaw, P. J. A. (2003). *Multivariate statistics for the environmental sciences*. Arnold, London.

Sheil, D. (1995). A critique of permanent plot methods and analysis with examples from Budongo Forest, Uganda. *Forest Ecology and Management*, 77, 11–34.

Sheil, D. (1996). Species richness, tropical forest dynamics and sampling: Questioning cause and effect. *Oikos*, 76(3), 587–590.

Sheil, D. (1999). Developing tests of successional hypotheses with size-structured populations, and an assessment using long-term data from a Ugandan rain forest. *Plant Ecology*, 140(1), 117–127.

Sheil, D. (2001a). Long-term observations of rain forest succession, tree diversity and responses to disturbance. *Plant Ecology*, 155(2), 183–199.

Sheil, D. (2001b). Conservation and biodiversity monitoring in the tropics: realities, priorities and distraction. *Conservation Biology*, 15(4), 1179–1182.

Sheil, D. (2003). Growth assessment in tropical trees: large daily diameter fluctuations and their concealment by dendrometer bands. *Canadian Journal of Forest Research*, 33(10), 2027–2035.

Sheil, D. and Ducey, M. (2002). An extreme-value approach to detect clumping and an application to tropical forest gap-mosaic dynamics. *Journal of Tropical Ecology*, 18, 671–686.

Sheil, D. and May, R. M. (1996). Mortality and recruitment rate evaluations in heterogeneous tropical forests. *Journal of Ecology*, 84(1), 91–100.

Sheil, D., Burslem, D. F. R. P., and Alder, D. (1995). The interpretation and misinterpretation of mortality-rate measures. *Journal of Ecology*, 83, 331–333.

Sheil, D., Jennings, S., and Savill, P. (2000). Long-term permanent plot observations of vegetation dynamics in Budongo, a Ugandan rain forest. *Journal of Tropical Ecology*, 16, 765–800.

Sheil, D., Ducey, M. J., and Sidiyasa, K. (2003). A new type of sample unit for the efficient assessment of diverse tree communities in complex forest landscapes. *Journal of Tropical Forest Science*, 15(1), 117–135.

Sheil, D., Salim, A., Chave, J., Vanclay, J., and Hawthorne, W. D. (2006). Illumination–size relationships of 109 coexisting tropical forest tree species. *Journal of Ecology*, 94, 494–507.

Shettigara, V. K. and Sumerling, G. M. (1998). Height determination of extended objects using shadows in SPOT images. *Photogrammetric Engineering and Remote Sensing*, 64, 35–44.

Shiver, B. D. and Borders, B. E. (1996). *Sampling techniques for forest resource inventory*. Wiley, New York.

Shortle, W. C. and Smith, K. T. (1987). Electrical properties and rate of decay in spruce and fir wood. *Phytopathology*, 77, 811–814.

Shugart, H. H. (1984). *A theory of forest dynamics: The ecological emplications of forest succession models*. Springer, New York.

Shugart, H. H. (1997). Plant and ecosystem functional types. In *Plant functional types*, eds. T. M. Smith, H. H. Shugart, and F. I. Woodward, pp. 20–43. Cambridge University Press, Cambridge.

Shugart, H. H. (1998). *Terrestrial ecosystems in changing environments*. Cambridge University Press, Cambridge.

Shugart, H. H. and Smith, T. M. (1996). A review of forest patch models and their application to global change research. *Climatic Change*, 34(2), 131–153.

Shugart, H. H. and West, D. C. (1977). Development of an Appalachian forest succession model and its application to assessment of the impact of the chestnut blight. *Journal of Environmental Management*, 5, 161–179.

Shugart, H. H. and West, D. C. (1980). Forest succession models. *BioScience*, 30, 308–313.

Silvertown, J. W. and Lovett Doust, J. (1993). *Introduction to plant population biology*. Blackwell Scientific, Oxford.

Silvertown, J., Franco, M., and Menges, E. (1996). Interpretation of elasticity matrices as an aid to the management of plant populations for conservation. *Conservation Biology*, 10, 591–597.

Sisk, T. D., Lanner, A. E., Switky, K. R., and Ehrlich, P. R. (1994). Identifying extinction threats: global analyses of the distribution of biodiversity and the expansion of the human enterprise. *Bioscience*, 44, 592–604.

Sisk, T. D., Haddad, N. M., and Ehrlich, P. R. (1997). Bird assemblages in patchy woodlands: modeling the effects of edge and matrix habitats. *Ecological Applications*, 7, 1170–1180.

Sjögren-Gulve, P. and Hanski, I. (2000). Metapopulation viability analysis using occupancy models. *Ecological Bulletins*, 48, 53–71.

Smith, D. M. (1986). *The practice of silviculture*, 8th edition. Wiley, New York.

Smith, J. L. (1986). Evaluation of the effects of photo measurement errors on predictions of stand volume from aerial photography. *Photogrammetric Engineering and Remote Sensing*, 52, 401–410.

Smith, N. J. H., Williams, J. T., Plucknett, D. P., and Talbot, J. P. (1992). *Tropical Forests and their Crops*. Cornell University Press, Ithaca, NY.

Snook, L. K. (1996). Catastrophic disturbance, logging and the ecology of mahogany (*Swietenia macrophylla* King): Grounds for listing a major tropical timber species on CITES. *Botanical Journal of the Linnean Society*, 122(1), 35–46.

Soehartono, T. and Newton, A. C. (2000). Conservation and sustainable use of tropical trees in the genus *Aquilaria*. I. Status and distribution in Indonesia. *Biological Conservation*, 96, 83–94.

Soehartono, T. and Newton, A. C. (2001). Conservation and sustainable use of tropical trees in the genus *Aquilaria*. II. The impact of gaharu harvesting in Indonesia. *Biological Conservation*, 97(1), 29–41.

Sokal, R. R. and Oden, N. (1978). Spatial autocorrelation in biology, 1. Methodology. *Biological Journal of the Linnean Society*, 10, 199–228.

Sollins, P. (1982). Input and decay of coarse woody debris in coniferous stands in western Oregon and Washington. *Canadian Journal of Forest Research*, 12, 18–28.

Sollins, P., Cline, S. P., Verhoeven, T., Sachs, D., and Spycher, G. (1987). Patterns of log decay in old-growth Douglas-fir forests. *Canadian Journal of Forest Research*, 17, 1585–1595.

Solomon, A. M. and Bartlein, P. J. (1992). Past and future climate change: response by mixed deciduous-coniferous forest ecosystems in northern Michigan. *Canadian Journal of Forest Research*, 22, 1727–1738.

Soltis, D. E. and Soltis, P. S. (eds.) (1990). *Isozymes in plant biology*. Advances in Plant Sciences Series, Vol. 4. Chapman & Hall, London.

Sørensen, L. L., Coddington, J. A., and Scarff, N. (2002). Inventorying and estimating sub-canopy spider diversity using semi-quantitative sampling methods in an Afromontane forest. *Environmental Entomology*, 31, 319–330.

Sork, V. L. (1984). Examination of seed dispersal and survival in red oak, *Quercus rubra* (Fagaceae), using metal-tagged acorns. *Ecology*, 65, 1020–1022.

Southwood, T. R. E. and Henderson, P. A. (2000). *Ecological methods*, 3rd edition. Blackwell Science, Oxford.

Spellerberg, I. F. (1991). *Monitoring ecological change*. Cambridge University Press, Cambridge.

Spetich, M. A., Shifley, S. R., and Parker, G. R. (1999). Regional distribution and dynamics of coarse woody debris in midwestern old-growth forests. *Forest Science*, 45, 302–313.

Spies, T. (1997). Forest stand structure, composition and function. In *Creating a forestry for the 21st century. The science of ecosystem management*, eds. K. A. Kohm and J. F. Franklin, pp. 11–30. Island Press, Washington, DC.

Spies, T. and Turner, M. G. (1999). Dynamic forest mosaics. In *Maintaining biodiversity in forest ecosystems*, ed. M. L. Hunter, pp. 95–160. Cambridge University Press, Cambridge.

Spies, T. A., Franklin, J. F., and Thomas, T. B. (1988). Coarse woody debris in Douglas-fir forests of Western Oregon and Washington. *Ecology*, 69, 1689–1702.

Ståhl, G., Ringvall, A., and Fridman, J. (2001). Assessment of coarse woody debris—a methodological overview. *Ecological Bulletins*, 49, 57–70.

Starfield, A. M., Smith, K. A., and Bleloch, A. L. (1990). *How to model it: Problem solving for the computer age*. McGraw-Hill, New York.

Staus, N., Strittholt, J., Dellasala, D., and Robinson, R. (2002). Rate and patterns of forest disturbance in the Klamath-Siskiyou ecoregion, USA between 1972 and 1992. *Landscape Ecology*, 17, 455–470.

Stellingwerf, D. A. and Hussin, Y. A. (1997). *Measurements and estimations of forest stand parameters using remote sensing*. VSP, Utrecht.

Stephenson, N. L. (1999). Reference conditions for Giant Sequoia forest restoration: structure, process and precision. *Ecological Applications*, 9(4), 1253–1265.

Stern, M. J. (1998). Field comparisons of two rapid vegetation assesment techniques with permanent plot inventory data in Amazonian Peru. In *Forest biodiversity research, monitoring and modeling: Conceptual background and Old World case studies*, eds. F. Dallmeier and J. A. Comiskey, pp. 269–283. UNESCO/Parthenon, Paris.

Stevenson, P. R., Quiñones, M. J., and Ahumada, J. A. (1998). Annual variation in fruiting pattern using different methods in a lowland tropical forest, Tinigua National Park, Colombia. *Biotropica*, 30(1), 129–134.

Stewart, G. B., Coles, C. F., and Pullin, A. S. (2005). Applying evidence-based practice in conservation management: Lessons from the first systematic review and dissemination projects. *Biological Conservation*, 126(2), 270–278.

Stewart, G. H. and Burrows, L. E. (1994). Coarse woody debris in old-growth temperate beech (*Nothofagus*) forests of New Zealand. *Canadian Journal of Forest Research*, 24, 1989–1996.

Stiles, F. G. (1977). Coadapted competitors: the flowering seasons of hummingbird-pollinated plants in a tropical forest. *Science*, 196, 1177–1178.

Stiles, F. G. (1978). Temporal organization of flowering among the hummingbird foodplants of a tropical wet forest. *Biotropica*, 10, 194–210.

Stockman, A. K., Beamer, D. A., and Bond, J. E. (2006). An evaluation of a GARP model as an approach to predicting the spatial distribution of non-vagile invertebrate species. *Diversity and Distributions*, 12, 81–89.

Stockwell, D. R. B. (1999). Genetic algorithms II. In *Machine learning methods for ecological applications*, ed. A. H. Fielding, pp. 123–144. Kluwer, Boston, MA.

Stockwell, D. R. B. and Peters, D. P. (1999). The GARP modelling system: Problems and solutions to automated spatial prediction. *International Journal of Geographic Information Systems*, 13, 143–158.

Stohlgren, T. J., Falkner, M. B., and Schell, L. D. (1995). A Modified-Whittaker nested vegetation sampling method. *Vegetatio*, 117, 113–121.

Stork, N. E., Boyle, T. J. B., Dale, V. et al. (1997). *Criteria and indicator for assessing the sustainability of forest management: conservation of biodiversity*. Working Paper no. 17. CIFOR, Jakarta, Indonesia.

Strickler, G. S. (1959). *Use of the densiometer to estimate density of canopy on permanent sample plots*. Research Note no. 180, USDA Forest Service, Washington, DC.

Summers, R. W. (1998). The lengths of fences in Highland woods: the measure of a collision hazard to woodland birds. *Forestry*, 71(1), 73–76.

Sutherland, W. (ed.) (1996). *Ecological census techniques, a handbook*. Cambridge University Press, Cambridge.

Sutherland, W. J. (2000). *The conservation handbook. Research, management and policy*. Blackwell Science, Oxford.

Sutherland, W. J., Newton, I., and Green, R. E. (eds.) (2004a). *Bird ecology and conservation. A handbook of techniques*. Techniques in Ecology and Conservation Series. Oxford University Press, Oxford.

Sutherland, W. J., Pullin, A. S., Dolman, P. M., and Knight. T. M. (2004b). The need for evidence-based conservation. *Trends in Ecology and Evolution*, 19(6), 305–308.

Sutton, S. L. (2001). Alice grows up: canopy science in transition from Wonderland to Reality. *Plant Ecology*, 153, 13–21.

Swaine, M. D., Lieberman, D., and Putz, F. E. (1987). The dynamics of tree populations in tropical forest: a review. *Journal of Tropical Ecology*, 3, 359–366.

Swaine, M. D. and Whitmore, T. C. (1988). On the definition of ecological species groups in tropical rain forests. *Vegetatio*, 75, 81–86.

Swartzman, G. L. and Kaluzny, S. P. (1987). *Ecological simulation primer (biological resource management)*. Macmillan, New York.

Tallent-Halsell, N. G. (ed.) (1994). *Forest health monitoring: 1994 field methods guide*. EPA/620/R-94/-027. U. S. Environmental Protection Agency, Washington, DC.

Taylor, A. H. and Halpern, C. B. (1991). The structure and dynamics of *Abies magnifica* forests in the Southern Cascade Range, USA. *Journal of Vegetation Science*, 2, 189–200.

ter Braak, C. J. F. (1986). Canonical correspondence analysis: a new eigenvector technique for multivariate direct gradient analysis. *Ecology*, 67, 1167–1179.

ter Braak, C. J. F. (1988). CANOCO: an extension of DECORANA to analyze species–environment relationships. *Vegetatio*, 75, 159–160.

ter Braak, C. J. F. and Šmilauer, P. (1998). *CANOCO reference manual and user's guide to Canoco for Windows: Software for canonical community ordination (version 4)*. Microcomputer Power, Ithaca, NY.

ter Steege, H. (1998). The use of forest inventory data for a National Protected Area Strategy in Guyana. *Biodiversity and Conservation*, 7, 1457–1483.

ter Steege, H. (1994). *HEMIPHOT, a programme to analyze vegetation indices, light and light quality from hemispherical photographs*. Tropenbos Documents 03, Tropenbos International, Wageningen.

Theis, J. and Grady, H. (1991). *Participatory rapid appraisal for community development*. Save the Children Fund, London.

Thomas, L. and Krebs, C. J. (1997). A review of statistical power analysis software. *Bulletin of the Ecological Society of America*, 78, 126–139.

Thompson, K. (1993). Persistence in soil. In *Methods in comparative plant ecology*, eds. G. A. F. Hendry and J. P. Grime, pp. 199–202. Chapman & Hall, London.

Thompson, K., Bakker, J. P., and Bekker, R. M. (1997). *The soil seed banks of north west Europe: Methodology, density and longevity*. Cambridge University Press, Cambridge.

Thompson, P. A. and Fox, D. J. C. (1971). A simple thermo-gradient bar designed for use in seed germination studies. *Proceedings of the International Seed Testing Association*, 36, 255–263.

Ticktin, T. (2004). The ecological implications of harvesting non-timber forest products. *Journal of Applied Ecology*, 41(1), 11–21.

Tischendorf, L. (2001). Can landscape indices predict ecological processes consistently? *Landscape Ecology*, 16(3), 235–254.

Tischendorf, L. and Fahrig, L. (2000a). How should we measure landscape connectivity? *Landscape Ecology*, 15, 633–641.

Tischendorf, L. and Fahrig, L. (2000b). On the usage and measurement of landscape connectivity. *Oikos*, 90, 7–9.

Tomppo, E. (1993). Multi-source national forest inventory of Finland. In *Proceedings of the Ilvessalo symposium on national forest inventories*, ed. A. Nyssönen, pp. 52–59. Research Paper 444, Finnish Forest Research Institute, Helsinki.

Toms, J. D. and Lesperance, M. L. (2003). Piecewise regression: a tool for identifying ecological thresholds. *Ecology*, 84, 2034–2041.

Trakhtenbrot, A., Nathan, R., Perry, G., and Richardson, D. M. (2005). The importance of long-distance dispersal in biodiversity conservation. *Diversity and Distributions*, 11, 173–181.

Trani, M. K. and Giles R. H. (1999). An analysis of deforestation: metrics used to describe pattern change. *Forest Ecology and Management*, 114, 459–470.

Trotter, C. M., Dymond, J. R., and Goulding, C. J. (1997). Estimation of timber volume in a coniferous forest using Landsat TM. *International Journal of Remote Sensing*, 18, 2209–2223.

Tuomisto, H., Linna, A., and Kalliola, R. (1994). Use of digitally processed satellite images in studies of tropical rain forest vegetation. *International Journal of Remote Sensing*, 15, 1595–1610.

Tuomisto, H., Ruokolainen, K., Kalliola, R., Linna, A., Danjoy, W., and Rodriguez, Z. (1995). Dissecting Amazonian biodiversity. *Science*, 269, 63–66.

Turner, I. M. (2001). *The ecology of trees in the tropical rain forest*. Cambridge University Press, Cambridge.

Turner, M. G., Gardner, R. H., and O'Neill, R. V. (2001). *Landscape ecology in theory and practice*. Springer, New York.

Turner, W., Spector, S., Gardiner, N., Fladeland, M., Sterling, E., and Steininger, M. (2003). Remote sensing for biodiversity science and conservation. *Trends in Ecology and Evolution*, 18(6), 306–314.

Ulanova, N. G. (2000). The effects of windthrow on forests at different spatial scales: a review. *Forest Ecology and Management*, 135, 155–167.

Underwood, A. J. (1990). Experiments in ecology and management: their logics, functions and interpretations. *Australian Journal of Ecology*, 15, 365–389.

Underwood, A. J. (1997). *Experiments in ecology: Their logical design and interpretation using analysis of variance*. Cambridge University Press, Cambridge.

Urban, D. L. (1990). *A versatile model to simulate forest pattern: a user's guide to ZELIG version 1.0*. Department of Environmental Sciences, University of Virginia, Charlottesville, VA.

Urban, D. L., Bonan, G. B., Smith, T. M., and Shugart, H. H. (1991). Spatial applications of gap models. *Forest Ecology and Management*, 42, 95–110.

Uusivuori, J., Lehto, E., and Palo, M. (2002). Population, income and ecological conditions as determinants of forest area variation in the tropics. *Global Environmental Change*, 12, 313–323.

USDA Forest Service. (2001). Down woody debris and fuels. In *Field Inventory and Analysis Field Methods for Phase 3 Measurements, 2001*. USDA Forest Service, Washington, DC ⟨http://fia.fs.fed.us/manuals/p3sec14_2001.doc⟩

Vales, D. J. and Bunnell, F. L. (1988). Comparison of methods for estimating forest overstory cover. I. Observer effects. *Canadian Journal of Forest Research*, 18, 606–609.

Vanclay, J. K. (1994). *Modelling forest growth and yield: Applications to mixed tropical forests*. CABI, Wallingford.

Vanclay, J. K. (1995). Growth models for tropical forests: a synthesis of models and methods. *Forest Science*, 41, 7–42.

Vanclay, J. K., Skovsgaard, J. P., and Hansen, C. P. (1995). Assessing the quality of permanent sample plot databases for growth modeling in forest plantations. *Forest Ecology and Management*, 71, 177–186.

Vanclay J. K., Gillison, A. N., and Keenan, R. J. (1997). Using plant functional attributes to quantify stand dynamics in mixed forests for growth modelling and yield prediction. *Forest Ecology and Management*, 94, 149–163.

Vandermeer, J. (1978). Choosing category size in a stage projection matrix. *Oecologia*, 32, 79–84.

Van der Pijl, L. (1982). *Principles of dispersal in higher plants*. Springer, New York.

Vander Wall, S. B., Kuhn, K. M., and Beck, M. J. (2005). Seed removal, seed predation, and secondary dispersal. *Ecology*, 86(3), 801–806.

van Dorp, D., Schippers, P., and Groenendael, J. M. (1997). Migration rates of grassland plants along corridors in fragmented landscape assessed with a cellular automation model. *Landscape Ecology*, 12, 39–50.

Veblen, T. T. (1979). Structures and dynamics of *Nothofagus* forests near timberline in South Chile. *Ecology*, 60, 937–945.

Veblen, T. T. (1985). Stand dynamics in Chilean *Nothofagus* forests. In *The ecology of natural disturbance and patch dynamics*, eds. S. T. A. Pickett and P. S. White, pp. 35–51. Academic Press, New York.

Veblen, T.T., Donoso, C., Schlegel, F. M., and Veblen, A. T. (1981). Forest dynamics in south-central Chile. *Journal of Biogeography*, 8, 211–247.

Veenendaal, E. M., Swaine, M. D., Lecha, R. T., Walsh, M. F., Abebrese, I. K., and Owusu-Afriyie, K. (1996). Responses of West African forest tree seedlings to irradiance and soil fertility. *Functional Ecology*, 10, 501–511.

Veldkamp, A. and Lambin, E. F. (2001). Predicting land-use change. *Agriculture, Ecosystems and Environment*, 85, 1–6.

Vellend, M. (2001). Do commonly used indices of beta-diversity measure species turnover? *Journal of Vegetation Science*, 12(4), 545–552.

Vera, F. W. M. (2000). *Grazing ecology and forest history*. CABI, Wallingford.

Verissimo, A., Barreto, P., Tarifa, R., and Uhl, C. (1995). Extraction of a high-value natural-resource in Amazonia—the case of mahogany. *Forest Ecology and Management*, 72(1), 39–60.

Villard, M.-A., Trzcinski, M. K., and Merriam, G. (1999). Fragmentation effects on forest birds: relative influence of woodland cover and configuration on landscape occupancy. *Conservation Biology*, 13, 774–783.

Vogler, P. and Desalle, R. (1994). Diagnosing units of conservation management. *Conservation Biology*, 8, 354–363.

Voysey, B. C., McDonald, K. E., Rogers, M. E., Tutin, C. E. G., and Pannell, R. J. (1999). Gorillas and seed dispersal in the Lopé Reserve, Gabon. I. Gorilla acquisition by trees. *Journal of Tropical Ecology*, 15, 23–28.

Wackernagel, H. (2003). *Multivariate geostatistics: An introduction with applications*. Springer-Verlag, Berlin.

Wade, P. R. (2000). Bayesian methods in conservation biology. *Conservation Biology*, 14(5), 1308–1316.

Waggoner, P. E. and Stephens, G. R. (1970). Transition probabilities for a forest. *Nature*, 225, 1160–1161.

Wagner, D. H. (1991). The '1-in-20' rule for plant collectors. *Plant Science Bulletin*, 37(2), 11.

Wagner, H. H. (2003). Spatial covariance in plant communities: integrating ordination, geostatistics, and variance testing. *Ecology*, 84, 1045.

Walter, J. and Soos, J. (1962). *The gimbal sight for the projection of crown radius*. Research Note No. 39, Faculty of Forestry, University of British Columbia, Vancouver, BC.

Ward, M., Dick, C. W., Gribel, R., Lemes, M., Caron, H., and Lowe, A. J. (2005). To self, or not to self: A review of outcrossing and pollen-mediated gene flow in neotropical trees. *Heredity*, 95, 246–254.

Watkinson, A. R. (1997). Plant population dynamics. In *Plant ecology*, ed. M. J. Crawley, 2nd edition, pp. 359–400. Blackwell Science, Oxford.

Watt, A. S. (1947). Pattern and process in the plant community. *Journal of Ecology*, 35, 1–22.

Weaver, P. L. (2000). Elfin woodland recovery 30 years after a plane wreck in Puerto Rico's Luquillo mountains. *Caribbean Journal of Science*, 36(1–2), 1–9.

Weber, R. O. (2001). Wildland fire spread models. In *Forest fires. Behaviour and ecological effects*, eds. E. A. Johnson and K. Miyanishi, pp. 151–169. Academic Press, San Diego, CA.

Webster, R. and Oliver, M. A. (2000). *Geostatistics for environmental scientists*. Wiley, New York.

Weihs, U., Dubbel, V., and Krummheuer, F. (1999). Die Elektrische Wiederstandstomographie— Ein vielversprechendes Verfahren zur Farbkerndiagnose am stehenden Rotbuchenstamm. *Forst und Holtz*, 6, 166–170.

Werner, P. A. (1975). A seed trap for determining patterns of seed deposition in terrestrial plants. *Canadian Journal of Botany*, 54, 1189–1197.

Wesseling, C. G., Karssenberg, D., Van Deursen, W. P. A., and Burrough, P. A. (1996). Integrating dynamic environmental models in GIS: the development of a dynamic modelling language. *Transactions in GIS*, 1, 40–48.

West, D. C., Shugart, H. H. and Botkin, D. B. (eds.) (1981). *Forest succession: Concepts and applications*. Springer, New York.

West, P. W. (2004). *Tree and forest measurement*. Springer-Verlag, Berlin.

Westoby, M. and Leishman, M. (1997). Categorizing plant species into functional types. In *Plant functional types*, eds. T. M. Smith, H. H. Shugart, and F. I. Woodward, pp. 104–121. Cambridge University Press, Cambridge.

Whelan, R. J. (1995). *The ecology of fire*. Cambridge University Press, Cambridge.

White, G. M. and Boshier, D. H. (2000). Fragmentation in Central American dry forests: genetic impacts on *Swietenia humilis* (Meliaceae). In *Genetics, demography and viability of fragmented populations*, eds. A. G. Young and G. M. Clarke, pp. 293–311. Cambridge University Press, Cambridge.

White, P. S. (1979). Pattern, process, and natural disturbance in vegetation. *Botanical Reviews*, 45, 229–299.

White, T. L., Adams, W. T., and Neale, D. B. (2002). *Forest genetics*. CABI, Wallingford.

Whitehead, C. J. (1969). *Oak mast yields on wildlife management areas in Tennessee*. Tennessee Wildlife Resource Agency, Nashville, TN.

Whitman, A. A., Brokaw, N. V. L., and Hagan, J. M. (1997). Forest damage caused by selection logging of mahogany (*Swietenia macrophylla*) in northern Belize. *Forest Ecology and Management*, 92(1–3), 87–96.

Whitmore, T. C. (1990). *An introduction to tropical rain forests*. Oxford University Press, Oxford.

Whitmore, T. C. and Brown, N. D. (1996). Dipterocarp seedling growth in rain forest canopy gaps during six and a half years. *Philosophical Transactions of the Royal Society of London*, B351, 1195–1203.

Whitmore, T. C., Brown, N. D., Swaine, M. D., Kennedy, D. K., Goodwin-Bailey, C. I., and Gong, W. K. (1993). Use of hemisphere photographs in forest ecology: measurement of gap size and radiation totals in a Bornean tropical rain forest. *Journal of Tropical Ecology*, 9, 131–151.

Whittaker, R. H. (1975). *Communities and ecosystems*, 2nd edition. Macmillan, New York.

Whitten, T., Holmes, D., and MacKinnon, K. (2001). Conservation biology: a displacement behaviour for academia? *Conservation Biology*, 15(1), 1–3.

Wiant, H. V., Wood, G. B., and Furnival, G. M. (1992). Estimating log volume using the centroid position. *Forest Science*, 38, 187–191.

Wiegand, T. and Moloney, K. A. (2004). Rings, circles, and null-models for point pattern analysis in ecology. *Oikos*, 104, 209–229.

Wiegand, T., Jeltsch, F., Hanski, I., and Grimm, V. (2003). Using pattern-oriented modeling for revealing hidden information: a key for reconciling ecological theory and application. *Oikos*, 100, 209–222.

Wiegert, R. G. (1962). The selection of an optimum quadrat size for sampling the standing crop of grasses and forbs. *Ecology*, 43, 125–129.

Wiersum, K. F. (1995). 200 years of sustainability in forestry: lessons from history. *Environmental Management*, 19(3), 321–329.

Wikramanayake, E., Dinerstein, E., Loucks C. J. *et al.* (2002). *Terrestrial ecoregions of the Indo-Pacific. A conservation assessment.* Conservation Science Program, WWF-US. Island Press, Washington, DC.

Williams, J. C., ReVelle, C. S., and Levin, S. A. (2004). Using mathematical optimization models to design nature reserves. *Frontiers in Ecology and the Environment*, 2(2), 98–105.

Williams, P., Gibbons, D., Margules, C., Rebelo, A., Humphries, C., and Pressey, R. (1996). A comparison of richness hotspots, rarity hotspots, and complementary areas for conserving diversity of British birds. *Conservation Biology*, 10, 155–174.

Williams, W. T., Lance, G. N., Webb, L. J., Tracey, J. G., and Connell, J. H. (1969). Studies in the numerical analysis of complex rain-forest communities. IV. The elucidation of small-scale forest pattern. *Journal of Ecology*, 57, 635–654.

Willson, M. F. (1983). *Plant reproductive ecology*. Wiley, New York.

Willson, M. F. (1993). Dispersal mode, seed shadows, and colonization patterns. *Vegetatio*, 108, 261–280.

Wilson, K., Newton, A., Echeverría, C., Weston, C., and Burgman, M. (2005a). A vulnerability analysis of the temperate forests of south central Chile. *Biological Conservation*, 122, 9–21.

Wilson, K., Pressey, B., Newton, A., Burgman, M., Possingham H., and Weston, C. (2005b). Measuring and incorporating vulnerability into conservation planning. *Environmental Management*, 35(5), 527–543.

Winn, A. A. (1989). Using radionuclide labels to determine the post-dispersal fates of seeds. *Trends in Ecology and Evolution*, 4, 1–2.

Woldendorp, G., Spencer, R. D., Keenan, R. J., and Barry, S. (2002). An analysis of sampling methods for coarse woody debris in Australian forest ecosystems. Bureau of Rural Sciences, Commonwealth of Australia, Canberra ⟨www.affa.gov.au/brs⟩

Wolf, P. R. and Dewitt, B. A. (2000). *Elements of photogrammetry with applications in GIS*, 3rd edition. McGraw-Hill, New York.

Wolter, P. and White, M. (2002). Recent forest cover type transitions and landscape structural changes in northeast Minnesota, USA. *Landscape Ecology*, 17, 133–155.

Woodward, F. I. and Yaqub, M. (1979). Integrator and sensors for measuring photosynthetically active radiation and temperature in the field. *Journal of Applied Ecology*, 16(2), 545–552.

Wright, S. (1951). The genetical structure of populations. *Annals of Eugenics*, 15, 323–354.

Wright, S. J., Muller-Landau, H. C., Condit, R., and Hubbell, S. P. (2003). Gap-dependent recruitment, realized vital rates, and size distributions of tropical trees. *Ecology*, 84(12), 3174–3185.

Wulder, M. A. and Franklin, S. E. (eds.) (2003). *Remote sensing of forest environments. Concepts and case studies*. Kluwer, Dordrecht.

Yaussy, D. A. (2000). Comparison of an empirical forest growth and yield simulator and a forest gap simulator using actual 30-year growth from two even-aged forests in Kentucky. *Forest Ecology and Management*, 126, 385–398.

Yoccoz, N. G., Nichols, J. D., and Boulinier, T. (2001). Monitoring of biological diversity in space and time. *Trends in Ecology and Evolution*, 16(8), 446–453.

Young, A., Boshier, D., and Boyle, T. (eds.) (2000). *Forest conservation genetics. Principles and practice*. CSIRO, Collingwood, Australia/CABI, Wallingford.

Young, A. G. and Clarke, G. M. (eds.) (2000). *Genetics, demography and viability of fragmented populations*. Cambridge University Press, Cambridge.

Yumoto, T. (1999). Seed dispersal by Salvin's curassow, *Mitu salvini* (Cracidae), in a tropical forest of Columbia: direct measurements of dispersal distance. *Biotropica*, 31, 654–660.

Zaniewski, A. E., Lehmann, A., and Overton, J. M. (2002). Predicting species distribution using presence-only data: a case study of native New Zealand ferns. *Ecological Modelling*, 157, 261–280.

Zapata, T. R. and Arroyo, M. T. K. (1978). Plant reproductive ecology of a secondary deciduous tropical forest in Venezuela. *Biotropica*, 10, 221–230.

Zeide, B. (1993). Analysis of growth equations. *Forest Science*, 39, 594–616.

Zhang, Q. F., Devers, D., Desch, A., Justice, C. O., and Townshend, J. (2005). Mapping tropical deforestation in Central Africa. *Environmental Monitoring and Assessment*, 101(1–3), 69–83.

Zhang, S.-Y. and Wang, L.-X. (1995). Comparison of three fruit census methods in French Guiana. *Journal of Tropical Ecology*, 11, 281–294.

Zimmerman, J. K., Everham, E. M., Waide, R. B., Lodge, D. J., Taylor, C. M. and Brokaw, N. V. L. (1994). Responses of tree species to hurricane winds in subtropical wet forest in Puerto Rico: implications for tropical tree life histories. *Journal of Ecology*, 82, 911–922.

Zobel, B. and Talbert, J. (1984). *Applied forest tree improvement*. Wiley, New York.

Zuidema, P. A. (ed.) (2000). *Demography of exploited tree species in the Bolivian Amazon*. PROMAB Scientific Series 2. Riberalta, Bolivia.

Zuidema, P. A. and Boot, R. G. A. (2002). Demography of the Brazil Nut Tree (*Bertholletia excelsa*) in the Bolivian Amazon: impact of seed extraction on recruitment and population dynamics. *Journal of Tropical Ecology* 18(1): 1–31.

Zuidema, P. A. and Zagt, R. J. (2000). Using population matrices for long-lived species: a review of published models for 35 woody plants. In *Demography of exploited tree species in the Bolivian Amazon*, ed. P. A. Zuidema, pp. 159–182. PROMAB Scientific Series 2. Riberalta, Bolivia.

Index of authors and names

Acevedo, M.F. *et al.* 227
Adams, J.B. *et al.* 62
Agee, J.K. 153
Akçakaya, H.R. 216
Akçakaya, H.R. and Sjögren-Gulve, P. 213–14
Alder, D. and Synnott, T.J. 182
Alves, D.S. *et al.* 58
Amarnath, G. *et al.* 67
Amaro, A. *et al.* 221
Anderson, M.C. 170
Anderson, R.P. *et al.* 323
Angelstam, P. and Dönz-Breuss M. 370
Angelstam, P. *et al.* 373
Arno, S.F. and Sneck, K.M. 153
Ashmole, Philip 378
Ashton, P.M.S. 192
Ashton, P.M.S. and Hall, P. 299
Asner, G.P. *et al.* 61, 63, 64
Augspurger, C.K. 201
Avery, T.E. 38
Avery, T.E. and Berlin, G.L. 38
Avery, T.E. and Burkhart, H.E. 86, 91

Bajracharya, S.B. *et al.* 334
Baker, P.J. and Wilson, J.S. 299
Baldocchi, D. and Collineau, S. 169
Ball, I.R. *et al.* 312
Balzter, H. *et al.* 228
Barabesi, L. and Fattorini, L. 120
Bardon, R.E. *et al.* 174
Barot, S. *et al.* 122
Baskin, C.C. and Baskin, J.M. 197, 250
Bawa, K.S. and Hadley, M. 235
Bebber, D. *et al.* 125
Becker, M. 170
Begon, M. and Mortimer, M. 205
Behera, M.D. *et al.* 68
Bekessy, S.A. *et al.* 216–18, 283
Bertault, J.G. and Sist, P. 160
Bhadresa, R. 155, 157, 193
Blasco, F. *et al.* 145
Boose, E.R. *et al.* 151
Boshier, D.H. *et al.* 275
Botkin, D.B. 231–4
Bradbeer, J.W. 197–8, 250
Bradstock, R.A. and Kenny, B.J. 200
Brändli, U.B. *et al.* 86, 303
Brandon, K. *et al.* 334
Bréda, N.J.J. 116, 117
Briggs, D. and Walters, S.M. 279
Brigham, C.A. and Schwartz, M.A. 213, 214

Brigham, C.A. and Thomson, D.M. 214
Brokaw, N.V.L. and Lent, R.A. 300
Brook, B.W. *et al.* 214
Brown, N.D. *et al.* 176–8
Brown, S.L. *et al.* 64
Brown, V.K. 155
Buckland S.T. *et al.* 95, 97
Bugmann, H. 226, 234
Bullock, J.M. 97, 313–15
Bullock, J.M. and Clarke, R.T. 259–61
Bunnell, F.L. and Vales, D.J. 176
Burgman, M.A. 318
Burkhart, H.E. 221
Byram, G.M. 153

Cahill, J.F. *et al.* 157
Cailliez, F. 112
Campbell, J.B. 32
Campbell, J.E. and Gibson, D.J. 256
Campbell, P. *et al.* 99
Cannell, M.G.R. and Grace, J. 170
Carter, R.E. *et al.* 144
Cavers, S. *et al.* 275–6
Cayuela, L. *et al.* 62
Chambers, R. 29
Chapin, F.S. *et al.* 374
Chapman, C.A. *et al.* 250
Charlesworth, D. 242
Chase, M.W. and Hills, H.H. 263
Chazdon, R.L. *et al.* 125, 127, 128
Chiariello, N.R. *et al.* 189
Clark, D.A. and Clark, D.B. 175, 194
Clark, D.F. *et al.* 294
Clarke, K.R. and Warwick, R.M. 135
Clark, J.S. *et al.* 260–1
Clark, N.A. 108
Clark, P.J. and Evans, F.C. 121
Cochran, W.G. 91
Cohen, W.B. *et al.* 62, 66
Cohen, W.B. and Goward, S.N. 32
Colwell, R.K. and Coddington, J.A. 125–8
Colwell, R.K. *et al.* 126
Comeau, P.G. *et al.* 178
Condit, R. *et al.* 100, 128, 132–3, 199
Cook, J.G. *et al.* 176
Cooney, R. 25
Cornelissen, J.H.C. *et al.* 201
Corona, P. *et al.* 32, 47
Crandall, K.A. *et al.* 278
Crawley, M.J. 157–9

Index of authors and names

Curran, P.J. and Williamson, H.D. 65
Cushman, S.A. and Wallin, D.O. 58
Czaplewski, R.L. 53

Dafni, A. 235
Dale, M.R.T. 121
Dallmeier, F. and Comiskey, J. 94
D'eca-Neves, F.F. and Morellato, L.P.C. 247
DeFries, R. *et al.* 65
Dennis, B. 8
de Vries, P.G. 87
Diamond, J. 194
Digby, P.G.N. and Kempton, R.A. 136
Diggle, P.J 121–2
Dow, B.D. and Ashley, M.V. 274
Durrett, R. and Levin, S.A. 228
Dytham, C. 8

Eberhardt, L.L. 95
Ebert, T.A. 205
Ek, A.R. *et al.* 221
Elith, J. and Burgman, M.A. 324
Ellner, S.P. *et al.* 214
Engeman, R.M. *et al.* 95, 96
Englund, S.R. *et al.* 177
Ennos, R.A. 270, 273
Erickson, A.A. *et al.* 157
Estabrook, G.F. *et al.* 248
Evans, G.C. 189
Evans, G.C. and Coombe, D.E. 170

Fahrig, L. 300, 325
Fairbanks, D.H.K. and McGwire, K.C. 67
Fassnacht, K.S. *et al.* 64
Hayes, D.J. and Sader, S.A. 58
Feisinger, P. and Busby, W.H. 244
Feldpausch, T.R. *et al.* 160
Fenner, M. 250
Ferment, A. *et al.* 178
Fieberg, J. and Ellner, S.P. 210
Fiorella, M. and Ripple, W.J. 62
Foody, G.M. and Cutler, M.E.J. 68
Foody, G.M. *et al.* 62, 64
Ford, E.D. 8, 203
Forget, P.-M. and Wenny, D. 257–8
Fortin, M.J. and Dale, M.R.T. 121
Fournier, R.A. *et al.* 64
Franklin, S.E. 32, 34, 41, 64
Fransson, J.E.S. *et al.* 66
Fraser, D.J. and Bernatchez, L. 278
Frazer, G.W. *et al.* 172
Freckleton, R.P. *et al.* 210
Frelich, L.E. 161, 162–3
Frelich, L.E. *et al.* 229
Fridman, J. and Walheim, M. 294
Friend, A. and Rapaport, D. 368
Friend, D.T.C. 174
Fritts, N.C. and Fritts, E.C. 184

Gardner, R.H. and Urban, D.L. 233
Gendron, F. *et al.* 169
Gentry, A.H. 99, 100
Ghazoul, J. and McAllister, M. 8
Gibbons, P. *et al.* 308–9
Gibson, D.J. 187, 230
Gillet, E. and Gregorius, H.R. 268
Gillison, A. 200
Gill, R.M.A. 159
Glaubitz, J.C. and Moran, G.F. 268
Gong, P. and Xu, B. 56
Gordon, J.E. 99
Gordon, J.E. *et al.* 101
Gotelli, N.J. and Colwell, R.K. 128–9
Gove, J.H. *et al.* 291, 292
Gray, A. 100 101
Greene, D.F. and Calogeropoulos, C. 261
Greene, D.F. and Johnson, E.A. 252
Guisan, A. and Zimmerman, N.E. 324

Hale, S.E. and Brown, N.D. 178
Hall, J.B. 96
Hall, P. *et al.* 99
Hall, R.J. 34
Hall, R.J. *et al.* 66
Hamrick, J.L. and Nason, J.D. 270
Hanski, I. 219
Haridasan, M. and de Araújo, G.M. 96
Harmon, M.E. *et al.* 290
Harmon, M.E. and Sexton, J. 287, 290
Hawthorne, W.D. 101, 143
Hawthorne, W.D. and Abu-Juam, M. 143
He, F.L. and Duncan, R.P. 210–11
Hemingway, C.A. and Overdorff, D.J. 247
Hendricks, W.A. 102
Henle, K. *et al.* 215
Heslop-Harrison, J. 279
Higman, S. *et al.* 340
Hill, M.O. 138
Hinsley, S.A. *et al.* 54
Hirata, Y. *et al.* 54
Hochberg, M.E. *et al.* 227
Hockings, M. *et al.* 334–5
Hohn, M.E. *et al.* 125
Holland, J. and Campbell, J. 29
Holmgren, P. and Thuresson, T. 40
Holvoet, B. and Muys, B. 339
Hope, J.C.E. 228
Horn, H.S. 227
Horn, H.S. *et al.* 261
Horning, N. 32, 61
Houle, A.C. *et al.* 103
House, S.M. 247
Hubbell, S.P. 11, 126
Hudson, W.D. 60
Humphrey, J. *et al.* 331

Hunter, M.L. 326
Hunt, R. 189
Hunt, R. and Parsons, I.T. 190
Husch, B. *et al.* 86, 97, 101, 107, 307
Hutchings, M.J. 187, 195, 196

Imbernon, J. and Branthomme, A. 84
Ingram, J.C. *et al.* 63
Iverson, L.R. *et al.* 152
Izhaki, I. *et al.* 256

Jennings, S.B. *et al.* 111–12, 174, 176, 178
Johnson, E.A. 153
Johnston, C.A. 68, 69
Jones, F.A. *et al.* 262
Jongman, R.H.G. *et al.* 136
Jordan, G.J. *et al.* 294

Kearns, C.A. and Inouye, D.W. 235
Keddy, P.A. and Drummond, C.G. 370–3
Keeley, J.E. and Fotheringham, C.J. 99, 100
Kenkel, N.C. *et al.* 98
Kent, M. and Coker, P. 136
Kesteven, J.L. *et al.* 64
Kimmins, J.P. 144
Kindt, R. and Coe, R. 129–30
Kint, V. *et al.* 100
Kleinn, C. 86
Kleinn, C. and Traub, B. 77
Kobe, R.K. *et al.* 201–2
Koenig, W.D. *et al.* 250
Köhl, M. and Gertner, G.Z. 125
Köhl, M. and Lautner, M. 61
Kohyama, T. 228
Krebs, C.J. 8, 9, 95, 97

Laffan, M. *et al.* 160
Lamb, D. and Gilmour, D. 350–2
Lambers, H. and Poorter, H. 189
Lambert, R.C. *et al.* 296
Lambin, E.F. 59
Lambin, E.F. *et al.* 363
Langlet, O. 279
Latham, P.A. *et al.* 299
Leck, M.A. *et al.* 195
Lefsky, M.A. *et al.* 54
Legendre, P. and Legendre, L. 136
Legg, C.J. and Nagy, L. 364–6
Leishman, M.R. and Westoby, M. 199
Lemmon, P.E. 176
Levin, S.A. *et al.* 261
Lieberman, M. *et al.* 176
Lilledsand, T.M. and Kiefer, R.W. 32
Lindenmayer, D.B. *et al.* 309
Lindner, M. *et al.* 224
Liu, J.G. and Ashton, P.S. 224, 229
Lobo, A. 62

Lockwood, M. *et al.* 336
Longino, J.T. *et al.* 126, 128
López-Barrera, F. *et al.* 257–9
López-Barrera, F. and Newton, A.C. 303–4
Lowe, A.J. *et al.* 275
Lowman, M.D. and Wittman, P.K. 103
Ludwig, D. 214
Lü, Y. *et al.* 335

Maathai, Wangari 378
MacArthur, R.H. and MacArthur, J.W. 297
McCarthy, M.A. *et al.* 215
McComb, W. and Lindenmayer, D. 312
McCracken, J.A. *et al.* 29
Mace, G.M. *et al.* 16
McElhinny, C. *et al.* 299–300
Machado, J.-L. and Reich, P.B. 178
McNab, W.H. *et al.* 144
Madany, M.H. *et al.* 153
Magurran, A.E. 125, 126, 131, 133
Manley, B.F.J. 366
Mansourian, S. *et al.* 350
Margoluis, R. and Salafsky, N. 354
Margules, C.R. and Pressey, R.L. 337
Maser, C. *et al.* 310
Matlack, G.R. and Litvaitis, J.A. 303
Matthews, E. 21
Mayaux, P. *et al.* 58
Mayers, J. and Bass, S. 17
Mayhew, J. and Newton, A.C. 194
Mertens, B. and Lambin, E.F. 362
Messier, C. and Puttonen, P. 169
Miles, L. *et al.* 145, 360
Mills, L.S. *et al.* 215
Mitchell, A.W. 103
Moffett, M.W. 103
Mueller-Dombois, D. and Ellenberg, H. 144
Munro, D.D. 221
Myers, N. 16

Nagendra, H. 67
Nanos, N. *et al.* 125
Nathan, R. *et al.* 261
Nei, M. 269
Newstrom, L.E. *et al.* 247
Newton, A.C. *et al.* 283–4
Newton, A.C. and Kapos, V. 87, 341, 374
Newton, A.C. and Oldfield, S. 25
Niemi, G.J. and McDonald, M.E. 367
Nilsson, M. *et al.* 65
Noble, I.R. and Slatyer, R.O. 199
Normand, F. *et al.* 249
Noss, R.F. 368–70, 372–3
Nussbaum, R. and Simula, M. 340, 341

Olson, D.N. *et al.* 145

O'Neill, R.V. *et al.* 77
Osawa, A. *et al.* 248–9

Parker, G.G. and Brown, M.J. 298–9
Parmenter, A.W. *et al.* 58
Parrado-Rosselli, A. *et al.* 251
Patil, G.P. *et al.* 120
Payandeh, B. and Ek, A.R. 96
Peng, C. 221
Pennanen, J. and Kuuluvainen, T. 231
Perrow, M. and Davy, A. 350
Petchey, O.L. and Gaston, K.J. 200
Peterken, G.F. 22
Peters, C.M. 346–7
Peterson, A.T. *et al.* 323
Peterson, G.D. *et al.* 374–6
Peters, R.H. 3, 8, 10–11, 19
Petit, R.J. *et al.* 278
Philip, M.S. 86
Phillips, O.L. *et al.* 98, 99, 182
Pillar, V.D. and Sosinski, E.E. 200–1
Poorter, H. 189
Pressey, R.L. *et al.* 335, 358
Pretty, J.N. *et al.* 29
Price, M.V. and Reichman, O.J. 196
Prins, E. and Kikula, I.S. 58
Putman, R.J. 159

Quine, C.P. and White, I.M.S. 151

Rackham, O. 161
Rametsteiner, E. and Simula, M. 341
Ramirez-Garcia, P. *et al.* 61–2
Ranneby, B. *et al.* 86
Redford, K.H. *et al.* 16
Reed, D.D. and Mroz, G.D. 86, 91
Reimoser, F. *et al.* 155
Rhodes, M. *et al.* 309
Ribbens, E.J. *et al.* 256
Richards, P.W. 102, 116
Rich, P.M. 170
Ries, L. *et al.* 304–5, 306
Riitters, K.H. *et al.* 84
Ringvall, A. and Ståhl, G. 294
Ripley, B.D 121–2
Ritland, K. 274–5
Roberts, H.A. 195, 196
Robinson, D. 357–8
Rorinson, I.H. and Robinson, D. 193
Rosenzweig, M.L. 126
Rubino, D.L. and McCarthy, B.C. 164
Runkle, J.R. 165

Sabersky, R.H. *et al.* 152
Sader, S.A. *et al.* 64
Salafsky, N. *et al.* 354, 357
Sala, O.E. *et al.* 374

Sample, V.A. and Cheng, A.S. 17
Schirone, B. *et al.* 249–50
Schnitzer, S.A. *et al.* 157
Schume, H. *et al* 125
Schwartz, M.W. 220
Schweingruber, F.H. 153
Shaw, P.J.A. 136
Sheil, D. 182, 184, 366–7
Sheil, D. *et al.* 96, 188, 201
Sheil, D. and May, R.M. 188
Shiver, B.D. and Borders, B.E. 87
Shugart, H.H. 231, 233
Smith, D.M. 345
Sørensen, L.L. *et al.* 133–5
Soehartono, T. and Newton, A.C. 87, 159
Soltis, D.E. and Soltis, P.S. 264
Southwood, T.R.E. and Henderson, P.A. 8
Spetich, M.A. *et al.* 310
Spies, T. *et al.* 288, 294
Ståhl, G. *et al.* 287, 289, 290, 293
Stellingwerf, D.A. and Hussin, Y.A. 34
Stern, M.J. 100
Stiles, F.G. 247
Stockman, A.K. *et al.* 323
Stork, N.E. *et al.* 368
Strickler, G.S. 176
Swaine, M.D. *et al.* 165
Swaine, M.D. and Whitmore, T.C. 201
Swartzman, G.L. and Kaluzny, S.P. 233

Tallent-Halsell, N.G. 112
ter Braak, C.J.F. 141
ter Steege, H. 87
Theis, J. and Grady, H. 29
Thomas, L. and Krebs, C.J. 9
Thompson, K. *et al.* 195–7
Trakhtenbrot, A. *et al.* 261
Turner, I.M. 201
Turner, M.G. *et al.* 77

Ulanova, N.G. 150
Underwood, A.J. 2, 6, 8

Vanclay, J.K. 221
Vander Wall, S.B. *et al.* 259
Veblen, T.T. 147
Velland, M. 135
Vera, F.W.M. 11, 159
Verissimo, A. *et al.* 161
Villard, M.-A. *et al.* 301
Voysey, B.C. *et al.* 256

Wade, P.R. 8
Waggoner, P.E. and Stephens, G.R. 227
Ward, M. *et al.* 275
Watkinson, A.R. 205
Watson, Alan 378

Watt, A.S. 7, 164
Weber, R.O. 153
West, P.W. 86
Whelan, R.J. 153
Whitman, A.A. *et al.* 161
Whitmore, T.C. *et al.* 170
Whittaker, R.H. 144
Whitten, T. *et al.* 16
Wiegand, T. 122
Wiegand, T. and Moloney, K.A. 122
Wiegert, R.G. 101
Willson, M.E. 235

Wilson, K. *et al.* 358, 360–2
Wolf, P.R. and Dewitt, B.A. 34, 36
Woodward, F.I. and Yaqub, M. 167
Wright, S. 269–70, 274
Wulder, M.A. Franklin, S.E. 32

Yaussy, D.A. 224
Yoccoz, N.G. *et al.* 363–4

Zhang, Q.F. *et al.* 58
Zhang, S.-Y. and Wang, L.-X. 251
Zuidema, P.A. and Zagt, R.J. 211–13

Subject index

Note: page numbers in *italics* refer to material in Figures, Tables and Boxes.

3D Mapper *37*

absence data 322–3
abundance, response to herbivory 158
abundance-based coverage estimator (ACE) *127*
AccuPAR *118*
accuracy 26
  of GPS systems 92–3
  of mapping 53
  of models 233
  of remote sensing 61, 65
adaptive cluster sampling 291, *293*
adaptive management 25, 347, *349*, 354–7
  forest restoration 353
adaptive management cycle *355*
additive variance 282–3
adult survival rate assessment 211–12
Aerial Image Corrector (AIC) *37*
aerial photography 33–4, 40
  forest cover changes assessment 55–6
  forest structure assessment 63
  forest type mapping 60
  image acquisition 34–6
  image processing 36–8
  timber volume estimation 64
  tree height estimation 63
age distribution, in assessment of disturbance history 162–4
age estimation 104–7
  gaps *166–7*
age structure of stands 113
aggregation *81*
airborne platform imaging 54
air temperature measurement 179
Alber's equal area projection 74
ALEX *215*
allowable annual cut (AAC) 345
allozymes 264
all taxa biodiversity inventory (ABTI) *328*
α-hulls 320
altitude measurement, GPS systems 93
AMI (SAR) *45*
amplified fragment length polymorphisms (AFLP) 265–8, 276
Analytica *318*
analytical methods, GIS 75–6
ancient forests 161–2
  Greenpeace campaign *14–15*
ancient trees
  felling 346
  habitat value 307–8
anemometers 150
anemophily, testing for 240
angle gauge *291*
annual rings 105–7
ANOVA (analysis of variance)
  estimation of heritability 283
  use with PCA 140
anthropogenic disturbances 148–9
  assessment *343*
  use of molecular markers 275–6
  harvesting 159–61
ANUCLIM *321*
APACK *83*
Aquilaria spp.
  A. malaccensis, life table and transition matrix *208*
  population size estimation 87
Araucaria araucana
  population viability analysis (PVA) *216–18*
  progeny tests 283
ArcGIS/ ArcView/ ArcInfo 72
ArcView Image Analysis *37*
area-limited species 372
area method, silviculture 345
area metrics 78, *79*, *80*
artificial forest restoration 352–3
artificial neural net classification *52*, 63
assumption testing 354
ASTER *45*
atmospheric conditions, in assessment of forest cover changes 59
ATTILA *83*
attractiveness index *245*
authenticity of forests 22
average canopy height 116
AVHRR (advanced very high resolution radiometer) sensor *43*, *45*, *46*, 48
  biomass estimation 64
  fire detection 66
  timber volume estimation 65
AWiFS *45*
azimuthal projections 74

bar graphs, phenology data 247, *249*
bark thickness measurement 108
bats, use of habitat trees 309
Bayesian methods 8
  of classification *52*
  of habitat modelling 324

Subject index

bayonet hollows *310*
below-ground herbivory, assessment 153
belt inventory, CWD 288–9
Berger–Parker index 131, *132*
beta diversity 133, 135
bias
   avoidance 26
   in distance-based sampling 97
   in searches 100–1
BIOCLIM 321
bioclimatic forest types 145
biodiversity
   conservation, CBD *18*, 19
   effects of fragmentation 300–2
   effects of sustainable forest management 341
   vertical stand structure, importance 297
   see also species diversity; species richness
biodiversity assessment 326–9
   UK Forestry Commission *329–31*
biodiversity hotspots *14*, 16
biodiversity indicators 341, *342–3*, *370–2*, 373–4
biodiversity information, use of national forest inventories 87
biodiversity mapping, remote sensing 66–8
biodiversity monitoring 363–7
biological characteristics, and potential for sustainable use 346–8
BIOMAP 321
BIOMAPPER 322
biomass estimation, remote sensing 63–4
biophysical classifications 144
BioRAP Toolbox 327
biplots 141–2
bird communities *371*
birds, use of tree cavities 308
blowdown 149–50
bole height 110
bootstrap estimator, species richness *127*
Borneo, satellite remote sensing *41*
boundaries, remote sensing imagery 53
branch-end hollows *310*
branch whorl counting 104
Braun–Blanquet classification, vegetation types 145
Braun–Blanquet scale, plant cover *313*
Brazil nuts, reproductive ecology 235
broad-sense heritability 282–3
browsing 154, *155*
browsing pressure indicators *156*
bryophytes *371*
   assessment *330*
bucket seed traps 255
buffering, GIS data 76
butt hollows *310*
buttressed trees, diameter measurement 108

C&I (criteria and indicators) processes 17–18, 20, 369, 370
   sustainable forest management 339
calipers 108
   measurement of diameter increments 184–5
camera angle, aerial photography 35–6
CANOCO *137*, 321
canonical correspondence analysis (CCA) 141, 321, 324
canopies, access techniques 103
canopy closure (density) 112
   measurement 174–8
canopy closure index 176
canopy composition, as biodiversity indicator *371*
canopy cover 111–12, 174, *175*
canopy loss estimation 150
canopy-scope *177*–8
canopy stratification 298–9
capacitance hygrometers 179
Carrifran Wildwood Project *271–3*, 378
Cartesian coordinates 75
catastrophic windthrow 149
categorical data analysis 259
category selection, matrix modelling 209
cavities 308–9, *310*
Cedrela odorata, progeny tests *281*, 283–4
cellular automata 227–8
   use in habitat modelling 324
census interval 188–9
census period 188
Center for International Forestry Research (CIFOR) 340
centroid sampling 119–20
CEPFOR Project *29–30*
ceptometers 168
certification of forests 18–19, 339, *340*, 341, 344
chance variables *319*
Chao 1, Chao 2 *127*
chloroform, use in assessment of herbivory 153
CI-110 Imager *173*
circular plots 94, 95, 99
CITES 143–4
classification 144–5
   landscape pattern studies 77
   of protected areas 334
   of satellite images 49–54
classification confusion matrix 53
clearcuts, remote sensing 66
climate change, Kyoto Protocol 19
climatic atlases 151
climatic classifications 144
climatic envelopes 321
climbing trees 102, *104*
clipping, GIS data 76
clonal plants, density measurement 120

Subject index | 439

'closed' forest edges *304*
cloud cover, satellite imagery 47
clumpiness *81*
clumping 121
cluster analysis 136–8
clustering, in simple random sampling 89
cluster sampling *86*, 90
coarse woody debris (CWD) 285–6
    as biodiversity indicator *371*
    decay class assessment 294–6, *311*
    decay rate estimation 296–7
    survey methods 287–8
        adaptive cluster sampling 291
        choice of method 293–4
        line intercept sampling *289*–90
        point and transect relascope sampling *291*–3
        sample plot inventory 288
        strip surveying 288–9
    volume assessment 286–7
    wood density assessment 296
co-dominant crown position 115
co-dominant marker systems 265
cohort life tables 205
colour allocation, satellite imagery 47, 48
column vectors 206
commission, errors of 53
common garden experiments 280
community involvement in forest management 27–31
Community Analysis Package (CAP) *137*
community-level models 220
comparative method 194
compass, use in location of sample units 91
composition metrics 78
computer software
    for conservation planning *337*–8
    for functional classification 200–1
    GIS 68–9, *70*, 71
        programs *72*
        selection 71, 73
    image processing packages
        aerial photography *37*
        satellite imagery 40, 48–9
    for landscape metric estimation 82, *83*
    for matrix calculations 207
    for outcrossing rates assessment 275
    patch occupancy models 218
    for PVA 214–*15*, 277
    for species richness estimation 127, 129–30
    for statistical analysis 122
        ANOVA 283
        cluster analysis 136
        F-statistics 270
        multivariate analysis 136, *137*
        spatial autocorrelation analysis 276
CONABIO, Mexico *326*

conceptual habitat models 317–20, *319*
conceptual level of models 232
conceptual modelling 369
    assessment of threats 357
configuration metrics 78
conformal projections 74
'confusion' matrix 233
connectivity 79, *81*, 82
conservation 332–3, 377–8
    adaptive management 354–7
    approaches 333
        forest restoration 347, 349, 350–4
        protected areas 334–8
        sustainable forest management 338–44
        sustainable use of tree species 344–7, *349*
    evidence-based 377
    indicators 367–74
    minimum habitat requirements 325
    modelling 231
    monitoring 363–7
    relevance of projects 12–16
    role of ecological understanding 1
    scenarios 374–6
    threats and vulnerability assessment 357–63
Conservation International 13, *14*, 16
conservation planning 336–8
    use of molecular marker data 276–9
conservation status, forest-dependent species *343*
consumer cameras, use for hemispherical photography 172, 174
contagion 79, *81*
contingency tables 233
continuous fields mapping 53
controls 9
Convention on Biological Diversity (CBD) *18*, 19
    ecosystem approach 25
convex hulls 320
coordinate systems, GIS 75
core area models 305
core extraction *105*–6
core metrics 79, *80*
correlation, landscape metrics 82, 84
Costa Rica, INBIO *326*
course of flowering *246*
cover 120
    understorey vegetation *313*
cover index (CI) 298
C-Plan *337*
criteria and indicators (C&I) 17–18, 20, 369, 370
    sustainable forest management 339
crossing experiments 268
crown length 110
crown measurement 112
crown position classification 115

crown position indices 175–6
cup anemometers 150
cylindrical form factor 307
cytoplasmic DNA (cpDNA) markers 274, 278

data analysis
    in adaptive management 356
    GIS 73, 75–6
data collection
    habitat models 325
    matrix modelling 209
data-defined approaches, functional
        classifications 199
data formats 71
data loggers 180–1
data presentation 374
data type selection, GIS 73–4
datums 75
Daubenmire classification 145
dead trees
    classification 310, 311
    volume assessment 287
    see also coarse woody debris (CWD)
deadwood, assessment 330
decay classes 311
    coarse woody debris 150, 294–6
    standing dead trees 310
decay rate estimation, coarse woody debris
        296–7
decay status assessment 310, 312
decision tree classification 52
decision variables 319
DECORANA 137, 141
deductive approaches, functional classifications
        199
deer
    assessment 330
    browsing 154
defoliation, remote sensing 66
deforestation
    assessment 362–3
    assessment of impact 274–5
    detection from satellite imagery 58
deforestation rate calculation 362
Delta-T devices 179
DEMON 118
Dempster–Shafer theory 52, 62
dendrochronology 104–7, 151
    assessment of disturbance history 162–4
dendrograms 136, 138
    in assessment of functional diversity 200
dendrographs 184
dendrometer bands 184, 185
density, understorey vegetation 312–13
density dependence
    PVA 215
    transition matrix models 210–11

density estimation 120
density of trees 96
depletion curves 185–6
depth of edge influence (DEI) 305
detection error 364
deterministic gap models 222
detrended correspondence analysis (DCA,
        DECORANA) 137, 141
dial-gauge micrometers 184
diameter at breast height (dbh) 107–8
    in assessment of disturbance history 162
    relationship to use of hollows 309
    remote sensing 64
diameter-exposed crown-area distribution 162
diameter growth measurement 184–5
diameter measurements 107–9, 113–15
diameter tapes 108
diazo paper, light measurement 174, 178
diffuse non-interceptance (DIFN) 168
digital elevation models (DEMs) 36, 37, 53,
        325
    use in classification of forest type 61
    use in habitat mapping 67
    use of lidar 54
Digital Globe 43, 45
digital images, aerial photography 35
digital number (DN), satellite imagery 47
dispersal-limited species 372
dispersal of pollen 237–9
dispersal of seeds 252–62
dispersion metrics 79, 81
distance-based sampling methods 93, 95–8
distance measurement 91–2
disturbances 148–9
    assessment 343
    fires 151–3
    harvesting 159–61
    herbivory 153–9
    historical analysis 161–4
    microclimate assessment 178–81
    removal 351
    risk assessment 360–3
    wind 149–51
diversity see biodiversity; species diversity
diversity indices 131–2
DNA markers see molecular markers
DOMAIN 321
dominant crown position 115
dominant eigenvalues 207, 209
dominant marker systems 265, 268
Domin scale, plant cover 313
DPSIR framework 368, 369
droppings traps 256
dry mass 190
dry mass estimation, coarse woody
        debris 296
duplication in conservation assessments 16

Subject index | 441

ecological communities, mapping 60–2
ecological models 221–2
   gap models 222–6
   transition models 226–8
ecological niche-factor analysis (ENFA) 322
economic sustainability 339
ecoregions 15, 16
ecosystem approach 23, 25
ecosystemic classifications 144
ecosystem management 25
edge characteristics and effects 302–6
edge density 79
edge effect, relationship to plot shape 95
edge metrics 78, 80
edge responses 305
edges, mechanistic model 306
edge-to-area ratio 79
effective area models 305
efficiency
   index of 91
   of sampling methods 99–100
elasticities, matrix modelling 209, 212
electronic rangefinders 92
elliptical stems, diameter measurement 108
elongation 80
endangered species, assessment of presence 142–4
environmental sustainability 339
equal area projections 74
equation of population flux 204–5
ERDAS IMAGINE 37, 48
ER Mapper 37
errors
   in combination of remote sensing with forest inventory data 65
   in mapping 53
   in sampling 88
establishment conditions, functional classification 200
EstimateS software 127, 129
ETM+ 45, 48
Euler equation 205
European Space Agency, provision of satellite data 42
evaluation
   of habitat models 325
   of protected areas 334–6
even-aged stands 113
evenness, measures of 131–2
evidence-based conservation 12, 377
evolutionarily significant units (ESUs) 277–8
experimental design 8–9
expert knowledge, use in habitat modelling 317–18
exposure 358
extinction risk assessment, use of molecular marker data 277

extinction risk curves 218
extinction vulnerability assessment 220
   see also population viability analysis (PVA)

faecal analysis 155, 157
Fauna and Flora International (FFI) 14
favourable conservation status (FCS) 373
fenced enclosures, use in assessment of herbivory 154, 193
field-based research, problems 193–4
film types, aerial photography 35
fire disturbance 153
fire response estimation 200
fires
   intensity 153
   remote sensing 66
   risk assessment 360
   temperature measurement 151–2
   velocity measurement 152
first-order Markov processes 226
Fisher's α 131, 132
'fish-eye' lenses 170
   see also hemispherical photography
fissures 310
fixation index (FST) 269–70, 274
fixed-area sampling methods 93, 94–5, 98–101
flagship species 373
flat seed traps 253
floods, remote sensing 66
floristic composition analysis 135–6
   importance values 142
   ordination 139–42
   TWINSPAN 138–9
floristic composition classifications 145
flower collection 247
flowering curves 246
flowering phenology 245–50
flower marking 236
fluorescent powdered dyes, use in pollen dispersal studies 238–9
fluorochromatic reaction (FCR) test, pollen viability 236
foliage height diversity (FHD) 297
Food and Agriculture Organization of the United Nations (FAO) 280, 340
   forest classifications 145
   forest-related definitions 23, 24
   national forest inventories 86–7
foraging behaviour, pollinators 244–5
foraging rate 245
forecasts 234
forest area, as biodiversity indicator 372
forest certification 18–19
forest cover, analysis of changes, remote sensing techniques 55–9
forest disturbances see disturbances

Subject index

forest dynamics 147
  assessment
    height and stem diameter growth 184–5
    natural regeneration surveys 182–4
    permanent sample plots (PSPs) 181–2
    plant growth analysis 189–90
    survival and mortality measurement 185–9
  growth and survival, influencing factors 191–5
forest extent, assessment *340*
ForestGALES 362
forest gaps 164–5
  characterization protocol *165–7*
forest habitat *see* habitat
forest harvesting
  impacts 159–61
  remote sensing 66
forest inventories
  sampling approaches 93–4
    distance-based sampling 95–8
    fixed area methods 94–5
    line intercept method 95
  sampling designs 87–90
  sampling intensity 91
  sampling unit location 91–3
  sampling unit selection 98–102
  types 85–7
forest inventory data, combination with remote sensing 65
forest landscape dynamics models (FLDM) 228–30
forest landscape restoration (FLR) 349, *350*
Forest Law Enforcement and Governance (FLEG) processes 19
forest management *24*
forest management unit (FMU) inventories 87
forest management unit (FMU) level 339
forest policy 16–21
Forest Policy Experts (POLEX) 17
Forest Principles, UNCED 338
forest recovery monitoring *353*
forest restoration 333, 347, 349, 351–4
  information sources *350*
Forest Restoration Information Service (FRIS) *350*
forest restoration methods, relative costs *352*
forestry practices, lessons for conservation management 332–3
forests, definition 21–2
forest standards 339
Forest Stewardship Council (FSC) 19, *340*
forest structure, assessment *343*
forest structure mapping, remote sensing 62–3
forest types, mapping, remote sensing methods 60–2
FORET 222, *225*, 230

form factors 307
FORMIND *225*, 228
FORMIX *225*, 228
form quotients 307
FORSKA 224
fossil evidence of disturbances 161
Foundations of Success 354
FPA solution 237
fractal dimension 79, *80*
fragmentation 76–7, *301, 342, 372*
  assessment of impact 274–5, 300–2
  edge characteristics and effects 302–6
  metric selection 78–82
  see also landscape pattern description
FRAGSTATS 82, *83*
frame quadrats 313–*14*
frameworks 368–9
frequency, understorey vegetation 312, 314
fresh mass 190
Frontier Forests campaign *15*
fruit, counting methods 250–2
fruit collection 247
fruiting curves 246
fruiting phenology 245–50
FSTAT 270
F-statistics 269–70
full-sib offspring 281
functional groups 198–202
fungal pathogens, effects 191
fungi *371*
  assessment *330*
funnel seed traps 255–6
fuzzy classification methods 52

gaharu collection, Indonesia 159, *160*
GALES 362
Galileo GPS system 92
gap analysis *327*
gap aperture *167*
gap-fraction-based methods, LAI measurement 117, *118*
gap markers *166*
gap models 222–6, 230
gap phase 164
GAPPS 277
gap size measurement *166*
GARP 323
Geary's c statistic 123, 124
genecology 280
gene flow estimation 270, 274
GENEPOP 270
generalized additive models (GAMs) 322
generalized linear models (GLMs) 322
general variables *319*
gene resource management units (GRMUs) 278

Subject index | 443

genetic algorithms 323
genetic differentiation measures 269
genetic diversity measures 268–74
genetic variation assessment
   molecular markers 262–3
      anthropogenic disturbance assessment 275–6
      choice of marker system 265, 268
      collection of material 263–4
      gene flow and mating system analysis 270, 274–5
      genetic diversity and differentiation assessment 268–70
      types of marker 264–5, *266–7*
      use in conservation planning 276–9
   quantitative variation 279–80
      heritability 282–3
      nursery and glasshouse experiments 283–4
      progeny test *281–2*
      provenance test 280–1
geographical information systems (GIS) 32, 48, 68–71
   analytical methods 75–6
   data type selection 73–4
   georeferencing 38
   geostatistics 124
   linkages to forest dynamics models 228–9, 230, 231
   mapping woodland cover *69*
   map projections 74
   software programs *72*
   software selection 71, 73
   use in assessment of threats 358, 360
   use in forest restoration projects 353
   use in habitat modelling 324
geographic envelopes 320–1
geometric processing, satellite imagery 47
GEOMOD 363
georeferencing (ground registration) 36, 38
geostatistics 124–5
germination, pollen 237
germination test 197–8
gimbal balance 112
glasshouse experiments 191–2
   provenance and progeny tests 283–4
Global Environment Outlook *376*
Global Forest Resources Assessment (GFRA) 18
Global Forest Watch *15*
Global Land Cover Facility (GLCF) *43*
Global Partnership on Forest Landscape Restoration *350*
Global Tree Campaign *14*
GLONASS 92
Google Earth 34
GPS (global positioning systems) 92–3

GRASS *72*
gravimetric water content of soil 179
grazing pressure indicators *156*
'greedy heuristic' algorithm, Sites *338*
Greenpeace *14–15*
'grey' literature 4
grey-scale format, satellite imagery 47
gridded quadrats 313–14, 366
ground control points (GCPs) 36, 38
ground registration (georeferencing) 36, 38
ground vegetation, assessment *330*
group definition, cluster analysis 137
growth
   influencing factors 191–5
   size-dependence *212*
growth analysis 189–90
growth chamber experiments 191
growth increment, gap models 223
growth increment models 107
growth patterns, in assessment of disturbance history 162–3
growth–yield models 220–1
   comparison with gap models 224
G_{ST} 269
Guatemala, national forest inventory 87
guided transect sampling 293
gut content analysis 157
Guyana, national forest inventory 87
gypsum resistance blocks 180

habitat 285
   coarse woody debris (CWD) 285–6
      decay class assessment 294–6
      decay rate estimation 296–7
      survey methods 287–94
      volume assessment 286–7
   edge characteristics and effects 302–6
   fragmentation 300–2
   understorey vegetation 312–16
   vertical stand structure 297–300
habitat management areas 334
habitat mapping, remote sensing 67
habitat models 316–*17*
   choice of method 324–6
   climatic envelopes 321
   conceptual models 317–20, *319*
   geographic envelopes 320–1
   machine learning methods 323–4
   multivariate association methods 321–2
   regression analysis 322–3
   tree-based methods 323
habitat suitability index (HSI) 318–20
habitat trees 307–12, *371*
half-sib offspring 281
hand pollination *243–4*
'hard' classification methods 51
'hard' forest edges 303, *304*

harvesting
  impacts 159–61
  remote sensing 66
hazard function 163–4
height of forest stands 115–16
height growth measurement 184
height measurement 109–11
Heip's index of evenness 131
HEMIPHOT 172
hemispherical photography 117, *170*–2, 178
  commercial systems *173*
HemiView *171*, *173*
herbaceous layer *371*
herbivore uncertainty principle 157
herbivory
  assessment of impact 153–9, 193, 362
  effects 191
heritability 282–3
heterogeneity, measures of 131–*2*
hierarchical classification schemes 146
hierarchical cluster analysis 136
high conservation value forest (HCVF) 23
'high-grading' 346
historical analysis, forest disturbances 161–4
historical ecology 161
holding times, semi-Markov models 227
holistic approach 27
hollows 308–9, *310*
HOLSIM 312
horizontal point sampling 291
Horvitz-Thompson (HT) estimation principle 287
hotspots *14*, 16
HRG *45*
HRS *45*
HRV *45*
Huber's formula 119, 287
Hugin Expert 318
human capital 31
human impacts 148–9
humidity measurement 179
HYBRID 224
hydrochory 252
hyperspectral sensors *43*, 46
hypothesis testing 5, 7–8, 11
hypothetico-deductive scientific method 7
hypsometers *110*–*11*, 184

Idrisi32 *72*
IDRISI 363
IDV *49*
Ikonos *43*, *45*, 46
  forest type assessment 61
illegal logging, policy developments 19
image acquisition
  aerial photography 34–6
  satellite remote sensing 42–7

image classification, satellite remote sensing 49–54
image comparison, assessment of forest change 56–9
image difference (ratio) 56
image processing
  aerial photography 36, 38
    computer software *37*, 71
  satellite remote sensing 47–9
image rectification, aerial photography 36
image segmentation *52*
image tone or colour, aerial photography 38
immigrant numbers, estimation 270
impact assessment, PVA *213*
impact of threatening processes 358, 360
impedance tomography 312
importance sampling 119–20
importance values 142
inbreeding coefficients (F_{IS}, F_{IT}) 269, 270
inbreeding depression 279
  incorporation into PVA 277
incidence-based coverage estimator (ICE), species richness *128*
increment analysis 162–4
increment borers *105*–6, 162
index of efficiency 91
index of flowering magnitude *246*
index of self-incompatibility (ISI) 241–2
indicator frameworks 368–9
indicators 367–8
  of biodiversity 329, 341
  selection and implementation 369–74
indicator species 372–3
indigocarmine test, seed viability 198
individual-based sampling 125, *129*
Indonesia, national forest inventory 87
influence diagrams 318, *319*
information sources
  aerial photography 34
  remote sensing and GIS technologies 32
  satellite remote sensing data 42–5
infrared gas analysers 179
infrared photography
  image colour 38
  temperature measurement 152
insect attack, remote sensing 66
insecticides, use in assessment of herbivory 154
insects, extraction of pollen 239
Instituto Nacional de Biodiversidad (INBIO), Costa Rica *326*
intensity of disturbance 148
intensity of vulnerability 358, 360
Intergovernmental Panel on Forests (IPF) *18*
intermediate crown position 115
International Forum on Forests (IFF) *18*
International Tropical Timber Organization (ITTO) *340*

International Union of Forest Research
    Organizations (IUFRO) 280
intersecting, GIS data 76
interspersion 79, *81*
inventories *see* forest inventories
inverse J structure *114*
invertebrates, assessment *330*
investigative frameworks 5–8, *6*
IRS *45*, 46
ISODATA *52*
isolation metrics 79, *81*
isozyme studies 264–8
    collection of material 264
IUCN (World Conservation Union) *15*
    authenticity of forests 22
    Forest Restoration Programme *350*
    Programme on Protected Areas (PPA) *336*
    Red List 142, 143, 320
    terminology 23

JABOWA 222, *225*, 230
Jaccard similarity index *134*, 135
jackknife estimators, species richness *127*
juvenile survivorship classification 201–2

kappa (κ) statistic 53, 233
kernel density estimators 320
keystone species 199, 373
K-function, Ripley 121–2
k nearest neighbour (*k*NN) method 65
kriging 124, 125
k values, leaves 184
Kyoto Protocol 19

LAI-2000 *118*
Lambert's projections 74
land cover, definition 51
landform classifications 144
LANDIS models *225*, *229*–31, 362
Landsat 32, *43*, *45*, 46, 48
    biomass estimation 64
    forest type assessment 61, 62
    image comparison 57, 58
    timber volume estimation 65
landscape ecology 68
landscape matrix 76
landscape pattern description 76–8
    metric estimation 82–4
    metric selection 78–82
landscape-scale modelling 231
landscape-scale studies, fragmentation 301, 302
LANDSIM 230
land use, definition 51
land use assessment *343*
land use change models 363
latitude–longitude system 75

leaf area index (LAI) 116–18, 168
    hemispherical photography *173*
    remote sensing 64
leaf area measurement 183
leaf area ratio (LAR) 189
leaf litter collection 117
leaf mass ratio (LMR) 190
leaf scar counting 104
leaves
    assessment for herbivore damage 154–5
    collection for genetic analysis 263–4
    staining for microscopy 155
Lefkovitch matrices 206, 207
Leslie matrices 206–7
lichens *371*
    assessment *330*
lidar (light detection and ranging) 54–5
    biomass estimation 64
life cycle stage 188
life history, use in functional classification 200
life tables 205, *208*
light availability, effects on growth and survival 191–3
light environment measurements
    canopy closure measurement 174–8
    choice of method 178
    hemispherical photography *170*–4
    light-sensitive paper 174
    light sensors 167–9
linear plant–herbivore interaction *158*
line intercept sampling
    CWD 289–90, *293*, 294
    understorey vegetation *316*
line transects 95
LISS *45*
logging
    impacts 159–61
    remote sensing 66
logistic growth function 158
logistic regression 322
logs
    decay scales 150, *311*
    volume assessment 286
    see also coarse woody debris (CWD)
long-distance seed dispersal 261–2
luxmeters 168–9

machine learning methods, habitat modelling 323–4
magnetic resonance imaging 312
mahogany harvesting *344*
    impact 161
mahogany silviculture, experimental trials 194
managed resource protected areas 334
management
    of habitat trees 312
    see also adaptive management

management cycle, protected areas 335
management interventions, prediction of response 213
management plans 356
management units (MUs) 278
MapInfo 72
Map Maker 72
mapping of vulnerability 358, 360
map projections, GIS 74
maps, accuracy 53
marginality 322
markers for plots 181
Markov models 226–7
MARXAN 337–8
mating systems 239–42
  analysis, use of molecular markers 274–5
matrices, assessment of threats 357–9
matrix graphs, phenology data 247, 249
matrix models 159, 203, 205–13
  PVA 214
matrix multiplication 207
maximum likelihood method, image classification 52
metal tagging, seeds 257
metapopulation models, PVA 215, 218–20
META-X 215
metric estimation, landscape pattern description 82–4
metric selection, landscape pattern description 78–82
Mexico, CONABIO 326
Michaelis–Menten equation 126
microclimate assessment 178–81
microhabitats, in gaps 166
micrometerological stations 180, 181
microsatellite markers (SSR) 265–8, 276
microwave scanning 312
midpoint flowering time 246
Millennium Ecosystems Assessment 333, 376
minimum habitat requirements 325
mirage method 288
MISR 45
mitochondrial DNA (mtDNA) 278
MLTR 275
modelling 203–4, 222, 228–34
  in adaptive management 354, 356
  community-level 220
  ecological models 221–2
    gap models 222–6
    transition models 226–8
  growth and yield models 220–1
  population dynamics 204
    equation of population flux 204–5
    life tables 205
    transition matrix models 205–7, 209–13
  population viability analysis (PVA) 213–20
  seed dispersal 259, 261

modified-Whittaker plot (MWP) 98–9
MODIS 41, 45–7
  fire detection 66
molecular markers 262–3
  anthropogenic disturbance assessment 275–6
  choice of marker system 265, 268
  collection of material 263–4
  gene flow and mating system analysis 270, 274–5
  genetic diversity and differentiation measurement 268–70
  marker types 264–7
  seed dispersal estimation 262
  use in conservation planning 276–9
molecular phylogeography 278
monitoring 3
  role in conservation 363–7
monitoring plans 356
monitoring programmes, design and implementation 365
Monte Carlo simulations 210
Moosehorn 176, 177
Moran's I statistic 123, 124
Morisita–Horn index of similarity 134–5
Morisita's index of dispersion (I_δ) 122
mortality rates 185, 188
MOSAIC 225
MSS 45
multiple-nearest-tree technique 96
MultiSpec 49
multispectral satellite imagery 43, 46
  image processing 47–8
multistemmed trees, diameter measurement 108
multivariate analysis, computer software 136, 137
multivariate association methods, habitat modelling 321–2

narrow endemic species 373
narrow-sense heritability 283
NASA, provision of satellite data 42
national forest inventories (NFIs) 86–7, 303
national parks 334
native species, use in forest restoration 349, 351
Natura 2000 network 373
natural experiments 194
natural forest 24
naturalness 22
natural regeneration surveys 182–4
natural seed traps 256
nature reserves 334
NAVSTAR-GPS system 92
nearest individual sampling method 96, 97
nearest-neighbour clustering 136

Subject index | 447

nearest-neighbour distances 121
nearest-neighbour functions 121–2
negative edge responses 305
nested plot designs 288
net multiplication rate of population (λ) 204, 207, 209, 212–13
net seed traps 253–5
neural networks 52, 63, 323–4
neutron probes 180
Newton's formula 119, 287
Neyman (optimal) allocation 89
NOAA 45, 46
non-governmental organizations (NGOs) 13–16, 20
  information on forest policy 17
non-hierarchical cluster analysis 136
non-native species, use in forest restoration 351
non-parametric estimators, species richness 127–8
non-timber forest products (NTFPs), CEPFOR study 29–30
normalized difference vegetative index (NDVI) 48
  relationship to species richness 67
null hypothesis 6, 7
nursery experiments, provenance and progeny tests 283–4

old-growth forest 24
  remote sensing 62
omission, errors of 53
OpenEV 49
'open' forest edges 304
open-pollinated offspring 281
open spaces, value 353
optical rangefinders 92
optimal (Neyman) allocation 89
Orbimage 43
Orbview 45
ordination 139–42
organelle markers, parentage analysis 274
organophosphates, use in assessment of herbivory 153
O-ring statistic 122
Orthoengine 37
orthophotos 36
orthorectification, aerial photographs 36
  computer software 37, 71
OSSIM 49
outcrossing rates, assessment 274–5
overall inbreeding coefficient (F_{IT}) 270
overdispersion 121
overlaying, GIS data 76
overtopped (suppressed) crown position 115

paint, temperature-sensitive 151–2
pair-correlation function (g) 121

parameters
  gap models 223–4
  PVA 214
parentage analysis 274
PAR (photosynthetically active radiation) quantum sensors 167–9, 178
participatory rural appraisal (PRA) 28
participatory threat mapping 358
passive microwave sensors 44
Patch Analyst 83
patch cohesion 81
patch identification 77–8
patch occupancy models 218
patch-scale studies, fragmentation 301, 302
patch size 79, 80
PC-ORD 137
PCRASTER 229
peak of flowering 246
penetrometers 296
percentage similarity 135
periodograms 250
permanent sample plots (PSPs) 181–2
persistence method, functional classification 200
phenology 245–50
  variables 246
phenophases 250
photogrammetry 34
photographic frame quadrats 314
photographic sensors, spectral resolution 43
photography
  aerial 33–40
    forest cover changes assessment 55–6
    forest structure assessment 63
    forest type mapping 60
    image acquisition 34–6
    image processing 36–8
    timber volume estimation 64
    tree height estimation 63
  hemispherical 117, 170–2, 178
    commercial systems 173
    value in location of sample units 91
photosensitive paper 174, 178
photosynthetic photon flux density (PPFD) 168, 169
physiognomic classifications 144–5
physiographic classifications 144
pitot tubes 152
pixel size 44
  aerial photography 35
  and landscape metric estimation 82
plantation forests 22
plant canopy analysers 168
plant growth analysis 189–90
plant–herbivore interactions 157–9
plant location techniques 183
plant-to-all-plants distance analysis 121

448 | Subject index

plot-based designs, landscape pattern studies 77
plot designs 94–5, 98–101
   biodiversity assessment 329–30
   coarse woody debris surveys 288
plot location 181
plot size 101–2
point-centred quarter method 96, 97
point quadrats 315
point relascope sampling 292–4
point-to-plant distances 120
point transect technique 316
policy 16–19
policy-relevance of projects 19–21
policy screening, scenario planning 375–6
pollen germination ability 237
pollination bags 239–41
pollination ecology 235–6
   flower marking 236
   hand pollination 243–4
   mating systems 239–42
   pollen dispersal 237–9
   pollen viability 236–7
   pollinator behaviour 244–5
pollinator movement studies 237–8
pollinators
   exclusion from flowers 239–40
   extraction of pollen 239
polylines 73
POPGENE 270
population ceilings, matrix models 210
population density 97
population dynamics modelling 204
   equation of population flux 204–5
   impact of harvesting 159
   life tables 205
   transition matrix models 205–7, 209–13
population recruitment curve (PRC) 182
population structure, and potential for sustainable use 348
population viability analysis (PVA) 213–18, 277
   avoiding mistakes 216
   example score sheet 219
   spatially explicit models 215, 218–20
position errors 53
positive edge responses 305
potted seedlings, use in experiments 193
power analysis 9
   monitoring programmes 365–6
practicality of projects 10
practical value of research 4
precautionary principle 25
precision 26–7, 91
precision dendrographs 184
predictions 234
pressure-state-response (PSR) framework 368

primary forest 24
Primer 5 137
principal components analysis (PCA) 139–40
principal coordinates analysis (PCO) 141
prior, Bayesian inference 8
probability density functions 114–15
process-limited species 372
process models 222
profile diagrams 116
Programme for the Endorsement of Forest Certification (PEFC) schemes 340
projection matrices 206–7
projections 234
proportional representation 89
protected areas 333, 334–8
   assessment 342
   Natura 2000 network 373
provenance test 280–1
proximity metrics 79, 81
pseudospecies, use in TWINSPAN 138
psychrometers 179, 180
pyranometers 168
pyrometers 151–2

quadratic paraboloid stem shape 119
quadrats 313–14, 366
   point quadrats 315
   sizes 314–15
quantitative genetic variation 279–80
   heritability 282–3
   nursery and glasshouse experiments 283–4
   progeny tests 281–2
   provenance test 280–1
quantitative trait loci (QTL) 279
querying, GIS data 75–6
Quickbird 43, 45, 46
   forest type assessment 61

rabbits, exclusion 154
radar, biomass estimation 64
radar imaging 312
radar sensors 44, 54
radial increment analysis 162–4
radiation measurement methods, LAI 117
radioactive tagging, seeds 257, 258
radiocarbon dating 107
radiometric processing, satellite imagery 47
radiometric resolution
   aerial photography 35
   satellite imagery 44
raked-ground surveys, fruits 251
RAMAS packages 214–16, 231
random amplified polymorphic DNA (RAPD) 264–7
random dispersion 121
randomization 9
random numbers 89

Subject index | 449

random sampling *see* sampling designs
rangefinders 92
rapid assessment programme (RAP) *328*
rapid biodiversity assessment (RBI) *328*
rapid rural appraisal (RRA) 28–9
RAPPAM methodology *335, 336*
rarefaction curves *129*
rare species, sampling methods 97
raster data 73–4
ratio-dependent plant-herbivore interaction 158
receiver-operator characteristic (ROC) curves 233, 325
reclamation 347
recognizable taxonomic units (RTUs) *328*
recording dendrographs 184
recruitment assessment 185
recruitment failure, identification 220
recruitment limitation 260–1
rectangular coordinates *183*
rectangular plots 94, 95, 100
rectification of images, aerial photography 36
   computer software 37
Red List, IUCN *15*, 142, 143, 320
reforestation 349
regeneration assessment 182–4
regeneration capacity 346–7, *348*
regeneration surveys 58
regression analysis, habitat modelling 322–4
rehabilitation 347
relascopes 109
relascope sampling *291*–3, 294
relative cover, understorey vegetation 313
relative flowering intensity *246*
relative growth rate (RGR) 189
relative growth rate of height (RGRH) 190
relative population viability assessment, score sheet example *219*
release identification 162–4
remote sensing 32, 54–5
   aerial photography 33–4
      forest cover changes assessment 55–6
      forest structure assessment 63
      forest type mapping 60
      image acquisition 34–6
      image processing 36–8
      timber volume estimation 64
      tree height estimation 63
   application 55
      analysis of forest cover changes 55–9
      biodiversity and habitat mapping 66–8
      forest structure mapping 62–3
      mapping different forest types 60–2
      mapping height, biomass, volume and growth 63–5
      mapping threats to forests 66
   biodiversity indicators *342*
   combination with forest inventory data 65
   satellite techniques 39–42
      biomass estimation 64
      forest cover change analysis 56–9
      forest structure assessment 63
      forest type mapping 60–2
      image acquisition 42–7
      image classification 49–54
      image processing 47–9
      raster data 73
      timber volume estimation 64
      tree height estimation 63
replication in experimental design 9, 26, 131, 133
reproductive characteristics, and potential for sustainable use *348*
reproductive ecology 235
   flowering and fruiting phenology 245–50
   pollination 235–6
      flower marking 236
      hand pollination 243–4
      mating systems 239–42
      pollen dispersal 237–9
      pollen viability 236–7
      pollinator behaviour 244–5
   seed ecology
      seed dispersal and predation 252–62
      seed production 250–2
reproductive success, variables *242*
resampling methods 210
research 3
   conservation relevance 12–16
   experimental design 8–9
   investigative framework 5–8
   objectives 2–5
   policy relevance 16–21
   precision and accuracy 26–7
   scientific value 9–11
resistance blocks 180
ResMap *43*
ResNet *338*
resolution, aerial photography 34
resource-limited species *372*
restoration of forest 333, 347, 349, 351–4
   information sources *350*
restriction fragment length polymorphism analysis (RFLP) 264, *266*–8
right eigenvector 207
ring counts 105–7
ring width changes, interpretation 162–3
r.le *83*
root competition, effects 191

sample-based assessment 125, *129*
sample designs, cluster sampling 90
sample plot inventory, CWD surveys 288, *293*

sample size 366
  species diversity estimation 133
sampling approaches 9, 93–4
  in biodiversity monitoring 364
  coarse woody debris (CWD) 288
  collection of phenology data 247
  distance-based sampling 95–8
  fixed-area methods 94–5
  for gaps 165–6
  for genetic analysis 263–4, 275, 276
  in increment analysis 163
  line intercept method 95
  seed bank studies 196
  for taxonomic determination 102–4
  understorey vegetation 316
  see also survey methods
sampling designs 87–8, 125
  simple random sampling 88–9
  stratified random sampling 89
  systematic sampling 90
sampling error 88
sampling intensity 91
sampling unit location 91–3
sampling unit selection 98–102
saplings, definition 182
satellite imagery 39–42
  biomass estimation 64
  forest cover change analysis 56–9
  forest structure assessment 63
  forest type mapping 60–2
  image acquisition 42–7
  image classification 49–54
  image processing 47–9
  raster data 73
  timber volume estimation 64
  tree height estimation 63
saturating plant–herbivore interaction 158
scale
  aerial photography 34, 39
  landscape pattern studies 77, 82
  satellite imagery 45–6
scanning, aerial photographs 35
scarring from fires 153
scenarios 234, 374–6
  information resources 376
Scenarios for Sustainability 376
seasonal variation, remote sensing 59
secondary forest 24
second-order statistics 121–2
sectional method, volume measurement 118–19, 287
seed banks, assessment 330
seed bank studies 195–8
seed dispersal 252
  dispersal curves 259–60
  long-distance 261–2
  movement observations 256–8
  recruitment limitation 260–1

seed traps 252–6
seeding, forest restoration 352
seedling experiments, relevance 194
seedling planting, forest restoration 352–3
seedling regeneration, assessment 331
seedlings
  definition 182
  height measurement 184
seed marking 257
seed predation 258–9
seed production 250–2
seed rain 252
seeds
  extraction from soil 196–7
  viability tests 198
seed sampling, trapless methods 256
seed shadow 259
seed transfer zones 273
seed traps 250–6
segmentation of images 52
self-incompatibility index (ISI) 241–2
self-pollination, testing for 240–2
self-pollination frequency 242
self-thinning, 3/2 law 120
semi-Markov models 226, 227
semi-natural forest 24
semivariograms (variograms) 124
sensitivity analysis 209, 233
sequential flowering 247, 249
severity of disturbance 148
shadehouse experiments 191–3
shade-tolerance 201–2
shadows, aerial photography 39
Shannon evenness measure 131
Shannon's index of genetic diversity 269
Shannon–Weiner function 132
shape metrics 78–80
shapes, aerial photography 39
Shell Global Scenarios 376
shifting-patch modelling 228
shigometers 312
sighting tubes 112
sigmoid plant-herbivore interaction 158
silica gel, use for preservation of material 263–4
similarity 81
similarity indices 134–5
  use in cluster analysis 136
simple random sampling 88–9
Simpson's index 131, 132
Simpson's measure of evenness 131, 132
'simulated annealing' algorithm, Sites 338
single-link clustering 136
Sites software 338
size-dependent relations, use in matrix models 211
Smalian's formula 118, 287
small mammals, assessment 330

small trees, sampling in subplots 94
SmartImage 37
snags, volume assessment 286
Sørensen's quantitative index *134*
Sørensen's similarity index *134*, 135
social capital 31
social issues 27–31
social survey techniques 28–9
social sustainability 339
Society for Ecological Restoration (SER) International 350
'soft' classification methods 51, 53, 62
'soft' forest edges 303, *304*
soil, effects of harvesting 160
soil corers *195*
soil microbes, assessment *330*
soil moisture, effect on remote sensing 59
soil moisture measurement 179–80
soil seed bank studies 195–8, *330*
soil seed traps 253–4
Solarcalc 172
solar energy measurement 168
solar illumination, in assessment of forest cover changes 59
solarimeters 168
songbirds, assessment *330*
SORTIE models *225*, 228, 230, 231
space agencies, provision of satellite data 42
Space Imaging 45
Spatial Analyst 83
spatial autocorrelation *81*, *123*–4, 276
spatial distribution of disturbances 148
spatiality of models 231–2
spatially explicit models 215, 217–20
spatial resolution
    aerial photography 34, 35
    in assessment of forest cover changes 59
    landscape pattern studies 77
    lidar systems 54
    satellite imagery 44–7
spatial structure of tree populations 121–5
species abundance distribution 126
species accumulation curves 126, *129*
species–area curves 126
species composition 85
species density 130–1
species diversity 125
    see also biodiversity
species diversity measurement 131–3
Species diversity and richness software 129
species: individual ratios, pitfalls 129, *130*
species management areas 334
species numbers, use in monitoring forest recovery 353
species richness
    measurement 125–31, *343*
    see also biodiversity
Species Survival Commission, IUCN 15

specific leaf area (SLA) 116, 190
specimen collection 102–3
spectral analysis, phenological patterns 250
spectral change vector analysis 56
spectral mixture analysis 61
spectral radiometers 168
spectral resolution
    in assessment of forest cover changes 59
    satellite imagery 44, 46
spectral sensitivity, aerial photography 35
spherical crown densiometers 176–7
splines 322
splitting *81*
S-Plus software 122
SPOMSIM 1.0 218
SPOT *43*, *45*, 46
SPOT VEGETATION 2 *45*, 46
    fire detection 66
SPRING 49
sprouts, definition 182
square plots 94, 95, 100
SSR (microsatellite markers) 265–8, 276
staining
    tree rings 107
    use in pollen dispersal studies 238
stakeholders, involvement 27–31
standard volume function 286
stand basal areas 113
stand density 120
stand scale modelling 231
stand structure 113
    age and size structure 113–15
    height and vertical structure 115–16
    leaf area 116–18
    vertical structure 297–300
stand tables 113
stand volume 118–20
statistical analysis 8–9, 11, 73
    categorical data 259
    monitoring programmes 365–6
    phenological data 248–50
statistical analysis software 122
    ANOVA 283
    cluster analysis 136
    F-statistics 270
    multivariate analysis 136, *137*
    spatial autocorrelation analysis 276
stem diameter measurement 107–9
stem form assessment 307
stem volume measurement 118–20
stereoscopic examination, aerial photographs 36
sticky seed traps 254–5
stigma, examination of pollen germination 237
stochastic gap models 222–3
stochasticity, incorporation into matrix models 210

Subject index

stochastic patch occupancy model 218
strategic management 232
stratification index 299
stratified random sampling 89
strips 94
strip surveying, CWD 288–9, *293*
structural characteristics, remote sensing 62–3
subjective approaches, functional classifications 199
subplots 94
successional changes, prediction 199–200
successional recovery 351–2
succession models 222
    gap models 222–6
    transition models 226–8
suitability indices (SI) 319
SunScan *118*
supervised classification, satellite images 50, 51, *52*
suppressed (overtopped) crown position 115
surface temperature measurement 179
survey methods, coarse woody debris (CWD) 287–8
    adaptive cluster sampling 291
    choice of method 293–4
    line intercept sampling *289*–90
    point and transect relascope sampling *291*–3
    sample plot inventory 288
    strip surveying 288–9
survival, influencing factors 191–5
survival rates 185, 187
survivorship curves 185–7
sustainable forest management (SFM) 23–4, 333, 338–9, 344
    assessment 341–3
    information sources *340*
    policies 17–19
Sustainable Forestry Initiative (SFI) Program 19
sustainable livelihoods approach (SLA) 29, *30*
sustainable rural livelihoods framework 29, 31
sustainable use of tree species 344–7, *349*
SYNCSA Minor software 201
systematic review, conservation evidence 377
systematic sampling 90

tagging
    of flowers 236
    of seeds 257
taper functions 287
tapes, use for diameter measurement 108
tatter flags 150–1
taxonomic determination, sampling techniques 102–4
Temiplaq 151
temperature, effects on seed germination 198
temperature measurement 179
    fires 151–2
Tempil tablets 152
temporal resolution, satellite imagery 44
tensiometers 180
terminology 19–24
    proposed definitions *24*
Terraserver *43*, *45*
tetrazolium test 198
texture, aerial photography 38
thematic errors 53
theories, identification 11
thermal imagery 54
thermistors 179
thermocouples 151, *152*, 179
threatened species, assessment of presence 142–4
threat matrices 357–9
threats
    assessment 357–8, 363
        scoring method *359*
    remote sensing 66
three-dimensional imaging, lidar 54, 55
threshold values, hemispherical photography 172
timber volume estimation 85
    remote sensing 64
time series graphs, phenology data 247, *248*
time series population counts 220
timing of disturbance 148
tip-up mounds 161
TM *45*
tone, aerial photography 38
total edge *80*
total edge contrast index *80*
total tree height 109
trait matrices 200
trampling 155
transect inventory, CWD 288–9
transect lengths, line intercept method 290
transect relascope sampling *292*–4
transects, assessment of edge characteristics 303
transition matrix models 159, 205–7, *208*, 209–13
    PVA 214
transition models 226–8
transition probabilities 226, 227
transmitters, pollinator movement studies 237–8
traversability *81*
tree-based methods, habitat modelling 323
tree climbing 102, *104*

Subject index | 453

tree crowns, access techniques 103
treefalls 150
tree height estimation, remote sensing 63
tree location techniques 183
tree markers 181–2
Tree Radar Unit (TRU) 312
tree size
   as biodiversity indicator *371*
   implications for plot design 100
   relationship to seed production 252
tree species 85
   sustainable use 344–7, *349*
tree species identification, remote sensing 60–1
tropical trees, functional classifications 201
T-square sampling method 96, 97, *98*
TSTRAT 299
t-test, power analysis 366
tube solarimeters 168
TWINSPAN (Two-Way INdicator SPecies ANalysis procedure) *137*–9
two-stage clustering 90

UK Forestry Commission, Biodiversity Assessment Project *329–31*
ultrasound rangefinders 92
uncertain values, management in matrix modelling 212
understorey vegetation 312–16
   impact of tree felling 159
uneven-aged stands 113
ungulate herbivory 154, 155
unions, GIS data 76
United Nations Conference on Environment and Development (UNCED) 17, *18*, 338
United Nations Environment Programme World Conservation Monitoring Centre UNEP-WCMC *14, 336*
   UNEP-WCMC Threatened Plants database 143
United Nations Forum on Forests (UNFF) *18*
United Nations Framework Convention on Climate Change (UNFCCC) 19
United Nations Millennium Development Goals 19
Universal Transverse Mercator (UTM) projection 74, 75
unsupervised classification, satellite images 50, *52*
UPGMA (Unweighted Pair Group Method with Arithmetic mean) 136
upper stem diameter measurement 109

validation of images 57
validation of indicators 373
validation procedures, modelling 233–4

vane anemometers 150
variability, assessment 26
variable-area transect sampling method 96
variable choice, modelling 232
variables 10–11
   conceptual models *319*
variograms 124
vector data 73, 74
VegClass software 200
vegetation classification 144–6
verification procedures, modelling 233
vertical air photographs 35–6
vertical stand structure 297–300
vertical structure, forest stands 115–16
veteran trees, habitat value 307–*8*
VI sensor *45*
visitation rates, pollinators 244–5
vital dyes, assessment of pollen viability 237
volume assessment, coarse woody debris (CWD) 286–7, 290
volume method, silviculture 345
volumetric water content of soil 179
VORTEX 214, *215*, 277
vulnerability assessment 357–63
   PVA *213*

water availability, effects 191
Weibull function 115, 164
wet seed traps 255
Whittaker's measure, beta diversity 133, 135
wide-angle relascope *291*
wilderness areas 334
wildfires
   remote sensing 66
   see also fires
wildlife tress 307–12, *371*
wind dispersal, seeds 261
wind disturbance 149–51
wind pollination, testing for 240
winds
   remote sensing 66
   risk assessment 360, 362
wind speed measurement 150
windthrow 149–50
WINPHOT 172
WinSCANOPY *173*
wood density measurement 296
woodland cover, application of GIS *69*
World Bank toolkit *327*
World Commission on Protected Areas (WCPA) *336*
World Conservation Union (IUCN) *15*
   authenticity of forests 22
   Forest Restoration Programme *350*

Programme on Protected Areas (PPA) *336*
Red List 142, 143, 320
  terminology 23
World Database of Protected Areas *336*
World List of Threatened Trees 143
WORLDMAP *338*
World Resources Institute (WRI) *15*
World Summit on Sustainable Development (WSSD) 19
World Wide Fund for Nature (WWF) 13, *15*, 16, 22
  Forest for Life Programme *350*
WWF and World Bank Alliance 335, *336*
wounding
  from fires 153
  from wind disturbance 150

X-ray tomography 312

yield models 220–1
  comparison with gap models 224

ZELIG 224, *225*, 227, 230
ZONATION *338*
zoochory 252, 256, 261

Printed and bound by CPI Group (UK) Ltd, Croydon, CR0 4YY